Low-Temperature Thermal and Vibrational Properties of Disordered Solids

A Half-Century of Universal "Anomalies" of Glasses

Low-Temperature Thermal and Vibrational Properties of Disordered Solids

A Half-Century of Universal "Anomalies" of Glasses

Editor

Miguel A Ramos

Universidad Autonóma de Madrid, Spain

World Scientific

NEW JERSEY • LONDON • SINGAPORE • BEIJING • SHANGHAI • HONG KONG • TAIPEI • CHENNAI • TOKYO

Published by

World Scientific Publishing Europe Ltd.
57 Shelton Street, Covent Garden, London WC2H 9HE
Head office: 5 Toh Tuck Link, Singapore 596224
USA office: 27 Warren Street, Suite 401-402, Hackensack, NJ 07601

Library of Congress Control Number: 2022942058

British Library Cataloguing-in-Publication Data
A catalogue record for this book is available from the British Library.

LOW-TEMPERATURE THERMAL AND VIBRATIONAL PROPERTIES OF DISORDERED SOLIDS
A Half-Century of Universal "Anomalies" of Glasses

ISBN 978-1-80061-257-0 (hardcover)
ISBN 978-1-80061-258-7 (ebook for institutions)
ISBN 978-1-80061-259-4 (ebook for individuals)

For any available supplementary material, please visit
https://www.worldscientific.com/worldscibooks/10.1142/Q0371#t=suppl

Desk Editor: Joseph Ang

Typeset by Stallion Press
Email: enquiries@stallionpress.com

Printed in Singapore

https://doi.org/10.1142/9781800612587_fmatter

Preface

About 50 years ago, a seminal paper by Zeller and Pohl established that specific heat and thermal conductivity at low temperatures in amorphous solids were unexpectedly very different from the behavior found in their crystalline counterparts. Only one year later, two independent works, one written by W.A. Phillips, another one by Anderson, Halperin, and Varma, postulated what is usually known as the Tunneling Model. A simple assumption about the general existence of a random distribution of two-level systems in non-crystalline solids appeared to successfully account for the "anomalous" behavior of thermal properties, as well as acoustic or dielectric properties investigated thereafter, at least below 1 K.

After the abovementioned publication by Zeller and Pohl in 1971, two decades of intensive research followed, especially focused on thermal and acoustic properties at very low temperatures. Then, the phenomenology of glasses (amorphous solids) seemed clear and well understood within the Tunneling Model. A few good review books summarized all those results and findings.

Nevertheless, in the second half of this "half-century of universal anomalies of glasses", some other research directions and unexpected results started to arise, especially around the feature known as the "boson peak", well beyond the range covered by the successful Tunneling Model. Furthermore, even the apparent success of this model in its range of applicability was seriously questioned.

To address all these most recent advances in a comprehensive way is the goal of this book. With the generous contributions of several reputed scientists in the field, and ranging from experimental to theoretical physics, this book aims to offer a timely review of low-temperature thermal and vibrational properties of glasses, and of disordered solids in general. From the very beginning, it has been my intention that the book should provide a coherent, but at the same time unbiased, broad survey of the topic. Most chapters were written by prominent researchers in each discipline, though sometimes with different views or approaches on the most controversial issues, such as the very nature of the "boson peak", the role of ultrastable glasses, or the glassy behavior found in crystalline solids with different degrees of disorder. The contents of the book are described in detail in the introductory first chapter and will not be repeated here.

It has not been an easy task to coordinate and edit this collective book, especially hampered by the global pandemic that we have suffered during the last two years. Some changes and unwanted decisions had to be made along the way. Finally, we reached the goal. For this, I am very grateful not only to all the contributors but also to those colleagues who encouraged me to move on and/or revised some chapters of the book, namely, Andrés Cano, Stephen Elliott, Anthony Leggett, Herbert Schober, Sebastián Vieira, and Alessio Zaccone. I also thank the editors of World Scientific Publications for their patience and understanding.

I do hope that this book will be useful for both experienced researchers in the field and younger researchers who are beginning to dive into the complex sea of low-temperature properties and low-frequency dynamics of glasses or non-crystalline solids.

Miguel Ángel Ramos
Universidad Autónoma de Madrid, Spain

https://doi.org/10.1142/9781800612587_fmatter

Contents

https://doi.org/10.1142/9781800612587_0001

Chapter 1

Introduction: About 50 Years of Two-Level Systems and Boson Peak

Miguel A. Ramos

Low Temperature Laboratory, Condensed Matter Physics Department
Condensed Matter Physics Center (IFIMAC) and Instituto Nicolás Cabrera
Universidad Autónoma de Madrid, Madrid 28049, Spain
miguel.ramos@uam.es

In this introductory chapter, the objective of the book is pointed out and its contents are briefly advanced. In short, the book aims to provide a balanced combination of an up-to-date review about the so-called universal anomalies of glasses at low temperature, thoroughly studied over the last half century, along with a comprehensive (albeit unavoidably incomplete) discussion on this controversial topic.

1. Introduction

Although there were antecedents and earlier experimental work,[1–8] it was the seminal paper by Zeller and Pohl,[9] published about 50 years before this review book, the one which opened up a new field of research in condensed matter physics (at that time likely called solid state physics). Unexpectedly, all studied amorphous solids were found to exhibit strikingly similar behavior in their specific heat and thermal conductivity at low temperatures, being in turn very different from that of their crystalline counterparts. Solid-state physicists would have however expected[7,10,11] a behavior similar for crystalline and amorphous (i.e. non-crystalline) solids, since low temperature properties should be dominated by long-wavelength acoustic vibrations following Debye's theory, and the presence or absence of long-range order should be essentially irrelevant. Thus, low-temperature specific heat (see Figure 1) and thermal conductivity (see Figure 2) of amorphous solids or glasses were soon considered as "anomalous".

Only one year after that work in 1971 by Zeller and Pohl,[9] two independent papers[12,13] came out, which essentially postulated the same theoretical framework to account for those low-temperature anomalies of amorphous solids. This approach became soon known as the Tunneling Model (TM).

In brief, it seemed reasonable that the excess of both low-temperature specific heat and defects scattering off thermal phonons was due to some additional excitations present in noncrystalline solids.[10] The TM proposed that in any glass (amorphous solids or even *spin glasses*) there should be many localized excitations of two-level systems (TLS), arising from two nearby equilibrium configurations of similar energy for small groups of atoms. Even at very low temperatures, those atoms or groups of atoms could move from one (metastable) equilibrium position to the other by quantum tunneling through the potential energy barrier. The statistical nature of this configurational disorder easily leads to a random distribution of energies for the TLS, which was shown[12,13] to produce both the observed linear temperature dependence for the specific heat and the quadratic one for the thermal conductivity.

Besides measurements of thermal properties of amorphous solids, many experiments on acoustic and dielectric properties of glasses (both metallic and nonmetallic) were conducted then, especially during the decade following the appearance of the TM.[10,14–16] The rich behavior of the sound velocity (dielectric constant) and the corresponding acoustic attenuation (dielectric loss) as a function of temperature, frequency or intensity, arose as a stringent test for the predictions of the TM, which passed most of them very successfully.

A very different story happened with the also anomalous properties of glasses above, say, 1 or 2 K. In this "intermediate" temperature range at $\sim 2-20\,\mathrm{K}$, the specific heat C_p of glasses exhibits a broad hump in C_p/T^3 (see Figure 1), the increasing thermal conductivity

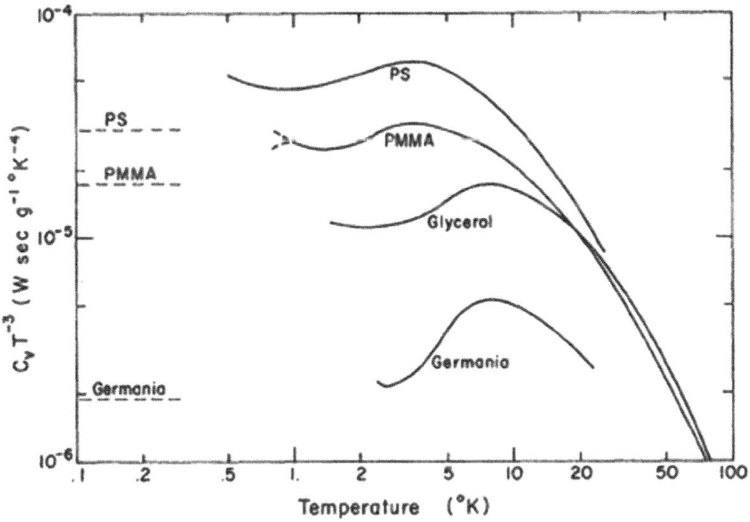

Fig. 1. Specific heat divided by T^3 of several amorphous materials: polystyrene (PS), poly(methyl methacrylate) or plexiglass (PMMA), glassy glycerol and vitreous germania. The horizontal dashed lines represent the low-temperature specific heat calculated from the elastic constants using Debye's theory. Reprinted with permission from Ref. [9]. Copyright 1971 by the American Physical Society.

with temperature reaches a universal *plateau* followed by a subsequent increase (see Figure 2), and the acoustic internal friction is also enhanced dramatically, among other features. Nowadays, it seems clear that all those additional glassy anomalies are related to the peculiar vibrational dynamics of noncrystalline solids, which possess low-energy excitations beyond textbook acoustic phonons, as presented in Debye's theory for crystals. Much more controversial is, however, the very nature of these ubiquitous low-frequency vibrational excitations in glassy systems, centred around the so-called *boson peak*, as it will be seen throughout this book.

As a matter of fact, the appearance of a broad and intense band at 30–120 cm^{-1} in the Raman spectrum of fused quartz (a-SiO_2) at room temperature had already been reported by Krishnan[17] in year 1953. This Raman band, with its maximum located more precisely at ~50 cm^{-1} for different specimens of vitreous silica — the paradigmatic glass for decades—, was reproduced in subsequent

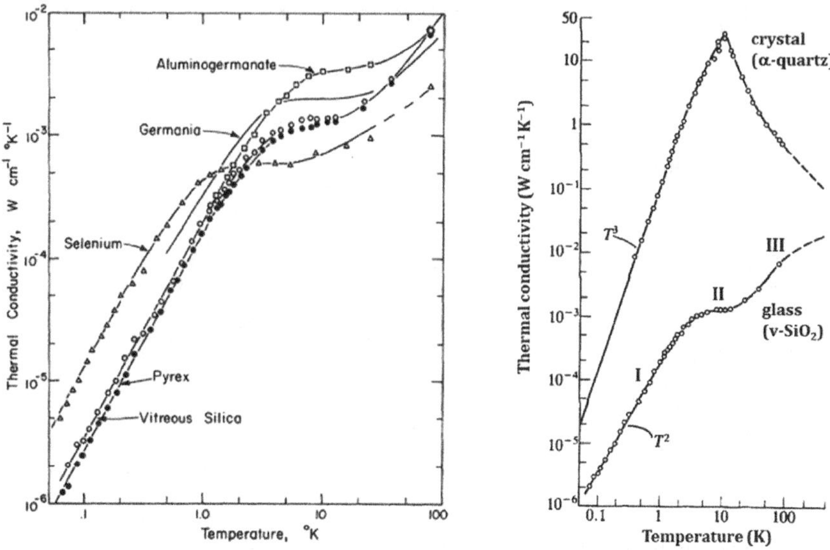

Fig. 2. (Left): Thermal conductivity of several amorphous materials at low temperatures: vitreous silica, Pyrex 7740, selenium, an aluminogermanate glass and vitreous germania. Note that no scaling has been done, all glasses exhibit very similar curves even quantitatively. Reprinted with permission from Ref. [9]. Copyright 1971 by the American Physical Society. (Right): Comparison of the thermal conductivity of SiO_2 in crystalline (α-quartz) and glass (vitreous silica) states. Reprinted with permission from Ref. [9]. Regions labelled I, II and III indicate the three typical temperature ranges observed for all glasses, as described in the text.

experiments.[2,18] This Raman feature, observed at low frequencies in this and in every other glass, and completely different from the null Raman spectrum of the corresponding crystals in that frequency range, was soon ascribed[19] to the breakdown of the wave-vector selection rules for crystalline (second order) Raman scattering, leading to a continuous (first order) vibrational spectrum in glasses. Furthermore, as early as in 1959, Flubacher *et al.*[2] measured Raman and Brillouin scattering concurrently with low-temperature specific heat of vitreous silica, and argued that the observed excess heat capacity over the Debye elastic contribution $\propto T^3$ was due to those *optical modes of very low frequency* seen by Raman scattering — what will be named "the boson peak" many years later. Leadbetter[7] interpreted the thermal properties of glasses at low temperatures in terms of the existence of localized vibrational modes of very low frequency.

Nonetheless, the historical paths of low-temperature properties and low-frequency vibrational spectroscopy in the field of glass science were divergent for many years. Since the 1970's, the abovementioned seminal paper by Zeller and Pohl[9] and the subsequent TM[12,13] triggered the interest in glassy properties at very low temperatures and tunneling TLS, which in turn were invisible by typical Raman scattering. On the other hand, many Raman-scattering experiments at low frequencies and varying temperature were conducted for different glasses, especially after Shuker and Gammon[19] established clearly that Raman scattering in noncrystalline solids was first-order, directly proportional the vibrational density of states $g(\nu)$ and to the Bose-Einstein thermal population factor $n(\nu, T)$, weighted by a photon-phonon coupling factor $C(\nu)$. Much more detail is provided in Section 4.2. of Chapter 7.

Indeed, the widely used term *boson peak* is something of a misnomer because it was coined to stress that the ubiquitous broad peak observed in glasses at low frequencies simply arose from the competition between a density of states $g(\nu)$ increasing with frequency ν and a decreasing $[n(\nu, T) + 1]$ Bose factor for typical temperatures — what is not the whole truth, as we will see soon.

Even worse, in 1974 an influential theoretical work by Martin and Brenig[20] postulated a definite expression for the unknown elasto-optical coefficient $C(\nu)$ between scattering light and an assumed Debye-like density of acoustic phonons, taking into account mechanical and electrical disorder in the amorphous solid. In this theoretical framework, the "boson peak" was mainly driven by a maximum in $C(\nu)$, ascribed to a correlation length 2σ of the various fluctuations involved. For at least two decades, many Raman experiments[21–25] were interpreted following that appealing theory.

Where was the problem? As Buchenau and coworkers[26,27] undoubtedly showed in the mid-1980's, a noteworthy excess in the vibrational density of states $g(\nu)$ over that predicted by Debye's theory $g(\nu) \propto \nu^2$ at low-enough frequencies occurs in glasses

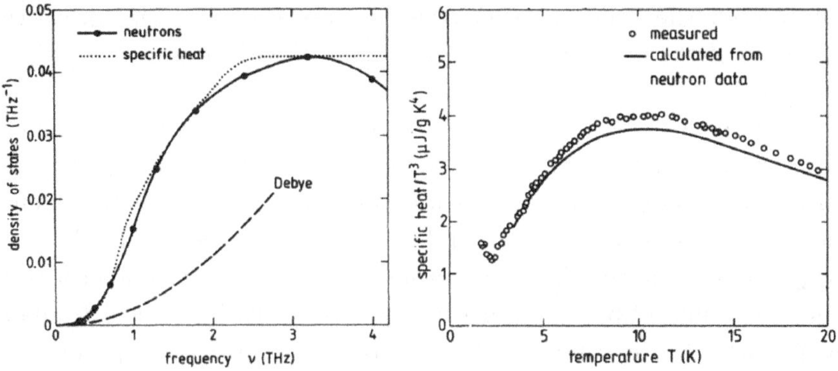

Fig. 3. (Left): Vibrational density of states in vitreous silica derived from inelastic neutron scattering (solid curve), compared to that obtained from specific heat data (dotted line). The dashed line shows the expected density of states due to acoustic phonons from Debye's theory. (Right): Measured specific heat C_p of vitreous silica plotted as C_p/T^3: T, compared to the curve calculated from inelastic neutron data, without any adjustable parameter. Reprinted with permission from Ref. [27]. Copyright 1986 by the American Physical Society.

(see Figure 3, for vitreous silica). Such an excess appeared as a broad peak in $g(\nu)/\nu^2$, which accounted fairly good for the broad maximum in C_p/T^3 always observed in glasses at a few K (as already shown in Figure 1), and which was indeed the real origin of the aforementioned "boson peak". The premise of Martin and Brenig theory that vibrations in amorphous solids follow Debye's law $g(\nu) \propto \nu^2$ was incorrect. It still took some more years until the destiny of the Raman-scattering *boson peak* and of the broad maximum in C_p/T^3 at a few K were brought back together.[28, 29]

After inelastic neutron scattering experiments[26–28] clearly unveiled that the boson peak originated from a broad maximum in the Debye-reduced vibrational density of states $g(\nu)/\nu^2$, many other experimental techniques have been employed in the last thirty years to study this central feature of glassy dynamics. Among them, we can mention far-infrared absorption,[30] hyper-Raman scattering,[31] inelastic X-ray scattering (IXS),[32–35] nuclear inelastic scattering (i.e., resonant inelastic scattering of X-rays via low-energy nuclear transitions),[36–38] single-molecule spectroscopy,[39] and terahertz-band time-domain spectroscopy.[40]

It should be emphasized, nevertheless, that despite the huge amount of accumulated experimental results about this ubiquitous "boson peak", its origin and microscopic nature remains a very controversial matter of discussion[41] by the community in the field (see Chapters 6–10).

2. How Universal Are "Universal Anomalies" of Glasses?

Two adjectives have pervaded the scientific literature regarding the behavior found generally in amorphous and other glassy solids: "anomalous" and "universal".

Given the common presence of non-crystalline solids (both natural and artificial) in real life, one may refuse to consider their behavior as *anomalous*, as opposed to the standard behavior of crystals in the framework of Debye's theory. This is even more true with the increasingly frequent finding of "glassy" properties in different crystals, as will be discussed throughout the book. As claimed by Leggett: *Glassy behavior is the norm, crystalline the anomaly!*[42]

On the other hand, it seems more appropriate indeed to refer to this ubiquitous set of unexpected low-temperature properties of glasses as *universal*, since they are very similar in such disparate materials (covalent amorphous solids, ionic glasses, polymers, metallic glasses, molecular and hydrogen-bonded glasses, plastic crystals, etc.).

However, after half a century of research on this topic, it may be pertinent to review the concept of universality. Are thermal properties of glasses (amorphous solids plus many disordered solids) as universal as so often assumed? To what extent are they universal? Which are the key magnitudes of that universal behavior and what is its ultimate reason?

We will briefly introduce this matter here, though more discussion about it will be found in following Chapters.

As already mentioned, the rich phenomenology widely found in glasses ranges from the low temperature thermal properties presented in Figures 1 and 2 to the particular temperature and frequency dependencies for the acoustic and dielectric properties.[10,14–16] This phenomenology was well accounted for by the standard TM, at least up to a few K. Probably, its best known fingerprint is the linear dependence of specific heat with temperature $C_p \sim T$, attributed to the presence everywhere of TLS with a density per unit volume and unit energy P_0, assumed constant. It is often heard that a characteristic of glasses at low temperatures is that they all systematically exhibit these tunneling TLS with a density $P_0 \sim 10^{45}$ $J^{-1}m^{-3}$ (see, for instance, the table in Ref. [43]). However, as it will be shown in Chapter 2, the range of variation for the linear coefficient of C_p, and hence for P_0, is larger than often believed, and the number of exceptions to the rule is not negligible.

On the contrary, others[42–45] have stressed that a more genuine (but unexplained), even *quantitative* universality is found in the behavior of the thermal conductivity (see Chapter 3) and the internal friction (see also Chapter 4) of glasses at low temperatures[9,10,14–16,42–45] — and hence of some related properties, such as sound velocity change with temperature. In Ref. [45], all available measurements on the low-temperature thermal conductivity and acoustic attenuation in amorphous solids as well as disordered crystals, on a total of over 60 different compositions, were exhaustively reviewed. It was confirmed that every kind of studied amorphous solid exhibited essentially the same thermal conductivity within one order of magnitude (see Figure 1 here, and also Figure 1 of Ref. [45] or Figure 2 in Chapter 3), as well as very similar internal friction for any frequency around the kHz range (see, e.g., Figure 2 of Ref. [45]). Let us remember that the corresponding behavior of their crystalline counterparts is dramatically different,[10] several orders of magnitude higher (thermal conductivity) or lower (internal friction), and with a distinct qualitative behavior. Specifically, both physical properties are featured in glasses by a central *plateau* (i.e., an almost temperature-independent region): (i) for thermal conductivity, the plateau typically ranges from 1 to 10 K and reaches a characteristic value $\kappa_{\text{plateau}} \sim 10^{-3} \text{Wcm}^{-1}\text{K}^{-1}$; (ii) for internal friction — the dimensionless acoustic attenuation — at low-enough frequencies, the plateau is in the range 0.1–10 K with a height $\sim 5 \times 10^{-4}$.

This more profound concept of universality can be traced back within the TM to the so-called tunneling strength $C \sim 3 \times 10^{-4}$, a dimensionless parameter defined as

$$C = \gamma^2 P_0 / \rho v^2 \tag{1}$$

where γ is the coupling energy between sound waves and TLS, ρ is mass density and v is sound velocity of the corresponding substance. (Strictly, one should use subscripts for C, γ, v, to distinguish longitudinal and transverse cases, but their differences are irrelevant here). Within the TM,[12,13] the thermal conductivity is inversely proportional to C (see Chapter 3), whereas the internal friction is linearly proportional to C.[10,14–16,45] In particular, the abovementioned plateau is predicted by the TM to be $Q^{-1} = (\pi/2)C$. Hence, the empirically observed universality of $Q^{-1} \sim 5 \times 10^{-4}$ in the wide plateau range directly translates into the universality of the TM tunneling strength $C \sim 3 \times 10^{-4}$. Therefore, the "universal" magnitude may not be the density of TLS P_0, but rather the product of the two TM fitting parameters scaled to the material elastic constants. It must be emphasized that the TM is basically consistent with all those experimental observations, provided we take $C \sim 3 \times 10^{-4}$, but it gives no reason at all why it is so. Since the four parameters entering C in Eq. (1) are fully independent, in principle, and vary significantly from substance to substance, it is quite surprising that C is so constant. This cannot be a mere coincidence, though the standard Tunneling Model does not provide any clue.

It is worth mentioning that Anderson and coworkers[46,47] were able to assess the behavior of the mean free path $l(\omega, T)$ of "phonons", after showing that phonons transport heat in glassy materials as in crystals. Thus, kinetic theory can also be applied to the thermal conductivity of amorphous solids, which can be expressed as

$$\kappa(T) = \frac{1}{3} \sum_j \int_0^{\omega_D} C_{Debye,j}(\omega, T) \nu_j(T) l_j(\omega, T) \tag{2}$$

where the sum is over the three acoustic phonon branches (one longitudinal and two degenerate transversal) up to the Debye frequency ω_D, and $C_{\text{Debye}}(\omega, T)$ is the heat capacity contribution of a phonon excitation of angular frequency $\omega = 2\pi\nu$ at temperature T.

In crystalline solids, Eq. (2) leads to the typical behavior shown in the right panel of Figure 2. Above ~10 K, phonons are scattered by intrinsic processes such as *umklapp* and phonon-phonon interactions, what produces a long monotonic decrease with temperature. Below a maximum at about 10 K, the phonon mean free path saturates by nonspecular scattering from the limiting surfaces of the sample. Since the sound velocities v_j are essentially constant, this implies a $\kappa(T) \propto T^3$ dependence.

By making use in Eq. (2) of the dominant-phonon approximation (where an average frequency $\nu = 100$ GHz is related to a temperature $T = 1$ K), Anderson and coworkers[46,47] deduced the general behavior of the phonon mean free path which led to that *universal* curve of thermal conductivity in glasses at low temperatures. Their conclusions are summarized in Figure 4, where three distinct frequency regions I, II and III can be observed, which correspond to the

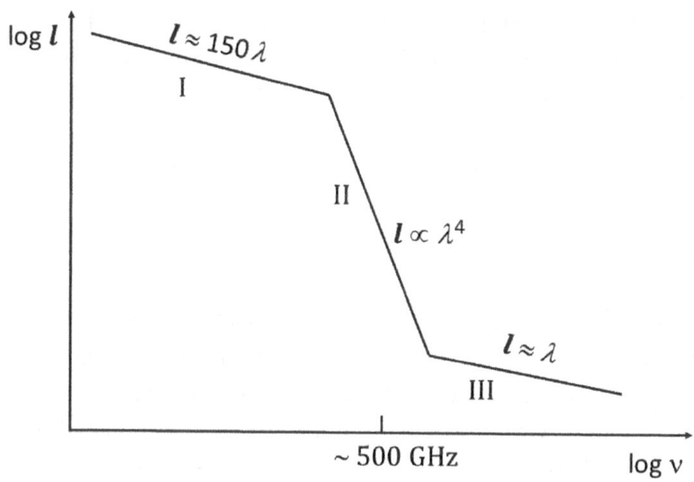

Fig. 4. Scheme of the frequency dependence of the intrinsic phonon mean free path l proposed by Anderson and coworkers[46,47] for glasses, widely accepted by the community in the field. λ is the phonon wavelength, inversely proportional to the frequency ν. The three frequency ranges labelled I, II and III are related to the corresponding three temperature ranges indicated in the right panel of Figure 2. Its physical meaning is explained in the text.

temperature ranges I, II and III identified in the right panel of Figure 2 for the thermal conductivity.

First, for temperatures below $\sim 2\,\mathrm{K}(\nu \leq 200\,\mathrm{GHz})$, all experimental data in glasses is consistent with a ratio of the mean free path over the wavelength of phonons $l/\lambda \approx 150$ (region I) to within a factor of 2. Within the Tunneling Model, the dominant mechanism that limits the mean free path of phonons at low temperatures is resonant scattering by TLS, which goes as $l(\omega, T) \propto \omega^{-1} \propto T^{-1}$. Since $C_{\mathrm{Debye}}(\omega, T) \propto T^3$, it is straightforwardly obtained that $\kappa(T) \propto T^2$, in excellent agreement with experiments.

Therefore, the equations found in the TM for the thermal conductivity at the lowest temperatures and for the acoustic attenuation in the kHz range are compatible and consistent[45] with the first $l \sim 150\,\lambda$ range, though its *quantitative* universality is not understood at all.

In the temperature range of about 2–20 K (region II), the thermal conductivity exhibits the ubiquitous *plateau*, whose center roughly coincides with the maximum in C_{p}/T^3 and with the end of the other plateau in the internal friction $Q^{-1} = (\pi/2)C$. In this range, the phonon mean free path drops rapidly with frequency, $l \propto \lambda^4$, that is, $l \propto \omega^{-4}$, in a frequency range centered at about $\nu \approx 500\,\mathrm{GHz}$, the frequency of the boson peak in $\mathrm{g}(\nu)/\nu^2$, again to within a factor of 2. On the one hand, this strong frequency dependence $l^{-1} \propto \omega^4$ of the phonon mean free path has often been attributed to Rayleigh scattering due to static disorder. On the other hand, one cannot rule out the possibility of a strong resonant scattering between phonons and localized excitations, which could lead to the abrupt decrease in $l(\omega)$. The former explanation was refuted by Zaitlin and Anderson,[46] who found that the magnitude of Rayleigh scattering in the studied glasses was not enough to account for region II. Nevertheless, at least for the case of vitreous silica, Raychaudhuri[48] showed that force constant fluctuations in SiO_2 along with the fluctuations in the Si-O-Si bond angle could indeed give enough Rayleigh scattering to account for the plateau in thermal conductivity. The latter possibility is the one assumed by the Soft-Potential Model (see Chapters 3 and 8), where the crossover from double-well to single-well potentials implies a resonant scattering passing from $l^{-1} \propto \omega$ (TLS) to $l^{-1} \propto \omega^4$

(*soft modes*), thus accounting for the thermal conductivity in the whole range below 20 K.

At still higher temperatures/frequencies (region III), the decreasing phonon mean free path eventually reaches its wavelength $l \sim \lambda$, a so-called Ioffe-Regel limit (see Chapters 6 and 7), when the acoustic vibrations cannot be considered as plane waves any longer. The corresponding Ioffe-Regel frequency when this limit is reached (the crossover between regions II and III) has been reported to coincide with the boson peak frequency in many glasses.[35]

For decades, exceptions to this general scheme had been observed only for the internal friction in amorphous films of covalently bonded, fourfold-coordinated Si, Ge, and C.[45] Further measurements of heat capacity and thermal conductivity in some evaporated thin films of amorphous silicon agreed with that suppression of glassy excitations at low temperature.[49] Nevertheless, more recent experiments by the same group showed that changing the preparation conditions of the a-Si thin films allowed the occurrence of typical low-energy glassy excitations.[50] On the other hand, a depletion of the TLS supposedly responsible for the typical glassy behavior at very low temperatures, have also been recently found experimentally in one ultrastable glass.[51] These apparent exceptions will be presented and discussed in Chapters 2 and 5.

Besides those few possible exceptions to the general rule of the universal behavior of glasses at low temperatures, a more fundamental criticism to the very concept of universality of tunneling states or TLS, and to the successful Tunneling Model itself, was raised first by Yu and Leggett[44] and thereafter by other theorists.[42,52–57] In brief (see more details in Chapter 4), it has been criticized the absent role of expected TLS-TLS interactions in the standard TM, which ignores them in most of situations. Although the TM is certainly consistent with most experimental observations at very low temperatures, it is not clear why this is so. It has been argued[42,44] that the TM may be just a subclass of a more general class of models for the many-body energy levels and matrix elements, that interact with each other by exchanging phonons. Nevertheless, the great success of the TM cannot be a coincidence, either. Furthermore, a simple mean-field scenario of phonon-mediated interactions between tunneling TLS would lead[57] to $C \sim 1$ instead of to the observed universal factor $C \sim 3 \times 10^{-4}$. Then, this puzzle hindered further progress to solve this enigma.

Also, from the experimental point of view, several measurements (mainly of acoustic attenuation and sound velocity variation at very low temperatures) reported[58–61] clear deviations from the basic TM predictions, both in insulating and metallic glasses, especially when varying the applied intensity or strain — hence probing the internal interactions. Only at low-enough intensity or strain, and/or at not so low temperatures, the picture of independent TLS inherent in the TM seemed to work.

Finally, we could also mention that the ubiquitous boson peak may be considered as another possible kind of "universal behavior" of glasses. This issue will be discussed in detail through Chapters 6–10.

Therefore, the issue of the *universal* properties of glasses — from amorphous solids to disordered crystals with glassy behavior — continues to be a matter of interest and vivid controversy. Hopefully, this book will shed some light on these open questions.

3. Aim and Contents of this Book

The search for understanding the very microscopic nature of the excitations responsible for all abovementioned "glassy anomalies", as well as for physical equations or models that account for them, has guided an exciting journey over the past 50 years. Several good reference books[10,14,16,62] and review articles[15,45] have been published about it during the last half century, leaving aside other more general review books[11,63–65] on glasses and the glass transition, which often include some section devoted to these topics.

However, it seems opportune — after this half century of studying universal "anomalies" in glasses — to update our knowledge on this matter from experimental, theoretical and computational points of view! And not only for amorphous solids, but also for other families of "glasses" such as "orientational glasses" and other crystals with distinct types of disorder, which have been incorporated more or less recently to the portfolio of glass phenomenology.

Certainly, not every theory or approach aiming to explain these ubiquitous properties of glasses at low temperature will be developed in this book. Let us mention, for instance, the Random First Order Transition (RFOT) theory,[66,67] which in fact does not specifically focus on the low-temperature properties, but rather on the glass transition phenomenon. This theory, including its explanations for the tunneling states and the boson peak (which are briefly summarized in Section 2.2 of Chapter 4 and in Section 2.1 of Chapter 5), is however covered exhaustively in a recent review book[65] and review articles,[67,68] to which the reader is referred. Nonetheless, the intention of the editor of the present book has been to provide a balanced combination of an updated overall review of the "anomalous" and "universal" low-temperature properties of glasses, with an open presentation and discussion of different state-of-the-art experiments, theories and models trying to account for them. Voice has been given to the main approaches currently used to understand the basic thermal, acoustic and vibrational properties of disordered solids, though unavoidably some others could be missed.

The content of the book is as follows. In the next chapters, the basic thermal properties, namely specific heat (Chapter 2) and thermal conductivity (Chapter 3) of amorphous solids and/or glasses at low temperatures, are reviewed. Some readers may miss an additional chapter devoted to thermal expansion measurements of glasses at low temperature. In fact, thermal expansion is much less significant here. As also occurs in crystals, thermal expansion in glasses[69] is very sensitive to microscopic material details and one cannot speak about "anomalous" nor "universal" behavior for this property.

Then, in Chapters 4 and 5, the universality of the behavior of amorphous solids at low temperature, with special emphasis on acoustic and dielectric properties, and the concept of a ubiquitous ensemble of *independent* TLS within the standard Tunneling Model, are critically discussed.

The boson peak, an essential though extremely controversial feature in the low-energy dynamics of glasses is addressed in Chapters 6–9. In Chapters 6 and 7, experimental results obtained from different techniques of vibrational spectroscopy are thoroughly discussed, aiming to shed light on the very microscopic nature of the boson peak and the low-frequency vibrations of glasses, in general. After this, in Chapters 8 and 9, some of the most popular models used specifically to account for the low-temperature/low-frequency properties of glasses are reviewed in more detail.

Once the most relevant experimental findings and models have been shown, Chapter 10 provides a comprehensive review on much recent work about powerful computer simulations that are currently being employed to investigate and test different hypotheses about the atomic dynamics of disordered solids.

Finally, Chapter 11 is a sort of *bonus*, which introduces the novel topic of topological matter in the context of amorphous solids. Although the field of topological insulators and topological states of matter was initially associated with symmetry and crystalline phases, the relevance of amorphous topological matter is increasing rapidly. This final chapter is expected to bridge the gap between the two research areas.

References

1. O. L. Anderson. (1959). The Debye temperature of vitreous silica. *J. Phys. Chem. Solids* **12**, 41–52.
2. P. Flubacher, A. J. Leadbetter, J. A. Morrison, and B. P. Stoicheff. (1959). The low-temperature heat capacity and the Raman and Brillouin spectra of vitreous silica. *J. Phys. Chem. Solids* **12**, 53–65.
3. A. C. Anderson, W. Reese, and J. C. Wheatley. (1963). Thermal conductivity of some amorphous dielectric solids below 1°K. *Rev. Sci. Instr.* **34**, 1386–1390.
4. W. Reese and J. E. Tucker. (1965). Thermal conductivity and specific heat of some polymers between 4.5° and 1°K. *J. Chem. Phys.* **43**, 105–114.
5. G. K. White and J. A. Birch. (1965). Thermal properties of silica at low temperatures. *Phys. Chem. Glasses* **6**, 85–89.

6. R. S. Craig, C. W. Massena, and R. M. Mallya. (1965). Thermal heat capacity of glycerol glass between 1.5° and 4°K. use of glycerol as a heat exchange medium in low-temperature calorimetry. *J. Appl. Phys.* **36**, 108–112.
7. A. J. Leadbetter. (1968). The thermal properties of glasses at low temperatures. *Phys. Chem. Glasses* **9**, 1–13.
8. C. L. Choy, R. G. Hunt, and G. L. Salinger. (1970). Specific heat of amorphous polymethyl methacrylate and polystyrene below 4°K. *J. Chem. Phys.* **52**, 3629–3633.
9. R. C. Zeller and R. O. Pohl. (1971). Thermal conductivity and specific heat of noncrystalline solids. *Phys. Rev. B* **4**, 2029–2041.
10. W. A. Phillips (ed.), (1981). *Amorphous Solids: Low-Temperature Properties.* Berlin-Heidelberg-New York: Springer.
11. S. R. Elliott. (1990). *Physics of Amorphous Materials.* 2nd ed., London: Longman.
12. P. W. Anderson, B. I. Halperin, and C. M. Varma. (1972). Anomalous thermal properties of glasses and spin glasses. *Philos. Mag.* **25**, 1–9.
13. W. A. Phillips. (1972). Tunneling states in amorphous solids. *J. Low Temp. Phys.* **7**, 351–360.
14. S. Hunklinger and W. Arnold. (1976). Ultrasonic properties of glasses at low temperatures. In W. P. Mason and R. N. Thurston (eds.), *Physical Acoustics.* New York: Academic, Vol. 12, pp. 155–215.
15. W. A. Phillips. (1987). Two-level states in glasses. *Rep. Prog. Phys.* **50**, 1657–1708.
16. P. Esquinazi (ed.), (1998). *Tunneling Systems in Amorphous and Crystalline Solids.* Berlin-Heidelberg-New York: Springer.
17. R. S. Krishnan. (1953). The scattering of light in fused quartz and its Raman spectrum. *Proc. Indian Acad. Sci. A* **37**, 377–384.
18. See p. 136 in Ref. [10].
19. R. Shuker and R. W. Gammon. (1970). Raman-scattering selection-rule breaking and the density of states in amorphous materials. *Phys. Rev. Lett.* **25**, 222–225.
20. A. J. Martin and W. Brenig. (1974). Model for brillouin scattering in amorphous solids. *Phys. Stat. Sol. (b)* **64**, 163–172.
21. R. J. Nemanich. (1977). Low-frequency inelastic light scattering from chalcogenide glasses and alloys. *Phys. Rev. B* **16**, 1655–1674.
22. See Chapter 8 in Ref. [10].
23. J. Lorösch, M. Couzi, J. Pelous, R. Vacher, and A. Levasseur. (1984). Brillouin and Raman scattering study of borate glasses. *J. Non-Cryst. Solids* **69**, 1–25.

24. G. Carini, M. Federico, A. Fontana, and G. A. Saunders. (1993). Low-frequency light scattering and structural defects in samarium phosphate glasses. *Phys. Rev. B* **47**, 3005–3010.
25. S. Kojima. (1993). Low-frequency Raman investigation of the liquid-glass transition in glycerol. *Phys. Rev. B* **47**, 2924–2927.
26. U. Buchenau, N. Nücker, and A. J. Dianoux. (1984). Neutron scattering study of the low-frequency vibrations in vitreous silica. *Phys. Rev. Lett.* **53**, 2316–2319.
27. U. Buchenau, M. Prager, N. Nücker, A. J. Dianoux, N. Ahmad, and W. A. Phillips. (1986). Low frequency modes in vitreous silica. *Phys. Rev. B* **34**, 5665–5673 .
28. V. K. Malinovsky, V. N. Novikov, P. P. Parshin, A. P. Sokolov, and M. G. Zemlyanov. (1990). Universal form of the low-energy (2 to 10 meV) vibrational spectrum of glasses. *Europhys. Lett.* **11**, 43–47.
29. M. A. Ramos. (1994). Comment on "Low-frequency light scattering and structural defects in samarium phosphate glasses". *Phys. Rev. B* **49**, 702–703.
30. T. Ohsaka and S. Oshikawa. (1998). Effect of OH content on the far-infrared absorption and low-energy states in silica glass. *Phys. Rev. B* **57**, 4995–4998.
31. B. Hehlen, E. Courtens, R. Vacher, A. Yamanaka, M. Kataoka, and K. Inoue. (2000). Hyper-Raman scattering observation of the Boson peak in vitreous silica. *Phys. Rev. Lett.* **84**, 5355–5358.
32. M. Foret, E. Courtens, R. Vacher, and J.-B. Suck. (1996). Scattering investigation of acoustic localization in fused silica. *Phys. Rev. Lett.* **77**, 3831–3834.
33. P. Benassi, M. Krisch, C. Masciovecchio, V. Mazzacurati, G. Monaco, G. Ruocco, F. Sette, and R. Verbeni. (1996). Evidence of high frequency propagating modes in vitreous silica. *Phys. Rev. Lett.* **77**, 3835–3838.
34. F. Sette, M. H. Krisch, C. Masciovecchio, G. Ruocco, and G. Monaco. (1998). Dynamics of glasses and glass-forming liquids studied by inelastic X-ray scattering. *Science* **280**, 1550–1555.
35. B. Rufflé, G. Guimbretière, E. Courtens, R. Vacher, and G. Monaco. (2006). Glass-specific behavior in the damping of acousticlike vibrations. *Phys. Rev. Lett.* **96**, 045502.
36. M. Seto, Y. Yoda, S. Kikuta, X.W. Zhang, and M. Ando. (1995). Observation of nuclear resonant scattering accompanied by phonon excitation using synchrotron radiation. *Phys. Rev. Lett.* **74**, 3828–3831
37. A. I. Chumakov, I. Sergueev, U. van Bürck, W. Schirmacher, T. Asthalter, R. Rüffer, O. Leupold, and W. Petry. (2004). Collective nature of the boson peak and universal transboson dynamics of glasses. *Phys. Rev. Lett.* **92**, 245508

38. A. I. Chumakov, A. Bosak, and R. Rüffer. (2009). Contribution of acoustic modes to the density of vibrational states measured by inelastic scattering techniques. *Phys. Rev. B* **80**, 94303.
39. Yu. G. Vainer, A. V. Naumov, and L. Kador. (2008). Local vibrations in disordered solids studied via single-molecule spectroscopy: Comparison with neutron, nuclear, Raman scattering, and photon echo data. *Phys. Rev. B* **77**, 224202.
40. M. Kabeya, T. Mori, Y. Fujii, A. Koreeda, B. W. Lee, J.-H. Ko, and S. Kojima. (2016). Boson peak dynamics of glassy glucose studied by integrated terahertz-band spectroscopy. *Phys. Rev. B* **94**, 224204.
41. For instance, in the last editions of the widely attended International Discussion Meeting on Relaxations in Complex Systems (IDMRCS) Conference Series, held in Lille (2005), Rome (2009), Barcelona (2013) and Wisła (2017), there were always specific sessions focused on the Boson Peak.
42. A. J. Leggett. (1991). Amorphous materials at low temperatures: Why are they so similar? *Physica B* **169**, 322–327.
43. J. F. Berret and M. Meissner. (1988). How universal are the low temperature acoustic properties of glasses? *Z. Phys. B — Condensed Matter* **70**, 65–72.
44. C. C. Yu and A. J. Leggett. (1988). Low temperature properties of amorphous materials: Through a glass darkly. *Comments Cond. Mat. Phys.* **14**, 231–251.
45. R. O. Pohl, X. Liu, and E. J. Thompson. (2002). Low-temperature thermal conductivity and acoustic attenuation in amorphous solids. *Rev. Mod. Phys.* **74**, 991–1013.
46. M. P. Zaitlin and A. C. Anderson. (1975). Phonon thermal transport in noncrystalline materials. *Phys. Rev. B* **12**, 4475–4486.
47. J. J. Freeman and A. C. Anderson. (1986). Thermal conductivity of amorphous solids. *Phys. Rev. B* **34**, 5684–5690.
48. A. K. Raychaudhuri. (1989). Origin of the plateau in the low-temperature thermal conductivity of silica. *Phys. Rev. B* **39**, 1927–1931.
49. B. L. Zink, R. Pietri, and F. Hellman. (2006). Thermal conductivity and specific heat of thin-film amorphous silicon. *Phys. Rev. Lett.* **96**, 055902.
50. D. R. Queen, X. Liu, J. Karel, T. H. Metcalf, and F. Hellman. (2013). Excess specific heat in evaporated amorphous silicon. *Phys. Rev. Lett.* **110**, 135901.
51. T. Pérez-Castañeda, C. Rodríguez-Tinoco, J. Rodríguez-Viejo, and M. A. Ramos. Suppression of tunneling two-level systems in ultrastable glasses of indomethacin. (2014). *Proc. Nat. Acad. Sci.* **111**, 11275–11280.

52. M. W. Klein. (1984). Density of states for interacting tunneling units in the absence of long-range order. *Phys. Rev. B* **29**, 5825–5835.
53. S. N. Coppersmith. (1991). Frustrated interactions and tunneling: Two-level systems in glasses. *Phys. Rev. Lett.* **67**, 2315–2318.
54. A. L. Burin, D. Natelson, D. D. Osheroff, and Yu. Kagan. (1998). Interactions between tunneling defects in amorphous solids. In P. Esquinazi (ed.), *Tunneling Systems in Amorphous and Crystalline Solids*. Berlin-Heidelberg-New York: Springer, Chapter 5.
55. D. C. Vural and A. J. Leggett. (2011). Universal sound absorption in amorphous solids: A theory of elastically coupled generic blocks. *J. Non-Cryst. Solids* **357**, 3528–3537.
56. M. Schechter and P. C. E. Stamp. (2013). Inversion symmetric two-level systems and the low-temperature universality in disordered solids. *Phys. Rev. B* **88**, 174202.
57. H. M. Carruzzo and C. C. Yu. (2020). Why phonon scattering in glasses is universally small at low temperatures. *Phys. Rev. Lett.* **124**, 075902.
58. P. Esquinazi, R. König, and F. Pobell. (1992). Acoustic properties of amorphous SiO_2 and PdSiCu, and of crystalline Ag, NbTi and Ta at very low temperatures. *Z. Phys. B: Condens. Matter* **87**, 305–321.
59. J. Classen, T. Burkert, C. Enss, and S. Hunklinger. (2000). Anomalous frequency dependence of the internal friction of vitreous silica. *Phys. Rev. Lett.* **84**, 2176–2179.
60. M. A. Ramos, R. König, E. Gaganidze, and P. Esquinazi. (2000). Acoustic properties of amorphous metals at very low temperatures: Applicability of the tunneling model. *Phys. Rev. B* **61**, 1059–1067.
61. R. König, M. A. Ramos, I. Usherov-Marshak, J. Arcas-Guijarro, A. Hernando-Mañeru, and P. Esquinazi. (2002). Strain dependence of the acoustic properties of amorphous metals below 1 K: Evidence for the interaction between tunneling states. *Phys. Rev. B* **65**, 180201.
62. M. I. Klinger. (2013). *Glassy Disordered Systems: Glass Formation and Universal Anomalous Low-Energy Properties*. Singapore: World Scientific.
63. E. Donth. (2002). *The Glass Transition*. Berlin: Springer.
64. K. Binder and W. Kob. (2005). *Glassy Materials and Disordered Solids*. Singapore: World Scientific.
65. P. G. Wolynes and V. Lubchenko (eds.). (2012). *Structural Glasses and Supercooled Liquids: Theory, Experiment, and Applications*. New Jersey: Wiley.
66. T. R. Kirkpatrick, D. Thirumalai, and P. G. Wolynes. (1989). Scaling concepts for the dynamics of viscous liquids near an ideal glassy state. *Phys. Rev. A* **40**, 1045–1054.

67. V. Lubchenko and P. G. Wolynes. (2007). Theory of structural glasses and supercooled liquids. *Annu. Rev. Phys. Chem.* **58**, 235–266.
68. V. Lubchenko and P. G. Wolynes. (2007). The microscopic quantum theory of low temperature amorphous solids. *Adv. Chem. Phys.* **136**, 95–206.
69. T. H. K. Barron and G. K. White. (1999). *Heat Capacity and Thermal Expansion at Low Temperatures.* New York: Kluwer/Plenum. See also Chapter 4 in Ref. [10].

https://doi.org/10.1142/9781800612587_0002

Chapter 2

Low-Temperature Specific Heat of Glasses and Disordered Crystals

Miguel A. Ramos

Low Temperature Laboratory, Condensed Matter Physics Department
Condensed Matter Physics Center (IFIMAC) and Instituto Nicolás Cabrera
Universidad Autónoma de Madrid, Madrid 28049, Spain
miguel.ramos@uam.es

The specific heat C_p of structural glasses (amorphous solids) and of many other disordered solids with glassy behavior, has been thoroughly studied in the past decades, as reviewed in this chapter. A quasilinear term in C_p dominant below 1 K (attributed to a constant density of *tunneling states* or two-level systems) and a ubiquitous hump in C_p/T^3 at about 3–10 K (due to the vibrational modes giving rise to the controversial *boson peak*), are the two main universal features of the low-temperature specific heat in glassy materials. To shed light on the very nature of these low-frequency excitations, investigating exceptions to the rule may be the key. Therefore, several relevant cases will be reviewed, from highly-stable amorphous solids devoid of these two-level systems to crystalline solids with minimal amounts of disorder that still exhibit those glassy features. Possible correlations between the magnitude of the contributions to the low-temperature specific heat and relevant parameters of the corresponding materials are also presented.

1. Introduction

The concept of heat capacity was introduced in the 18th century by the Scottish scientist Joseph Black, who also measured its magnitude for many substances. Nowadays, the specific heat (the extensive heat capacity scaled to the amount of substance) is likely the most significant thermodynamic coefficient, both at practical ambient conditions and at low temperatures. Experimentally, the heat capacity is just the gained energy (heat) per unit temperature increase. From a fundamental point of view, the heat capacity (at constant volume V) is directly the temperature derivative of the internal energy of the system: $C_{\mathrm{V}} = (\partial U/\partial T)_{\mathrm{V}}$, hence its leading role in thermodynamics. The corresponding specific heat is most usefully expressed per mole of atoms or molecules.

Indeed, the main advantage and, at the same time, drawback of the specific heat is precisely that it accounts for any kind of energy contribution (i.e., excitations of quantum energy levels) in the system. Further, these energy excitations are smeared out by thermal factors as we will see below. Thus, it is often difficult to draw conclusions about the microscopic origin of those measured contributions to the internal energy and discriminate among different theories without additional studies.

Since the early 19th century, the specific heat of solids (typically at room temperature) was successfully accounted for using the *classical* Dulong-Petit limit,[1] $C_{\mathrm{V}} = 3R = 3N_{\mathrm{A}}k_{\mathrm{B}} = 24.94\,\mathrm{J mol^{-1} K^{-1}}$, where R is the so-called gas constant, N_{A} is Avogadro number and k_{B} is Boltzmann constant. Strictly speaking, it is not C_{V} what is experimentally measured, but rather the heat capacity at constant pressure: $C_{\mathrm{P}} = (\partial H/\partial T)_p$, where H is enthalpy. Fortunately, C_{P}–$C_{\mathrm{V}} = TV\beta^2/\kappa$, with β the volume thermal expansion coefficient and κ the isothermal compressibility coefficient. Therefore, $C_{\mathrm{P}} \approx C_{\mathrm{V}}$ is fulfilled in every solid material, especially at low temperatures.

Of course, with the advent of *Quantum Physics* it soon became of great interest to study the behavior of C_{P} at low temperatures, like with any other physical property. In addition, both C_{P} and C_{V} should tend to zero for $T \to 0$ according to the Third Law of Thermodynamics, which manifested the clear failure of the classical Dulong-Petit law when decreasing the temperature. Except for the additional linear term of free electrons in metals (though we deal mostly with insulating glass-forming substances) and for some particular impurities (producing Schottky peaks), specific heat in solids was found to be directly proportional to T^3 at low enough temperature.[2–4]

By assuming that atoms vibrate in solids around their equilibrium lattice positions, Einstein[5] postulated a model of an ensemble of $3N$ harmonic oscillators (N being the number of present atoms), all of them vibrating with an averaged "Einstein frequency". Using the basic hypotheses of the quantum theory already introduced by Planck, he correctly obtained the Dulong-Petit classical limit at high temperature, but the heat capacity was predicted to decrease exponentially with decreasing temperature for $T \to 0$, that is, too fast compared to experiments.

A few years later, Debye[6] proposed his much more successful model. As can be studied in any textbook on solid state physics,[2–4] the Debye theory considers the energy contribution of the vibrational modes in a solid (*phonons*, the quanta or quasiparticles of the elastic field) as the result of the collective vibrations of the entire atomic lattice, a set of coupled harmonic oscillators. In brief, the total contribution of these phonons to the internal energy in a solid (in principle, crystalline) at a certain temperature is given by the sum of the energies $\hbar\omega$ over all vibrational modes, appropriately weighted by the Planck thermal distribution factor. Given the huge number of infinitesimally near vibrational modes, the summation can be replaced by an integral, and the total energy reads

$$U = \sum_{j=1}^{3} \int d\omega D_j(\omega) \frac{\hbar\omega}{\exp(\frac{\hbar\omega}{k_B T}) - 1} \tag{1}$$

where the density of states $D_j(\omega)$ is the number of vibrational modes per unit frequency range for each one of the three acoustic branches j.

The Debye model assumes a linear dispersion relation $\omega = v_j k$ (where k stands for the wavevector and v_j for the constant sound velocity of the corresponding polarization branch), up to a cutoff wave vector in the reciprocal space $k_D = (6\pi^2 N/V)^{1/3}$. This cutoff is reached when the number of acoustic modes is exhausted for a solid with N primitive cells, that is, the number of atoms in a monoatomic lattice. Hence the vibrational density of states (VDOS) is obtained to depend quadratically on frequency, $D_j(\omega) = (V/2\pi^2 v_j^3)\omega^2$ in the low-frequency (acoustic) range. Only acoustic phonons contribute at low temperatures, the contribution of higher-frequency optical phonons is negligible here.

To perform the sum over the three branches in Eq. (1), it is useful to define an averaged "Debye sound velocity" v_D:

$$\frac{1}{v_D^3} = \frac{1}{3}\sum_{j=1}^{3}\frac{1}{\nu_j^3} \tag{2}$$

In simple isotropic solids, including glasses, the summation runs over two degenerate transverse branches (with the same v_T) and one longitudinal branch (with v_L).

Then, the total contribution of acoustic phonons to the internal energy in the Debye model is:

$$\begin{aligned} U &= \frac{3\hbar}{2\pi^2 v_D^3} V \int_0^{\omega_D} d\omega \frac{\omega^3}{\exp\left(\frac{\hbar\omega}{k_B T}\right) - 1} \\ &= 9Nk_B T \left(\frac{T}{\theta_D}\right)^3 \int_0^{x_D} dx \frac{x^3}{(e^x - 1)^2} \end{aligned} \tag{3}$$

where a cutoff Debye frequency is given by $\omega_D = k_D v_D$, and hence a Debye temperature θ_D by $\hbar\omega_D = k_B\theta_D$. The Debye heat capacity is obtained by differentiation of the energy with respect to temperature:

$$\begin{aligned} C_v &= \frac{3V\hbar^2}{2\pi^2 v_D^3 k_B T^2} \int_0^{\omega_D} d\omega \frac{\omega^4 \exp(\frac{\hbar\omega}{k_B T})}{\left[\exp\left(\frac{\hbar\omega}{k_B T}\right) - 1\right]^2} \\ &= 9Nk_B \left(\frac{T}{\theta_D}\right)^3 \int_0^{x_D} dx \frac{x^4 e^x}{(e^x - 1)^2} \end{aligned} \tag{4}$$

In the right-side members of Eqs. (3) and (4), the useful change of variable $x \equiv \hbar\omega/k_B T$, with $x_D \equiv \hbar\omega_D/k_B T \equiv \theta_D/T$, has been made.

In the limit of high temperatures ($T \gg \theta_D$), Eq. (4) recovers asymptotically the Dulong-Petit value $C_{\mathrm{v}} = 3N_{\mathrm{A}}k_{\mathrm{B}}$. However, we are interested in the low-temperature limit (compared to the characteristic Debye temperature), $T \ll \theta_D$. There, one can replace the upper limit of the integral for infinity with negligible effects, so that it can solved analytically:

$$C_v \approx \frac{12\pi^4}{5} N k_B \left(\frac{T}{\theta_D}\right)^3 \tag{5}$$

Thus, the cubic coefficient of the specific heat of (crystalline) solids at low-enough temperatures can be successfully[2–4] predicted if we know the Debye temperature of the material, which is directly obtained from the elastic constants of the solid, without any fitting parameter or unknown constant, via

$$\theta_D = \frac{\hbar v_D}{k_B} \left(\frac{6\pi^2 N}{V}\right)^{1/3} \tag{6}$$

Alternatively, θ_D can be determined from the measured cubic coefficient C_v/T^3 in the low-temperature limit by means of Eq. (5).

A note of warning: strictly speaking, N/V should be taken as the number of primitive cells[2,3] per unit volume, each one providing three acoustic branches. This coincides with the atomic density only when there is one atom per primitive cell. However, N/V is often taken as the atomic density also in cases with non-monoatomic lattices. This amounts to include optical phonons in the vibrational spectrum, thus modifying (increasing) both ω_D and θ_D. Furthermore, in glasses, where the concept of primitive cells is ill-defined, sometimes the number of molecules per unit volume has been used. Therefore, one should take care of which interpretation for N/V is used in order to assess or compare different θ_D values in the literature.

The Debye theory for the vibrational and thermal properties of solids can be considered one of the most successful highlights of the quantum theory of solids developed during the first half of the 20th century. In particular, the cubic dependence of the specific heat at low temperatures has long been regarded as a key fingerprint of the quantum behavior of (crystalline) solids.

2. The Specific Heat of Glasses vs Crystals

Before, say, 1970, it was expected that Debye's theory should explain also the thermal properties of glasses or amorphous solids at low temperatures. The idea of $3N$ normal modes describing the vibrational excitations did not rely on the assumption of a periodic lattice with long-range translational order. Moreover, the wavelength and the mean free path of the contributing phonons should increase at low temperature, and hence the structural disorder would become less and less important with decreasing temperature. In the dominant phonon approximation, $\hbar\omega \approx 4 - 5\, k_B T$. Then, at 1 K, the dominant phonon frequency is $\approx 10^{11}$ Hz and for typical sound velocities, the wavelength of the relevant acoustic phonons is typically $\lambda > 300$ Å for $T < 1$ K. Therefore, a glass can be considered as a perfect, isotropic elastic continuum and its thermal properties should therefore be similar to those of crystals, following the Debye laws. As already commented in the previous chapter, this view unexpectedly turned out to be wrong, as there were other energy excitations in glasses (such as the TLS postulated by the Tunneling Model) that dominated their low-temperature properties. In particular, the seminal paper by Zeller and Pohl[7] and other works[8] (see the Introduction in Chapter 1) clearly showed that glasses exhibited thermal properties very different from those of their crystalline counterparts.

On the one hand, the specific heat of glasses was found to be much higher than that in the corresponding crystal state, decreasing much more slowly than the expected T^3-dependence at the lowest temperatures (see Figure 1 for vitreous germania and selenium). On the other hand, the thermal conductivity of glasses was found to be orders of magnitude lower than that of their crystalline counterparts and also exhibited a very distinct qualitative behavior (see Chapters 1 and 3).

The universal "anomalies" of glasses are not restricted to thermal properties. Indeed, both their acoustic and dielectric behavior are also "anomalous" (in the sense of very different to that of crystals) and, even more strikingly, universal (i.e., very similar among themselves irrespective of the type of material, chemical bonding, etc.), as anticipated in Section 2 of Chapter 1 and developed in more detail in following chapters. All this "glassy" behavior of acoustic and dielectric properties at low temperature was soon ascribed to the same new excitations responsible for the "anomalous" specific heat and thermal conductivity below a few K.

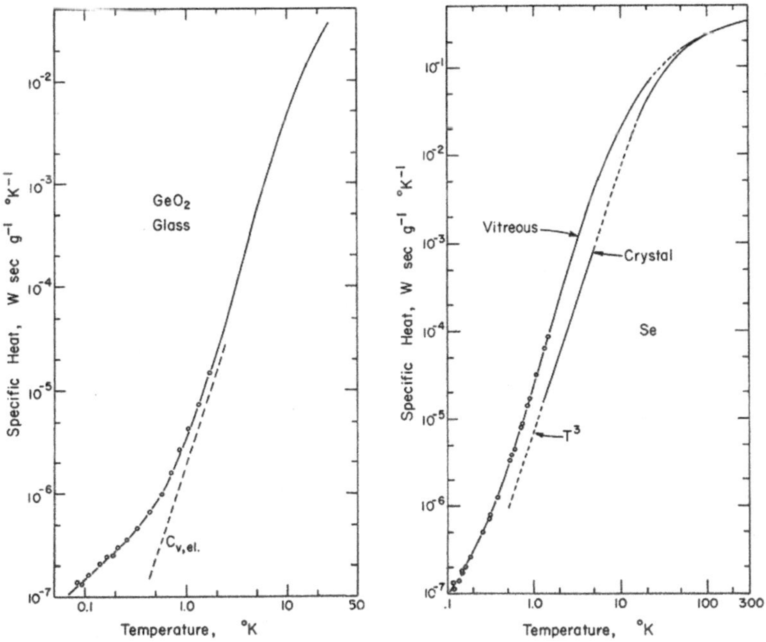

Fig. 1. Specific heat as a function of temperature in a logarithmic scale of vitreous germanium dioxide (left) and of selenium for both glass and crystal states (right). The dashed lines indicate the cubic dependence expected from Debye's theory for GeO_2 glass and for Se crystal. Reprinted with permission from Ref. [7]. Copyright 1971 by the American Physical Society.

2.1. *The specific heat below 1 K*

Since the reference for the specific heat of an insulating solid at low temperature is the Debye law, $C_{\mathrm{p}} \approx C_{\mathrm{v}} = C_{\mathrm{Deb}}T^3$ (see Eq. (5)), it is convenient to plot the "Debye-reduced" specific heat C_{p}/T^3 as a function of temperature (often in a logarithmic scale to amplify the low-temperature range). This was done by Zeller and Pohl in their abovementioned paper (see Figure 1 of Chapter 1) and will be done here in most cases (see Figure 2 and many figures thereafter). In the crystalline state of the studied substances, when available, a flat curve can be observed at low-enough temperatures, as expected. However, for amorphous solids or glasses, the specific heat is always above the predicted Debye contribution, as indicated by the dashed lines in the left panels of Figure 1 (germanium dioxide) and of Figure 2 (glycerol), respectively.

In general, one can distinguish two relevant temperature ranges: below or above of, say, 1–2 kelvin. Below 1–2 K, glasses exhibit an upturn in C_p/T^3 above the Debye level. Moreover, they also show a hump (the so-called boson peak) in C_p/T^3 typically around 3–10 K, reminiscent of — but quantitatively and qualitatively different from — the van Hove singularity seen in many crystals at higher temperatures.

Thus, at $T < 1$ K, the specific heat is clearly dominated by an unexpected linear dependence on temperature, attributed to the TLS of the Tunneling Model already mentioned in Chapter 1 and further developed in this and following chapters. Strictly speaking, a slightly *supralinear* dependence $C_p \sim T^{1+\delta}$ has often been found,[8–10] consistent with a corresponding $\kappa \sim T^{2-\delta}$ dependence reported for the thermal conductivity in the same temperature range, with $\delta \approx 0.1 - 0.2$. Nevertheless, for easy data handling, a purely linear dependence is usually assumed. Therefore, when representing C_p/T vs T^2, a simple least-squares linear fit at the lowest temperatures provides the TLS linear term (the intercept with the ordinate axis). This is exemplified in the right panel of Figure 2 for glasses of $Ca_{0.4}K_{0.6}(NO_3)_{1.4}$ (CKN) and glycerol, to be compared with the crystal of the latter devoid of any linear term.

2.2. *The specific heat above 1 K*

Above 1–2 K, the linear contribution of TLS to the specific heat of glasses becomes more and more negligible, but still $C_p(T)$ remains over the Debye level. In principle, the slopes of the linear fits of C_p/T vs T^2 as in Figure 2 (right panel) should provide the cubic Debye coefficient. This is true for the crystal, but it is not straightforward for the glass. The permanent excess of $C_p(T)$ over the Debye level, characterized by the broad maximum in C_p/T^3 (the boson peak), makes the cubic coefficient obtained from the fit to be higher than the C_{Deb} expected from the Debye's theory, determined via Eqs. (5) and (6). In fact, it has been stated[8,9] for many years that the *calorimetric* Debye coefficients in glasses exceeded the *elastic*

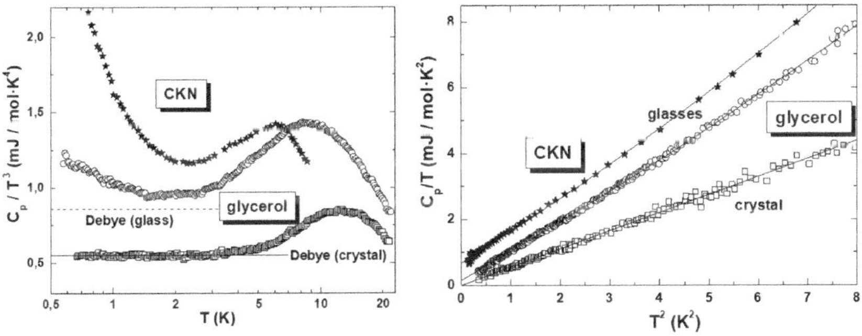

Fig. 2. Specific heat as a function of temperature for CKN glass[11] and for glycerol glass and crystal.[12] (Left): Data plotted in a C_p/T^3 representation with a logarithmic scale for temperature. Horizontal dashed and solid lines indicate the Debye level calculated from elastic constants for glass and crystal of glycerol, respectively. The Debye coefficient for CKN is 1.0 mJ/mol·K (not shown). (Right): Lowest-temperature data data plotted in a C_p/T vs T^2 representation. The solid lines are fits to the Soft-Potential Model (glasses) and to the Debye's theory (crystal). The glasses show a positive intercept with the ordinate axis (the linear coefficient in C_p), whereas the crystal tends to zero, as expected.

coefficients obtained from Debye's theory in those cases where sound-velocity measurements were available. We'd better say that a simple linear fit of C_p/T vs T^2 data in glasses does not determine correctly the cubic Debye coefficient because other contributions should be considered too. We will discuss this point more quantitatively in Section 5.3.

Interestingly, the temperature range at about 3–12 K where the C_p/T^3 hump occurs, is the same where the *plateau* in the thermal conductivity appears (see Chapters 1 and 3 for further detail). Although other alternative explanations have arisen along past decades, such as Rayleigh scattering by structural heterogeneities or through extensions of the Tunneling Model with more complicated distributions of TLS, it seems reasonable to assume that the *plateau* in the thermal conductivity is due to the scattering of acoustic phonons by the same additional excitations responsible for the broad maximum in C_p/T^3.

In the 1980's, inelastic neutron scattering experiments by Buchenau and coworkers (see the discussion and Figure 3 in Chapter 1) clearly demonstrated that those additional excitations

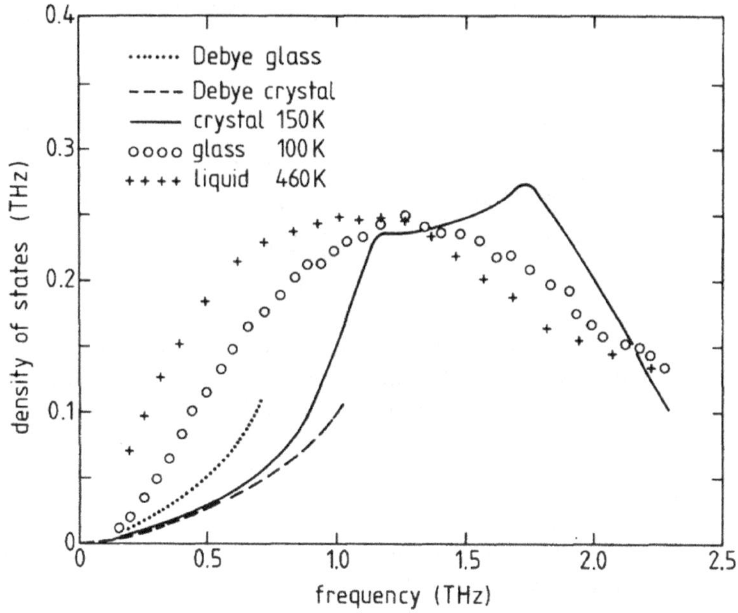

Fig. 3. The low-frequency density of states on an absolute scale for glassy, liquid, and polycrystalline selenium, obtained from inelastic neutron scattering. The limiting Debye form is also shown. Reprinted with permission from Ref. [13]. Copyright 1997 by the American Physical Society.

producing the *boson peak* were indeed vibrational modes of rather harmonic character (as opposed to the anharmonic TLS, dominant at lower energies or frequencies, but invisible in these experiments), though different from ordinary sound waves. We show here in Figure 3 the vibrational density of states (VDOS) so obtained for selenium, in its different states: liquid, crystal and glass. As can be seen, in contrast to the agreement with the Debye prediction observed for the crystal, in the glass even at low temperature a clear excess in the VDOS is observed at low frequencies, reminiscent of the behavior found in the liquid state.

One should not confuse the van Hove peaks for the crystal or the broad maximum for the glass in the *total* VDOS, shown in Figure 3 for the different states of selenium, with the proper boson peak. The boson peak producing the ubiquitous hump in C_p/T^3 is a maximum in the *reduced* density of states $g(\nu)/\nu^2$, that is, it takes place at the frequency $\sim 0.5\,\text{THz}$ in Se, where there exists the largest separation between the expected Debye curve (dotted line, Debye glass, in Figure 3) and the measured VDOS (open circles, glass).

In the 1990's, Sokolov *et al.*[11,14] found an interesting (though still controversial) inverse correlation of the strength of the boson peak relative to the Debye background with the so-called fragility[15,16] of the glassforming liquid. The fragility index m, or alternative definitions of the fragility also employed through the literature,[17] have been used to characterize the deviation of an averaged structural relaxation time of the liquid (and supercooled liquid) from a standard Arrhenius dependence on reciprocal temperature. Sokolov *et al.*[11,14] assessed the height of the boson peak either from reduced Raman-scattering data or from C_p/T^3 data, scaled to the observed Debye backgrounds, and found that the boson peak of the glass was higher, the less fragile the substance was. In Figure 4, their data for *structural*

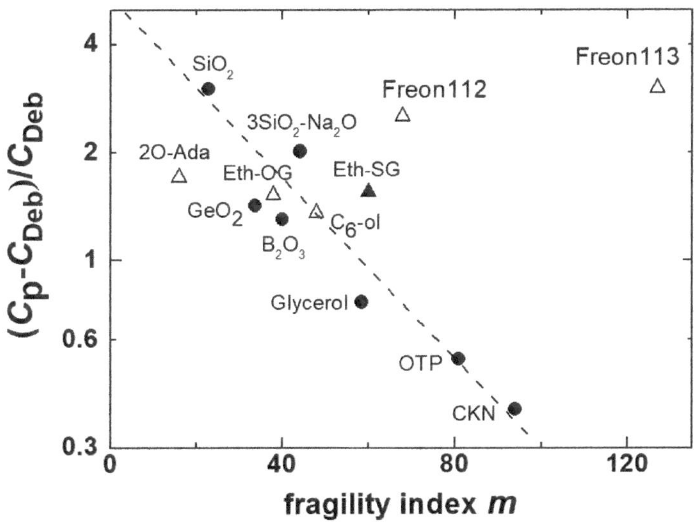

Fig. 4. The ratio of the excess specific heat to the Debye value at the "boson peak" maximum in C_p/T^3 (at $T \sim$ 3–10 K) vs the fragility index m (for definitions, see Ref. [17]). Circles are data taken from Sokolov *et al.*[11,14] for different glasses. Solid triangle: structural glass of ethanol (Eth-SG).[18] Open triangles: data for several *orientational* glasses, after Ref. [17]: ethanol (Eth-OG), adamantanone (2O-Ada), cyclohexanol (C_6−ol), Freon 112 and Freon 113. In general, solid symbols indicate structural glasses (amorphous solids) and open symbols indicate "orientational glasses" (i.e. orientationally-disordered crystals). The dashed line is a guide to the eye intended for structural glasses.

glasses are depicted, together with additional data for *orientational glasses*,[17] that is, molecular crystals with orientational disorder which exhibit glassy properties (see Section 3.2). These orientational glasses (frozen-in plastic crystals) do not follow the posited correlation of the boson peak for structural glasses.

3. Glassy Features in Disordered Crystals

3.1. *Mixed crystals*

The finding of "anomalous" and universal behavior for many low-temperature properties in amorphous materials was initially thought to be a consequence of the lack of translational order in non-crystalline solids. Nonetheless, with the time passing, more and more cases of crystalline solids with some kind of disorder exhibiting *glassy* behavior have been reported. As a matter of fact, the current use of the term "glassy" anomalies or "glassy" behavior is more general and accurate than referring to amorphous solids only, as it was often done a few decades ago. Indeed, it has been natural to investigate those poorly understood glassy excitations in systems that preserve at least a periodic crystalline lattice, with some more-or-less controlled amount of disorder, what should facilitate their study.

The first systematic experimental and theoretical studies, aiming to elucidate the microscopic nature of the low-energy excitations responsible for the low-temperature anomalies of glasses, were focused on disordered alkali-halide alkali-cyanide crystals[19–23] and other mixed crystals such as $Ba_{1-x}La_2F_{2+x}$.[24]

De Yoreo *et al.*[19] doped KBr crystals with different x concentrations of nonspherical CN^- ions that replace Br^- ions in the lattice. Their specific-heat data in a C_p/C_{Debye} representation is reproduced in Figure 5 (left panel). Of course, pure KBr or KCN crystals followed Debye's theory. For dilute concentrations $0 < x < 0.05$, isolated 180° tunnelling flips of the elongated CN^- ions produced Schottky-like contributions to the specific heat. However, for higher concentrations, stronger interactions between them and with the lattice led to the typical glassy behavior of both specific heat and

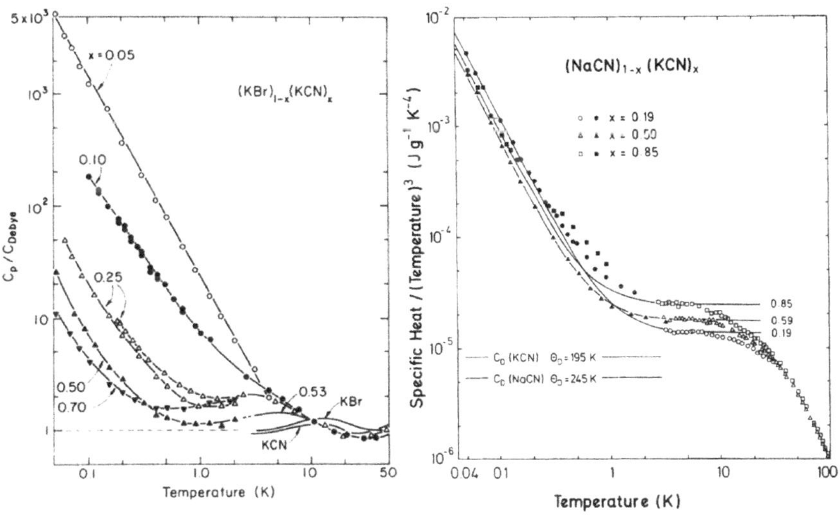

Fig. 5. (Left): Low-temperature specific heat of crystalline $(\mathrm{KBr})_{1-x}(\mathrm{KCN})_x$ for different CN^- concentrations x, plotted as $C_\mathrm{p}/C_\mathrm{D}$, where C_D is the Debye specific heat calculated from the elastic constants. Reprinted with permission from Ref. [19]. Copyright 1986 by the American Physical Society. (Right): Low-temperature specific heat of different crystals of $(\mathrm{NaCN})_{1-\mathrm{x}}(\mathrm{KCN})_x$, plotted as C_p/T^3. Solid lines are best fits to a quasilinear plus cubic curve $C_\mathrm{p}(T)$. Debye contributions at lowest temperatures for both pure NaCN and KCN are indicated. Reprinted with permission from Ref. [21]. Copyright 1986 by the American Physical Society.

thermal conductivity. In brief, the excess specific heat was thought to arise from CN^- librational excitations coupled to the lattice.

A slightly different case (right panel in Figure 5) of alkali-halide alkali-cyanide mixed crystal is that of $(\mathrm{NaCN})_{1-\mathrm{x}}(\mathrm{KCN})_\mathrm{x}$. Here, static random strains are generated by the volume difference of the K^+ and Na^+ ions, which are distributed randomly through the cationic sublattice, whereas the cyanide sublattice remains undiluted.[21] As in the former case, a large quasilinear contribution to C_p attributed to TLS is observed below 1 K as an upturn in C_p/T^3, but with only a shallow maximum (boson peak) just below 10 K.

All these results in mixed crystals pointed out to the relevance of random strain fields developed in the atomic lattice as the possible source of TLS.[23]

3.2. *Orientationally-disordered crystals ("glassy crystals")*

Beyond abovementioned stoichiometrically-mixed disordered crystals, another kind of "orientational glass" exists that furthermore exhibits a *glass transition* between a homogeneous glassy (nonergodic) state and a liquidlike (ergodic) state. They are orientationally-disordered crystals (ODC) or "glassy crystals" — as they are also often called[25] — which are produced by quenching molecular plastic crystals, with a *rotational* disorder that becomes frozen-in into *orientational* disorder below a "glass transition" temperature T_g. Some of these "orientational glasses" were included before in Figure 4. It appears that the inverse correlation there proposed between the boson peak strength and the fragility does not work for these ODC.

Specially interesting is the paradigmatic case of ethanol, where a fully-ordered monoclinic crystal phase, a genuine glass (amorphous) phase with $T_g = 97\,\mathrm{K}$, and a cubic crystal phase with random orientational disorder (presumably of the methyl groups of ethanol molecules) after supercooling the plastic rotator-phase also below $T_\mathrm{g} = 97$ K, are all experimentally accessible at low temperatures.[26–28]

Astonishingly, the specific heat at low temperature of the ODC of ethanol shows, even quantitatively, the same density of TLS and almost the same *boson peak* as the structural glass, in clear contrast to the orthodox Debye behavior of the monoclinic crystal, as can be seen in Figure 6 for fully-deuterated ethanol.[29] Exactly the same phenomenology was observed for normal, fully-hydrogenated ethanol.[29,30] Subsequent thermal conductivity measurements[31] in ethanol confirmed those findings, since the ODC-phase curve was indistinguishable from that of the structural glass, both showing the universal plateau, and very different from the typical polycrystalline behavior of the monoclinic crystalline phase. The case of ethanol is unique is the sense of being a simple molecule which allows to select the different degrees of order (or lack of it) in a solid, either translational or orientational, or even rotational, allowing a direct *quantitative* comparison among them. The robust conclusion to be

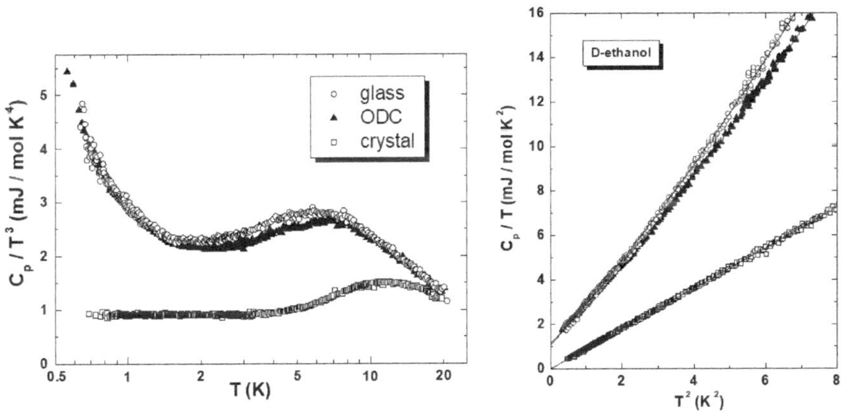

Fig. 6. Specific heat of fully-deuterated ethanol in its three possible states at low temperature: glass (open circles), orientationally-disordered crystal (ODC, solid triangles) and stable monoclinic crystal (open squares).[29] Data are plotted as C_p/T^3 in the whole temperature range measured (left) and in a C_p/T vs T^2 plot for the same data at the lowest temperatures (right). The solid lines (fits) emphasize a very similar intercept with the ordinate axis, and hence linear coefficient in C_p, for glass and ODC, which is null in the fully-ordered crystal.

drawn is that the lack of long-range translational order characterizing amorphous solids is not a key ingredient for "glassy behavior", neither for low-temperature "anomalies" nor for the glass transition itself.

3.3. *Other disordered crystals with glassy behavior*

Some of these glassy anomalies at low temperature (in specific heat, but also in thermal conductivity or acoustic attenuation) have also been reported in other crystals with different types of disorder. It would not be fair to say that *any* kind of disordered solid exhibits glassy behavior, but certainly new examples of them, beyond abovementioned *orientational glasses*, are continuously found.

As early as forty years ago, some *ferroelectric* substances were shown[32,33] to present the typical phenomenology of glasses at low temperatures, not only in their specific heat, but also in thermal conductivity or dielectric dispersion, what is an important support to disregard other possible contributions and alternative explanations. Interestingly, glasslike behaviour was most pronounced in polycrystalline samples of relaxor ferroelectrics, with a diffuse phase transition (such as in PLZT, SBN or PMN), whereas those ferroelectrics with a sharp phase transition ($BaTiO_3$, KDP or TGS) lack glassy anomalies.[33]

In addition, the specific heat of different incommensurate modulated crystals with broken translational periodicity have been found to present features at low temperatures similar to those of glasses. On the one hand, there are quasi one-dimensional conductors[34,35] presenting a charge density wave (CDW) such as the $(TaSe_4)_2I$ compound,[34] which exhibit a quasilinear term at low temperatures and a bump in C_p/T^3 over the Debye behavior. In this case, glassy dynamics has been ascribed to CDW phase excitations above their pinning frequency. On the other hand, in order to avoid possible ambiguity due to *metallic-like* contributions, more attention has been paid to 3-dimensional incommensurate *insulators* such as BCPS $[(ClC_6H_4)_2SO_2]$[36] or, more recently, $ThBr_4$[35] (Figure 7, left panel). In the latter case, the authors explain the maximum in C_p/T^3 ("boson peak") from the gaps occurring in the phase and amplitude branches, while the lowest-temperature upturn is ascribed to the finite lifetime of the phason excitations.

An interesting, related system is that of *quasicrystals*, which are neither perfect crystals (at least, not in 3D) nor amorphous solids. They are not typical "disordered crystals" either. Naturally, the question arose whether their lattice vibrations resemble more closely those of crystals (Debye theory at low temperatures) or of amorphous solids (glassy anomalies with TLS). Few available results of specific heat[37] are however ambiguous, again due to alternative contributions arising from the semimetallic or superconducting behavior of most quasicrystals at low temperatures. Nevertheless, at least for an icosahedral AlPdMn quasicrystal, universal glassy anomalies in thermal conductivity and internal friction were clearly observed.[38]

More recently, several interesting cases of crystals with minimal amount of disorder have also been reported to exhibit standard glassy behavior in their low-temperature specific heat and related properties. In the right panel of Figure 7, low-temperature specific-heat

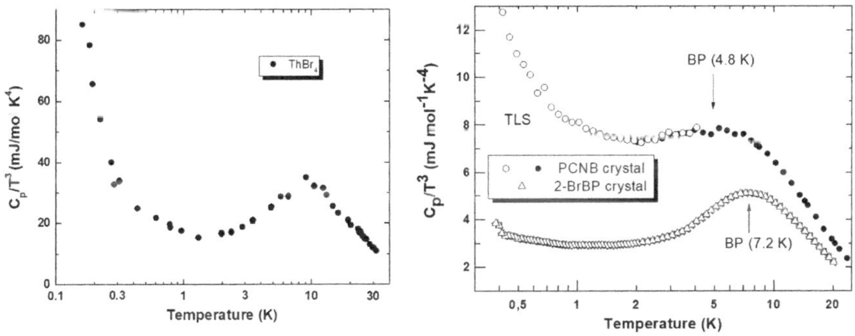

Fig. 7. (Left): Specific heat of an incommensurate modulated crystal of insulating $ThBr_4$ with broken translational periodicity.[35] Data are plotted as C_p/T^3 in the whole temperature range. (Right): Specific heat of pentachloronitrobenzene (PCNB), a highly anisotropic crystal[39] (open and solid circles correspond to different experimental runs), and of a single monoclinic crystal of 2-bromobenzophenone (2-BrBP)[40] (open triangles). Both crystals exhibit the upturn at low temperatures characteristic of TLS, and a "boson peak" at 4.8 K and 7.2 K, respectively.

data for two different substances are shown. The most clear case of glassy behavior at low temperatures here is pentachloronitrobenzene (PCNB, $C_6Cl_5NO_2$), with a large density of TLS and a boson peak at 4.8 K.[39] PCNB is a quasi-planar hexasubstituted benzene derivative that exhibits a layered structure of rhombohedral parallel planes in which the molecules can rotate around a six-fold-like axis. Pentachloronitrobenzene is therefore a *highly anisotropic* solid that can be considered a crystal with a low-dimensional orientational disorder. Moreover, its thermal conductivity at low temperatures was also found to be very similar to that of noncrystalline solids.[39] The other example here shown is ortho-bromobenzophenone (2-BrBP), an organic crystal with low-energy intramolecular modes, capable of distorting the low-energy density of states.[40] The low-temperature specific heat of a single-crystal sample (presenting a monoclinic lattice) of 2-BrBP, grown with a very low concentration of lattice defects, is also seen to exhibit a modest quasilinear contribution and a boson peak at 7.2 K.

Another interesting system to be mentioned is that of $CBr_nCl_{4-n}(n = 0, 1, 2)$ halomethanes, studied at low temperature by Moratalla *et al.*[41] where, despite being translationally-ordered stable monoclinic crystals, *glassy* dynamical features had been previously reported. For both Br-substituted substances, a sort of glass-transition feature appearing at about 90 K in their C_p curves suggested the freezing of a molecular motion, associated with an "occupational disorder" introduced by the exchange of halide ions. This may be the origin of their low-frequency excitations for $n = 1, 2$. Nonetheless, the three studied CBr_nCl_{4-n} crystals were found to exhibit the characteristic glassy features at low temperatures, namely a broad *boson peak* in both C_p/T^3 and $g(\omega)/\omega^2$ VDOS (from inelastic-neutron-scattering data), as well as a nonzero linear term in C_p, all features increasing with the number n of Br atoms per molecule, $n = 0, 1, 2$. Thus, also pure tetrachloromethane, CCl_4, a fully-ordered monoclinic crystal (its stable phase below 225 K) devoid of any occupational disorder, exhibited a similar glassy behavior, though to a lesser extent. Interestingly, taking advantage of the fully periodic structure of the CCl_4 crystal, the authors[41] were able to determine and study the dispersion relations for acoustic and optical phonons over the whole Brillouin zone, by means of Density Functional Theory (DFT) molecular calculations. In brief, the low-energy (*soft*) modes at 3–4 meV, responsible for the "boson peak", were found to arise from a *piling up* of the lowest optical modes that merged with Brillouin-zone-boundary acoustic phonons.

4. Low-temperature Specific Heat of Highly-stable Glasses

A long-standing question during this last half-century of universal "anomalies" of glasses, has been whether those glassy excitations at low temperatures are intrinsic properties of any non-crystalline solid, or rather will they gradually disappear if we get glasses of much higher stability. In other words, whether the TLS density or the boson-peak strength are robust dynamic properties of glasses and other disordered solids, or are they simply proportional to the degree of configurational disorder. In the latter case, the magnitude of glassy features could be drastically reduced, even suppressed, by allowing the system to descend very deep into its energy landscape approaching the *ideal glass* state, thus reaching states of very low enthalpy or negligible configurational entropy (see Ref. [42] and references therein). In principle, there are two ways to seek these highly-stable glasses: (i) identify particular cases of amorphous solids with a very compact, rigid network, such as silicon or other covalently-bonded elements with 4-fold coordination (see Section 2 in Chapter 1, as well as Chapter 5); (ii) obtain canonical glasses by extremely slow cooling or after extremely efficient annealing or aging processes below the glass-transition temperature T_g.

Indeed, there are only a few papers in the literature claiming the absence of TLS (likely, the most characteristic fingerprint of glassy behavior) following the option (i). Pohl and coworkers[43] observed a dramatic reduction of the internal friction plateau, which is proportional to the amount of TLS weighted by the TLS–phonon coupling energy, in 1%-hydrogenated amorphous silicon (*a*-Si), whereas they did find TLS in pure *a*-Si. The suppression of TLS in hydrogenated *a*-Si was attributed by the authors to a more compact fourfold coordination due to the passivation of the dangling bonds by the hydrogen, hence producing an overconstrained vibrational network.[43] Later on, Hellman's group has reported[44,45] specific heat and thermal conductivity experiments that show negligible TLS in some *a*-Si thin films but not others, all depending on mass densities and methods of preparation (see Chapter 5). In general, measurements in these films of amorphous semiconductors have often given contradictory results.[45] Another interesting case is that of (polyamorphic) amorphous water. Agladze and Sievers[46] reported absence of far-infrared resonant absorption by TLS in low-density amorphous (LDA) ice at low temperature, whereas its counterpart high-density amorphous (HDA) ice did exhibit the typical TLS response of other glasses.

Nevertheless, one of the most important recent advances in the scientific research of glasses, both experimentally and theoretically, has been the discovery by Ediger's group of a novel way to obtain higher-quality, more dense and lower-entropy glasses, the so-called *ultrastable glasses*.[47,48] They are prepared by physical vapor deposition (PVD) of (usually organic) glasses at temperatures $\sim 0.85T_g$.

Following this idea, several groups in the world are currently able to obtain glassy thin films with extraordinary kinetic and thermodynamic stability within the laboratory time scale. A controlled evaporation rate in combination with an optimal substrate temperature have proved to drastically favour two-dimensional mobility and, as a consequence, low-energy-minima states in the potential-energy landscape. It has been estimated that their stability is equivalent to have aged a liquid-cooled glass for thousands of years. For a review on these PVD highly-stable glasses, see Ref. [48].

Up to date, only one experiment of specific-heat measurements at very low temperatures has been reported for these PVD ultrastable glasses (USG), namely that by Pérez-Castañeda *et al.*[49] in indomethacin (IMC). Their data plotted as C_p/T^3 for two slightly different USG, compared to similar samples of IMC but in a conventional glass state or in the crystal state, are shown in the left panel of Figure 8. As can be seen there, USG films of IMC exhibit a full suppression of TLS within experimental error (following the Debye law as in crystals), whereas glassy films prepared in conditions other than ideal, show a typical linear term in C_p and hence a

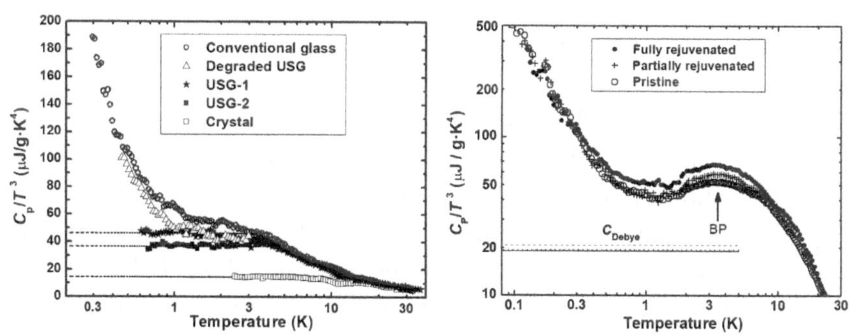

(a) (b)

Fig. 8. Specific heat data at low temperatures for two types of highly-stable glass: ultrastable physical-vapor-deposited indomethacin (IMC) glasses[49] (a), and pristine 110-million-year amber glass[42] (b), both compared to their standard glassy states. For IMC, the crystal state is also available and shown. For amber glass, an intermediate case of partially-rejuvenated glass is also presented. Horizontal dashed lines indicate corresponding Debye contributions. See text for more details.

significant density of TLS.[49] Only a very shallow boson peak is suggested below 5 K for both ordinary and ultrastable glasses.

At first glance, one might thus ascribe the found suppression of TLS for the USG samples to the extraordinary stability (either thermodynamic or structural) of these particular glasses. Moreover, recent computational studies[50] have supported this finding. By employing an advanced (*swap*) Monte Carlo method, a strong depletion of TLS in ultrastable computer-generated glasses has been found, in stark contrast to the usual quenched glasses. However, earlier work on a very different type of highly-stable glass casts some doubt on such apparently simple explanation, or at least on its possible generality, as will be discussed next.

As pointed out above, a second way (ii) to get highly-stable glasses would be by means of patient structural-relaxation processes, such as extremely-long aging or annealing below T_g. Of course, this method was more obvious and many attempts of studying the dependence of glassy properties on its thermal history can be found in the literature since the 1970's, but only by employing the very modest annealing or aging processes that could be performed in a laboratory.[42] Depending on the glass studied and several other factors, these works have given conflicting or ambiguous results. The answer was thus not clear, though most authors tended to think that if a drastic reduction of the frozen-in configurational disorder could be realized, these glassy anomalies should be significantly depleted.

To overcome, however, that practical difficulty of requiring geological times for glass stabilization, one can resort to the help of nature indeed. Specifically, Dominican amber glasses, geologically aged for some 20 million years, were firstly identified and studied.[51,52] Calorimetric and mechanical characterizations demonstrated that those glasses have really experienced a very significant stabilization, as shown by its densification, enthalpy release at the glass-transition temperature T_g, or noticeable reduction in the so-called fictive temperature.[51–53] In contrast to standard annealing processes, here the pristine (hyperaged) sample is the highly-stable glass. After performing all desired measurements and the necessary characterization, the very same sample is partially or totally rejuvenated — as also usually done with conventional aged polymers, though with a very different temporal scale.

Nevertheless, most clear-cut and striking results on the specific heat at very low temperature for these amber glasses, were reported on an even more stable Spanish amber,[42] recently discovered within El Soplao territory (Cantabria, Spain), dated to be 110–112 million years old. As shown in Figure 8 (right panel), the glassy anomalies in the low-temperature specific heat were found to persist essentially unchanged in these extremely stabilized amber glasses, supposedly approaching the "ideal glass" state of negligible configurational entropy, when compared to the very same sample after being subsequently rejuvenated.[42,53] C_p data down to 0.1 K all exhibited the same quasilinear behavior $C_p \sim T^{1.27}$, even quantitatively, whereas the C_p/T^3 maximum was found to increase ~26% in height after fully rejuvenation of the pristine amber glass, as later confirmed by inelastic X-ray scattering experiments.[54] All these experiments clearly showed that, at least in amber glasses, the glassy anomalies (especially, the TLS excitations) persisted as robust features of noncrystalline solids, independently of the degree of thermodynamic and kinetic stability, and in clear contrast to the case of PVD ultrastable glasses of IMC.

As anticipated above, recent developments in swap Monte Carlo methods[50,55] are allowing to investigate computer-generated glasses, which can also access glassy states very deep in the energy landscape, analogous to PVD ultrastable glasses, whereas in the past computer glasses typically were obtained only after very fast quenching from liquid configurations. This kind of computer studies have found that the low-frequency part of the VDOS $g(\omega)$ splits between extended and quasi-localized modes, the latter ones obeying $g(\omega) \sim \omega^4$, irrespective of the stability, in agreement with the Soft-Potential Model (see below). The prefactor of this quartic law decreases with increasing stability and the corresponding modes become more localized and sparser.[55] These results reasonably agree with experimental data on amber. However, when searching lower energy TLS, computer simulations showed a dramatic depletion of TLS with increasing stability in these computer glasses,[50] in agreement with the experimental observations in IMC glasses, but opposite to those in amber.

It might be hypothesized that highly-stable "ideal glasses" are associated with a negligible excess in *configurational* entropy at T_g, whereas noncrystalline solids lacking low-energy excitations (TLS, boson peak ...) could be associated with a low *vibrational* entropy. Both features can be interrelated sometimes, but they do not have to be connected automatically, as the case of hyperaged amber apparently demonstrates. In any case, it would be desirable to perform further experiments on other different non-anisotropic PVD ultrastable glasses, as well as computer simulations either on anisotropic or polymeric-like systems, to test these hypotheses.

5. Theories and Models

As said at the beginning of this chapter, looking at the $C_\mathrm{p}(T)$ curves alone is often not enough to deduce the nature of the excitations contributing. One usually needs complementary experimental information, as well as to make use of theories or models to verify or refute the different hypotheses. In this section, we will provide a very brief (and unavoidably incomplete) overview of some of the most employed approaches or models to analyze specific-heat data at low temperatures in glasses.

5.1. *Tunneling Model*

Although the Tunneling Model (TM) is discussed and developed throughout the first half of this book, a brief summary of its main assumptions and most relevant equations for the specific heat is presented here, for the sake of completeness.

As is well known, the TM was independently proposed by Phillips[56] and by Anderson, Halperin and Varma[57] only one year after the paper by Zeller and Pohl.[7] Its fundamental hypothesis is the ubiquitous existence of atoms (or better, groups of atoms) in amorphous solids, which can oscillate between two configurational states of similar energy by quantum-mechanical tunneling through a potential-energy barrier, as illustrated in Figure 9. In the TM, the

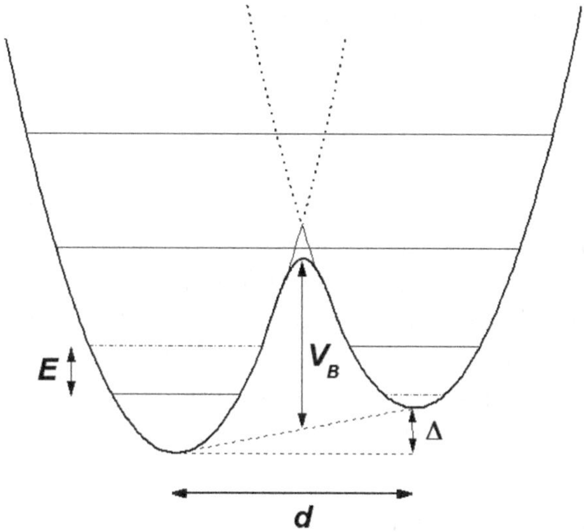

Fig. 9. Schematic illustration of a typical asymmetric double-well potential in the TM. The two-level system is given by the two lowest energy levels with an energy difference E. These two states are separated a distance d (generalized coordinate) by the tunneling barrier of height V_B. Higher energy levels corresponding to usual vibrational modes are also depicted.

abscissa is a configurational generalized coordinate, describing the collective motion of all participating atoms. Although this coordinate is not well defined (to put it politely), the success of the TM and its apparent universality rely precisely on how far it can go without knowing details such as the very microscopic nature of the tunneling state.

Specifically, any double-well potential (DWP) is characterized in the TM by its barrier height V_{B} and by its asymmetry energy Δ, both of which are expected to be randomly distributed in a disordered solid. At low enough temperatures, only the two lowest energy levels will contribute significantly to thermodynamic quantities, its energy excitation E being

$$E = \sqrt{\Delta_0^2 + \Delta^2} \tag{7}$$

where Δ_0 is the energy splitting arising from quantum-mechanical tunneling for the case of a symmetric DWP, i.e. $\Delta = 0$. Assuming a high enough potential barrier V_B, it is found that

$$\Delta_0 = \hbar\omega_0 \exp\left(-d\sqrt{2mV_B/\hbar^2}\right) \tag{8}$$

where ω_0 can be taken as the angular frequency of oscillation within one single well, m is the effective mass of the tunneling entity, and d is the separation between the two minima along the generalized coordinate. The argument of the exponential is the so-called tunneling parameter λ:

$$\lambda = -d\sqrt{2mV_B/\hbar^2} \tag{9}$$

Assuming random disorder, it is natural to postulate a constant distribution function in terms of Δ and λ as independent variables, that is $P(\Delta, \lambda) = P_0$. By a simple Jacobian transformation, this is equivalent to $f(\Delta, \Delta_0) = P_0/\Delta_0$.

The second basic assumption of the TM is related to the interaction of these tunneling states with the sound waves or acoustic "phonons" of the atomic lattice (see more details in Chapter 4). In brief, the strain field ε_j induced by the elastic sound wave in the TLS system changes the asymmetry Δ of the DWP (neglecting small variations in Δ_0) by an amount characterized by a coupling energy

$$\gamma_j = \frac{1}{2}\frac{\partial\Delta}{\partial\epsilon_j} \tag{10}$$

where $j = l,t$ stand for longitudinal and transverse, respectively. The essential hypothesis of the TM is to assume an averaged constant coefficient γ_j for a given substance.

Then, the relaxation rate of a tunneling state in the thermal bath of phonons is obtained to be[10]

$$\tau^{-1} = \left[\frac{\gamma_l^2}{v_l^5} + 2\frac{\gamma_t^2}{v_t^5}\right]\frac{E^3}{2\pi\rho\hbar^4}\left(\frac{\Delta_0}{E}\right)^2 \coth\left(\frac{E}{2k_BT}\right) \tag{11}$$

with $\nu_{l,t}$ and ρ being the corresponding sound velocity and the mass density of the material, respectively. Thus, the minimum relaxation time $\tau_{\min}$ (i.e. the maximum relaxation rate) occurs when the DWP is symmetric ($\Delta = 0, \text{hence}\, E = \Delta_0$).

For calculating the contribution of these tunneling TLS to the specific heat, it is more convenient to use the TLS excitation energy E and its relaxation time τ as independent variables. By making the appropriate Jacobian transformation from $P(\Delta, \lambda) = P_0$, it is readily obtained

$$g(E, \tau) = \frac{P_0}{2\tau(1 - \tau_{\min}(E)/\tau)^{1/2}} \tag{12}$$

By integrating Eq. (12) over all relaxation times from its minimum possible value $\tau_{\min}$ up to a given experimental time $t_{\exp}$ of the heat-capacity measurement, the density of tunneling states per unit volume and energy interval is calculated to be

$$n(E) = \frac{1}{2} P_0 \ln\left(\frac{4t_{\exp}}{\tau_{\min}}\right) \tag{13}$$

and hence

$$C_p \approx C_v = k_B \int_0^\infty n(E) \frac{(E/2k_BT)^2}{\cosh^2(E/2k_BT)} dE \approx \frac{\pi^2}{12} P_0 k_B^2 T \ln\left(\frac{4t_{\exp}}{\tau_{\min}}\right) \tag{14}$$

Therefore, the TM is shown to predict a quasilinear temperature dependence with a small logarithmic factor that increases with measuring time, what was later observed indeed in experiments,[58] and that roughly accounts for the typical $C_p \sim T^{1+\delta}$ dependence mentioned in Section 2.

For practical purposes, the specific distribution $g(E, \tau)$ of the two chosen variables and the corresponding relaxation times are often not very important. In fact, the influence of the coupling coefficients γ_j in the specific heat is essentially irrelevant: it would only affect the parameter $\tau_{\min}$ in the logarithmic factor. On the contrary, these TLS-phonon interaction coefficients γ_j are essential to address the behavior of thermal conductivity and acoustic properties[8, 10] through the dimensionless tunneling strength $C = \gamma_j^2 P_0/\rho v_j^2$, as commented in the precedent chapter. Therefore, when considering the specific heat, the genuine tunneling states (and the truly TM) are often replaced by a simplified distribution of TLS, only characterized by their energy E, whose density can be taken as approximately constant, $n(E) \approx n_{\mathrm{TLS}} \approx 10P_0$. In this case, Eq. (14) becomes $C_p = (\pi^2/6)n_{\mathrm{TLS}}k_{\mathrm{B}}^2T$. This is the simpler, linear temperature dependence usually ascribed to the specific heat in glasses.

In general, the great success of the Tunneling Model has relied on both the simplicity of its hypotheses and the apparent capability of prediction for many distinct properties (heat capacity, thermal conductivity, acoustic and dielectric properties at different frequencies...), at least below 1–2 K.

5.2. *The maximum in C_p/T^3 in glasses and in crystals*

As shown above, in the intermediate range of temperatures 2–30 K the specific heat of glasses still goes over the expected Debye-like values, what is characterized by a hump or broad maximum in C_p/T^3, actually associated with the so-called boson peak at ω_{BP} in the Debye-reduced density of states $g(\omega)/\omega^2$. As a matter of fact, a maximum in the Debye-reduced specific heat or C_p/C_{Debye}, is not that new. It is well known that many crystals (though not all of them) also present a maximum in C_p/T^3. Nevertheless, in textbook crystals that maximum in C_p/T^3 is undoubtedly originated by a van Hove singularity, i.e. a flattening of the $\omega(q)$ dispersion curve for acoustic phonons near the Brillouin zone boundary, which produces an increase in $g(\omega)$ over the expected $g(\omega) \propto \omega^2$ at those frequencies and hence an increase in C_p over the expected $C_p \propto T^3$.

For decades, similarities and differences between the "boson peak" in glasses and the "van Hove" peaks in crystals have been discussed. Liu and von Löhneysen[59] found a rough correlation between the height P_{c} of the maximum in C_p/T^3 and the temperature of that maximum T_{max} for a large collection of both amorphous and crystalline solids: $P_{\mathrm{c}} \sim T_{\mathrm{max}}^{-1.6}$. Moreover, Chumakov *et al.*[60,61] have argued that the boson peak of glasses also arises from van Hove singularities for transverse acoustic phonons, therefore suggesting an identical nature of the two features. The quantitative differences in their values between glasses and crystals would disappear when the curves are properly normalized by the corresponding mass densities or elastic constants.

However, one should take into account several considerations:

(i) Below the peak at ~10–50 K in crystals, their specific heat follow the Debye law, whereas in glasses C_p never reaches the Debye level.
(ii) When performing a scaling plot[59] $(C_p/T^3)/P_c$:T/T_{max}, the maximum of crystalline solids is always narrower than that in their amorphous counterparts.[41,59]
(iii) There is something tricky when, by comparing different curves, an increase in the boson peak intensity is observed or claimed with decreasing frequencies: with ω^2 or T^3 in the denominator, it is trivial that peaks at lower frequency or temperature show higher $g(\omega)/\omega^2$ or C_p/T^3 values. As Caponi *et al.*[62] correctly pointed out, when using a normalized frequency $\Omega \equiv \omega/\omega_{BP}$, conservation of the total number of vibrational states should be imposed, $g(\Omega)d\Omega = g(\omega)d\omega$, and hence one obtains $g(\Omega)/\Omega^2 = [g(\omega)/\omega^2] \times \omega_{BP}^3$. By performing this consistent scaling, it was found that many apparent inverse correlations of the boson peak intensity with ω_{BP} disappeared, since they were only due to the ω_{BP} shift produced by changes in the elastic constants.[62]
(iv) As has been already mentioned, $C_p(T)$ is a smooth function, where any kind of energy excitation is weighted by thermal factors. To elucidate the possible same nature of the vibrational states producing the boson peak in glasses and van Hove peaks in crystals, it is convenient to inspect their influence on other properties such as thermal conductivity at low temperatures. As shown in Figure 1 of Chapter 7, where crystalline and amorphous states of SiO_2 are compared, the overall thermal behavior in glasses cannot be mapped to that of their crystalline counterparts with just an appropriate scaling.

5.3. *Soft potential model*

For a long time (see the Introduction in Chapter 1), the glassy behavior below 1–2 K (the TLS) and that above 1–2 K (the maximum in C_p/T^3 or boson peak) were treated separately. The first serious attempt to unify all those glassy features in one single picture was the Soft Potential Model (SPM), introduced by the St. Petersburg's group[63,64] on the 1980's and further developed[65–72] along next decades. The SPM will be presented more exhaustively in Chapter 8. Nonetheless, it will be useful to summarize here some pertinent equations related to the specific heat within this model to facilitate the following discussions. In fact, beyond the higher or lower degree of confidence one could give to this model in comparison with others in the literature, some specific approaches[69,70,73] of the SPM are particularly useful since they provide a straightforward method to evaluate quantitatively, from the experimental data, the different contributions to the specific heat and other low-temperature properties.

In brief, the SPM posits the presence in glasses of a variety of quasilocalized low-energy excitations, in addition to standard acoustic phonons. These "soft modes" range from very anharmonic two-level systems (TLS), appearing in double-well potentials (DWP) such as that in Figure 9, to more or less quasiharmonic vibrational modes in single-well potentials (SWP). All these low-frequency modes can be represented by energy potentials up to a quartic coefficient (positive, to stabilize the mode). The original terminology of the model can be reformulated into a simpler one,[66,67,70] where the potential of any soft mode is represented by

$$V(x) = W(D_1 x + D_2 x^2 + x^4) \tag{15}$$

x being a dimensionless generalized spatial coordinate. Thus, each mode has a linear coefficient D_1 (providing the asymmetry) and a quadratic one D_2 (the restoring force), which can be either positive or negative.[67] The basic parameter of the model is the energy W, the stabilizing fourth-order term, which is assumed to be the same for all atomic potentials. An additional cubic coefficient is unnecessary, since it can always be eliminated by a simple change of variable (a shift) in the spatial coordinate x.

Similarly to the Tunneling Model, the basic SPM assumes a random distribution of potentials: $P(D_1, D_2) = P_s$. Below 1 K, the specific heat is dominated by the contribution of TLS in DWP, with analogous results to those from the TM, though with some slightly different equations, as discussed in Chapter 8. Specifically, its contribution is found[70] to be:

$$C_{p,TLS}(T) = \frac{\pi^2}{6}\left(\frac{1}{9}\right)^{1/3} P_s k_B \left(\frac{k_B T}{W}\right) \ln^{1/3}\left[\frac{t_{\rm exp}}{\tau_{\rm min}(T)}\right] \tag{16}$$

where $t_{\rm exp}$ and $\tau_{\rm min}$ have the same meaning as in the TM above.

Above 1–2 K (strictly speaking, above the minimum in C_p/T^3 occurring at $T_{\rm min} \approx W/1.8k_{\rm B}$), the dominant contribution is that from vibrational excitations in SWP (*soft modes*), which is found to be proportional to ω^4 (see Eq. (26) of Chapter 8). In the quasiharmonic approximation, it leads to the analytical expression for the specific heat:

$$C_{p,sm}(T) = \frac{2\pi^6}{21} P_s k_B \left(\frac{k_B T}{W}\right)^5 \tag{17}$$

Obviously, this basic assumption of an unlimited distribution of soft modes $P(D_1, D_2) = P_s$ giving $g(\omega) \propto \omega^4$ cannot be valid at higher frequencies without an end. So, it cannot account for the boson peak maximum in $g(\omega)/\omega^2$, only for its low-temperature tail. However, several approaches to assess the true VDOS at higher frequencies including the boson peak have been proposed, as briefly discussed in Section 4.1. of Chapter 8.

Nonetheless, a very practical assessment of the different contributions to the specific heat at low temperature in glasses can be performed by using that most basic SPM. In short, the whole of the specific heat below the boson peak can be approximately written as:

$$C_p \approx C_{\rm TLS} T + C_{\rm Debye} T^3 + C_{\rm sm} T^5 \tag{18}$$

where the cubic term $C_{\rm Debye} T^3$ should be that given by Debye's theory. Hence by plotting $C_p/T{:}T^2$ a least-squares fit to a quadratic polynomial directly provides the three coefficients. In fact, the cubic coefficient could be fixed from the elastic constants of the glass, when available. Therefore, only two free C_p coefficients remain, which provide the two SPM fitting parameters, namely the density of soft modes P_s and their average stabilization energy W. To be consistent, the fit should be conducted for data ranging from zero up to some temperature between the minimum T_{min} and the boson peak T_{max}, since above this range the approximation of Eq. (17) is no longer valid. In particular, it was argued[73] to perform the fits for data in the range $0 < T < (3T_{\text{min}} + T_{\text{max}})/4$.

Interestingly, by following this procedure for C_p data analysis, an old puzzle related to the Debye coefficients in glasses or amorphous solids was solved. For decades, it has been stated[8,9,19,44,45] that the *calorimetric* Debye coefficient (the cubic term obtained from fitting C_p data) is always in excess over the *elastic* Debye coefficient (obtained from the measured elastic constants) in glasses. The reason for such discrepancy seems clear within the SPM analysis of Eq. (18). Typical linear fits of $C_p/T{:}T^2$ plots at low temperatures could give a reasonable value for C_{TLS} but systematically overestimate the cubic term if one neglects the contribution of the "soft modes" (i.e. the low-temperature tail of the boson peak). However, when a quadratic fit in the abovementioned temperature range is performed,[73,74] the discrepancy disappears and an excellent agreement is found between *calorimetric* and *elastic* Debye coefficients (see Table 1 in Ref. [73]). This procedure has been followed to determine all the specific-heat coefficients of Tables 1 and 2 in Section 6.

5.4. *Other theories and approaches*

We will briefly mention here other theories and approaches about the low-temperature specific heat of glasses, at least providing the pertinent references for the reader.

One of the most considered theories to understand the vibrational density of states in glasses or strongly disordered solids is the heterogeneous-elasticity theory put forward by Schirmacher and co-workers.[75,76] By employing mean-field techniques, this heterogeneous-elasticity theory was shown to be able to explain the boson peak or the plateau in thermal conductivity in terms of spatially fluctuating transverse elastic constants. The glassy behaviour at the lowest temperatures, however, is beyond the reach of this theory, which simply assumes the existence of additional two-level system excitations following the Tunneling Model. A direct quantitative check of this theory by experimental data is not as easy as with the SPM. Nonetheless, it has been at least fruitfully employed to analyse variations of the boson peak with geological ageing in amber.[54] A thorough update of this theory is provided in the Chapter 9 of this book.

As mentioned in the Chapter 1 and briefly summarized in Section 2.2 of Chapter 4 and in Section 2.1 of Chapter 5, another relevant theory about glasses and the glass transition is the Random First Order Transition (RFOT) theory.[77–80] RFOT is not specifically intended to account for the low-temperature properties of glasses, but this matter is nonetheless widely discussed in Refs. [77–80], to which the reader is referred for more detail.

It is also worth mentioning the extensive experimental studies carried out by the Messina's group on alkali borate glasses,[81–83] combining specific heat measurements with light and neutron scattering, aiming to understand the influence of the atomic structure — including densification at high pressures — on the low-temperature anomalies of these glasses.

Finally, a new paradigm[84] of the vibrational and thermal properties of solids (both amorphous and crystals) has been recently introduced by Zaccone and coworkers.[85–88] By relying on the above-mentioned findings of glassy anomalies in minimally disordered crystals and in incommensurate structures, the authors question the old picture of low-temperature anomalies arising from structural disorder. They link the linear term in the specific heat to anharmonic low-energy modes resulting from fundamental symmetries and random matrix theory,[85,86] whereas the boson peak is attributed to gapped low-energy optical phonons in crystals and/or to anharmonic diffusive damping.[87,88]

6. Discussion: How Universal is the Linear Term in the Specific Heat of Glassy Materials

As briefly discussed in Chapter 1, the most genuine (and unexplained) universalities in the low-temperature properties of glasses are the *plateaux* in both thermal conductivity $\kappa(T)$ and internal friction $Q^{-1}(T)$ centred at about 1–10 K. They exhibit even almost *quantitative* universality: $\kappa(T) \approx 10^{-1}\,\mathrm{Wm^{-1}K^{-1}}$ and $Q^{-1} \approx (2{-}5)\times 10^{-4}$ within one order of magnitude for dozens of different substances of any kind.[89]

The total specific heat could be expected, however, to vary among different materials depending on atomic density, elastic constants, etc., like in crystals. Nevertheless, it is also often stated that the dominant contribution at the lowest temperatures $C_p = (\pi^2/6)n_{\mathrm{TLS}}k_{\mathrm{B}}^2T$ (see Section 5.1.), is associated to a constant density of TLS which is always of the order $n_{\mathrm{TLS}} \sim 10^{46}\,\mathrm{J^{-1}m^{-3}}$, with variations of one order of magnitude at most. Indeed, the more exact Eq. (13) relates the approximately constant n_{TLS} obtained from the specific heat to the TLS spectral density P_0, which has been exhaustively obtained from acoustic measurements and compiled for many glasses.[90–92] It has been generally found that $P_0 \approx (0.5-3)\times 10^{45}\,\mathrm{J^{-1}m^{-3}}$. Estimates of the temperature-dependent logarithmic factor in Eq. (13) for typical glasses give[90,93] $n_{\mathrm{TLS}} \approx 10P_0$, in good agreement with reported experimental data. Since n_{TLS} and P_0 were found to vary little among different substances, there have hardly been attempts to seek for any correlation with some physical parameter. Let us mention that a slight decrease of P_0 with increasing value of the glass transition temperature T_g has been pointed out by some authors.[91,92]

In contrast, the strength or height of the C_p/T^3 broad maximum is accepted to be much more dependent on the studied substance. This is partly due to the inherent contribution of the Debye-like background of the material to the vibrational density of states in this range. As discussed in Section 2.2., Sokolov and coworkers[14,94] showed an inverse correlation of the boson-peak height relative to the Debye level with the fragility index of the glass-forming substance (see Figure 4), though several failures of this correlation have been reported.[17,95]

The abovementioned suggested correlation between the boson-peak relative height and the fragility led Zhu[96] to seek and find an apparent correlation between the TLS density and the fragility, in this case *direct* instead of *inverse*. The correlation was a bit better if the density of tunneling states was expressed per unit mol instead of per unit mass, what seems reasonable.

Another proposed correlation comes from the interstitialcy model proposed by Granato.[97,98] According to Granato, liquids could be considered as crystals containing a few interstitials in thermal equilibrium, which become frozen in the glassy state. Following a simple approximation, and by putting a single frequency for all the resonance modes, the maximum in C_p/T^3 was predicted[99] to occur for $T_{\text{max}} = \theta_D/35$. This general prediction has been tested by different authors,[95,100,101] finding empirically that $\theta_D \approx (35 - 40)T_{\text{max}}$.

Let us however return to the central question: to what extent the TLS density is quantitatively universal? After those abovementioned new data and relevant findings in highly stable glasses or in different disordered crystals, it is deemed pertinent to review the situation and take a fresh look into this issue. Older and more recent specific-heat data C_p at low temperatures, both for structural glasses (amorphous solids) and for disordered crystals with glassy properties, are compiled and presented in Tables 1 and 2 respectively, updating and expanding on previous work.[74] The data analysis to determine the linear and Debye coefficients has been generally carried out by recourse to the SPM methodology described in Section 5.3., using Eq. (18). The found linear coefficient C_1 is presented, as usual, in $\mu\text{J/g} \cdot \text{K}^2$. Those rows marked with an asterisk mean that only C_p data above 1 K are available, therefore giving for those materials less accurate, extrapolated C_1 values.

In principle, one may ask which is the most reasonable scale, to assess and compare the linear coefficients C_1 and the corresponding density of TLS, n_{TLS}, whether per unit volume, per unit mass, per mol of molecules or of atoms... It comes out that changing the normalization scale does not make much difference. In any case, a sensible comparison of the amount of (presumable) TLS contributing to the specific heat between different materials could be scaling its linear contribution C_1T to either the total C_p, or to the Debye background $C_{\text{Deb}}T^3$, evaluated at, say, 1 K. Both ratios assessing the relative contribution of TLS to the specific heat for a given material are also displayed in Tables 1 and 2. The former ratio (more direct

Table 1. Specific-heat linear coefficients and corresponding densities of two-level systems for many different structural glasses, together with their relevant material parameters (see text for more details). Most of the original data sources can be found in Ref. [74], as well as in Ref. [8].

Structural Glasses								
Material	density (g/cm^3)	M_{mol}; M_{at} (g/mol)	C_1 ($\mu J/g \cdot K^2$)	C_1T/C_p (1 K)	$C_1T/C_{Deb}T^3$ (1 K)	n_{TLS} ($J^{-1}m^{-3}$) $\times 10^{45}$	n_{TLS} ($meV^{-1}atom^{-1}$) $\times 10^{-6}$	T_{BP} (K)
$(B_2O_3)_{0.75}(Na_2O)_{0.25}$	2.288	68; 15.1	3.24	0.69	2.25	23.6	41.5	11
$(B_2O_3)_{0.84}(Na_2O)_{0.16}$	2.122	68.7; 14.7	3.50	0.61	1.6	23.7	43.7	10
SiO_2	2.20	60; 20	1.34	0.61	1.7	9.4	22.7	10
glycerol	1.42	92.1; 6.58	1.83	0.16	0.20	8.3	10.2	8.7
GeO_2	3.6	105; 35	0.9	0.26	0.47	10.3	26.7	8
$(B_2O_3)_{0.94}(Na_2O)_{0.06}$	1.960	69.5; 14.2	3.75	0.49	1.0	23.4	45.2	7.5
1-propanol	0.99	60.1; 5.0	7.05	0.19	0.24	22.3	29.9	6.7
H-ethanol	0.96	46.1; 5.12	26.0	0.43	0.77	79.6	113	6.1
D-ethanol	0.96	52.1; 5.79	20.2	0.36	0.58	61.8	99.2	6.0
$Ca_{0.4}K_{0.6}(NO_3)_{1.4}$	2.1	126.3; 19.13	5.39	0.42	0.73	36.1	87.5	6.0
B_2O_3 (annealed)	1.85	70; 14	1.59	0.22	0.35	9.4	18.9	5.5
*n-butanol	0.951	74.12; 4.94	11.6	0.32	0.49	35.2	48.6	5.4
B_2O_3 (quenched)	1.81	70; 14	2.76	0.23	0.34	15.9	32.8	5.2
PB	0.96	54.1; 5.41	6.0	0.22	0.30	18.4	27.5	5.1
2-propanol	0.91	60.1; 5.0	8.59	0.16	0.20	24.9	36.4	5.0
As_2S_3	3.2	246; 49.2	0.44	0.19	0.18	4.5	18.4	4.9
*sec-butanol	0.948	74.12; 4.94	28.5	0.51	1.16	86.2	119	4.8
*Isobutanol	0.908	74.12; 4.94	55.7	0.64	2.1	161.3	233	4.8
*Toluene	0.9	92.14; 6.14	0 ± 16	0 ± 0.31	0 ± 0.32	0 ± 46	0 ± 83	4.5
PMMA	1.2	100.1; 6.67	5.71	0.22	0.30	21.9	32.3	3.6
IMC (ultrastable)	1.336	357.8; 8.73	0 ± 0.8	0 ± 0.02	0 ± 0.02	0 ± 3.4	0 ± 5.9	3.5
IMC (conventional)	1.31	357.8; 8.73	13.7	0.22	0.28	57.2	101	3.5
Amber (hyperaged)	1.055	—	5.8	0.13	0.31	19.5	—	3.4
Amber (rejuvenated)	1.035	—	5.8	0.11	0.28	19.1	—	3.4
Se	4.3	79; 79	0.58	0.026	0.033	8.0	38.9	3.1
PS	1.06	104.1; 6.51	5.69	0.13	0.16	19.2	31.4	3.0
Lexan	1.2	272.3; 7.56	3.8	0.085	0.13	14.5	24.4	2.7

Table 2. Same data columns as in Table 1, but here for some disordered crystals with glassy properties. Most of the original data sources can be found in Ref. [74], as well as in Refs. [19, 21, 35].

Disordered Crystals								
Material	density (g/cm^3)	M_{mol}; M_{at} (g/mol)	C_1 ($\mu J/g \cdot K^2$)	C_1T/C_p (1 K)	$C_1T/C_{Deb}T^3$ (1 K)	n_{TLS} ($J^{-1}m^{-3}$) $\times 10^{45}$	n_{TLS} ($meV^{-1}atom^{-1}$) $\times 10^{-6}$	T_{BP} (K)
$ThBr_4$	5.84	551.7; 110.3	2.39	0.077	0.085	44.5	224	9.2
*CCl_4	1.94	153.8; 30.76	29.3	0.70	2.5	181	765	9.2
*$CBrCl_3$	2.45	198.3; 39.65	27.2	0.68	2.3	213	915	7.7
*CBr_2Cl_2	2.97	242.7; 48.54	32.1	0.59	1.6	304	1320	7.5
2-BrBP	1.45	261.1; 10.88	0.46	0.042	0.044	2.13	4.25	7.2
H-ethanol-OG	0.96	46.1; 5.12	27.5	0.46	0.88	84.2	119	6.8
D-ethanol-OG	0.96	52.1; 5.79	21.7	0.39	0.66	66.4	107	6.4
*Freon-113	2.11	187.4; 23.42	9.6	0.29	0.47	64.6	191	5.0
$(NaCN)_{0.81}(KCN)_{0.19}$	1.64	52.07; 17.36	13.5	0.49	1.2	70.6	199	4.8
PCNB	1.8	295.3; 21.09	3.59	0.13	0.16	20.6	64.2	4.8
*Freon-112	2.14	203.8; 25.48	12.3	0.33	0.56	83.9	266	4.5
$(NaCN)_{0.41}(KCN)_{0.59}$	1.62	58.51; 19.5	8.5	0.32	0.53	43.9	141	4.5
$(KBr)_{0.75}(KCN)_{0.25}$	2.50	105.5; 46.9	2.4	0.10	0.18	19.1	95.5	4
$(KBr)_{0.30}(KCN)_{0.70}$	1.95	81.28; 30.1	0.43	0.018	0.029	2.67	11.0	3

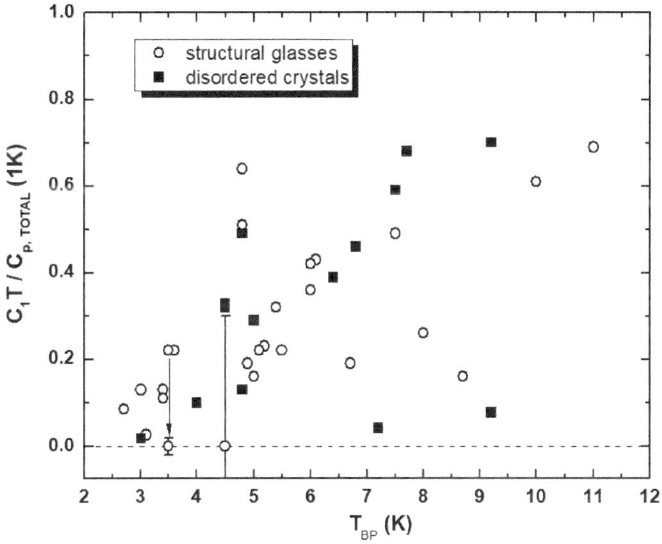

Fig. 10. Relative contribution of the linear term (ascribed to TLS) to the total specific heat, evaluated at 1 K, for both structural glasses (open circles, Table 1) and disordered crystals (solid squares, Table 2), as a function of the boson-peak temperature. Error bars for the reported null-C_1 cases of toluene and ultrastable IMC are shown. For the latter, the arrow indicates the change from conventional to ultrastable glass state.

and model independent than the latter) is plotted for both structural glasses and disordered crystals in Figure 10 as function of the boson peak temperature $T_{\rm BP}$ (defined as the temperature of the maximum in C_p/T^3). Besides looking for a possible correlation, at least $T_{\rm BP}$ runs as an index to display a wide range of relative TLS contributions for many different substances.

In Figure 10, the ratio $C_1T/C_{\rm p,TOTAL}$ evaluated at 1 K seems to increase with $T_{\rm BP}$, as a general rule. Let us notice, however, that materials with a C_p/T^3 maximum at higher temperatures will tend to have a weaker contribution of the boson peak tail at 1 K. Thus, the apparent correlation between $C_1T/C_{\rm p,TOTAL}$ and the boson peak may not be very meaningful.

From the collected C_1 coefficients, the density of two-level systems $n_{\rm TLS}$ following the standard Tunneling Model is readily obtained (see Section 5.1) from $C_1 = (\pi^2/6)n_{\rm TLS}k_{\rm B}^2T$.

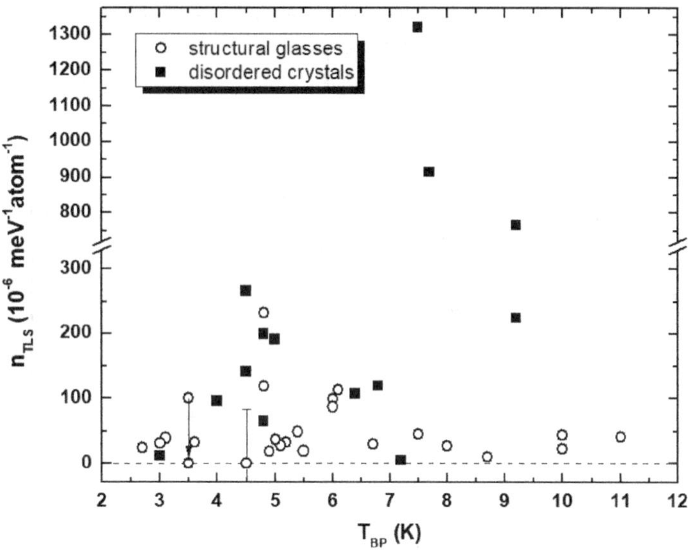

Fig. 11. Density of TLS determined from the linear term in the specific heat, for both structural glasses (open circles, Table 1) and disordered crystals (solid squares, Table 2), as a function of the boson-peak temperature. Error bars for the reported null-C_1 cases of toluene and ultrastable IMC are shown (that is negligible for the latter). The arrow indicates the change from conventional to ultrastable glass state of IMC. Please note the break in the ordinate axis to encompass some high values from halomethane crystals.

In Tables 1 and 2, n_{TLS} is expressed both in the usual units of $\mathrm{J}^{-1}\mathrm{m}^{-3}$ ($n_{\mathrm{TLS}} \sim 10^{46} - 10^{47}\,\mathrm{J}^{-1}\mathrm{m}^{-3}$ is typically obtained) and in units of meV^{-1}atom^{-1}. This latter column ($\times 10^{-6}$) offers the insight to give roughly the ppm amount of TLS per atom, since the energy range of relevant tunneling TLS can be considered to be 0–1 meV. This density of TLS is plotted in Figure 11. No clear correlation with the C_p/T^3 maximum T_{BP} is really observed. Interestingly, some disordered crystals exhibit the highest density of TLS (~1000 ppm). Although those cases (marked with an asterisk in Tables 1 and 2) where C_p data above 1 K have been extrapolated to determine C_1 and n_{TLS} are less reliable, it is clear nevertheless that the obtained densities of TLS are far from being quantitatively universal.

Figures 10 and 11 also help to quantify what null or a very small number of TLS (i.e., a small linear term in C_p) really means.

For instance, Selenium has an apparently small linear contribution to C_p, well determined with measurements down to very low temperatures. Nevertheless, when its materials parameters are properly taken into account, the atomic density of TLS, n_{TLS}, lies in the average. However, the reported lack of a linear term in toluene (with data only down to 1.8 K) is compatible with "normal" values of C_1T/C_p or n_{TLS}, when the corresponding error bar from the fit is considered. On the contrary, the error bars for C_1T/C_p and n_{TLS} in the case of the IMC ultrastable glass, which was measured down to very low temperatures, imply an essentially zero value, in clear contrast to its conventional glass state. This confirms the reported suppression of TLS (or whatever excitations are responsible for these low-temperature anomalies) for this highly-stable glass, in striking opposition to the case of amber glasses.

7. Conclusion and Outlook

Along the last half century, many specific-heat C_p data of glasses and disordered solids have been compiled and analyzed. Different overall trends, correlations and theories about the apparent universality found in glasses have been postulated for this core thermal property at low temperature. Some of them have been reviewed in this chapter, focused on the linear term ascribed to the existence of tunneling TLS and on the broad maximum in C_p/T^3(the boson peak).

Although the C_p/T^3 peak seems more controversial and its magnitude varies more widely, quantitatively and qualitatively, among different substances, the linear term (i.e. the density of TLS) has appeared as a robust universal property of noncrystalline solids for decades.

Nevertheless, some careful studies with a wider scope, including recent experiments on new kinds of glasses and disordered crystals, cast now doubts even on its assumed phenomenological universality, as discussed in the previous section. Furthermore, some authors[84–86, 102, 103] have questioned the very existence of an ensemble of independent TLS tunneling states in any amorphous solid, which is the basis of the standard Tunneling Model.

At least, the following questions remain: (i) Is the linear coefficient of C_p really similar between different glassy solids — in some appropriate units — as often stated? (ii) Are both this linear term and the C_p/T^3 boson peak truly fingerprints of long-range atomic disorder in solids (either amorphous solids or "glassy" crystals)? (iii) If so, do glasses consistently exist without those low-temperature "anomalies" (e.g. glasses possessing high thermodynamic and kinetic stability)? (iv) Which kind of disorder is really needed to make a crystal *glassy*?

After about 50 years of intensive research (theoretical, experimental and computational) on these issues, we have got many data and many theories, but also have more open questions than answers, in my view. Hopefully, it will take less than other 50 years to solve these enigmas, especially with the help of new theoretical approaches and models with predictive capability, the rapidly increasing power of computer simulations on glasses, and with new crucial experiments on key glassy systems, all aiming to verify or refute the different hypotheses on this matter.

References

1. A. T. Petit and P. L. Dulong. (1819). Recherches sur quelques points importants de la Théorie de la Chaleur. *Annales de Chimie et de Physique* **10**, 395–413.
2. N. W. Ashcroft and N. D. Mermin. (1976). *Solid State Physics.* Philadelphia: Saunders.
3. C. Kittel. (2005). *Introduction to Solid State Physics.* 8th edn. USA: Wiley.
4. C. Enss and S. Hunklinger. (2005). *Low-Temperature Physics.* Berlin-Heidelberg: Springer.
5. A. Einstein. (1907). Planck's theory of radiation and the theory of the specific heat. *Ann. d. Physik* **22**, 180–190.
6. P. Debye. (1912). Zur Theorie der spezifischen Wärmen. *Ann. d. Physik* **39**, 789–807.
7. R. C. Zeller and R. O. Pohl. (1971). Thermal conductivity and specific heat of noncrystalline solids. *Phys. Rev. B.* **4**, 2029–2041.
8. W. A. Phillips (ed.), (1981). *Amorphous Solids: Low-Temperature Properties.* Berlin-Heidelberg-New York: Springer.

9. R. B. Stephens. (1976). Intrinsic low-temperature thermal properties of glasses. *Phys. Rev. B.* **13**, 852–865.
10. W. A. Phillips. (1987). Two-level states in glasses. *Rep. Prog. Phys.* **50**, 1657–1708.
11. A. P. Sokolov, R. Calemczuk, B. Salce, A. Kisliuk, D. Quitmann, and E. Duval. (1997). Low-temperature anomalies in strong and fragile glass formers. *Phys. Rev. Lett.* **78**, 2405–2408.
12. C. Talón, Q. W. Zou, M. A. Ramos, R. Villar, and S. Vieira. (2002). Low-temperature specific heat and thermal conductivity of glycerol. *Phys. Rev. B.* **65**, 012203.
13. W. A. Phillips, U. Buchenau, N. Nücker, A.-J. Dianoux, and W. Petry. (1989). Dynamics of glassy and liquid selenium. *Phys. Rev. Lett.* **63**, 2381–2384.
14. A. P. Sokolov, E. Rössler, A. Kisliuk, and D. Quitmann. (1993). Dynamics of strong and fragile glass formers: Differences and correlation with low-temperature properties. *Phys. Rev. Lett.* **71**, 2062–2065.
15. C. A. Angell. (1988). Structural instability and relaxation in liquid and glassy phases near the fragile liquid limit. *J. Non-Cryst. Solids* **102**, 205–221.
16. R. Böhmer, K. L. Ngai, C. A. Angell, and D. J. Plazek. (1993). Non-exponential relaxations in strong and fragile glass formers. *J. Chem. Phys.* **99**, 4201–4209.
17. A. Vispa, M. Romanini, M. A. Ramos, L. C. Pardo, F. J. Bermejo, M. Hassaine, A. I. Krivchikov, J. W. Taylor, and J. Ll. Tamarit. (2017). Thermodynamic and kinetic fragility of Freon 113: The most fragile plastic crystal. *Phys. Rev. Lett.* **118**, 105701.
18. C. Talón, M. A. Ramos, and S. Vieira. (2002). Low-temperature specific heat of amorphous, orientational glass, and crystal phases of ethanol. *Phys. Rev. B.* **66**, 012201.
19. J. J. De Yoreo, W. Knaak, M. Meissner, and R. O. Pohl. (1986). Low-temperature properties of crystalline $(\mathrm{KBr})_{1-\mathrm{x}}(\mathrm{KCN})_x$: A model glass. *Phys. Rev. B.* **34**, 8828–8842.
20. E. R. Grannan, M. Randeria, and J. P. Sethna. (1990). Low-temperature properties of a model glass. II. Specific heat and thermal transport. *Phys. Rev. B.* **41**, 7799–7821.
21. B. Mertz, J. F. Berret, R. Böhmer, A. Loidl, M. Meissner, and W. Knaak. Calorimetric investigations of $(\mathrm{NaCN})_{1-x}(\mathrm{KCN})_x$ glasses. (1990). *Phys. Rev. B.* **42**, 7596–7603.
22. S. K. Watson and R. O. Pohl. (1995). Low-temperature glasslike properties in $(\mathrm{NaCl})_{1-x}(\mathrm{NaCN})_x$. *Phys. Rev. B.* **51**, 8086–8101.
23. S. K. Watson. Tunneling states in crystals with large random strains. (1995). *Phys. Rev. Lett.* **75**, 1965–1968.

24. D. G. Cahill and R. O. Pohl. (1989). Low-energy excitations in the mixed crystal $Ba_{1-x}La_2F_{2+x}$. *Phys. Rev. B.* **39**, 10477–10480.
25. H. Suga and S. Seki. (1974). Thermodynamic investigations on glassy states of pure simple compounds. *J. Non-Cryst. Solids* **16**, 171–194.
26. O. Haida, H. Suga, and S. Seki. (1977). Calorimetric study of the glassy state. XII. Plural glass-transition phenomena of ethanol. *J. Chem. Thermodyn.* **9**, 1133–1148.
27. M. A. Ramos, S. Vieira, F. J. Bermejo, J. Dawidowski, H. E. Fischer, H. Schober, M. A. González, C. K. Loong, and D. L. Price. (1997). Quantitative assessment of the effects of orientational and positional disorder on glassy dynamics. *Phys. Rev. Lett.* **78**, 82–85.
28. M. A. Ramos, I. Shmyt'ko, E. Arnautova, R. Jiménez-Riobóo, V. Rodriguez-Mora, S. Vieira, and M. J. Capitán. (2006). On the phase diagram of polymorphic ethanol: Thermodynamic and structural studies. *J. Non-Cryst. Solids.* **352**, 4769–4775.
29. C. Talón, M. A. Ramos, and S. Vieira. (2002). Low-temperature specific heat of amorphous, orientational glass, and crystal phases of ethanol. *Phys. Rev. B.* **66**, 012201.
30. M. A. Ramos, C. Talón, R. Jiménez-Riobóo, and S. Vieira. (2003). Low-temperature specific heat of structural and orientational glasses of simple alcohols. *J. Phys.* **15**, S1007.
31. A. I. Krivchikov, A. N. Yushchenko, V. G. Manzhelii, O. A. Korolyuk, F. J. Bermejo, R. Fernández-Perea, C. Cabrillo, and M. A. González. (2006). Scattering of acoustic phonons in disordered matter: A quantitative evaluation of the effects of positional versus orientational disorder. *Phys. Rev. B.* **74**, 060201R.
32. D. A. Ackerman, D. Moy, R. C. Potter, A. C. Anderson, and W. N. Lawless. (1981). Glassy behavior of crystalline solids at low temperatures. *Phys. Rev. B.* **23**, 3886–3893.
33. E. Hegenbarth. (1995). Glasslike behaviour of ferroelectrics at low temperatures. *Ferroelectrics* **168**, 25–37.
34. K. Biljaković, J. C. Lasjaunias, F. Zougmore, P. Monceau, F. Levy, L. Bernard, and R. Currat. (1986). Contributions of phasons and of low-energy excitations to the specific heat of the quasi one-dimensional compound $(TaSe_4)_2I$. *Phys. Rev. Lett.* **57**, 1907.
35. G. Reményi, S. Sahling, K. Biljaković, D. Starešinić, J. C. Lasjaunias, J. E. Lorenzo, P. Monceau, and A. Cano. (2015). Incommensurate systems as model compounds for disorder revealing low-temperature glasslike behavior. *Phys. Rev. Lett.* **114**, 195502, and references therein.

36. J. Etrillard, J. C. Lasjaunias, K. Biljaković, B. Toudic, and G. Coddens. (1996). Excess low temperature specific heat and related phonon density of states in a modulated incommensurate dielectric. *Phys. Rev. Lett.* **76**, 2334–2337.
37. J. E. Graebner and H. S. Chen. (1987). Specific heat of an icosahedral superconductor, $Mg_3Zn_3Al_2$. *Phys. Rev. Lett.* **58**, 1945–1948.
38. E. J. Thompson, P. D. Vu, and R. O. Pohl. (2000). Glasslike lattice vibrations in the quasicrystal $Al_{72.1}Pd_{20.7}Mn_{7.2}$. *Phys. Rev. B.* **62**, 11437–11443.
39. J. F. Gebbia, M. A. Ramos, D. Szewczyk, A. Jezowski, A. I. Krivchikov, Y. V. Horbatenko, T. Guidi, F. J. Bermejo, and J. Ll. Tamarit. (2017). Glassy anomalies in the low-temperature thermal properties of a minimally disordered crystalline solid. *Phys. Rev. Lett.* **119**, 215506.
40. A. Jeżowski, M. A. Strzhemechny, A. I. Krivchikov, N. A. Davydova, D. Szewczyk, S. G. Stepanian, L. M. Buravtseva, and O. O. Romantsova. (2018). Glassy anomalies in the heat capacity of an ordered 2-bromobenzophenone single crystal. *Phys. Rev. B.* **97**, 201201(R).
41. M. Moratalla, J. F. Gebbia, M. A. Ramos, L. C. Pardo, S. Mukhopadhyay, S. Rudić, F. Fernandez-Alonso, F. J. Bermejo, and J. L. Tamarit. (2019). Emergence of glassy features in halomethane crystals. *Phys. Rev. B.* **99**, 024301.
42. T. Pérez-Castañeda, R. J. Jiménez-Riobóo, and M. A. Ramos. Two-level systems and boson peak remain stable in 110-million-year-old amber glass. (2014). *Phys. Rev. Lett.* **112**, 165901.
43. X. Liu, B. E. White, Jr., R. O. Pohl, E. Iwanizcko, K. M. Jones, A. H. Mahan, B. N. Nelson, R. S. Crandall, and S. Veprek. (1997). Amorphous solid without low energy excitations. *Phys. Rev. Lett.* **78**, 4418–4421.
44. B. L. Zink, R. Pietri, and F. Hellman. (2006). Thermal conductivity and specific heat of thin-film amorphous silicon. *Phys. Rev. Lett.* **96**, 055902.
45. D. R. Queen, X. Liu, J. Karel, T. H. Metcalf, and F. Hellman. (2013). Excess specific heat in evaporated amorphous silicon. *Phys. Rev. Lett.* **110**, 135901.
46. N. I. Agladze and A. J. Sievers. (1998). Absence of an isotope effect in the two level spectrum of amorphous ice. *Phys. Rev. Lett.* **80**, 4209–4212.
47. S. F. Swallen, K. L. Kearns, M. K. Mapes, Y. S. Kim, R. J. McMahon, M. D. Ediger, T. Wu, L. Yu, and S. Satija. (2007). Organic glasses with exceptional thermodynamic and kinetic stability. *Science.* **315**, 353–356.

48. C. Rodríguez-Tinoco, M. González-Silveira, M. A. Ramos, and J. Rodríguez-Viejo. (2022). Ultrastable glasses: New perspectives for an old problem. *Riv. Nuovo Cim.* **45**, 325–406.
49. T. Pérez-Castañeda, C. Rodríguez-Tinoco, J. Rodríguez-Viejo, and M. A. Ramos. (2014). Suppression of tunneling two-level systems in ultrastable glasses of indomethacin. *PNAS* **111**, 11275–11280.
50. D. Khomenko, C. Scalliet, L. Berthier, D. R. Reichman, and F. Zamponi. (2020). Depletion of two-level systems in ultrastable computer-generated glasses. *Phys. Rev. Lett.* **124**, 225901.
51. T. Pérez-Castañeda, R. J. Jiménez-Riobóo, and M. A. Ramos. (2013). Low-temperature thermal properties of a hyperaged geological glass. *J. Phys.: Cond. Matter.* **25**, 295402.
52. J. Zhao, S. L. Simon, and G. B. McKenna. (2013). Using 20-million-year-old amber to test the super-Arrhenius behaviour of glass-forming systems. *Nat. Commun.* **4**, 1783.
53. T. Pérez-Castañeda, R. J. Jiménez-Riobóo, and M. A. Ramos. (2016). Do two-level systems and boson peak persist or vanish in hyperaged geological glasses of amber? *Philos. Mag.* **96**, 774–787.
54. E. A. A. Pogna, A. I. Chumakov, C. Ferrante, M. A. Ramos, and T. Scopigno. (2019). Tracking the connection between disorder and energy landscape in glasses using geologically hyperaged amber. *J. Phys. Chem. Lett.* **10**, 427–432.
55. L. Wang, A. Ninarello, P. Guan, L. Berthier, G. Szamel, and E. Flenner. (2019). Low-frequency vibrational modes of stable glasses. *Nat. Commun.* **10**, 26.
56. W. A. Phillips. Tunneling states in amorphous solids. (1972). *J. Low Temp. Phys.* **7**, 351–360.
57. P. W. Anderson, B. I. Halperin, and C. M. Varma. (1972). Anomalous thermal properties of glasses and spin glasses. *Philos. Mag.* **25**, 1–9.
58. M. Meissner and K. Spitzmann. (1981). Experimental evidence on time-dependent specific heat in vitreous silica. *Phys. Rev. Lett.* **46**, 265–268.
59. X. Liu and H. v. Löhneysen. (1996). Specific-heat anomaly of amorphous solids at intermediate temperatures (1 to 30 K). *Europhys. Lett.* **33**, 617–622.
60. A. I. Chumakov *et al.* (2011). Equivalence of the boson peak in glasses to the transverse acoustic van hove singularity in crystals. *Phys. Rev. Lett.* **106**, 225501.
61. A. I. Chumakov *et al.* (2014). Role of disorder in the thermodynamics and atomic dynamics of glasses. *Phys. Rev. Lett.* **112**, 025502.
62. S. Caponi, A. Fontana, F. Rossi, G. Baldi, and E. Fabiani. (2007). Effect of temperature on the vibrational density of states in vitreous SiO_2: A Raman study. *Phys. Rev. B.* **76**, 092201.

63. V. G. Karpov, M. I. Klinger, and F. N. Ignatiev. (1982). Atomic tunneling states and low-temperature anomalies of thermal properties in amorphous materials. *Solid State Commun.* **44**, 333–337.
64. M. A. Il'in, V. G. Karpov, and D. A. Parshin. (1987). Parameters of soft potentials in glasses. *Sov. Phys. JETP.* **65**, 165–174.
65. U. Buchenau, Yu. M. Galperin, V. L. Gurevich, D. A. Parshin, M. A. Ramos, and H. R. Schober. (1992). Interaction of soft modes and sound waves in glasses. *Phys. Rev. B.* **46**, 2798–2808.
66. L. Gil, M. A. Ramos, A. Bringer, and U. Buchenau. (1993). Low temperature specific heat and thermal conductivity of glasses. *Phys. Rev. Lett.* **70**, 182–185.
67. M. A. Ramos, L. Gil, A. Bringer, and U. Buchenau. (1993). The density of tunneling and vibrational states of glasses within the soft-potential model. *Phys. Stat. Sol.* (a) **135**, 477–492.
68. D. A. Parshin. (1994). Model of soft atomic potentials and universality of low-temperature properties of glasses. *Phys. Solid State.* **36**, 991–1024.
69. M. A. Ramos and U. Buchenau. (1997). Low-temperature thermal conductivity of glasses within the soft-potential model. *Phys. Rev. B.* **55**, 5749–5754.
70. M. A. Ramos and U. Buchenau. (1998). Beyond the standard tunneling model: The soft-potential model. In P. Esquinazi (ed.), *Tunneling Systems in Amorphous and Crystalline Solids*, pp. 528–569. New York: Springer.
71. V. L. Gurevich, D. A. Parshin, and H. R. Schober. (2003). Anharmonicity, vibrational instability, and the Boson peak in glasses. *Phys. Rev. B.* **67**, 094203.
72. D. A. Parshin, H. R. Schober, and V. L. Gurevich. (2007). Vibrational instability, two-level systems, and the Boson peak in glasses. *Phys. Rev. B.* **76**, 064206.
73. M. A. Ramos. (2004). Are the calorimetric and elastic Debye temperatures of glasses really different? *Phil. Mag.* **84**, 1313–1321.
74. M. A. Ramos. (2020). Are universal "anomalous" properties of glasses at low temperatures truly universal? *Low Temp. Phys./Fizika Nizkikh Temperatur.* **46**, 130–137.
75. W. Schirmacher. (2006). Thermal conductivity of glassy materials and the "boson peak". *Europhys. Lett.* **73**, 892–898.
76. W. Schirmacher, T. Scopigno, and G. Ruocco. (2015). Theory of vibrational anomalies in glasses. *J. Non-Cryst. Solids.* **407**, 133–140, and references therein.

77. P. G. Wolynes and V. Lubchenko (eds.), (2012). *Structural Glasses and Supercooled Liquids: Theory, Experiment, and Applications.* New Jersey: Wiley.
78. T. R. Kirkpatrick, D. Thirumalai, and P. G. Wolynes. (1989). Scaling concepts for the dynamics of viscous liquids near an ideal glassy state. *Phys. Rev. A* **40**, 1045–1054.
79. V. Lubchenko and P. G. Wolynes. (2007). Theory of structural glasses and supercooled liquids. *Annu. Rev. Phys. Chem.* **58**, 235–266.
80. V. Lubchenko and P. G. Wolynes. (2007). The microscopic quantum theory of low temperature amorphous solids. *Adv. Chem. Phys.* **136**, 95–206.
81. G. D'Angelo, G. Carini, C. Crupi, M. Koza, G. Tripodo, and C. Vasi. (2009). Boson peak in alkaline borate glasses: Raman spectroscopy, neutron scattering, and specific-heat measurements. *Phys. Rev. B.* **79**, 014206.
82. G. Carini, Jr., G. Carini, G. D'Angelo, G. Tripodo, G. Di Marco, C. Vasi, and E. Gilioli. (2013). Influence of packing on low energy vibrations of densified glasses. *Phys. Rev. Lett.* **111**, 245502.
83. C. Crupi, G. Carini Jr., G. Carini, M. Federico, A. Mandanici, V. Romano, and G. D'Angelo. (2020). Structure and low energy vibrations in silver borate glasses with low silver content. *J. Phys. Chem. Solids.* **138**, 109304.
84. M. Baggioli and A. Zaccone. (2021). New paradigm for glassy-like anomalies in solids from fundamental symmetries. *Int. J. Mod. Phys. B.* **35**, 2130002.
85. M. Baggioli and A. Zaccone. (2019). Hydrodynamics of disordered marginally stable matter. *Phys. Rev. Res.* **1**, 012010.
86. M. Baggioli, R. Milkus, and A. Zaccone. (2019). Vibrational density of states and specific heat in glasses from random matrix theory. *Phys. Rev. E.* **100**, 062131.
87. M. Baggioli and A. Zaccone. (2019). Universal origin of boson peak vibrational anomalies in ordered crystals and in amorphous materials. *Phys. Rev. Lett.* **122**, 145501.
88. M. Baggioli and A. Zaccone. (2020). Low-energy optical phonons induce glassy-like vibrational and thermal anomalies in ordered crystals. *J. Phys.: Mater.* **3**, 015004.
89. R. O. Pohl, X. Liu, and E. J. Thompson. (2002). Low-temperature thermal conductivity and acoustic attenuation in amorphous solids. *Rev. Mod. Phys.* **74**, 991–1013.
90. J. Y. Duquesne and G. Bellessa. (1985). Ultrasonic study of tunnelling defects in aorphous Se, Ge and Se-Ge compounds. *Philos. Mag. B.* **52**, 821–842.

91. J. F. Berret and M. Meissner. (1988). How universal are the low temperature acoustic properties of glasses? *Z. Phys. B — Condensed Matter.* **70**, 65–72.
92. S. Hunklinger. (1987). The universality of the density of states of low-energy excitations in glasses. *Phil. Mag. B.* **56**, 199–211.
93. J. M. Grace and A. C. Anderson. (1989). Low-temperature specific heat and thermal conductivity of a glassy polymer under applied pressure. *Phys. Rev. B.* **40**, 1901–1917.
94. A. P. Sokolov, R. Calemczuk, B. Salce, A. Kisliuk, D. Quitmann, and E. Duval. (1997). Low-temperature anomalies in strong and fragile glass formers. *Phys. Rev. Lett.* **78**, 2405–2408.
95. D.-M. Zhu and H. Chen. (1998). Low temperature specific heat and fragility of glasses. *J. Non-Cryst. Solids.* **224**, 97–101.
96. D.-M. Zhu. (1996). Correlation between density of tunnelling states and fragility of glasses. *Phys. Rev. B.* **54**, 6287–6291.
97. A. V. Granato. (1992). Interstitialcy model for condensed matter states of face-centered-cubic metals. *Phys. Rev. Lett.* **68**, 974–977.
98. A. V. Granato. (1994). Self-interstitials as basic structural units of liquids and glasses. *J. Phys. Chem. Solids.* **55**, 931–939.
99. A. V. Granato. (1996). An interstitialcy model for the boson peak. In M. Giordano, D. Leporini, and M. P. Tosi (eds.), *Non Equilibrium Phenomena in Supercooled Fluids, Glasses and Amorphous Materials*, pp. 361–362. Singapore: World Scientific.
100. G. Carini, G. D'Angelo, G. Tripodo, and G. A. Saunders. (1995). Specific heat and low-energy excitations of samarium phosphate glasses. *Philos. Mag.* **71**, 539–545.
101. M. A. Ramos, C. Talón, and S. Vieira. (2002). The Boson peak in structural and orientational glasses of simple alcohols: Specific heat at low temperatures. *J. Non-Cryst. Solids.* **307–310**, 80–86.
102. C. C. Yu and A. J. Leggett. (1988). Low temperature properties of amorphous materials: Through a glass darkly. *Comments Cond. Mat. Phys.* **14**, 231–251.
103. A. Kumar, I. Procaccia, and M. Singh. (2021). Density of quasi-localized modes in athermal glasses. EPL **135**, 66001.

https://doi.org/10.1142/9781800612587_0003

Chapter 3

Thermal Conductivity of Glasses and Disordered Crystals

A. I. Krivchikov*,‡ and A. Jeżowski†,§

*B. Verkin Institute for Low Temperature Physics and Engineering of the NAS of Ukraine, 47 Nauky Ave., Kharkiv 61103, Ukraine
†Institute of Low Temperature and Structure Research, PAS, 2 Okolna Str., 50-422 Wrocław, Poland
‡krivchikov@ilt.kharkov.ua
§a.jezowski@intibs.pl

Among the many physical properties, the amorphous state manifests itself in the most spectacular way in heat transport. Anomalously low thermal conductivity, its low-temperature dependence as a function of temperature, and the presence of a plateau of thermal conductivity are all definitely different when compared with the ordered state of crystals. For this reason, the microscopic mechanisms from which these surprising behaviors may emerge have been the subject of intensive theoretical research and even more extensive experimental research. These investigations have led to the detection of a huge range of disordered materials with somewhat similar properties, which have become a rich base for various physical interpretations. In this chapter, we present the last 50 years of the history of experimental research together with the key theoretical physics scenarios.

1. Introduction

This chapter deals with experimental studies of the thermal conductivity κ of solids, which is characterized by a unique temperature dependence of the same type, the so-called glass-like thermal conductivity, in a wide range from ultralow to high (1,000 K) temperatures. The turning point in the understanding of glassy behavior of the thermal conductivity was the pioneering work[1] of Zeller and Pohl in 1971. For the first time, they undertook a joint measurement of low-temperature thermal properties (heat capacity and thermal conductivity) of several structural glasses such as SiO_2, GeO_2, multi-component structural glass Pyrex, and aluminogermanate, as well as amorphous Se, in a temperature range from 0.05 to 100 K. A comparative analysis of the obtained results and literature data for a large number of amorphous substances of various nature allowed the authors to reach the fundamental conclusion that the low-temperature behavior of the thermal properties of amorphous substances is rather insensitive to the chemical composition of the material and is quite different from the behavior of the corresponding crystals, and that cannot therefore be explained in the framework of collective excitations of a phonon gas in a crystal lattice. They found a direct correlation between the additional linear-in-temperature contribution to the heat capacity and the thermal conductivity quadratic dependence $\kappa \sim T^2$. This allowed them to formulate the problem of low-temperature anomalies of disordered solids, which was the inspiration for plenty of new experimental research not only of thermal properties but also of others, among them dielectric, acoustic, structural, and mechanical ones. Schematic dependences of reduced heat capacity C/T^3 and glass-like thermal conductivity of a disordered solid on its reduced temperature, around the temperature of the boson peak (see Chapters 2, 8 and 9) $T_{\max}$, are shown in Figure 1.

2. Low-Temperature Thermal Conductivity Data

To date, thermal conductivity has been investigated in more than 100 amorphous solids and in more than 80 crystalline substances, in which glass-like behavior takes place. Studies performed after the work of Zeller and Pohl can be divided into several stages that we separate by decades as follows.

2.1. *Studies in the decade 1971–1980*

The initial period (from 1971 to 1980) was primarily associated with experimental work on the measurement of the thermal conductivity

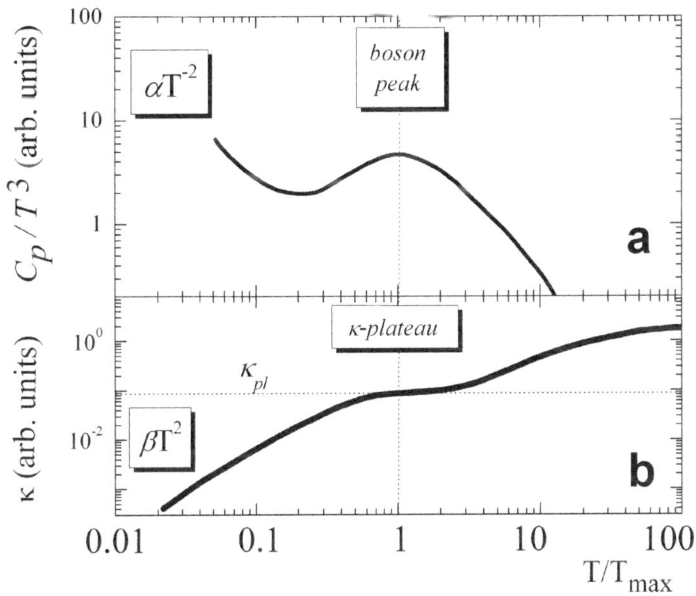

Fig. 1. Schematic dependences of reduced heat capacity C/T^3 (a) and glass-like thermal conductivity (b) of a disordered solid on its reduced temperature T/T_{max}, below and above the temperature of the boson peak T_{max}. The thermal conductivity quadratic factor is β and the plateau thermal conductivity is κ_{pl}.

of typical amorphous substances and glasses: *polymers*;[1–12] *chalcogenide glasses*;[13,14] *structural glasses*;[14–19] *metallic glasses*;[20–22] *amorphous solids*;[23–25] *radiation damaged crystals*;[26] and *mixtures of glass oxides*.[27]

The above cited works show that thermal conductivity of amorphous solids (obtained in various ways) exhibits some universal behavior, which is due to the interaction (resonant scattering) of acoustic excitations with a set of localized low-energy excitations according to the well-known phenomenological tunneling model of Anderson, Halperin, and Varma,[28] and of Phillips,[29] at least at the lowest temperatures (see Section 3.1).

In the same period, in the first review of the thermal conductivity of polymers,[12] it was noted that *the temperature dependence for all amorphous polymers is approximately equal in magnitude and characterized by a* T^2 *dependence below 0.5 K, a plateau region between 5 and 15 K and a slow increase at yet higher temperatures.*

2.2. Studies in the decade 1981–1990

In the 1980s — the second important period — an exciting event emerged, namely, the detection of glass-like behavior not only in new amorphous substances but also in crystalline structures, such as ferroelectrics, ionic conductors, crystals with orientational disorder, as well as in radiation-damaged crystals. Many experimental investigations revealed new features of glass-like behavior in the κ-plateau and high-temperature increase in molecular glasses, mixture of glass oxides, chalcogenide glasses, disordered crystals, as well as in a variety of metal glasses and in many modern — at that time — polymers and structural glasses. The search also began for an explanation of which external factors (pressure, electric field, radiation, variation in composition of polymeric and quasicrystalline materials) affect the glass-like behavior in a wide temperature range. The thermal activation character of the high-temperature dependence of the thermal conductivity of clathrate hydrates was reported for the first time.[30] Among others, the following works were published in this decade: *ferroelectric crystals*;[31–37] $Pb(Sc_{0.5}Nb_{0.5})O_3$ under electric field;[38] tourmaline in pyroelectric state;[39] $PbZrO_3$ in pyroelectric state, antiferroelectric;[40] *ionic conductors and ceramics*;[31,41–44] *orientational glasses*;[45–49] *mixtures of glass oxides*;[50–54] *molecular glasses*;[55,56] *clathrate compounds*;[30,57–59] *quasicrystals*;[60] *radiation damaged solids*;[61,62] *amorphous solids*;[63–67] *structural glasses*;[53,56,68,69] SiO_2 porous;[70,71] SiO_2 aerogel;[72] SiO_2 + Ne;[73,74] *chalcogenide glasses*;[75] *polymers*;[56,76–85] electron-irradiated epoxy resin,[86] epoxy resin under pressure;[87] and *metallic glasses*.[88–97]

In the same period, a large number of reviews[75,85,98–109] were published, in which the results of many experimental investigations of different properties including thermal conductivity of glasses, disordered crystals, and glassy crystals were presented. In Ref. [102], new data on the glass-like behavior of thermal conductivity in metallic and insulating glasses, and in crystalline disordered solids, were discussed. The glass-like thermal conductivity of complex inclusion crystals with orientational disorder, clathrate hydrates at high temperatures and under pressure, was discussed in Ref. [103].

Another review[104] stressed that the low-temperature glassy properties of disordered crystals exhibit the same anomalous behavior as found in amorphous solids. However, these did not provide a definitive answer to the main problem. *There is no final answer to the main problem. The major problem is of course the microscopic nature of the tunneling entity.*[105] In Ref. [106], some inorganic ferroelectrics, antiferroelectrics, and pyroelectrics were reviewed, which display glass-like thermal properties at low and ultralow temperatures. It was found[107] that glass-like behavior of thermal conductivity in ferroelectric single crystals of the relaxor type is sensitive to an electric field.

A review by Phillips[108] covered a wide range of experimental and theoretical studies of two-level or tunneling states in glasses and concluded the following: *A convincing microscopic description of a tunneling state in any amorphous material remains a problem to tax the solid state physicist.* Cahill and Pohl[75] showed that in amorphous and highly disordered solids, the picture of quantized vibrations exhibits serious deficiencies at high frequencies. Greig[85] noted that there is no *universal* thermal conductivity behavior in semi-crystalline polymers. Also, several general reviews of experimental works were published, concerning anomalous very-low-energy properties of nonmetallic solids and their theoretical interpretation.[98–102, 105, 110]

2.3. *Studies in the decade 1991–2000*

In the following stage (during the 1990s), it was first discovered that a complex polycrystalline filled skutterudite[111] $Ir_4NdGe_3Sb_9$ containing weakly bound atoms in some crystal nodes, and almost-crystalline three-component metal structures called quasicrystals, showed glass-like behavior. Pioneering works were done in the study of the extreme effects of pressure, radiation, and electric field on thermal conductivity of glasses, polymers, and crystals. At that time, evidence emerged that the thermal conductivity of porous SiO_2 aerogel anomalously depends on its density. It was also revealed that thermal conductivity of metallic glasses at high temperatures presents the same regularity as structural glasses and amorphous films. Crystalline clathrate compounds, significantly different in chemical bonding, display the usual glassy behavior. Further, it was found that orientationally disordered crystals of both inorganic and organic substances, as well as molecular glasses, present all the characteristic features of the glass-like behavior of thermal conductivity. Studies of a large number of both single crystals and mixed crystals revealed only new quantitative features of the glass-like behavior of the universal temperature dependence of thermal conductivity in a wide range of temperatures. For example, thermal conductivity of disordered solids slightly increases with increasing density. Data on the small Bridgman parameter $g \approx 2 \div 3$ for a group of polymer substances were obtained[112–115] and later confirmed.[116,117]

The following works were published in this decade: *quasicrystals*;[118–120] *skutterudites*;[111] *amorphous solids*,[121,122] a-Si porous;[123] *radiation damaged crystals:*neutron-irradiated C;[124] *structural glasses:* SiO_2 film,[125,126] Al_2O_3 film,[127] aerogel SiO_2,[128,129] densified SiO_2,[130] SiO_2 under pressure;[131] *mixtures of glass oxides*;[132,133] *polymers*;[112–115,134–146] *molecular glasses*;[147] *orientational glasses*;[134,149,150] *clathrate compounds*;[151–157] *metallic glasses*;[158,159] and *ferroelectric crystals*.[160–165]

As in the previous period, several general reviews of experimental works were published.[132,166–178] These reviews were related to anomalous very-low-energy properties of metallic and non-metallic glass and disordered crystals, either under normal conditions or under pressure. They also dealt with the theoretical interpretation within the framework of the soft-potential model,[171,179,180] which is able to explain the glassy anomalies in the specific heat and the thermal conductivity in a much wider low-temperature range. Other works considered the phonon-glass electron-crystal concept,[175] or a lower limit to the thermal conductivity of solids.[181] This experimentally observed lower limit to the thermal conductivity of solids was ascribed to the minimum thermal conductivity earlier proposed by Einstein, in which the elastic energy propagates in a random walk among the atoms which are vibrating with random phases.

2.4. *Studies in the decade 2001 2010*

In the fourth stage, i.e., in the first decade of the new millennium, considerable efforts were directed to experimental and theoretical studies of crystals with a different kind of dynamic, static, orientational, positional, polaronic, and other kinds of disorder (see Figure 2). At this stage, researchers revealed various new crystals with both weak and strong interatomic forces, in which glass-like behavior of thermal conductivity takes place. It became clear that the κ-plateau can manifest itself not only in the low-temperature region (4–10 K) but also in the medium-temperature region (20–50 K) in the case of metallic glasses, quasicrystals, in some semiconductor clathrate crystals, and in some ferroelectrics and ceramics. In the temperature range above the "plateau", the thermal conductivity of quasicrystals is described by the Arrhenius-type equation. At that time, it had been suggested that avoiding the crossing between heat-carrying acoustic modes and the low-lying optical

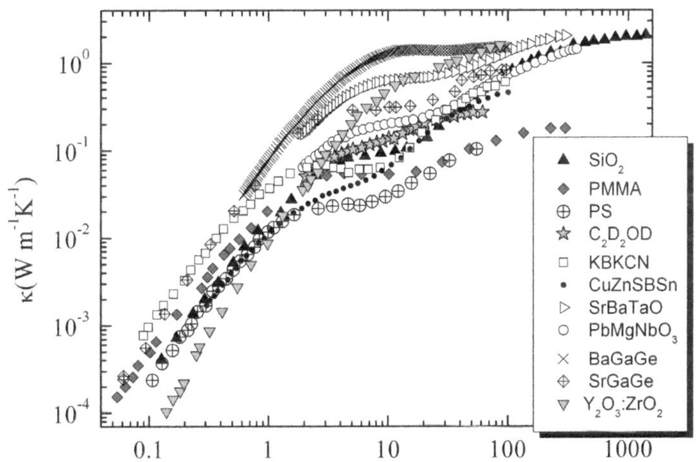

Fig. 2. Glass-like thermal conductivity of different glasses and disordered crystals: SiO_2;[16,56] polymers PMMA, PS;[15] molecular glass C_2D_5OD;[243] orientational glass $(KBr)_{0.75}(CN)^{47}_{0.25}$; ferroelectric crystal PMN;[244] SrBaTaO;[296] clathrate compounds SGG;[153] BGG[245] and CuZnSbSn;[246] single-crystal Y_2O_3:ZrO_2.[247]

modes[182–184] can influence glass-like behavior of thermal properties. The following works were published in this decade: *clathrate compounds*;[185–202, 245] *quasicrystals*;[203–208] *orientational glasses*;[209–211] *skutterudites and other rattling structures*;[212–214] *ferroelectric crystals*;[215, 216, 244] *ceramics*;[217] *molecular glasses*;[213, 217–220] *amorphous solids*;[221–225] *structural glasses*;[226–229] *chalcogenide glasses*;[230, 231] *polymers*;[119, 232–235] *films*;[236] *metallic glasses*;[237–239] and *others:* ZrHfTiNiSn,[240] zeolite NaX,[241] and $(La_{1-x}Yb_x)_2Zr_2O_7$.[242]

Also, several general reviews of experimental and theoretical works were published.[182, 214, 248–254] In Ref. [250], attention was drawn to ... *the fact that the low-energy excitations in numerous disordered crystals resemble very closely those found in amorphous solids also strongly suggests a picture in which lack of long-range order per se is not the cause of the low-energy excitations. But in that case, what is important?* Using the example of over 60 different compositions, it was shown that T^2 dependence of thermal conductivity is universal and is dominated by phonons that are resonantly scattered by the low-energy excitations in the frame of the tunneling model.

2.5. *Studies in the decade 2011–2020*

In the fifth stage, which comprises this last decade, the study of glass-like behavior in new organic and inorganic crystals of various thermoelectric materials, deuterated substances, porous amorphous films and glasses, unusual ceramics, and ferroelectrics is actively continuing. In this period, new representatives appeared in the discussed large family of "glass-like" behavior, for example, BSA, myoglobin proteins, $Tb_2Ti_2O_7$ spin-liquid, disordered layered structures and organic metal anisotropic single crystals, cyanate ether resin, HMS ribbon, nanofiber, and nanocrystals. It has been established that deuteration leads to an increase in the high-temperature thermal conductivity of clathrate hydrates,[154] amorphous water,[255] and molecular glasses.[243] A relevant line of current research was noted in the review of Chang and Zhao[256] where the relationship between anharmonicity and low thermal conductivity in thermoelectrics was summarized. Several strategies which yield anharmonicity were also suggested, including lone pair electron, resonant bonding, and the rattling model found in clathrates and skutterudites with cage-like structure.

The following main works were published in this decade: *biomaterials*;[257] *clathrate compounds*;[246,258–264] *quasicrystals*;[265] *orientational glasses*;[243,266–268] *molecular glasses*;[243,269–272] *amorphous solids*;[253,273–277] *structural glasses*;[278–280] *polymers*;[281–292] *metallic glasses*;[293,294] *ferroelectric crystals*;[295–298] *ceramics*;[247,299–304] and *others*: layered crystal structure of $NaZn_4Sb_3$,,[305] $Tb_2Ti_2O_7$ spin liquid,[306] silica zeolite,[307] Si nanowires,[308] boron carbide nanowires,[309] lead halide perovskite nanowires,[310] amorphous nanowires,[311,312] SiN films,[313] a-Si_3N_4 nanowires,[314] Si nanocrystals,[315] HMS ribbon,[316] θ-$(BEDT\text{-}TTF)_2MZn(SCN)_4$,[317] $Ln_3NbO_7(Ln{=}Dy, Y, Er, Yb)$,[318] $CH_3NH_3PbBr_3$ nanowire,[319] hydroxyapatite nanoparticles,[320] and *nanostructures*.[321–325]

Also, several general reviews of experimental and theoretical works have been published.[288,315,326–335] In Refs. [326, 328], important correlations between the properties at very low temperatures and at moderately low temperatures in glasses are discussed in the frame of soft-potential model and other theoretical approaches. In Ref. [327], the authors discussed recent advances in materials engineering to control thermal conductivity and stated that ... *good progress has been made with materials possessing rattling atoms or complex unit cells.* A similar statement can be made on the basis of another review,[329] which presents a large variety of properties of phonon-glass electron-crystal thermoelectric clathrates. The thermal properties under pressure and the transition behavior of several host–guest inclusion compounds (urea, thiourea, Dianin's compound, clathrate hydrates, and hydroquinone) have been reviewed in Ref. [330]. There, the unusual glass-like thermal conductivity of inclusion compounds was found to be of great technological and fundamental importance to seek for improved thermoelectrical materials, despite the fact that the origin of their glass-like thermal conductivity is not understood. In another review,[331] the following can be read: *Recently, however, a growing number of studies have re-examined the thermal properties of amorphous semiconductors, such as amorphous Si. These studies, which included both computational and experimental work, have revealed that phonon transport in amorphous materials is perhaps more complicated than previously thought.* On the other hand, De Angelis *et al.*[333] have reviewed the theory and computational modeling approaches that can be applied to disordered materials and discussed corresponding experimental techniques. They concluded that ... *important questions lay on the horizon for better understanding thermal transport when disorder is present and tremendous progress is possible in the next few decades.*

3. Theoretical Descriptions

Disordered solids have no translation symmetry and therefore Peierls's theoretical picture that heat carriers are the crystalline excitation of propagating vibrational waves seems to break. To explain this anomalous behavior of an amorphous solid, various phenomenological models were developed that introduced a number of low-energy excitations: tunneling two-level systems (TLS), relaxation systems, and low-frequency quasilocal vibrational states, in addition to acoustic continuous-medium excitations. The temperature dependence of the thermal conductivity is linked to the nature of the energy transport which is quantified by ballistic propagation and thermal diffusivity of collective excitations in a disordered solid. According to the Standard Tunneling Model and the Soft-Potential Model, in a ballistic regime or in a quasiparticle picture, continuous-medium acoustic excitations are strongly resonant scattered by TLS excitations and quasi-local vibrations. A harmonic theory of thermal transport[336] in glasses was proposed by Allen and Feldman in the year 1989: An additional thermal diffusive regime or wave interference picture was implemented where heat is carried by new carriers, the coupling of high-energy vibration modes being named diffusons and locons. Allen-Feldman approaches, however, do not always provide physical insight into how the modes are actually contributing to the thermal conductivity.[336,337] The thermal conductivity of complex ordered crystals displays a glass-like low-temperature behavior[338–340] (the universal T^2 dependence predicted by the tunneling model) and a glass-like high-temperature behavior[219,260,266,316,341–343] (decay milder than the universal T^{-1} dependence predicted by the phonon gas model). It should be noted that one review[333] outlined theoretical computational modern approaches, summarized the progress in understanding materials with disorder, and highlighted open questions.

3.1. *Standard Tunneling Model*

The Standard Tunneling Model was proposed by Anderson, Halperin, and Varma,[28] and by Phillips[29] in 1972 to describe many aspects of the universal low-temperature properties of amorphous solids. Two-level tunneling states (TLSs) with a constant density of states as defects of an elastic continuum were postulated. For an elastic continuum such as a glass, the thermal conductivity $\kappa(T)$ is given by the standard expression obtained from the well-known phonon-gas kinetic equation, and using the Debye approximation for the density of states of the sound waves transporting heat. The thermal conductivity quadratic dependence $\kappa = \beta T^2$ is evaluated on the assumption of strongly resonant scattering of continuous-medium excitations by TLS excitations. One obtains the following[28,29,250]:

$$\kappa(T) = \frac{k_B^3}{6\pi\hbar^2}\left(\sum_\alpha \frac{\rho s_\alpha}{\overline{P}\gamma_\alpha^2}\right)T^2 = \frac{k_B^3}{6\pi\hbar^2}\left(\sum_\alpha \frac{1}{C_\alpha s_\alpha}\right)T^2, \tag{1}$$

with

$$C_\alpha = \frac{\overline{P}\gamma_\alpha^2}{\rho s_\alpha^2} \tag{2}$$

where α is the polarization of phonons, $\overline{P}$ is the spectral density of the tunneling states, and γ_α is the sound excitation-TLS coupling. ρ and s_α, respectively, represent the mass density and the corresponding sound velocity.

3.2. Soft-potential model

For temperatures between 1 and 10 K, the anomalous behavior of thermal conductivity that manifests itself by the appearance of a temperature-independent plateau, as well as a concomitant increase in heat capacity well above that corresponding to the Debye model, was the subject of conflicting explanations. The explanations have been now rationalized on phenomenological grounds by means of a generalization of the tunneling model known as the soft-potential model (SPM)[180, 344, 345] and also as soft-mode model.[346]

In the model of soft potentials, glass is considered as an elastic medium with specific structural defects, the particle motion of which can be not only harmonic but also anharmonic. It has been shown to be able to account for the plateau of the thermal conductivity on the basis of the assumption of resonant scattering of sound waves from localized low-frequency vibrations. The soft-potential model describing elementary excitations in a wide range of energies introduces a characteristic energy W (often expressed in K), which assigns the energy scale in the classification of elementary excitations in the harmonic soft potential. According to Ref. [180], $\kappa(T)$ is described by a universal function in the representation of a normalized temperature (reduced temperature) $u = T/W$, which can also be read as

$$\kappa(T) = \frac{18k_B^3}{\pi h^2}\frac{W^2}{\bar{C}\cdot s}\frac{T^2}{1.1\,W^2 + 0.7\,T\cdot W + 3\,T^2} \tag{3}$$

The dimensionless parameter $\bar{C}$ takes on a fairly narrow interval 10^{-3} $\div 10^{-4}$ and characterizes the strength of the interaction of TLS with sound excitations at a properly averaged sound velocity s.

The expression (3) allows us to relate the observed three characteristics — the plateau in thermal conductivity (κ_{pl}), the thermal conductivity quadratic factor (β), and the characteristic energy of the SPM (W) — to each other in the following form:

$$\frac{\kappa_{\text{pl}}}{\beta W^2} \approx \frac{1}{3} \tag{4}$$

The ratio (4) is in good agreement with experimental data for disordered solids of a different nature.

4. Recapitulation

The large amount of materials described above, in which glass-like thermal conductivity is observed, raises the question of finding common physical patterns, and also raises the question of the factors influencing the character of this property.

A lot of work has clearly demonstrated a correlation[31] between the maximum temperature $T_{\max}$ (boson peak) in the reduced heat capacity C/T^3 and the temperature range of the glass-like plateau in the thermal conductivity.[15, 31, 179, 326, 347–350, 361] This unambiguous correlation was observed both in joint and independent measurements of the heat capacity and thermal conductivity of glasses and amorphous solids of a different nature, as well as in complex crystals with disorder, such as the following examples: *amorphous solids*;[1, 24, 222, 224] *chalcogenide glasses*;[54, 75, 231, 351, 352] *clathrate compounds*;[151, 188, 189, 199, 201, 202, 245, 246, 258, 259, 262, 264] *ferroelectric crystals*;[31, 36, 38, 164, 217, 244] *ionic conductors*;[31, 247, 300] *metallic glasses*;[237, 238] *mixture of glass oxides*;[353] *molecular glasses*;[217, 269, 270] *orientational glasses*;[45, 47] *radiation solids*;[19, 354] *polymers*;[78, 79, 140, 143, 355] (also under pressure);[87] *structural glasses*;[1, 356] and *other materials*.[240]

The correlation Boson peak–plateau persists after external and internal effects on a solid when varying factors such as pressure,[87, 130, 140, 141, 144, 145] annealing,[73] electric field,[161, 162] neutrons irradiation,[26, 62] or electrons irradiation[19] express their influence. The position and size of the plateau depend on the porosity of the solid[129, 227, 356] and on the concentration of the solvent in the two-component solid solution.[50, 53, 97, 199, 216, 353, 357]

However, it should be noted that there is also a skeptical view on the existence of a direct correlation between the specific features of heat capacity and thermal conductivity, which can be expressed in the following quote: *Our most important conclusion is that the "reincrease" of thermal conductivity above the plateau region is attributable to heat carried by "diffuson" modes in much the way imagined by Birch, Clark, and Kittel, and that the plateau is a simple crossover region, not requiring any new physics to explain. In particular, we believe that "excess modes" (also known as a "Boson peak") is not a necessary ingredient to explain the plateau. Amorphous silicon seems to lack these "excess modes" but still to have a plateau.*[358] But, the following fact seems to contradict the above statements. Namely, the correlation Boson peak–plateau can be clearly seen if we represent the dependence of thermal conductivity κ on the temperature normalized to the temperature of the maximum of the boson peak $T_{\max}$ (see Figure 3). In the $T \approx T_{\max}$ region, the $\kappa(T/T_{\max})$ systematically exhibits a plateau. This circumstance can be used for a new definition of the term "glass-like thermal conductivity".

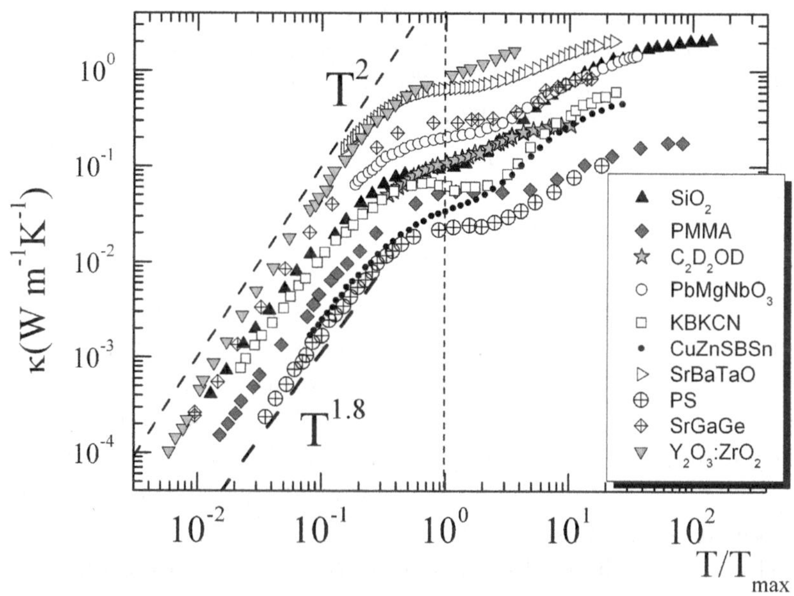

Fig. 3. Glass-like temperature thermal conductivity depending on temperature scaled by $T_{\max}$ for glasses and crystalline materials (see the symbols in the caption of Figure 2).

The glass-like thermal conductivity can be characterized by the following four features:

(i) $\kappa(T)$ is proportional to $\beta T^{2-\delta}(\delta = 0 \div 0.2)$ at $T \ll T_{\max}$. (ii) There is a plateau region with the plateau thermal conductivity $\kappa_{\rm pl}(T) \approx const$ near $T_{\max}$. (iii) $\kappa(T)$ has a rising inflection point above the plateau from $T > T_{\max}$. (iv) A saturating regime where $\kappa(T)$ is independent of T for $T > 10T_{\max}$.

It is to be emphasized that the κ-plateau is very sensitive to external factors. From a large set of experimental results, we can distinguish in principle the following effects that affect the thermal conductivity:

- Low-temperature $T^{2-\delta}$ effect — most of the data in the literature in the temperature range 0.1–1 K report a $\delta > 0$, but for $T < 0.1$K $\delta \approx 0$: SiO_2,[17,228] SBN,[173] B_2O_3,[228] PVC,[232] $Zr_{52.5}Ti_5Cu_{17.9}Ni_{14.6}Al_{10}$,[293] Polyimide.[292] A nearly linear specific heat and a thermal conductivity roughly proportional to T^2 ($\kappa(T) = \beta T^2$) are common features of glass-like behavior. As thermal conductivity is concerned, not only is this universality qualitative but also quantitative.[250] For many different glassy solids, the universal T^2 temperature dependence of the thermal conductivity is in reasonable agreement with the prediction of the tunneling model. The values of the coefficient β for the T^2 dependence are located in a rather confined interval of the thermal conductivity quadratic factor $\beta \approx 0.01 \div 0.1\,\mathrm{Wm^{-1}K^{-3}}$. The experimental results of thermal conductance of amorphous nanowires between 0.05 K and 5K show a uniform quadratic dependence on the temperature.[311,312]
- The effect of pressure — increase of heat conductivity in the κ-plateau area with increase in external pressure, as observed for epoxy resin,[87] SBN,[160] epoxy resin,[140] PC,[145] PS.[141]
- The effect of irradiation — a change in the thermal conductivity at the κ-plateau with increasing dose: falling for epoxy resin;[88] rising for SiO_2.[19,27]

- Electric field effect (poling) — increase of thermal conductivity in the κ-plateau region with an increase in the magnitude of the electric field: $Pb(Sc_{0.5}Nb_{0.5})O_3$, SBN,[35,38,162] $Pb(Mg_{1/3}Nb_{2/3})O_3$.[162,359]
- The "shoulder effect" — a transformation of the plateau into a curve with a slightly increasing slope is observed in two-component substances: $0.9SiO_2 \cdot 0.1GeO_2$ and $0.925SiO_2 \cdot 0.075Ti_2O$,[133] $(1-x)B_2O_3 \cdot xNa_2O$,[52] $Ge_xAs_{0.4-x}S_{0.6}$,[231] $Fe_{0.4}Ni_{0.4}P_{0.14}B_{0.06}$,[91] and $PMN_{1-x}[PbTiO_3]_x$.[215]
- Porosity effect — decrease of heat conductivity in the κ-plateau area with an increase in porosity: see Refs. [11,70,129,227,278,356].
- Cross-link density effect — increase of thermal conductivity in the κ-plateau area with a decrease of cross-link density: epoxy resin,[78,79,81,136,357] PS.[84]
- Annealing effect — a change of thermal conductivity in the κ-plateau under the action of annealing: falling for PC;[360] rising for SiO_2.[73]
- The effect of deuteration — an increase in the value of high-temperature thermal conductivity in deuterated substances (above the plateau): THF $+17D_2O$[154], C_2D_5OD,[243] a-ice.[255]

This huge amount of work on glass-like behavior over the last 50 years shows how fascinating the problem was and still is. Many common features and analogies have been found, e.g., the resonant phonon scattering by localized excitations and the concept of a lower limit to the thermal conductivity. But, despite the fact that so many interesting results, including thermal conductivity data, have been collected, the question about the nature of the observed anomalies in significantly different objects remains open.

Acknowledgments

This work was supported by the National Research Foundation of Ukraine (Grant No.197/02.2020).

References

1. R. C. Zeller and R. O. Pohl. (1971). Thermal conductivity and specific heat of noncrystalline solids. *Phys. Rev. B.* **4**(6), 2029–2041.
2. R. B. Stephens, G. S. Cieloszyk, and G. L. Salinger. (1972). Thermal conductivity and specific heat of non-crystalline solids: Polystyrene and polymethyl methacrylate. *Phys. Lett. A.* **38**(3), 215–217.
3. T. Scott and M. Giles. Dislocation scattering in Teflon at low temperatures. (1972). *Phys. Rev. Lett.* **29**(10), 642–643.
4. H. Abe and K. I. Koga. (1977). Thermal conductivity of some practical insulators around 100 mK. *Jpn. J. Appl. Phys.* **16**(9), 1583–1587.
5. T. A. Scott, J. de Bruin, M. M. Giles, and C. Terry. (1973). Low-temperature thermal properties of nylon and polyethylene. *J. Appl. Phys.* **44**(3), 1212–1216.
6. S. Burgess and D. Greig. (1975). The low-temperature thermal conductivity of polyethylene. *J. Phys. C.* **8**(11), 1637–1648.
7. C. L. Choy and D. Greig. (1975). The low-temperature thermal conductivity of a semi-crystalline polymer, polyethylene terephthalate. *J. Phys. C.* **8**(19), 3121–3130.
8. G. S. Cieloszyk, M. T. Cruz, and G. L. Salinger. (1973). Thermal properties of dielectric solids below 4 K l—polycarbonate. *Cryogenics.* **13**(12), 718–721.
9. K. W. Garrett and H. M. Rosenberg. (1974). The thermal conductivity of epoxy-resin/powder composite materials. *J. Phys. D.* **7**(9), 1247–1258.
10. D. S. Matsumoto, C. L. Jr. Reynolds, and A. C. Anderson. (1977). Thermal boundary resistance at metal-epoxy interfaces. *Phys. Rev. B.* **16**(8), 3303–3307.
11. M. P. Zaitlin and A. C. Anderson. (1975). Phonon thermal transport in noncrystalline materials. *Phys. Rev. B.* **12**(10), 4475–4486.
12. C. L. Choy. Thermal conductivity of polymers. (1977). *Polymer.* **18**(10), 984–1004.
13. R. Flasck and H. K. Rockstad. (1973). The thermal conductivity of some chalcogenide glasses. *J. Non-Cryst. Solids.* **12**(3), 353–356.
14. A. J. Leadbetter, A. P. Jeapes, C. G. Waterfield, and R. Maynard. (1977). Conduction thermique des verres aux basses temperatures. *J. de Physique* **38**(1), 95–99.
15. R. B. Stephens. (1973). Low-temperature specific heat and thermal conductivity of noncrystalline dielectric solids. *Phys. Rev. B.* **8**(6), 2896–2905.

16. D. H. Damon. (1973). Thermal conductivity of vitreous silica at low temperatures. *Phys. Rev. B.* **8**(12), 5860–5865.
17. J. C. Lasjaunias, A. Ravex, M. Vandorpe, and S. Hunklinger. (1975). The density of low energy states in vitreous silica: specific heat and thermal conductivity down to 25 mK. *Solid State Commun.* **17**(9), 1045–1049.
18. R. B. Stephens. (1976). Intrinsic low-temperature thermal properties of glasses. *Phys. Rev. B.* **13**(2), 852–865.
19. H. V. Löhneysen and M. Platte. (1979). Low temperature thermal properties and intrinsic defects in vitreous silica. *Zeitschrift für Physik B Condensed Matter.* **36**(2), 113–120.
20. J. R. Matey and A. C. Anderson. (1977). Low-temperature thermal conductivity of an amorphous palladium silicon alloy. *J. Non-Cryst. Solids.* **23**(1), 129–137.
21. J. E. Graebner, B. Golding, R. J. Schutz, F. S. L. Hsu, and H. S. Chen. (1977). Low-temperature properties of a superconducting disordered metal. *Phys. Rev. Lett.* **39**(23), 1480–1483.
22. J. R. Matey and A. C. Anderson. (1977). Phonon transport in glassy metals below 100 K. *Phys. Rev. B.* **16**(8), 3406–3410.
23. H. V. Löhneysen and F. Steglich. (1977). Low-temperature thermal conductivity of amorphous germanium. *Phys. Rev. Lett.* **39**(22), 1420–1423.
24. D. P. Jones, N. Thomas, and W. A. Phillips. (1978). The low-temperature thermal properties of amorphous arsenic. *Philos. Mag. B.* **38**(3), 271–288.
25. M. Meissner and D. Wobig. (1979). Specific heat and thermal conductivity of trigonal and vitreous selenium in the temperature range 3–300 K. In E. Gerlach, P. Grosse (eds.), The Physics of Selenium and Tellurium, pp. 68–73. Berlin-Heidelberg-New York: Springer.
26. T. L. Smith, J. Anthony, and A. C. Anderson. (1978). Effect of neutron irradiation on the density of low-energy excitations in vitreous silica. *Phys. Rev. B.* **17**(12), 4997–5008.
27. F. Canal, M. C. Schmidt, J. P. Redoules, and P. Carrara. (1976). Influence of phase separation on the thermal conductivity of potassium borosilicate glasses between 1 and 10 K. *J. Non-Cryst. Solids.* **21**(1), 73–84.
28. P. W. Anderson, B. I. Halperin, and C. M. Varma. (1972). Anomalous low-temperature thermal properties of glasses and spin glasses. *Philos. Mag.* **25**(1), 1–9.
29. W. A. Phillips. (1972). Tunneling states in amorphous solids. *J. Low Temp. Phys.* **7**(3–4), 351–360.

30. P. Andersson and R. G. Ross. (1983). Effect of guest molecule size on the thermal conductivity and heat capacity of clathrate hydrates. *J. Phys. C.* **16**(8), 1423–1432.
31. D. A. Ackerman, D. Moy, R. C. Potter, A. C. Anderson, and W. N. Lawless. (1981). Glassy behavior of crystalline solids at low temperatures. *Phys. Rev. B.* **23**(8), 3886–3893.
32. J. J. De Yoreo, R. O. Pohl, and G. Burns. (1985). Low-temperature thermal properties of ferroelectrics. *Phys. Rev. B.* **32**(9), 5780–5784.
33. E. Fischer, W. Hässler, E. Hegenbarth, and V. I. Fritsberg. (1981). Amorphous behaviour of PLZT demonstrated by thermal conductivity measurements. *phys. status solidi (a).* **66**(2), K169–K171.
34. E. Fischer, W. Hässler, and E. Hegenbarth. (1982). Glass-like behaviour in the thermal conductivity of $Sr_{1-x}Ba_x(Nb_2O_6)$ single crystal. *phys. status solidi (a).* **72**(2), K169–K171.
35. W. Hässler and E. Hegenbarth. (1985). Glasslike behaviour of thermal conductivity at ferroelectric single crystals op relaxor-type. *Ferroelectr. Lett. Sec.* **4**(4), 117–121.
36. W. N. Lawless. (1982). Glasslike properties of ferroelectrics at low temperatures. *Ferroelectrics* **43**(1), 223–228.
37. E. Fischer. (1986). Dielectric and thermal properties of relaxation ferroelectrics at low temperatures. *phys. status solidi (a).* **97**(1), 121–128.
38. H. E. Zimmer, J. Engert, P. Frach, I. Henning, D. Chat, R. Oszolinsk, and E. Hegenbarth. (1987). Glasslike behaviour of polycrystalline $Pb(Sc_{0.5}Nb_{0.5})O_3$ at low temperatures. *Ferroelectr. Lett. Sec.* **7**(5), 113–120.
39. W. N. Lawless and R. K. Pandey. (1984). Glasslike thermal conductivity of tourmaline at low temperatures. *Solid State Commun.* **52**(10), 833–835.
40. W. N. Lawless. (1984). Glasslike thermal properties of antiferroelectric $PbZrO_3$ at low temperatures. *Phys. Rev. B.* **30**(11), 6555–6559.
41. W. N. Lawless and T. K. Gupta. (1983). Thermal properties of tetragonal ZrO_2 at low temperatures. *Phys. Rev. B.* **28**(10), 5507–5510.
42. F. J. Walker and A. C. Anderson. (1984). Low-energy excitations in yttria-stabilized zirconia. *Phys. Rev. B.* **29**(10), 5881–5890.
43. M. N. Regueiro, D. Castello, M. A. Izbizky, D. Esparza, and C. D'ovidio. (1987). Anomalous thermal conductivity of $La_{2-x}Sr_xCuO_4$: Possibility of a polaronic glass. *Phys. Rev. B.* **36**(16), 8813–8815.

44. T. Kopte and E. Hegenbarth. (1990). Thermal conductivity of high-Tc YBCO ceramics in magnetic fields up to 10 T. *phys. status solidi (a).* **120**(2), 547–555.
45. E. Bonjour, R. Calemczuk, R. Lagnier, and B. Salce. (1981). Low temperature thermal properties of cyclohexanol: A glassy crystal system. *J. Phys. Colloques* **42**(C6), 63–65.
46. D. Moy, J. N. Dobbs, and A. C. Anderson. (1984). Low-temperature properties of the orientational glass $KBr_{1-x}(CN)_x$. *Phys. Rev. B.* **29**(4), 2160–2165.
47. J. J. De Yoreo, W. Knaak, M. Meissner, and R. O. Pohl. (1986). Low-temperature properties of crystalline $(KBr)_{1-x}(KCN)_x$: A model glass. *Phys. Rev. B.* **34**(12), 8828–8842.
48. D. G. Cahill and R. O. Pohl. (1989). Low-energy excitations in the mixed crystal $Ba_{1-x}La_xF_{2+x}$. *Phys. Rev. B.* **39**(14), 10477–10480.
49. L. N. Yadon, C. I. Nicholls, and D. G. Haase. (1989). Thermal conductivity of solid N_2-Ar alloys. *Phys. Rev. B.* **40**(7), 5215–5217.
50. K. A. McCarthy, H. H. Sample, and W. G. D. Dharmaratna. (1984). The thermal conductivity of three fluorozirconate glasses. *J. Non Cryst. Solids* **64**, 445–448.
51. W. M. MacDonald, A. C. Anderson, and J. Schroeder. (1985). Low-temperature behavior of potassium and sodium silicate glasses. *Phys. Rev. B.* **31**(2), 1090–1101.
52. R. Villar, M. L. de la Torre, and S. Vieira. (1986). Low-temperature thermal conductivity of sodium borate glasses. *Phys. Rev. B.* **34**(10), 7394–7395.
53. B. Hanna and R. G. Bohn. (1988). The thermal conductivity of $Li_2O \cdot Al_2O_3 \cdot n\,SiO_2$ glasses between 5 and 100 K. *J. Appl. Phys.* **64**(8), 3911–3914.
54. A. Avogadro, S. Aldrovandi, G. Carini, and A. Siri. (1989). Specific heat and thermal conductivity of ionic conductors and chalcogenide glasses at low temperatures. *Philos. Mag. B.* **59**(1), 33–42.
55. N. O. Birge. (1986). Specific-heat spectroscopy of glycerol and propylene glycol near the glass transition. *Phys. Rev. B.* **34**(3), 1631–1642.
56. D. G. Cahill and R. O. Pohl. (1987). Thermal conductivity of amorphous solids above the plateau. *Phys. Rev. B.* **35**(8), 4067–4073.
57. R. G. Ross and P. Andersson. (1982). Clathrate and other solid phases in the tetrahydrofuran–water system: Thermal conductivity and heat capacity under pressure. *Can. J. Chem.* **60**(7), 881–892.
58. N. Ahmad and W. A. Phillips. (1987). Thermal conductivity of ice and ice clathrate. *Solid State Commun.* **63**(2), 167–171.

59. J. S. Tse and M. A. White. (1988). Origin of glassy crystalline behavior in the thermal properties of clathrate hydrates: A thermal conductivity study of tetrahydrofuran hydrate. *J. Phys. Chem.* **92**(17), 5006–5011.
60. J. J. Freeman, K. J. Dahlhauser, A. C. Anderson, and S. J. Poon. (1987). Low-temperature thermal conductivity of glassy and icosahedral Pd-U-Si alloys. *Phys. Rev. B.* **35**(5), 2451–2452.
61. J. W. Gardner and A. C. Anderson. (1981). Low-temperature specific heat and thermal conductivity of neutron-irradiated crystalline quartz. *Phys. Rev. B.* **23**(2), 474–482.
62. A. K. Raychaudhuri and R. O. Pohl. (1982). Thermal conductivity of neutron-irradiated silica. *Solid State Commun.* **44**(5), 711–714.
63. J. E. Graebner and L. C. Allen. (1983). Tunneling systems in amorphous germanium. *Phys. Rev. Lett.* **51**(17), 1566–1569.
64. M. Mertig, G. Pompe, and E. Hegenbarth. (1984). Specific heat of amorphous silicon at low temperatures. *Solid State Commun.* **49**(4), 369–372.
65. G. Pompe and E. Hegenbarth. (1988). Thermal conductivity of amorphous Si at low temperatures. *phys. status solidi (b).* **147**(1), 103–108.
66. J. E. Graebner and L. C. Allen. (1984). Thermal conductivity of amorphous germanium at low temperatures. *Phys. Rev. B.* **29**(10), 5626–5633.
67. J. E. Graebner, B. Golding, L. C. Allen, J. C. Knights, and D. K. Biegelsen. (1984). Thermal properties of a–Si:H at low temperatures. *Phys. Rev. B.* **29**(6), 3744–3746.
68. R. O. Pohl. (1985). Some facts, old and new, about the low temperature thermal properties of noncrystalline solids. *Phase Transit.* **5**(4), 239–260.
69. V. I. Sviridenko, V. A. Medvedev, N. P. Rybkin, and V. G. Gorbunova. (1987). The thermal conductivity of KV fused silica at 2–300° K. Izmer. Tekhn. No. 5, 34–36 (in Russian) [*Meas. Tech.* **30**(5), 454–458].
70. T. C. Hsieh, W. M. MacDonald, and A. C. Anderson. (1981). Low-temperature thermal conductivity of porous Vycor. *J. Non-Cryst. Solids.* **46**(3), 437–440.
71. W. N. Lawless, A. C. Anderson, and F. Walker. (1981). Glasslike thermal and dielectric properties of $Cd_2Nb_2O_7$ at low temperatures. *Ferroelectrics.* **37**(1), 627–629.
72. R. Calemczuk, L. Jaqmin, G. P. Singh, K. Dransfeld, and R. Vacher. (1984). Experimental evidence for the existence of two-level-tunneling systems in the glassy crystalline state of cyclohexanol. *Phys. Rev. B.* **29**(6), 3767–3769.

73. H. V. Löhneysen, H. Rüsing, and W. Sander. (1985). Structural relaxation effects on the low-temperature properties of vitreous silica. *Z. Phys. B.* **60**(2-4), 323–330.
74. G. X. Mack and A. C. Anderson. (1985). Low-temperature behavior of vitreous silica containing neon solute. *Phys. Rev. B.* **31**(2), 1102–1106.
75. D. G. Cahill and R. O. Pohl. (1988). Lattice vibrations and heat transport in crystals and glasses. *Ann. Rev. Phys. Chem.* **39**(1), 93–121.
76. D. M. Finlayson, P. Mason, J. N. Rogers, and D. Greig. (1980). The thermal conductivity and specific heat of extruded polyethylene below 1K. *J. Phys. C.* **13**(9), L185–L188.
77. A. K. Hasen and K. S. Dubey. (1981). Lattice thermal conductivity of polyvinyl acetate at low temperatures. *Solid State Commun.* **38**(12), 1185–1187.
78. C. I. Nicholls and H. M. Rosenberg. (1981). The effects of cross-link length on the thermal properties of epoxy-resins from 1.5 to 80 K. *Physica B+C.* **108**(1–3), 1015–1016.
79. S. Kelham and H. M. Rosenberg. (1981). The thermal conductivity and specific heat of epoxy-resin from 0.1–80K. *J Phys. C.* **14**(12), 1737–1749.
80. F. J. Walker and A. C. Anderson. (1981). Thermal conductivity and specific heat of a glass–epoxy composite at temperatures below 4 K. *Rev. Sci. Instrum.* **52**(3), 471–472.
81. C. I. Nicholls and H. M. Rosenberg. (1984). The excitation spectrum of epoxy resins; neutron diffraction, specific heat and thermal conductivity at low temperatures. *J. Phys. C.* **17**(7), 1165–1178.
82. W. Scheibner and M. Jäckel. (1985). Thermal conductivity and specific heat of an epoxy resin/epoxy resin composite material at low temperatures. *phys. status solidi (a).* **87**(2), 543–547.
83. J. D. Boyer, J. C. Lasjaunias, R. A. Fisher, and N. E. Phillips. (1983). The low-temperature specific heat of PTFE (Teflon) at pressures to 5.2 kbar. *J. Non-Cryst. Solids.* **55**(3), 413–432.
84. J. X. Mack, J. J. Freeman, A. C. Anderson, and D. Greig. (1987). Low temperature thermal conductivity of polystyrene. *J. Non-Cryst. Solids.* **91**(3), 391–401.
85. D. Greig. (1988). Low temperature thermal conductivity of polymers. *Cryogenics.* **28**(4), 243–247.
86. K. Jahn, M. Jäckel, and W. Meyer. (1983). Thermal conductivity and dielectric properties at low temperatures of an epoxy-resin after electron-irradiation. *Cryogenics.* **23**(3), 160–162.
87. J. M. Grace and A. C. Anderson. (1989). Low-temperature specific heat and thermal conductivity of a glassy polymer under applied pressure. *Phys. Rev. B.* **40**(3), 1901–1916.

88. D. M. Herlach, E. F. Wassermann, and J. Kästner. (1981). Heat conductivity of amorphous PdCuSi alloys containing Fe and Mn impurities. *Phys. Rev. B.* **24**(6), 3204–3210.
89. J. J. Freeman and A. C. Anderson. (1986). Thermal conductivity of amorphous solids. *Phys. Rev. B.* **34**(8), 5684–5690.
90. R. Van den Berg, S. Grondey, J. Kästner, and H. Löhneysen. (1983). Low temperature thermal properties of amorphous $Mg_{0.7}Zn_{0.3}$. *Solid State Commun.* **47**(2), 137–140.
91. G. Pompe, M. Gaafar, P. Büttner, and T. Francke. (1983). Low temperature thermal conductivity of amorphous (Fe, Ni, Co)(P, B, Si) alloys and their change by heat treatment. *phys. status solidi (b).* **119**(2), 579–587.
92. R. Van den Berg, S. Grondey, J. Kästner, and H. Löhneysen. (1983). Low temperature thermal properties of amorphous $Mg_{0.7}Zn_{0.3}$. *Solid State Commun.* **47**(2), 137–140.
93. J. C. Lasjaunias, A. Ravex, O. Laborde, and O. Bethoux. (1984). Thermal and superconducting properties of Zr-based amorphous alloys; effect of structural relaxation. *Physica B+C.* **126**(1–3), 126–133.
94. A. Ravex, J. C. Lasjaunias, and O. Bethoux. (1984). Heat treatment effect on thermal, superconducting and structural properties of amorphous sputtered $Zr_{76}Cu_{24}$ alloy. *J. Phys. F.* **14**(2), 329–346.
95. A. Jeżowski, J. Mucha, and G. Pompe. (1987). Thermal conductivity of the amorphous alloy $Fe_{40}Ni_{40}P_{14}B_6$ between 80 and 300 K. *J. Phys. D.* **20**(11), 1500–1506.
96. C. Sürgers and H. V. Löhneysen. (1988). Low-temperature properties of amorphous $(Mo_{1-x}Ru_x)_{0.8}P_{0.2}$ alloys. *Z. Phys. B.* **70**(3), 361–369.
97. P. Svoboda and P. Vašek. (1988). Low temperature thermal conductivity of metallic glasses $Fe_xNi_{80-x}B_{20}$. *phys. status solidi (a).* **106**(1), 115–121.
98. G. Bellessa. (1980). Low temperature properties, low-energy excitations in amorphous metals. *J. Phys. Colloques* **41**(C8), 723–730.
99. W. A. Phillips. (1981). *Properties, A.S.L.T. Topics in Current Physics,* Vol. 24. New York: Springer.
100. B. Golding, J. E. Graebner, and L. C. Allen. (1986). The thermal conductivity plateau in disordered systems. *Phonon Scattering in Condensed Matter V.* Berlin, Heidelberg: Springer.
101. A. Bhattacharyya. (1981). A tunnelling-state model for the specific heat capacity and thermal conductivity of amorphous materials at low temperatures. *Contemp. Phys.* **22**(1), 117–127.
102. H. V. Löhneysen. (1981). Low energy excitations in amorphous metals. *Phys. Rep.* **79**(3), 161–212.

103. R. G. Ross, P. Andersson, B. Sundqvist, and G. Backstrom. (1984). Thermal conductivity of solids and liquids under pressure. *Rep. Progr. Phys.* **47**(10), 1347–1402.
104. A. C. Anderson. (1985). Low temperature glassy properties of disordered crystals. *Phase Transit.* **5**(4), 301–316.
105. S. Hunklinger and A. K. Raychaudhuri. (1986). Thermal and elastic anomalies in glasses at low temperatures. *Progr. Low Temp. Phys.* **9**, 265–344.
106. W. N. Lawless. (1985). Recent topics in ferroelectrics at low temperatures: Quantum, vibronic, and glasslike properties. *Jpn. J. Appl. Phys.* **24**(S2), 94–98.
107. W. Hässler and E. Hegenbarth. (1985). Glasslike behaviour of thermal conductivity at ferroelectric single crystals op relaxor-type. *Ferroelectr. Lett. Sec.* **4**(4), 117–121.
108. W. A. Phillips. (1987). Two-level states in glasses. *Rep. Progr. Phys.* **50**(12), 1657–1708.
109. Y. M. Galperin, V. G. Karpov, and V. I. Kozub. (1989). Localized states in glasses. *Adv. Phys.* **38**(6), 669–737.
110. R. G. Ross, P. Andersson, and G. Bäckström. (1981). Unusual PT dependence of thermal conductivity for a clathrate hydrate. *Nature.* **290**(5804), 322–323.
111. G. S. Nolas, G. A. Slack, D. T. Morelli, T. M. Tritt, and A. C. Ehrlich. (1996). The effect of rare-earth filling on the lattice thermal conductivity of skutterudites. *J. Appl. Phys.* **79**(8), 4002–4008.
112. S. P. Andersson and O. Andersson. (1997). Thermal conductivity, heat capacity, and compressibility of atactic poly (propylene) under high pressure. *Int. J. Thermophys.* **18**(3), 845–864.
113. S. P. Andersson, O. Andersson, and G. Bäckström. (1997). Thermal conductivity of amorphous Teflon (AF 1600) at high pressure. *Int. J. Thermophys.* **18**(1), 209–219.
114. S. P. Andersson and O. Andersson. (1998). Volume dependence of thermal conductivity and isothermal bulk modulus up to 1 GPa for poly (vinyl acetate). *J. Polym. Sci. Part B.* **36**(9), 1451–1463.
115. S. P. Andersson and O. Andersson. (1998). Volume dependence of thermal conductivity and bulk modulus for poly (propylene glycol). *J. Polym. Sci. Part B.* **36**(2), 345–355.
116. T. Kikuchi, T. Takahashi, and K. Koyama. (2003). Temperature and pressure dependence of thermal conductivity measurement of polystyrene and polycarbonate. *J. Macromol. Sci. Part B.* **42**(5), 1097–1110.
117. O. Andersson and A. Inaba. (2005). Thermal conductivity of crystalline and amorphous ices and its implications on amorphization and glassy water. *Phys. Chem. Chem. Phys.* **7**(7), 1441–1449.

118. M. A. Chernikov, A. Bianchi, and H. R. Ott. (1995). Low-temperature thermal conductivity of icosahedral $Al_{70}Mn_9Pd_{21}$. *Phys. Rev. B.* **51**(1), 153–158.
119. K. Edagawa, M. A. Chernikov, A. D. Bianchi, E. Felder, U. Gubler, and H. R. Ott. (1996). Low-temperature thermodynamic and thermal-transport properties of decagonal $Al_{65}Cu_{20}Co_{15}$. *Phys. Rev. Lett.* **77**(6), 1071–1074.
120. A. D. Bianchi, F. Bommeli, E. Felder, M. Kenzelmann, M. A. Chernikov, L. Degiorgi, H. R. Ott, and K. Edagawa. (1998). Low-temperature thermal and optical properties of single-grained decagonal Al-Ni-Co quasicrystals. *Phys. Rev. B.* **58**(6), 3046–3056.
121. D. G. Cahill, M. Katiyar, and J. R. Abelson. (1994). Thermal conductivity of a-Si:H thin films. *Phys. Rev. B.* **50**(9), 6077–6081.
122. D. G. Cahill, M. Katiyar, and J. R. Abelson. (1995). Heat transport in micron thick a-Si:H films. *Philos. Mag. B.* **71**(1), 677–682.
123. G. Gesele, J. Linsmeier, V. Drach, J. Fricke, and R. Arens-Fischer. (1997). Temperature-dependent thermal conductivity of porous silicon. *J. Phys. D.* **30**(21), 2911–2916.
124. D. T. Morelli, T. A. Perry, J. W. Vandersande, and C. Uher. (1993). Glasslike thermal transport in heavily irradiated diamond. *Phys. Rev. B.* **48**(5), 3037–3041.
125. D. G. Cahill. (1990). Thermal conductivity measurement from 30 to 750 K: The 3ω method. *Rev. Sci. Instrum.* **61**(2), 802–808.
126. S. M. Lee and D. G. Cahill. (1997). Heat transport in thin dielectric films. *J. Appl. Phys.* **81**(6), 2590–2595.
127. D. G. Cahill, S. M. Lee, and T. I. Selinder. (1998). Thermal conductivity of κ-Al_2O_3 and α-Al_2O_3 wear-resistant coatings. *J. Appl. Phys.*, **83**(11), 5783–5786.
128. D. Posselt, J. K. Kjems, A. Bernasconi, T. Sleator, and H. R. Ott. (1991). The thermal conductivity of silica aerogel in the phonon, the fracton and the particle-mode regime. *Europhys. Lett.* **16**(1), 59–65.
129. P. Scheuerpflug, H. J. Morper, and G. Neubert. (1991). Low-temperature thermal transport in silica aerogels. *J. Phys. D.* **24**(8), 1395–1403.
130. D. M. Zhu and K. I. Cheong. (1996). Thermal transport in densified amorphous solids. *Phys. B: Conden. Matter.* **219**, 317–319.
131. M. Jäckel, K. Wagner, and E. Hegenbarth. (1996). Low-temperature thermal conductivity of vitreous silica under high pressure. *Phys. B: Conden. Matter.* **219**, 308–310.
132. D. G. Cahill, S. K. Watson, and R. O. Pohl. (1992). Lower limit to the thermal conductivity of disordered crystals. *Phys. Rev. B.* **46**(10), 6131–6140.

133. D. M. Zhu and T. Kosugi. (1996). Thermal conductivity of $GeO_2 - SiO_2$ and $TiO_2 - SiO_2$ mixed glasses. *J. Non-Cryst. Solids.* **202**(1–2), 88–92.
134. D. M. Zhu and A. C. Anderson. (1990). Low-temperature thermal conductivity of a glassy epoxy-epoxy composite. *J. Low Temp. Phys.* **80**(3–4), 153–160.
135. A. Licea-Claverie, M. Müller, M. Jäckel, and K. F. Arndt. (1990). Low-temperature thermal conductivity of diolmodified epoxies. *Polym. Bull.* **24**(4), 421–427.
136. M. Jäckel, M. Müller, A. L. Claverie, and K. F. Arndt. (1991). Thermal conductivity of modified epoxy resins with different cross-link densities at low temperatures. *Cryogenics.* **31**(4), 228–230.
137. M. Jäckel. (1995). Thermal properties of polymer/particle composites at low temperatures. *Cryogenics.* **35**(11), 713–716.
138. A. Nittke, M. Scherl, P. Esquinazi, W. Lorenz, J. Li, and F. Pobell. (1995). Low temperature heat release, sound velocity and attenuation, specific heat and thermal conductivity in polymers. *J. Low Temp. Phys.* **98**(5–6), 517–547.
139. J. E. De Oliveira, A. Azevedo, and F. L. A. Machado. (1995). Thermal and structural properties of the MY750 Epoxy diluted with xylene. *Solid State Commun.* **95**(11), 781–785.
140. K. I. Cheong, D. M. Zhu, and E. W. Hellmuth. (1996). Thermal transport in a glassy polymer after being compressed at high pressure. *J. Non-Cryst. Solids.* **204**(2), 158–163.
141. R. Geilenkeuser, T. Porschberg, M. Jackel, and A. Gladun. (1999). Influence of high pressure on thermal properties of amorphous polystyrene. *Phys. B: Conden. Matter.* **263**, 276–279.
142. S. P. Andersson and R. G. Ross. (1994). Thermal conductivity and heat capacity per unit volume of poly (methyl methacrylate) under high pressure. *Int. J. Thermophys.* **15**(5), 949–962.
143. E. Duval, T. Achibat, A. Boukenter, B. Varrel, R. Calemczuk, and B. Salce. (1995). Low-energy excitations in polyvinyl chloride: Raman scattering and thermal properties. *J. Non-Cryst. Solids.* **190**(3), 258–263.
144. M. Jäckel, F. V. Schoenebeck, U. Escher, and A. Gladun. (1996). Low-temperature thermal properties of amorphous polycarbonate. *Czechoslovak J. Phys.* **46**(1), 2249–2250.
145. M. Jäckel, F. Weise, J. Opitz, and R. Geilenkeuser. (1998). Influence of hydrostatic pressure on the thermal properties of polymers at low temperatures. *Cryogenics.* **38**(1), 105–108.

146. G. Ventura, E. Gottardi, I. Peroni, A. Peruzzi, and G. Ponti. (1999). Low temperature thermal conductivity of polyamide-imide. *Nucl. Phys. B Proc. Suppl.* **78**, 571–572.
147. M. S. Love and A. C. Anderson. (1991). Influence of structural relaxation on the low-temperature thermal conductivity of ancient natural glasses. *J. Low Temp. Phys.* **84**(1–2), 19–36.
148. S. K. Watson and R. O. Pohl. (1995). Low-temperature glasslike properties in $(NaCl)_{1-x}(NaCN)_x$. *Phys. Rev. B.* **51**(13), 8086–8101.
149. S. Vieira, M. A. Ramos, Q. W. Zou, and C. Talon. (1997). Low-temperature thermal properties of molecular glasses and crystals. *Phase Transit.* **64**(1–2), 87–102.
150. J. F. Berret, M. Meissner, S. K. Watson, R. O. Pohl, and E. Courtens. (1991). Glasslike thermal properties and isotope effect in $Rb_{1-x}(NH_4)_xH_2PO_4$ mixed crystals. *Phys. Rev. Lett.* **67**(1), 93–96.
151. N. Ahmad, C. J. Adkins, and W. A. Phillips. (1993). Heat capacity and thermal conductivity of $LiCl.7H_2O$. *J. Phys.: Condens. Matter* **5**(24), 4007–4012.
152. G. S. Nolas, J. L. Cohn, and G. A. Slack. (1998). Effect of partial void filling on the lattice thermal conductivity of skutterudites. *Phys. Rev. B.* **58**(1), 164–170.
153. J. L. Cohn, G. S. Nolas, V. Fessatidis, T. H. Metcalf, and G. A. Slack. (1999). Glasslike heat conduction in high-mobility crystalline semiconductors. *Phys. Rev. Lett.* **82**(4), 779–782.
154. O. Andersson and H. Suga. (1996). Thermal conductivity of normal and deuterated tetrahydrofuran clathrate hydrates. *J. Phys. Chem. Solids.* **57**(1), 125–132.
155. M. Zakrzewski and M. A. White. (1992). Thermal conductivities of a clathrate with and without guest molecules. *Phys. Rev. B.* **45**(6), 2809–2817.
156. D. Michalski and M. A. White. (1995). Thermal conductivity of a clathrate with restrained guests: The CCl_4 clathrate of Dianin's compound. *J. Phys. Chem.* **99**(11), 3774–3780.
157. D. Michalski and M. A. White. (1997). Thermal conductivity of an organic clathrate: Possible generality of glasslike thermal conductivity in crystalline molecular solids. *J. Chem. Phys.* **106**(14), 6202–6203.
158. C. L. Choy, K. W. Tong, H. K. Wong, and W. P. Leung. (1991). Thermal conductivity of amorphous alloys above room temperature. *J. Appl. Phys.* **70**(9), 4919–4925.
159. R. Schmidt, T. Franke, and P. Häussler. (1999). Thermal conductivity of Cu_xSn_{100-x} films at low temperatures. *Phys. B.* **263**, 296–298.

160. K. Wagner and E. Hegenbarth. (1993). Thermal conductivity of Strontium-Barium-Niobat under high pressure and at low temperatures. *Ferroelectr. Lett. Sec.* **16**(3–4), 95–102.
161. B. Salce, J. L. Gravil, and L. A. Boatner. (1994). Disorder and thermal transport in undoped $KTaO_3$. *J. Phys. Conden. Matter* **6**(22), 4077–4092.
162. E. Hegenbarth. (1995). Glasslike behaviour of ferroelectrics at low temperatures. *Ferroelectrics.* **168**(1), 25–37.
163. M. Fahland, G. Mattausch, and E. Hegenbarth. (1995). Thermal conductivity measurements on $(Pb_{1-x}Ba_x)(Sc_{0.5}Nb_{0.5})O_3$. *Ferroelectrics.* **168**(1), 9–16.
164. J. Suchanicz, A. Jeżowski, and R. Poprawski. (1998). Low-temperature thermal and dielectric properties of $Na_{0.5}Bi_{0.5}TiO_3$. *Phys. Status Solidi (a).* **169**(2), 209–215.
165. H. Fujishiro and M. Ikebe. (1999). Two-level-like anomalous phonon scattering in $La_{2-x}Sr_xCuO_4$ and $La_{1-x}Sr_xMnO_3$. *Phys. B.* **263**, 691–694.
166. U. T. Höchli, K. Knorr, and A. Loidl. (1990). Orientational glasses. *Adv. Phys.* **39**(5), 405–615.
167. H. Löhneysen. (1991). Thermal properties of amorphous metals. *Mater. Sci. Eng. A.* **133**, 51–58.
168. R. G. Ross. (1991). Thermal conductivity and disorder in nonmetallic materials. *Phys. Chem. Liq.* **23**(4), 189–210.
169. Y. K. Godovsky. (1992). *Thermophysical Properties of Polymers.* Heidelberg: Springer-Verlag.
170. I. A. Smirnov and V. S. Oskotski. (1993). Thermal conductivity of rare earth compounds. *Handb. Phys. Chem. Rare Earths.* **16,** 107–224.
171. M. A. Ramos, L. Gil, A. Bringer, and U. Buchenau. (1993). The density of tunneling and vibrational states of glasses within the soft-potential model. *Phys. Status Solidi (a).* **135**(2), 477–492.
172. B. C. Sales, D. Mandrus, and R. K. Williams. (1996). Filled skutterudite antimonides: A new class of thermoelectric materials. *Science.* **272**(5266), 1325–1328.
173. G. Mattausch, T. Felsner, E. Hegenbarth, B. Kluge, and S. Sahling. (1996). Glassy properties of the relaxor ferroelectric strontium barium niobate at low temperatures. *Phase Transitions.* **59**(4), 189–223.
174. F. J. Bermejo, G. J. Cuello, E. Courtens, R. Vacher, and M. A. Ramos. (1998). Comment on "high frequency dynamics of glass forming liquids at the glass transition". *Phys. Rev. Lett.* **81**(17), 3801.
175. G. S. Nolas, D. T. Morelli, and T. M. Tritt. (1999). Skutterudites: A phonon-glass-electron crystal approach to advanced thermoelectric energy conversion applications. *Ann. Rev. Mater. Sci.* **29**(1), 89–116.

176. M. A. White. (1999). Thermal conductivities of molecular materials. *J. Therm. Anal. Calorim.* **57**(3), 765–771.
177. A. D. Bianchi. (1999). *Thermal and Transport Properties of Quasicrystals* (Doctoral dissertation, ETH Zurich).
178. D. G. Cahill. (1997). Heat transport in dielectric thin films and at solid-solid interfaces. *Microscale Thermophys. Eng.* **1**(2), 85–109.
179. D. A. Parshin. (1993). Soft potential model and universal properties of glasses. *Phys. Scripta.* **T49A**, 180–185.
180. M. A. Ramos and U. Buchenau. (1997). Low-temperature thermal conductivity of glasses within the soft-potential model. *Phys. Rev. B.* **55**(9), 5749–5754.
181. D. G. Cahill, R. B. Stephens, R. H. Tait, S. K. Watson, and R. O. Pohl. (1990). Thermal conductivity and lattice vibrations in glasses. In C. J. Cremers and H. A. Fine (eds.), *Thermal Conductivity*, Vol. 21, pp. 3–16, New York: Plenum Press.
182. M. I. Klinger and A. M. Kosevich. (2001). Soft-mode-dynamics model of acoustic-like high-frequency excitations in boson-peak spectra of glasses. *Phys. Lett. A.* **280**(5–6), 365–370.
183. E. Duval, A. Mermet, and L. Saviot. (2007). Boson peak and hybridization of acoustic modes with vibrations of nanometric heterogeneities in glasses. *Phys. Rev. B.* **75**(2), 024201–9.
184. M. Christensen, A. B. Abrahamsen, N. B. Christensen, F. Juranyi, N. H. Andersen, K. Lefmann, J. Andreasson, C. R. Bahl, and B. B. Iversen. (2008). Avoided crossing of rattler modes in thermoelectric materials. *Nature Mater.* **7**(10), 811–815.
185. B. C. Sales, B. C. Chakoumakos, R. Jin, J. R. Thompson, and D. Mandrus. (2001). Structural, magnetic, thermal, and transport properties of $X_8Ga_{16}Ge_{30}$ (X= Eu, Sr, Ba) single crystals. *Phys. Rev. B.* **63**(24), 245113–8.
186. S. Paschen, W. Carrillo-Cabrera, A. Bentien, V. H. Tran, M. Baenitz, Y. Grin, and F. Steglich. (2001). Structural, transport, magnetic, and thermal properties of $Eu_8Ga_{16}Ge_{30}$. *Phys. Rev. B.* **64**(21), 214404–11.
187. G. S. Nolas, J. L. Cohn, J. S. Dyck, C. Uher, and J. Yang. (2002). Transport properties of polycrystalline type-I Sn clathrates. *Phys. Rev. B.* **65**(16), 165201–5.
188. S. Paschen, V. Pacheco, A. Bentien, A. Sanchez, W. Carrillo-Cabrera, M. Baenitz, B. B. Iversen, Y. Grin, and F. Steglich. (2003). Are type-I clathrates Zintl phases and 'phonon glasses and electron single crystals? *Phys. B: Condens. Matter.* **328**(1–2), 39–43.
189. A. Bentien, M. Christensen, J. D. Bryan, A. Sanchez, S. Paschen, F. Steglich, G. D. Stucky, and B. B. Iversen. (2004). Thermal conductivity of thermoelectric clathrates. *Phys. Rev. B.* **69**(4), 045107–5.

190. L. Qiu, I. P. Swainson, G. S. Nolas, and M. A. White. (2004). Structure, thermal, and transport properties of the clathrates $Sr_8Zn_8Ge_{38}$, $Sr_8Ga_{16}Ge_{30}$, and $Ba_8Ga_{16}Si_{30}$. *Phys. Rev. B.* **70**(3), 035208–8.
191. O. Andersson, V. Murashov, and M. A. White. (2002). Thermal conductivity and heat capacity of dianin's clathrates under pressure. *J. Phys. Chem. B.* **106**(1), 192–196.
192. G. S. Nolas, M. Beekman, J. Gryko, G. A. Lamberton Jr, T. M. Tritt, and P. F. McMillan. (2003) Thermal conductivity of elemental crystalline silicon clathrate Si 136. *Appl. Phys. Lett.* **82**(6), 910–912.
193. A. I. Krivchikov, V. G. Manzhelii, O. A. Korolyuk, B. Y. Gorodilov, and O. O. Romantsova. (2005). Thermal conductivity of tetrahydrofuran hydrate. *Phys. Chem. Chem. Phys.* **7**(5), 728–730.
194. A. I. Krivchikov, B. Y. Gorodilov, O. A. Korolyuk, V. G. Manzhelii, H. Conrad, and W. Press. (2005). Thermal conductivity of methane-hydrate. *J. Low Temp. Phys.* **139**(5-6), 693–702.
195. I. Fujita, K. Kishimoto, M. Sato, H. Anno, and T. Koyanagi. (2006). Thermoelectric properties of sintered clathrate compounds $Sr_8Ga_xGe_{46-x}$ with various carrier concentrations. *J. Appl. Phys.* **99**(9), 093707–8.
196. A. I. Krivchikov, B. Y. Gorodilov, O. A. Korolyuk, V. G. Manzhelii, O. O. Romantsova, H. Conrad, W. Press, J. S. Tse, and D. D. Klug. (2006). Thermal conductivity of Xe clathrate hydrate at low temperatures. *Phys. Rev. B.* **73**(6), 064203–6.
197. M. Beekman, W. Schnelle, H. Borrmann, M. Baitinger, Y. Grin, and G. S. Nolas. (2010). Intrinsic electrical and thermal properties from single crystals of Na 24 Si 136. *Phys. Rev. Lett.* **104**(1), 018301–4.
198. E. J. Rosenbaum, N. J. English, J. K. Johnson, D. W. Shaw, and R. P. Warzinski. (2007). Thermal conductivity of methane hydrate from experiment and molecular simulation. *J. Phys. Chem. B.* **111**(46), 13194–13205.
199. K. Suekuni, M. A. Avila, K. Umeo, and T. Takabatake. (2007). Cage-size control of guest vibration and thermal conductivity in $Sr_8Ga_{16}Si_{30-x}Ge_x$. *Phys. Rev. B.* **75**(19), 195210–6.
200. J. V. Zaikina, W. Schnelle, K. A. Kovnir, A. V. Olenev, Y. Grin, and A. V. Shevelkov. (2007). Crystal structure, thermoelectric and magnetic properties of the type-I clathrate solid solutions $Sn_{24}P_{19.3(2)}Br_xI_{8-x}(0 \leq x \leq 8)$ and $Sn_{24}P_{19.3(2)}Cl_yI_{8-y}(y \leq 0.8)$. *Solid State Sci.* **9**(8), 664–671.
201. M. A. Avila, K. Suekuni, K. Umeo, H. Fukuoka, S. Yamanaka, and T. Takabatake. (2008). $Ba_8Ga_{16}Sn_{30}$ with type-I clathrate structure: Drastic suppression of heat conduction. *Appl. Phys. Lett.* **92**(4), 041901–3.

202. K. Suekuni, M. A. Avila, K. Umeo, H. Fukuoka, S. Yamanaka, T. Nakagawa, and T. Takabatake. (2008). Simultaneous structure and carrier tuning of dimorphic clathrate $Ba_8Ga_{16}Sn_{30}$. *Phys. Rev. B*. **77**(23), 235119–8.
203. E. Thompson, P. D. Vu, and R. O. Pohl. (2000). Glasslike lattice vibrations in the quasicrystal $Al_{72.1}Pd_{20.7}Mn_{7.2}$. *Phys. Rev. B*. **62**(17), 11437–11443.
204. K. Giannó, A. V. Sologubenko, L. Liechtenstein, M. A. Chernikov, and H. R. Ott. (2001). Low-temperature thermoelectric power and thermal conductivity of two icosahedral Al-Pd-Re quasicrystals. *Ferroelectrics*. **250**(1), 249–252.
205. Y. K. Kuo, J. R. Lai, C. H. Huang, W. C. Ku, C. S. Lue, and S. T. Lin. (2004). Thermoelectric properties of binary Cd-Yb quasicrystals and Cd_6Yb. *J. Appl. Phys.* **95**(4), 1900–1905.
206. J. Dolinšek, S. Vrtnik, M. Klanjšek, Z. Jagličić, A. Smontara, I. Smiljanić, A. Bilušić, Y. Yokoyama, A. Inoue, and C. V. Landauro. (2007). Intrinsic electrical, magnetic, and thermal properties of single-crystalline $Al_{64}Cu_{23}Fe_{13}$ icosahedral quasicrystal: Experiment and modeling. *Phys. Rev. B*. **76**(5), 054201–9.
207. I. Smiljanić, A. Smontara, A. Bilušić, N. Barišić, D. Stanić, J. Lukatela, J. Dolinšek, M. Feuerbacher, and B. Grushko. (2008). Thermal and electrical conductivities in Al-based complex metallic alloys. *Philos. Mag.* **88**(13–15), 2155–2162.
208. Y. K. Kuo, N. Kaurav, W. K. Syu, K. M. Sivakumar, U. T. Shan, S. T. Lin, Q. Wang, and C. Dong. (2008). Transport properties of Ti-Zr-Ni quasicrystalline and glassy alloys. *J. Appl. Phys.* **104**(6), 063705–3.
209. A. I. Krivchikov, A. N. Yushchenko, V. G. Manzhelii, O. A. Korolyuk, F. J. Bermejo, R. Fernández-Perea, C. Cabrillo, and M. A. González. (2006). Scattering of acoustic phonons in disordered matter: A quantitative evaluation of the effects of positional versus orientational disorder. *Phys. Rev. B*. **74**(6), 060201–4.
210. C. Chiritescu, D. G. Cahill, N. Nguyen, D. Johnson, A. Bodapati, P. Keblinski, and P. Zschack. (2007). Ultralow thermal conductivity in disordered, layered WSe2 crystals. *Science*. **315**(5810), 351–353.
211. W. Wang, Z. Wang, J. Tang, S. Yang, H. Jin, G. L. Zhao, and Q. Li. (2009). Seebeck coefficient and thermal conductivity in doped C_{60}. *J. Renewable Sustainable Energy*. **1**(2), 023104–8.
212. G. S. Nolas, T. J. R. Weakley, J. L. Cohn, and R. Sharma. (2000). Structural properties and thermal conductivity of crystalline Ge clathrates. *Phys. Rev. B*. **61**(6), 3845–3850.
213. K. Takahata, Y. Iguchi, D. Tanaka, T. Itoh, and I. Terasaki. (2000). Low thermal conductivity of the layered oxide (Na, Ca)Co_2O_4:

Another example of a phonon glass and an electron crystal. *Phys. Rev. B.* **61**(19), 12551–12555.

214. C. Uher. Skutterudites: Prospective novel thermoelectrics. (2001). *Semicond. Semimetals.* **69**, 139–253.
215. M. Tachibana, K. Sasame, H. Kawaji, T. Atake, and E. Takayama-Muromachi. (2009). Thermal signatures of nanoscale inhomogeneities and ferroelectric order in $[PbZn_{1/3}Nb_{2/3}O_3]_{1-x}[PbTiO_3]_x$. *Phys. Rev. B.* **80**(9), 094115–7.
216. M. Tachibana and E. Takayama-Muromachi. (2009). Thermal conductivity and heat capacity of the relaxor ferroelectric $[PbMg_{1/3}Nb_{2/3}O_3]_{1-x}[PbTiO_3]_x$. *Phys. Rev. B.* **79**(10), 100104–4.
217. C. Talon, Q. W. Zou, M. A. Ramos, R. Villar, and S. Vieira. (2001). Low-temperature specific heat and thermal conductivity of glycerol. *Phys. Rev. B.* **65**(1), 012203–4.
218. A. I. Krivchikov, A. N. Yushchenko, O. A. Korolyuk, F. J. Bermejo, R. Fernández-Perea, I. Bustinduy, and M. A. Gonzalez. (2008). Effects of resonant phonon scattering from internal molecular modes on the thermal conductivity of molecular glasses. *Phys. Rev. B.* **77**(2), 024202–7.
219. A. I. Krivchikov, I. V. Sharapova, O. A. Korolyuk, O. O. Romantsova, and F. J. Bermejo. (2009). Experimental evidence of the role of quasilocalized phonons in the thermal conductivity of simple alcohols in orientationally ordered crystalline phases. *Fiz. Nizk. Temp.* **35**(11), 1143–1150 [*Low Temp. Phys.* **35**(11), 891–897].
220. J. H. Pöhls, M. B. Johnson, and M. A. White. (2016). Origins of ultralow thermal conductivity in bulk [6, 6]-phenyl-C 61-butyric acid methyl ester (PCBM). *Phys. Chem. Chem. Phys.* **18**(2), 1185–1190.
221. O. Andersson and H. Suga. (2002). Thermal conductivity of amorphous ices. *Phys. Rev. B.* **65**(14), 140201–4(R).
222. B. L. Zink and F. Hellman. (2004). Specific heat and thermal conductivity of low-stress amorphous Si–N membranes. *Solid State Commun.* **129**(3), 199–204.
223. B. Revaz, B. L. Zink, and F. Hellman. (2005). Si-N membrane-based microcalorimetry: Heat capacity and thermal conductivity of thin films. *Thermochim. Acta.* **432**(2), 158–168.
224. B. L. Zink, R. Pietri, and F. Hellman. (2006). Thermal conductivity and specific heat of thin-film amorphous silicon. *Phys. Rev. Lett.* **96**(5), 055902.
225. X. Liu, J. L. Feldman, D. G. Cahill, R. S. Crandall, N. Bernstein, D. M. Photiadis, M. J. Mehl, and D. A. Papaconstantopoulos. (2009). High thermal conductivity of a hydrogenated amorphous silicon film. *Phys. Rev. Lett.* **102**(3), 035901–4.

226. D. Rosenberg, D. Natelson, and D. D. Osheroff. (2000). Thermal conductivity in glasses below 1K: New technique and results. *J. Low Temp. Phys.* **120**(3–4), 259–268.
227. S. K. Watson and R. O. Pohl. (2003). Low-energy lattice vibrations of porous silica glass. *Phys. Rev. B.* **68**(10), 104203–11.
228. H. Y. Hao, M. Neumann, C. Enss, and A. Fleischmann. (2004). Contactless technique for thermal conductivity measurement at very low temperature. *Rev. Sci. Instrum.* **75**(8), 2718–2725.
229. A. Netsch, A. Fleischmann, and C. Enss. (2007). Thermal conductivity in glasses with a phononic crystal like structure. *J. Phys. Conf. Ser.* **92**(1), 012130.
230. B. Terziyska, H. Misiorek, E. Vateva, A. Jeżowski, and D. Arsova. (2005). Low-temperature thermal conductivity of $Ge_xAs_{40-x}S_{60}$ glasses. *Solid State Commun.* **134**(5), 349–353.
231. B. Terziyska, E. Vateva, and D. Arsova. (2007). Low-temperature contributions to the specific heat of $Ge_xAs_{40-x}S_{60}$ glasses. In *AIP Conf. Proc.* **899**(1), 589–589.
232. L. Risegari, M. Barucci, E. Olivieri, and G. Ventura. (2006). Low temperature thermal conductivity of PVC. *J. Low Temp. Phys.* **144**(1–3), 49–59.
233. A. L. Woodcraft and A. Gray. (2009). A low temperature thermal conductivity database. In *AIP Conf. Proc.* **1185**(1), 681–684.
234. M. Barucci, E. Olivieri, E. Pasca, L. Risegari, and G. Ventura. (2005). Thermal conductivity of Torlon between 4.2 and 300 K. *Cryogenics.* **45**(4), 295–299.
235. M. Barucci, E. Gottardi, E. Olivieri, E. Pasca, L. Risegari, and G. Ventura. (2002). Low-temperature thermal properties of polypropylene. *Cryogenics.* **42**(9), 551–555.
236. C. Hu, M. Kiene, and P. S. Ho. (2001). Thermal conductivity and interfacial thermal resistance of polymeric low *k* films. *Appl. Phys. Lett.* **79**(25), 4121–4123.
237. Y. Tian, Z. Q. Li, and E. Y. Jiang. (2009). Low temperature specific heat and thermal conductivity of bulk metallic glass $(Cu_{50}Zr_{50})_{94}Al_6$. *Solid State Commun.* **149**(37–38), 1527–1530.
238. Z. Zhou, C. Uher, D. Xu, W. L. Johnson, W. Gannon, and M. C. Aronson. (2006). On the existence of Einstein oscillators and thermal conductivity in bulk metallic glass. *Appl. Phys. Lett.* **89**(3), 031924–3.
239. Y. K. Kuo, K. M. Sivakumar, C. A. Su, C. N. Ku, S. T. Lin, A. B. Kaiser, J. B. Qiang, Q. Wang, and C. Dong. (2006). Measurement of low-temperature transport properties of Cu-based Cu-Zr-Ti bulk metallic glass. *Phys. Rev. B.* **74**(1), 014208–7.

240. M. B. Tang and J. T. Zhao. (2009). Low temperature transport and thermal properties of half-Heusler alloy $Zr_{0.25}Hf_{0.25}Ti_{0.5}NiSn$. *J. Alloys Compd.* **475**(1–2), 5–8.
241. M. B. Jakubinek, B. Z. Zhan, and M. A. White. (2007). Temperature-dependent thermal conductivity of powdered zeolite NaX. *Microporous Mesoporous Mater.* **103**(1–3), 108–112.
242. C. Wan, W. Zhang, Y. Wang, Z. Qu, A. Du, R. Wu, and W. Pan. (2010). Glass-like thermal conductivity in ytterbium-doped lanthanum zirconate pyrochlore. *Acta Mater.* **58**(18), 6166–6172.
243. A. I. Krivchikov, F. J. Bermejo, I. V. Sharapova, O. A. Korolyuk, and O. O. Romantsova. (2011). Deuteration effects in the thermal conductivity of molecular glasses. *Fiz. Nizk. Temp.* **37**(6), 651–658 [*Low Temp. Phys.* **37**(6), 517–523].
244. M. Tachibana, T. Kolodiazhnyi, and E. Takayama-Muromachi. (2008). Thermal conductivity of perovskite ferroelectrics. *Appl. Phys. Lett.* **93**(9), 092902–3.
245. M. A. Avila, K. Suekuni, K. Umeo, H. Fukuoka, S. Yamanaka, and T. Takabatake. (2006). Glasslike versus crystalline thermal conductivity in carrier-tuned $Ba_8Ga_{16}X_{30}$ clathrates (X= Ge, Sn). *Phys. Rev. B.* **74**(12), 125109–8.
246. K. Suekuni, H. I. Tanaka, F. S. Kim, K. Umeo, and T. Takabatake. (2015). Glasslike versus crystalline thermophysical properties of the Cu–S based minerals: Tetrahedrite and colusite. *J. Phys. Soc. Jpn.* **84**(10), 103601–4.
247. E. I. Salamatov, A. V. Taranov, E. N. Khazanov, E. V. Charnaya, and E. V. Shevchenko.(2017). Transport characteristics of phonons and the specific heat of Y_2O_3: ZrO_2 solid solution single crystals. *Zh. Eksp. Teor. Fiz.* **152**(5), 910–916 (in Russian) [*J. Exp. Theor. Phys.* **125**(5), 768–774].
248. S. Hunklinger and C. Enss. (2000). Tunneling systems in crystalline and amorphous solids. *Insulating and Semiconducting Glasses* [Series on Directions in Condensed Matter Phys: Vol. 17], pp. 499–551. Singapore: World Scientific.
249. G. S. Nolas, G. A. Slack, and S. B. Schujman. (2001). Semiconductor clathrates: A phonon glass electron crystal material with potential for thermoelectric applications. Recent Trends in Thermoelectric Materials Research [Series: Semiconductors and Semimetals, Vol. 69], pp. 255–300. San Diego: Academic Press Inc.
250. R. O. Pohl, X. Liu, and E. Thompson. (2002). Low-temperature thermal conductivity and acoustic attenuation in amorphous solids. *Rev. Mod. Phys.* **74**(4), 991–1013.

251. V. Murashov and M. A. White. (2004). Thermal conductivity of insulators and glasses. In T. M. Tritt (ed.), Thermal conductivity: Theory, properties, and applications, pp. 93–104. New York: Kluwer Acad./Plenum Publ.
252. T. M. Tritt (ed.), (2004). Thermal conductivity: Theory, properties, and applications. New York: Kluwer Acad./Plenum Publ.
253. R. O. Pohl. (2006). Lattice vibrations of glasses. *J. Non-Cryst. Solids.* **352**(32–35), 3363–3367.
254. G. P. Johari and O. Andersson. (2007). Vibrational and relaxational properties of crystalline and amorphous ices. *Thermochim. Acta.* **461**(1–2), 14–43.
255. O. Andersson. (2018). Thermal conductivity of normal and deuterated water, crystalline ice, and amorphous ices. *J. Chem. Phys.* **149**(12), 124506–8.
256. C. Chang and L. D. Zhao. (2018). Anharmoncity and low thermal conductivity in thermoelectrics. *Mater. Today Phys.* **4**, 50–57.
257. B. M. Foley, C. S. Gorham, J. C. Duda, R. Cheaito, C. J. Szwejkowski, C. Constantin, B. Kaehr, and P. E. Hopkins. (2014). Protein thermal conductivity measured in the solid state reveals anharmonic interactions of vibrations in a fractal structure. *J. Phys. Chem. Lett.* **5**(7), 1077–1082.
258. U. Aydemir, C. Candolfi, A. Ormeci, Y. Oztan, M. Baitinger, N. Oeschler, F. Steglich, and Y. Grin. (2011). Low-temperature thermoelectric, galvanomagnetic, and thermodynamic properties of the type-I clathrate $Ba_8Au_xSi_{46-x}$. *Phys. Rev. B.* **84**(19), 195137–14.
259. S. Mano, T. Onimaru, S. Yamanaka, and T. Takabatake. (2011). Off-center rattling and thermoelectric properties of type-II clathrate $(K, Ba)_{24}(Ga, Sn, \square)_{136}$ single crystals. *Phys. Rev. B.* **84**(21), 214101–6.
260. W. Li and N. Mingo. (2015). Ultralow lattice thermal conductivity of the fully filled skutterudite $YbFe_4Sb_{12}$ due to the flat avoided-crossing filler modes. *Phys. Rev. B.* **91**(14), 144304–6.
261. F. Sui, H. He, S. Bobev, J. Zhao, F. E. Osterloh, and S. M. Kauzlarich. (2015). Synthesis, structure, thermoelectric properties, and band gaps of alkali metal containing type I clathrates: $A_8Ga_8Si_{38}$ (A= K, Rb, Cs) and $K_8Al_8Si_{38}$. *Chem. Mater.* **27**(8), 2812–2820.
262. S. Christensen, M. S. Schmøkel, K. A. Borup, G. K. Madsen, G. J. McIntyre, S. C. Capelli, M. Christensen, and B. B. Iversen. (2016). "Glass-like" thermal conductivity gradually induced in thermoelectric $Sr_8Ga_{16}Ge_{30}$ clathrate by off-centered guest atoms. *J. Appl. Phys.* **119**(18), 185102–11.

263. P. F. Lory, S. Pailhèès, V. M. Giordano, H. Euchner, H. D. Nguyen, R. Ramlau, H. Borrmann, M. Schmidt, M. Baitinger, M. Ikeda, and P. Tomeš. (2017). Direct measurement of individual phonon lifetimes in the clathrate compound $Ba_{7.81}Ge_{40.67}Au_{5.33}$. *Nature Commun.* **8**(1), 491.
264. P. Tomeš, T. Himmelbauer, A. Sidorenko, X. Yan, A. Prokofiev, and S. Paschen. (2017). Physical properties of the type-I clathrate phase $Ba_{8-x}Eu_xAu_ySi_{46-y}$. *Acta Mater.* **129**, 521–531.
265. P. Popčević, D. Stanić, Ž. Bihar, A. Bilušić, and A. Smontara. (2011). Heat transport in aluminum-based quasicrystals i-AlPdMn, i-AlCuFe, and d-AlCoNi. *Israël J. Chem.* **51**(11–12), 1340–1348.
266. A. I. Krivchikov, O. A. Korolyuk, I. V. Sharapova, J. L. Tamarit, F. J. Bermejo, L. C. Pardo, M. Rovira-Esteva, M. D. Ruiz-Martin, A. Jezowski, J. Baran, and N. A. Davydova. (2012). Effects of internal molecular degrees of freedom on the thermal conductivity of some glasses and disordered crystals. *Phys. Rev. B.* **85**(1), 014206–10.
267. J. F. Gebbia, M. A. Ramos, D. Szewczyk, A. Jezowski, A. I. Krivchikov, Y. V. Horbatenko, T. Guidi, F. J. Bermejo, and J. L. Tamarit. (2017). Glassy anomalies in the low-temperature thermal properties of a minimally disordered crystalline solid. *Phys. Rev. Lett.* **119**(21), 215506–6.
268. G. P. Johari, O. Andersson, and B. Sundqvist. (2018). Instability and thermal conductivity of pressure-densified and elastically altered orientational glass of Buckminsterfullerene. *J. Chem. Phys.* **148**(14), 144502–8.
269. M. Hassaine, M. A. Ramos, A. I. Krivchikov, I. V. Sharapova, O. A. Korolyuk, and R. J. Jiménez-Riobóo. (2012). Low-temperature thermal and elastoacoustic properties of butanol glasses: Study of position isomerism effects around the boson peak. *Phys. Rev. B,* **85**(10), 104206–11.
270. T. Pérez-Castañeda, R. J. J. Riobóo, and M. A. Ramos. (2013). Low-temperature thermal properties of a hyperaged geological glass. *J. Phys. Condens. Matter.* **25**(29), 295402–10.
271. A. I. Krivchikov and O. Andersson. (2016). Thermal conductivity of triphenyl phosphite's liquid, glassy, and glacial states. *J. Phys. Chem. B.* **120**(10), 2845–2853.
272. O. Andersson and G. P. Johari. (2016). Thermal conductivity of Glycerol's liquid, glass, and crystal states, glass-liquid-glass transition, and crystallization at high pressures. *J. Chem. Phys.* **144**(6), 064504–10.
273. H. S. Yang, D. G. Cahill, X. Liu, J. L. Feldman, R. S. Crandall, B. A. Sperling, and J. R. Abelson. (2010). Anomalously high thermal conductivity of amorphous Si deposited by hot-wire chemical vapor deposition. *Phys. Rev. B.* **81**(10), 104203–7.

274. X. Liu, J. L. Feldman, D. G. Cahill, H. S. Yang, R. S. Crandall, N. Bernstein, D. M. Photiadis, M. J. Mehl, and D. A. Papaconstantopoulos. (2011). Anomalously high thermal conductivity of amorphous silicon films prepared by hot-wire chemical vapor deposition. *Chin. J. Phys.* **49**(1), 359 368.
275. K. Valalaki and A. G. Nassiopoulou. (2013). Low thermal conductivity porous Si at cryogenic temperatures for cooling applications. *J. Phys. D.* **46**(29), 295101–9.
276. K. Valalaki and A. G. Nassiopoulou. (2014). Thermal conductivity of highly porous Si in the temperature range 4.2 to 20 K. *Nanoscale Res. Lett.* **9**(1), 318–6.
277. O. Andersson and Y. Nakazawa. (2015). Transitions in pressure collapsed clathrate hydrates. *J. Phys. Chem. B.* **119**(9), 3846–3853.
278. Z. G. Cheng and M. H. Chan. (2013). Threefold reduction in thermal conductivity of Vycor glass due to adsorption of liquid 4He. *New J. Phys.* **15**(6), 063030–8.
279. A. Gurlo, E. Ionescu, R. Riedel, and D. R. Clarke. (2016). The thermal conductivity of polymer-derived amorphous Si–O–C compounds and nano-composites. *J. Am. Ceram. Soc.* **99**(1), 281–285.
280. S. N. Zhang, J. He, T. J. Zhu, X. B. Zhao, and T. M. Tritt. (2009). Thermal conductivity and specific heat of bulk amorphous chalcogenides $Ge_{20}Te_{80-x}Se_x$ (x= 0, 1, 2, 8). *J. Non-Cryst. Solids* **355**(2), 79–83.
281. G. Ventura and V. Martelli. (2010). Low temperature thermal conductivity of PVC. In C. Leroy *et al.* (ed.), *Astroparticle, Particle and Space Physics, Detectors and Medical Physics Applications*, Vol. 5, pp. 145–149. Singapore: World Scientific.
282. A. L. Woodcraft, V. Martelli, and G. Ventura. (2010). Thermal conductivity of Tecamax® SRP from millikelvin temperatures to room temperature. *Cryogenics.* **50**(2), 66–70.
283. T. Seidl, A. Plotnikov, E. Mustafin, R. Lopez, D. Severin, E. Floch, C. Trautmann, A. Golubev, A. Smolyakov, D. Tommasini, and W. Ensinger. (2012). Influence of swift heavy ion beams and protons on the dielectric strength of polyimide. *Polym. Degrad. Stab.* **97**(11), 2396–2402.
284. O. Andersson and G. P. Johari. (2011). Effect of pressure on thermal conductivity and pressure collapse of ice in a polymer-hydrogel and kinetic unfreezing at 1 GPa. *J. Chem. Phys.* **134**(12), 124903–10.
285. J. C. Duda, P. E. Hopkins, Y. Shen, and M. C. Gupta. (2013). Exceptionally low thermal conductivities of films of the fullerene derivative PCBM. *Phys. Rev. Lett.* **110**(1), 015902–5.

286. X. Wang, V. Ho, R. A. Segalman, and D. G. Cahill. (2013). Thermal conductivity of high-modulus polymer fibers. *Macromolecules.* **46**(12), 4937–4943.
287. V. Singh, T. L. Bougher, A. Weathers, Y. Cai, K. Bi, M. T. Pettes, S. A. McMenamin, W. Lv, D. P. Resler, T. R. Gattuso, and D. H. Altman. (2014). High thermal conductivity of chain-oriented amorphous polythiophene. *Nature Nanotechnol.* **9**(5), 384–390.
288. G. Ventura and M. Perfetti. (2014). *Thermal Properties of Solids at Room and Cryogenic Temperatures.* Springer Netherlands.
289. P. Limelette, B. Schmaltz, D. Brault, M. Gouineau, C. Autret-Lambert, S. Roger, V. Grimal, and F. Tran Van. (2014). Conductivity scaling and thermoelectric properties of polyaniline hydrochloride. *J. Appl. Phys.* **115**(3), 033712–6.
290. S. Nakamura, T. Fujii, S. Matsukawa, M. Katagiri, and H. Fukuyama. (2018). Specific heat, thermal conductivity, and magnetic susceptibility of cyanate ester resins – an alternative to commonly used epoxy resins. *Cryogenics.* **95**, 76–81.
291. L. Dong, Q. Xi, D. Chen, J. Guo, T. Nakayama, Y. Li, Z. Liang, J. Zhou, X. Xu, and B. Li. (2018). Dimensional crossover of heat conduction in amorphous polyimide nanofibers. *Nat. Sci. Rev.* **5**(4), 500–506.
292. M. Daal, N. Zobrist, N. Kellaris, B. Sadoulet, and M. Robertson. (2019). Properties of selected structural and flat flexible cabling materials for low temperature applications. *Cryogenics.* **98**, 47–59.
293. D. S. Rothfuß. (2013). Thermische Eigenschaften von supraleitenden massiven metallischen Gläsern bei ultratiefen Temperaturen (Doctoral dissertation Universität Heidelberg).
294. V. V. E. Pryadun, D. V. Louzguine-Luzgin, L. V. Shvanskaya, and A. N. Vasiliev. (2015). Thermoelectric properties of Au-based metallic glass at low temperatures. *JETP Lett.* **101**(7), 465–468.
295. T. Kolodiazhnyi, H. Sakurai, O. Vasylkiv, H. Borodianska, and Y. Mozharivskyj. (2014). Abnormal thermal conductivity in tetragonal tungsten bronze $Ba_{6-x}Sr_xNb_{10}O_{30}$. *Appl. Phys. Lett.* **104**(11), 111903–3.
296. M. Tachibana. (2015). Thermal conductivity of Aurivillius compounds Bi_2WO_6, $SrBi_2Ta_2O_9$, and $Bi_4Ti_3O_{12}$. *Solid State Commun.* **211**, 1–3.
297. B. M. Foley, E. A. Paisley, C. DiAntonio, T. Chavez, M. Blea-Kirby, G. Brennecka, J. T. Gaskins, J. F. Ihlefeld, and P. E. Hopkins. (2017). Phonon scattering mechanisms dictating the thermal conductivity of lead zirconate titanate ($PbZr_{1-x}Ti_xO_3$) thin films across the compositional phase diagram. *J. Appl. Phys.* **121**(20), 205104–15.

298. B. Ramachandran, K. K. Wu, Y. K. Kuo, and M. R. Rao. (2015). Phonon thermal transport and phonon–magnon coupling in polycrystalline $BiFeO_3$ systems. *J. Phys. D.* **48**(11), 115301–7.
299. F. Drymiotis, T. Drye, D. Rhodes, Q. Zhang, J. C. Lashey, Y. Wang, S. Cawthorne, B. Ma, S. Lindsey, and T. Tritt. (2009). Glassy thermal conductivity in the two-phase $Cu_xAg_{3-x}SbSeTe_2$ alloy and high temperature thermoelectric behavior. *J. Phys. Condens. Matter.* **22**(3), 035801–5.
300. M. D. Nielsen, V. Ozolins, and J. P. Heremans. (2013). Lone pair electrons minimize lattice thermal conductivity. *Energy Environ. Sci.* **6**(2), 570–578.
301. P. Wu, M. Y. Hu, X. Y. Chong, and J. Feng. (2018). The glass-like thermal conductivity in ZrO_2-Dy_3TaO_7 ceramic for promising thermal barrier coating application. *Appl. Phys. Lett.* **112**(13), 131903–5.
302. T. J. Zhu, S. N. Zhang, S. H. Yang, and X. B. Zhao. (2010). Improved thermoelectric figure of merit of self-doped $Ag_{8-x}GeTe_6$ compounds with glass-like thermal conductivity. *Phys. Status Solidi.* **4**(11), 317–319.
303. D. Hitchcock. (2014). Unusually low thermal conductivity in the argyrodite Ag_8GeTe_6 attributed to strong anharmonicity. All Dissertations1414, *Clemson University.*
304. J. L. Braun, C. M. Rost, M. Lim, A. Giri, D. H. Olson, G. N. Kotsonis, G. Stan, D. W. Brenner, J. P. Maria, and P. E. Hopkins. (2018). Charge-induced disorder controls the thermal conductivity of entropy-stabilized oxides. *Adv. Mater.* **30**(51), 1805004.
305. V. Gvozdetskyi, B. Owens-Baird, S. Hong, T. Cox, G. Bhaskar, C. Harmer, Y. Sun, F. Zhang, C. Z. Wang, K. M. Ho, and J. V. Zaikina. (2019). From $NaZn_4Sb_3$ to HT-$Na_{1-x}Zn_{4-y}Sb_3$: Panoramic hydride synthesis, structural diversity, and thermoelectric properties. *Chem. Mater.* **31**(21), 8695–8707.
306. Q. J. Li, Z. Y. Zhao, C. Fan, F. B. Zhang, H. D. Zhou, X. Zhao, and X. F. Sun. (2013). Phonon-glass-like behavior of magnetic origin in single-crystal $Tb_2Ti_2O_7$. *Phys. Rev. B.* **87**(21), 214408–6.
307. J. Fang, Y. Huang, C. M. Lew, Y. Yan, and L. Pilon. (2012). Temperature dependent thermal conductivity of pure silica MEL and MFI zeolite thin films. *J. Appl. Phys.* **111**(5), 054910–6.
308. C. Blanc, A. Rajabpour, S. Volz, T. Fournier, and O. Bourgeois. (2013). Phonon heat conduction in corrugated silicon nanowires below the Casimir limit. *Appl. Phys. Lett.* **103**(4), 043109–5.
309. Q. Zhang, Z. Cui, Z. Wei, S. Y. Chang, L. Yang, Y. Zhao, Y. Yang, Z. Guan, Y. Jiang, J. Fowlk, J. Yang, D. Xu, Y. Chen, T. Xu, and

D. Li. (2017). Defect facilitated phonon transport through kinks in boron carbide nanowires. *Nano Lett.* **17**(6), 3550–3555.

310. Y. Wang, R. Lin, P. Zhu, Q. Zheng, Q. Wang, D. Li, and J. Zhu. (2018). Cation dynamics governed thermal properties of lead halide perovskite nanowires. *Nano Lett.* **18**(5), 2772–2779.
311. A. Tavakoli, C. Blanc, H. Ftouni, K. J. Lulla, A. D. Fefferman, E. Collin, and O. Bourgeois. (2017). Universality of thermal transport in amorphous nanowires at low temperatures. *Phys. Rev. B.* **95**(16), 165411–6.
312. A. Tavakoli, K. Lulla, T. Crozes, N. Mingo, E. Collin, and O. Bourgeois. (2018). Heat conduction measurements in ballistic 1D phonon waveguides indicate breakdown of the thermal conductance quantization. *Nature Commun.* **9**(1), 4287–8.
313. H. Ftouni, C. Blanc, D. Tainoff, A. D. Fefferman, M. Defoort, K. J. Lulla, J. Richard, E. Collin, and O. Bourgeois. (2015). Thermal conductivity of silicon nitride membranes is not sensitive to stress. *Phys. Rev. B.* **92**(12), 125439–7.
314. A. Tavakoli, C. Blanc, H. Ftouni, K. J. Lulla, A. D. Fefferman, E. Collin, and O. Bourgeois. (2017). Universality of thermal transport in amorphous nanowires at low temperatures. *Phys. Rev. B.* **95**(16), 165411–6.
315. T. Hori and J. Shiomi. (2019). Tuning phonon transport spectrum for better thermoelectric materials. *Sci. Technol. Adv. Mater.* **20**(1), 10–25.
316. A. Weathers, J. Carrete, J. P. DeGrave, J. M. Higgins, A. L. Moore, J. Kim, N. Mingo, S. Jin, and L. Shi. (2017). Glass-like thermal conductivity in nanostructures of a complex anisotropic crystal. *Phys. Rev. B.* **96**(21), 214202–7.
317. T. Nomoto, S. Imajo, S. Yamashita, H. Akutsu, Y. Nakazawa, and A. I. Krivchikov. (2019). Construction of a thermal conductivity measurement system for small single crystals of organic conductors. *J. Therm. Anal. Calorim.* **135**(5), 2831–2836.
318. J. Yang, Y. Wang, H. Yang, W. Tang, J. Yang, L. Chen, and W. Zhang. (2019). Thermal transport in thermoelectric materials with chemical bond hierarchy. *J. Phys. Condens. Matter.* **31**(18), 183002–14.
319. Y. Wang, R. Lin, P. Zhu, Q. Zheng, Q. Wang, D. Li, and J. Zhu. (2018). Cation dynamics governed thermal properties of lead halide perovskite nanowires. *Nano Lett.* **18**(5), 2772–2779.
320. R. V. Nikonkov, P. Stachowiak, T. V. Romanova, A. Jeżowski, and V. V. Sumarokov. (2015). Investigations of thermal conductivity of

simple van der Waals crystal-based nanocomposites. *Fiz. Nizk. Temp.* **41**(6), 625–628 [*Low Temp. Phys.* **41**(6), 492–494].

321. M. T. Buscaglia, F. Maglia, U. Anselmi-Tamburini, D. Marrè, I. Pallecchi, A. Ianculescu, G. Canu, M. Viviani, M. Fabrizio, and V. Buscaglia. (2014). Effect of nanostructure on the thermal conductivity of La-doped SrTiO3 ceramics. *J. Eur. Ceram. Soc.* **34**(2), 307–316.
322. E. Dechaumphai, D. Lu, J. J. Kan, J. Moon, E. E. Fullerton, Z. Liu, and R. Chen. (2014). Ultralow thermal conductivity of multilayers with highly dissimilar debye temperatures. *Nano Lett.* **14**(5), 2448–2455.
323. J. Yang, H. Tang, Y. Zhao, Y. Zhang, J. Li, Z. Ni, Y. Chen, and D. Xu. (2015). Thermal conductivity of zinc blende and wurtzite CdSe nanostructures. *Nanoscale.* **7**(38), 16071–16078.
324. B. M. Foley, H. J. Brown-Shaklee, M. J. Campion, D. L. Medlin, P. G. Clem, J. F. Ihlefeld, and P. E. Hopkins. (2015). Glass-like thermal conductivity of (010)-textured lanthanum-doped strontium niobate synthesized with wet chemical deposition. *J. Am. Ceram. Soc.* **98**(2), 624–628.
325. L. Li, X. J. Yan, S. T. Dong, Y. Y. Lv, X. Li, S. H. Yao, Y. B. Chen, S. T. Zhang, J. Zhou, Y. H. Lu, M. H. Lu, and Y. F. Chen. (2017). Ultra-low thermal conductivities along c-axis of naturally misfit layered $Bi_2[AE]_2Co_2O_y$ (AE= Ca, $Ca_{0.5}Sr_{0.5}$, Sr, Ba) single crystals. *Appl. Phys. Lett.* **111**(3), 033902–4.
326. M. I. Klinger. (2010). Soft atomic motion modes in glasses: Their role in anomalous properties. *Phys. Rep.* **492**(4–5), 111–180.
327. E. S. Toberer, L. L. Baranowski, and C. Dames. (2012). Advances in thermal conductivity. *Ann. Rev. Mater. Res.* **42**, 179–209.
328. M. I. Klinger. (2013). Glassy disordered systems: Glass formation and universal anomalous low-energy properties. Singapore: World Scientific. *Nanoscale Microscale Thermophys. Eng.* **23**(2), 81–116.
329. T. Takabatake, K. Suekuni, T. Nakayama, and E. Kaneshita. (2014). Phonon-glass electron-crystal thermoelectric clathrates: Experiments and theory. *Rev. Mod. Phys.* **86**(2), 669–716.
330. O. Andersson and Y. Nakazawa. (2014). Thermal properties and transition behavior of host-guest compounds under high pressure. *Curr. Inorg. Chem.* **4**(1), 2–18.
331. M. C. Wingert, J. Zheng, S. Kwon, and R. Chen. (2016). Thermal transport in amorphous materials: A review. *Semiconductor Sci. Technol.* **31**(11), 113003–16.
332. M. Beekman and D. G. Cahill. (2017). Inorganic crystals with glass-like and ultralow thermal conductivities. *Cryst. Res. Technol.* **52**(10), 1700114–13.

333. F. DeAngelis, M. G. Muraleedharan, J. Moon, H. R. Seyf, A. J. Minnich, A. J. McGaughey, and A. Henry. (2018). Thermal transport in disordered materials. *Nanoscale Microscale Thermophys. Eng.* 1–36.
334. X. Chen, A. Weathers, J. Carrete, S. Mukhopadhyay, O. Delaire, D. A. Stewart, N. Mingo, S. N. Girard, J. Ma, D. L. Abernathy, and J. Yan. (2015). Twisting phonons in complex crystals with quasi-one-dimensional substructures. *Nature Commun.* **6**, 6723–9.
335. D. G. Cahill, P. V. Braun, G. Chen, D. R. Clarke, S. Fan, K. E. Goodson, P. Keblinski, W. P. King, G. D. Mahan, A. Majumdar, and H. J. Maris. (2014). Nanoscale thermal transport. II. 2003–2012. *Appl. Phys. Rev.* **1**(1), 011305.
336. P. B. Allen and J. L. Feldman. (1989). Thermal conductivity of glasses: Theory and application to amorphous Si. *Phys. Rev. Lett.* **62**(6), 645–648.
337. P. B. Allen and J. L. Feldman. (1993). Thermal conductivity of disordered harmonic solids. *Phys. Rev. B.* **48**(17), 12581–12588.
338. T. Romanova, P. Stachowiak, A. Jeżowski, A. I. Krivchikov, and G. A. Vdovychenko. (2015). A universal T^2 behavior of low temperature thermal conductivity of some simple molecular polycrystals. *Physica B.* **459**, 93–96.
339. A. I. Krivchikov, G. A. Vdovichenko, O. A. Korolyuk, F. J. Bermejo, L. C. Pardo, J. L. Tamarit, and D. Szewczyk. (2015). Effects of site-occupation disorder on the low-temperature thermal conductivity of molecular crystals. *J. Non-Cryst. Solids.* **407**, 141–148.
340. D. Szewczyk, A. Jeżowski, G. A. Vdovichenko, A. I. Krivchikov, F. J. Bermejo, J. L. Tamarit, L. C. Pardo, and J. W. Taylor. (2015). Glassy dynamics versus thermodynamics: The case of 2-adamantanone. *J. Phys. Chem. B.* **119**(26), 8468–8474.
341. C. H. Lee, A. Nishida, T. Hasegawa, H. Nishiate, H. Kunioka, S. Ohira-Kawamura, M. Nakamura, K. Nakajima, and Y. Mizuguchi. (2018). Effect of rattling motion without cage structure on lattice thermal conductivity in $LaOBiS_{2-x}Se_x$. *Appl. Phys. Lett.* **112**(2), 023903–4.
342. M. Simoncelli, N. Marzari, and F. Mauri. (2019). Unified theory of thermal transport in crystals and glasses. *Nature Phys.* **15**(8), 809–813.
343. G. Kumar, F. G. VanGessel, D. C. Elton, and P. W. Chung. Phonon lifetimes and thermal conductivity of the molecular crystal α-RDX (2019). *arXiv preprint arXiv*:1904.12038.
344. V. G. Karpov, M. I. Klinger, and P. N. Ignatiev. (1982). Atomic tunneling states and low-temperature anomalies of thermal properties in amorphous materials. *Solid State Commun.* **44**(3), 333–337.

345. U. Buchenau, Y. M. Galperin, V. L. Gurevich, D. A. Parshin, M. A. Ramos, and H. R. Schober. (1992). Interaction of soft modes and sound waves in glasses. *Phys. Rev. B.* **46**(5), 2798–2808.
346. V. G. Karpov, M. I. Klinger, and P. N. Ignatiev. (1982). Atomic tunneling states and low temperature anomalies of thermal properties in amorphous materials. *Solid State Commun.* **44**(3), 333–337.
347. V. G. Karpov, I. Klinger, and F. N. Ignat'ev. (1983). Theory of the low-temperature anomalies in the thermal properties of amorphous structures. *Zh. Eksp. Teor. Fiz.* **84**(2), 760–770 [*J. Exp. Theor. Phys.* **57**(2), 439–448].
348. B. Golding, J. E. Graebner, and L. C. Allen. (1986). The thermal conductivity plateau in disordered systems. *Phonon Scattering in Condensed Matter.* **V**, [Proc. SSSOL, vol. 68], pp. 23–25. Heidelberg: Springer.
349. C. Y. Clare and J. J. Freeman. (1987). Thermal conductivity and specific heat of glasses. *Phys. Rev. B.* **36**(14), 7620–7624.
350. L. Gil, M. A. Ramos, A. Bringer, and U. Buchenau. (1993). Low-temperature specific heat and thermal conductivity of glasses. *Phys. Rev. Lett.* **70**(2), 182–185.
351. X. Liu and H. V. Löhneysen. (1993). Low-temperature thermal properties of amorphous As_xSe_{1-x}. *Phys. Rev. B.* **48**(18), 13486–13494.
352. J. L. Liu, A. Khitun, K. L. Wang, W. L. Liu, G. Chen, Q. H. Xie, and S. G. Thomas. (2003). Cross-plane thermal conductivity of self-assembled Ge quantum dot superlattices. *Phys. Rev. B.* **67**(16), 165333–6.
353. G. D'Angelo, C. Crupi, G. Tripodo, and G. Salvato. (2010). Relation between low-temperature thermal conductivity and the specific heat of cesium borate glasses. *J. Phys. Chem. B.* **114**(7), 2467–2475.
354. J. W. Gardner and A. C. Anderson. (1981). Low-temperature specific heat and thermal conductivity of neutron-irradiated crystalline quartz. *Phys. Rev. B.* **23**(2), 474–482.
355. R. Geilenkeuser, T. Porschberg, M. Jackel, and A. Gladun. (1999). Influence of high pressure on thermal properties of amorphous polystyrene. *Phys. B. Condens. Matter.* **263**, 276–279.
356. J. M. Grace and A. C. Anderson. (1986). Low-temperature thermal properties of a sol-gel glass. *Phys. Rev. B.* **33**(10), 7186–7191.
357. A. Licea-Claverie, M. Müller, M. Jäckel, and K. F. Arndt. (1990). Low-temperature thermal conductivity of diolmodified epoxies. *Polym. Bull.* **24**(4), 421–427.
358. P. B. Allen, J. L. Feldman, J. Fabian, and F. Wooten. (1999). Diffusons, locons and propagons: Character of atomic vibrations in amorphous Si. *Philos. Mag. B.* **79**(11–12), 1715–1731.

359. M. Fahland and E. Hegenbarth. (1993). Thermal conductivity of $Pb(Mg_{1/3}Nb_{2/3})O_3$ under the influence of high electric field. *Ferroelectr. Lett. Sec.* **15**(3–4), 89–96.
360. W. Scheibner, K. Jahn, D. Jehnichen, and M. Jäckel. (1986). Heterogeneous structure of amorphous polycarbonate as determined by low-temperature thermal conductivity measurements. *Acta Polym.* **37**(1), 56–59.
361. O. A. Korolyuk, A. I. Krivchikov, and O. O. Romantsova. (2020). Universal temperature dependence of the thermal conductivity of clathrate compounds, molecular crystals and glasses at low temperatures. *Fiz. Nizk. Temp.* **46**(2) 138–146 [*Low Temp. Phys.* **46**(2), 111–117].

https://doi.org/10.1142/9781800612587_0004

Chapter 4

Two-Level Systems and the Tunneling Model: A Critical View

Clare C. Yu* and Hervé M. Carruzzo

Department of Physics and Astronomy, University of California, Irvine CA, 92697, USA
**cyu@uci.edu*

The thermal and acoustic properties displayed by a wide variety of glasses at low temperatures are well described by the model of tunneling two-level systems (TLS). We review the standard TLS model as well as developments that have occurred since the earlier work. In spite of half a century of effort and impressive successes, mysteries still remain. We discuss some of these long-standing puzzles about glasses at low temperatures and speculate about future directions.

1. Introduction

The goal of this chapter is to raise questions about the properties of glasses at low temperatures.[1,2] So let us start with this one, "Why are the thermal properties of glasses different from those of crystals at low temperatures?" Crystals have a phonon (Debye) specific heat and a thermal conductivity that both go as T^3 at low temperatures, independent of the crystal structure. This is duc to phonons whose wavelengths λ are much larger than the lattice constant a; as a result the phonons are insensitive to differences in crystal structures. This insensitivity to crystal structure could lead one to predict that the specific heat and thermal conductivity of all insulators should behave similarly at low temperatures. But this prediction would be wrong.

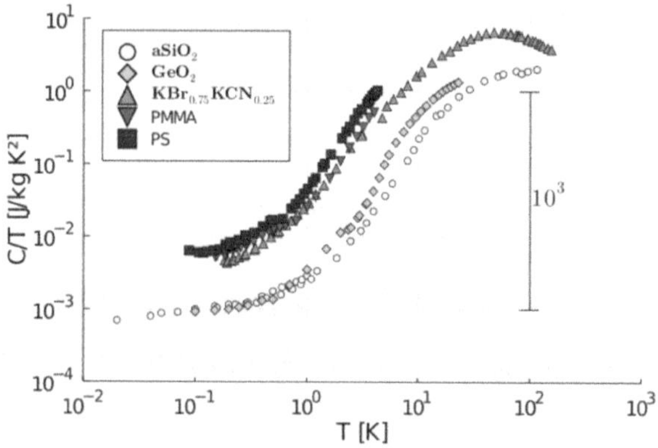

Fig. 1. C/T vs. T for amorphous SiO_2 ($aSiO_2$) [3, 4], GeO_2 [3, 5], $KBr_{0.75}$ $CN_{0.25}$ [6], PMMA [7–9], and PS [8, 9] shows a steep rise by two to three orders of magnitude between 3–10 K.

Amorphous materials below 1 K have a specific heat that is linear in temperature and a thermal conductivity that is quadratic in temperature (see Chapter 1, Figures 1 and 2).[3] Between 3 and 10 K, the thermal conductivity displays a plateau followed by a rise at higher temperatures.[3] In the temperature range of the plateau, the specific heat displays a bump when C/T^3 is plotted versus T on log–log plot.[3] This bump is associated with the so-called 'boson peak'. If C/T vs. T is plotted, there is a steep rise by two to three orders of magnitude between 1 and 20 K (Figure 1).

It is remarkable that the amorphous structure, rather than chemical composition, determines the thermal, acoustic, and dielectric properties of glassy materials. What is it about the disordered structure that produces such universal behavior?

1.1. *Standard model of two-level systems*

To explain the behavior below 1 K, the standard model of tunneling two-level systems (TLS) was proposed by Anderson, Halperin, and Varma[10] and independently by Phillips.[11] The basic idea is that there are dynamic defects consisting of an atom or group of atoms that can sit more or less equally well in either of two energy minima of a double-well potential. At low temperatures the atom or group of atoms tunnels back and forth between the two configurations represented by the two minima. The energy difference between the two minima is given by the energy asymmetry ε. The potential barrier height is V while d is the distance between the minima. In the right-well–left-well basis, the Hamiltonian is given by

$$H = \frac{1}{2}\begin{pmatrix} \varepsilon & \Delta_o \\ \Delta_o & -\varepsilon \end{pmatrix} \tag{1}$$

The tunneling matrix element, Δ_o, is given by the WKB formula

$$\Delta_o = \omega_o e^{-W} \tag{2}$$

where

$$W \sim \frac{\sqrt{2MV}d}{\hbar} \tag{3}$$

M is the mass of the tunneling entity, and the attempt frequency ω_o is on the order of the Debye frequency since atomic motion is involved in the tunneling. The two level systems that tunnel fast enough to be seen experimentally must have low barrier heights ($V \leq 0.1\,\text{eV}$). In order to conserve energy, a phonon must be emitted or absorbed when a transition is made between energy levels given by the following energy eigenvalues:

$$E_{\pm} = \pm\frac{1}{2}\sqrt{\varepsilon^2 + \Delta_o^2} \tag{4}$$

This implies an energy splitting, $E = E_+ - E_- \leq k_BT$. A typical estimate for the density of two-level systems is 100 ppm, which comes from integrating the specific heat of SiO_2 up to 10 K. The TLS model assumes that the distribution $N(\varepsilon)$ of asymmetry energies is flat and the distribution $P(W)$ of the WKB exponent is flat.[10,11] Even though $N(\varepsilon)$ probably varies on the scale of 0.1 eV, it is reasonable to expect it to be flat for energies of order k_BT where $T \sim 1\,K$.[10] On the other hand, the distribution for the WKB exponent W does not have such a natural explanation besides the fact that it leads to a flat distribution of energy splittings. Regardless, the assumption is then that $P(\varepsilon, W) = \bar{P}$ where $\bar{P}$ is a constant. Changing variables from W to Δ_o yields

$$P(\varepsilon, \Delta_o) = \frac{\bar{P}}{\Delta_o} \tag{5}$$

As a result, the distribution $n(E)$ of energy splittings E in Eq. (4) is a constant n_o. It is straightforward to show (see Chapter 2) that this uniform density of states produces a linear specific heat using an approach similar to that used to calculate the linear heat capacity of an electron gas.

To calculate the thermal conductivity, we note that up to about 50 to 60 K, Zaitlin and Anderson showed that heat is transported by phonons in insulating glasses.[12] The thermal conductivity is given by

$$\kappa(T) = \frac{1}{3}\int_0^{\omega_D} C_{ph}(T,\omega) v \ell(T,\omega)\, d\omega \tag{6}$$

where ω_D is the Debye frequency, C_{ph} is the phonon specific heat which goes as T^3, v is the sound velocity, and $\ell(T,\omega)$ is the phonon mean free path. The phonon mean free path is limited by resonant scattering from TLS in which a phonon with energy $\hbar\omega$ is absorbed by a TLS with energy splitting $E \simeq \hbar\omega$. This requires that phonons couple to the strain field as described by the following term in the Hamiltonian:

$$H_{TLS-Ph} = \gamma e \sigma_z \tag{7}$$

where γ is the coupling constant, e is the strain field, and σ_z is the z-component Pauli matrix. Calculating the resonant scattering rate yields $\ell \sim \omega^{-1} \sim T^{-1}$, which, when combined with $C_{ph} \sim T^3$, produces $\kappa(T) \sim T^2$. In fitting the thermal conductivity data to Eq. (6), Freeman and Anderson[13] noted the remarkable result that $\ell \sim 150\lambda$ below about 200 GHz and for temperatures around and below 1 K for a wide variety of glasses.

1.2. *Crossover region*

As we mentioned earlier, between 3–10 K there is a plateau in the thermal conductivity and a bump in C/T^3 that corresponds to the boson peak. In fitting the thermal conductivity to Eq. (6) to find the frequency dependence of the mean free path ℓ below about 10 K, Freeman and Anderson[13] found that $\ell \sim 150\lambda$ at low frequencies, drops precipitously around 200 GHz, and then $\ell \sim \lambda$ at higher frequencies (see Figure 4 in Chapter 1). One can regard the plateau as a crossover from a long mean free path to a regime with a short mean free path. One way to account for these observations is to ascribe them to a steep rise at an energy E_o in the density of states of dynamic defects that scatter phonons.[12,14,15] This is consistent with the bump in C/T^3, i.e., the rise in C/T. Fitting the specific heat data results in E_o between 10 and 40 K.[15] This is consistent with the rise in the density of states observed in Raman scattering[16] and neutron scattering.[17,18] (E_o/k_B being greater than the temperature of the bump in C/T^3 is due to excitations being able to contribute to the specific heat at temperatures well below their characteristic energy.) Yu and Freeman[15] used this rise in the density of states of phonon scatterers to fit the specific heat and thermal conductivity of various glasses up to roughly 100 K. In order to fit both the plateau in the thermal conductivity and the bump in C/T^3 with the same density of states, they had to introduce strong (Rayleigh) scattering at low frequencies ($\hbar\omega < E_o$) but not at high frequencies ($\hbar\omega > E_o$).

Zaitlin and Anderson were unable to find physically realistic values for the coefficient of Rayleigh scattering due to density fluctuations.[12] However, it is possible that Rayleigh scattering could arise from other sources, e.g., fluctuations in local force constants,[19] though they did not consider this. Zaitlin and Anderson were able to fit the thermal conductivity without invoking Rayleigh scattering by adding a term quadratic in the energy to the density of states of the phonon scatterers and cutting off the steep drop in the phonon mean free path at high frequencies with a minimum mean free path of order a few angstroms.[12,14] However, unlike Yu and Freeman,[15] they did not try to do a detailed fit to the specific heat with this same density of states.

These findings raise several questions: First, what are the excess excitations responsible for the boson peak? This question is still not settled, though it has been discussed extensively in the literature[20] and is reviewed in Chapters 7–11. Second, what picks out the energy scale of E_o for the rise in the density of states? It is much less than the other energy scales in the problem, e.g., the Debye energy is of order 300 K and electronic energies are of order thousands of degrees. Third, what is the source of the strong (Rayleigh) scattering and why does the strong (Rayleigh) scattering appear to stop at high frequencies? If Rayleigh scattering of phonons were allowed to occur at all frequencies, the thermal conductivity would not rise above the plateau. However, it is quite likely that above the plateau, where the phonon mean free path is of order the wavelength, ballistic phonons are no longer operating. Rather, phonon diffusion or hopping of quasi-localized vibrations may be transporting heat.[21,22] These are questions worth pondering on, but let us return to the subject of two-level systems.

2. Beyond the Standard Model of Two-Level Systems

2.1. *Interacting two-level systems*

Since TLS couple to the strain field, TLS can interact with each other by exchanging phonons.[23,24] Using second-order perturbation theory, one can show that this interaction has a dipolar form that goes as g/r^3 where $g \sim \gamma^2/\rho v^2$, r is the distance between the two interacting TLS, and ρ is the mass density.[23,24] Yu and Leggett argued that this interaction should not be neglected.[25] To see why, let us estimate g/r^3 as follows. If we take $\gamma \sim 1\,\text{eV}$ (a typical experimentally deduced value), then $g \sim 5 \times 10^4$ K-Å^3. For a TLS concentration of 100 ppm, $r^3 \sim 10^4$ Å^3, assuming 1 atom/Å^3. Putting this all together yields $g/r^3 \sim 5\,\text{K} \gg 0.05\,\text{K} \sim 1\,\text{GHz}$, the typical TLS energy splitting ΔE seen in ultrasound experiments.[26,27] Thus, interactions between TLS cannot be ignored. Indeed, some experiments can only be explained by including interactions.

Historically, TLS–TLS interactions were largely ignored in the standard TLS model except to account for spectral diffusion.[28] For example, in acoustic hole burning experiments,[29] a high power acoustic pulse is resonantly absorbed by those TLS whose energy splitting matches the frequency of the pulse, leading to an increase in the population of the upper energy level and to a decrease in the attenuation. A probe pulse reveals a "hole" in the resonant attenuation centered at the frequency of the high power pulse. The width of the hole is much wider than one would expect from the typical decay rate of excited TLS.[29] To understand this,[28,29] we note that the energy splitting of any given TLS will fluctuate adiabatically due to the fluctuating strain field produced by nearby fluctuating TLS that are undergoing thermal excitations and relaxations back to their ground state. Thus, the broadening of the width of the hole in the attenuation is due to fluctuations in the energy splittings of the excited TLS. While the explanation of spectral diffusion experiments requires TLS–TLS interactions, other types of experiments, e.g., specific heat and thermal conductivity, do not, and so, such interactions have beeen largely ignored.

Interactions between TLS were not widely incorporated into the description of glasses at low temperatures until the work of Yu and Leggett[25] and other theoretical efforts,[30–42] together with subsequent nonequilibrium dielectric experiments from the Osheroff group.[43,44] In those experiments, an ac capacitance bridge was used to study the dielectric response of glasses such as SiO_2, SiO_x, and polymers. Slow dc sweeps showed a minimum in the real part of the ac dielectric susceptibility χ' at the electric field $\boldsymbol{E_o}$ in which the sample was cooled. In the absence of an applied bias voltage ($\boldsymbol{E_o} = 0$), the hole in χ' is centered at zero bias. Upon switching the dc electric field to a new value $\boldsymbol{E_{new}}$ different from $\boldsymbol{E_o}$, χ' increased abruptly and then decayed logarithmically in time. If the electric field is held fixed at $\boldsymbol{E_{new}}$, a second minimum develops at that field as revealed by subsequent dc field sweeps. The abrupt increase of χ' upon the sudden application of $\boldsymbol{E_{\mathbf{new}}}$ is due to χ' jumping out of the bottom of the hole, followed by the logarithmic relaxation in time corresponding to digging a new hole.

An analysis of these experiments by Carruzzo, Grannan, and Yu[45] showed that while many features could be explained with the model of non-interacting TLS, the frequency dependence of the slope, $dC/d(ln\, t)$, of the logarithmic relaxation of the capacitance C required interacting TLS. In particular, interactions lead to the formation of clusters of TLS, with large slow clusters contributing to the dielectric response only at low frequencies. Interactions also provide a natural explanation for the hole in χ' which reflects the hole in $P(h)$, the distribution of local fields h, produced by neighboring TLS. (These fields could be local strain and/or electric fields.) Stability arguments[46–48] require that there be a hole in $P(h)$ as $h \to 0$. To see this, consider a spin glass with long-range interactions. Suppose there is a ground state spin configuration with $P(h = 0) \neq 0$. This implies that there are spins that have zero local field and that, therefore, can flip without any change in energy. But if these spins flip, this would change the local fields of other spins, and so, some of them would flip. This in turn would cause still other spins to flip, leading to a proliferation of flipping spins. This avalanche means that the supposed ground state is not stable. So in order to have a stable ground state, $P(h)$ must go to zero as $h \to 0$. This hole in $P(h)$ is sometimes referred to as the dipole gap. Carruzzo *et al.* used Monte Carlo simulations of an Ising spin glass to show that upon abrupt application of a dc field, the susceptibility jumps up and then relaxes logarithmically, with interactions leading to frequency dependence of the slope of the logarithmic relaxation.[45] An Ising spin glass could be considered as the limit of densely interacting TLS, though Carruzzo *et al.* did not specify the distance between nearest neighbor Ising spins.

Burin[44, 49] also analyzed the Osheroff experiments. Like Carruzzo *et al.*,[45] he found that these experiments could not be explained by non-interacting TLS and that interactions between TLS lead to a dipole gap in the density of states. He considered the limit of dilute interacting TLS and argued that slowly relaxing pairs of TLS lead to slow, logarithmic relaxation following an abrupt change in the dc field.

2.2. *Universally small phonon scattering*

Not only do glasses exhibit qualitatively universal behavior, measurements also reveal quantitatively universal values. Phonon scattering is an example. We have already mentioned that fitting the thermal conductivity below 1 K leads to the ratio $\ell/\lambda \approx 150$, where ℓ is the phonon mean free path and λ is the phonon wavelength.[13] This is equivalent to the universal value found for the inverse Q factor, sometimes called internal friction, which is a measure of ultrasonic attenuation. Recall that if one mounts a sample on an oscillator, e.g., a double paddle oscillator, and measures the amplitude as a function of frequency, then Q^{-1} is the ratio of the resonant line width to the resonant frequency f_o, i.e., $Q^{-1} = \Delta f / f_o$. Q^{-1} is proportional to λ/ℓ:

$$Q^{-1} = \frac{\lambda}{4\pi\ell} \tag{8}$$

Measurements on a wide variety of glasses find that Q^{-1} lies in the range between 10^{-4} and 10^{-3}, with a typical value of 3×10^{-4}. The universality of this value has been a long-standing puzzle. Why is phonon scattering so insensitive to the composition and structure of different amorphous materials?

Within the context of the standard TLS model, phonons scatter from TLS and the measured quantities can be expressed in terms of C, a dimensionless scattering rate, as follows:

$$C = \frac{\bar{P}\gamma^2}{\rho v^2} \tag{9}$$

where γ is the strength of the TLS–phonon coupling, ρ is the mass density of the glass, and v is the velocity of sound given by

$$\frac{1}{v^3} = \frac{1}{3}\sum_s \frac{1}{v_s^3} \tag{10}$$

where v_s is the sound velocity for polarization s. In Eq. (9), $\bar{P}$ is the same $\bar{P}$ found in Eq. (5) and corresponds to the density of states of TLS that tunnel fast enough to scatter phonons, a small subset of all the TLS that contribute to the specific heat. The TLS density of states, n_o, that enters into the specific heat is typically about ten times larger than $\bar{P}$.

Various measured quantities can be expressed in terms of C. For example,

$$Q^{-1} = \frac{\pi}{2}C \tag{11}$$

The change in sound velocity in the relaxation regime is given by

$$\frac{\Delta v}{v} = -\frac{1}{2} C \ln\left(\frac{T}{T_o}\right) \tag{12}$$

where T_o is an arbitrary reference temperature. In the relaxation regime, phonons perturb the TLS energy-level separation, and as a result, the level population must readjust to a new Boltzmann equilibrium. In terms of C, the ratio of the mean free path to the wavelength can be expressed as

$$\frac{\ell}{\lambda} = \frac{1}{2\pi^2 C} \tag{13}$$

Measurements of these quantities yield values of C between 2×10^{-4} and 5×10^{-4} for a wide variety of amorphous materials. This universal value of C is quite surprising given the large variation from glass to glass of the parameters entering into C in Eq. (9).[50] It is also odd that a dimensionless number should result in a number so far from unity. Fits to experimental measurements on insulating glasses yield TLS–phonon couplings γ on the order of 1 eV,[51] a puzzling energy scale since it does not match any other in the problem.

Yu and Leggett[25] (YL) made the first attempt to explain the universal value of C using a mean field approach in which the interactions between TLS determined the density of states, resulting in

$$\bar{P} \sim \frac{\rho v^2}{\gamma^2} \tag{14}$$

Plugging this in Eq. (9) yields $C \sim 1$ which is universal but of the wrong order of magnitude, though it is consistent with $\ell \sim \lambda$ found at temperatures above the plateau in the thermal conductivity.

Subsequently, a number of other authors have proposed a variety of rather complicated approaches to achieve the right order of magnitude for C.[52–56] Burin and Kagan[52] obtain a renormalized value of $\bar{P}$ in Eq. (5) by considering a dilute system of TLS with two-body interactions, $J_{ij} = g_{ij}/r_{ij}^3$, between TLS at sites i and j. The coupling g_{ij} has a typical value of g_o. As the temperature T decreases, the physics is dominated by increasingly large TLS clusters with constituent pairs that satisfy $T < J_{ij}$. Using their renormalized value of $\bar{P}$, Burin and Kagan find that $C = \bar{P}g_o$ is of order 10^{-3}. Vural and Leggett[53] used a block scaling renormalization group approach in which TLS were replaced by blocks represented by stress tensors that interacted with one another via elastic dipolar interactions. In going to longer length scales by combining smaller blocks to form larger blocks, they find that Q^{-1} vanishes in the thermodynamic limit. They rescue this by postulating an ansatz for the form of Q^{-1}. Lubchenko and Wolynes[54] related their theory of the glass transition as a random first-order (phase) transition (RFOT) to two-level systems at low temperatures. In their RFOT theory, a cluster of cooperatively rearranging particles forms a droplet with a typical volume of ξ^3. At low temperatures, the lowest two energy levels of a droplet constitute a two-level system. Lubchenko and Wolynes[54] argue that

$$C \sim \left(\frac{\xi}{a}\right)^3 \sim \frac{1}{200} \tag{15}$$

where a defines the lattice constant. This is certainly the right order of magnitude, but identifying which particles are in a droplet and which are not, in a molecular dynamics simulation, say, is a challenging proposition.

Schechter and Stamp[56] proposed two types of TLS. One set has weak interactions with phonons due to inversion symmetry, and hence, weak elastic interactions between TLS, while the other type has strong interactions with phonons, and hence, strong elastic interactions between TLS. The ratio of the TLS–phonon coupling constants associated with these two types of interactions naturally gives rise to the right order of magnitude for the universally small value of the ultrasonic attenuation as well as to the small energy scale of 3 K.

Parshin[41] invoked the soft potential model to try to explain the universal value of C. The soft potential model postulates that the excitations in glasses at low temperature are due to quasilocal soft anharmonic oscillators described by quartic polynomials.[57–59] Some of these are double-well potentials that correspond to TLS while others are anharmonic single potential wells. By making assumptions and estimates of various parameters that enter the model, e.g., those that govern the distribution of the coefficients of the fourth-order polynomials describing the anhamonic potentials, Parshin was able to arrive at a small value for C consistent with experiment. Sometime later, Parshin, Schober, and Gurevich[55] considered localized harmonic oscillators coupled to high frequency modes such as phonons. This coupling leads to a vibrational instability that produces quasilocal harmonic vibrations that populate the boson peak as well as double-well potentials that correspond to tunneling two-level systems. In their model, C is a sensitive function of the coupling of the localized harmonic oscillators to the high frequency modes. They point out that only a small fraction of TLS are able to tunnel fast enough to contribute to experimental measurements, and it is this fraction that determines the small value of C. Note that while the aforementioned models have been able to obtain the right order of magnitude for C, their use of a variety of assumptions and estimates have precluded their ability to make quantitative predictions of the value of C for specific glasses.

To avoid the cancellation that Yu and Leggett (YL)[25] found, Carruzzo and Yu[60] treated $\bar{P}$ and $\gamma^2/\rho v^2$ separately in Eq. (8). They revisited the standard model of TLS and exploited aspects that had not been previously appreciated. First, they pointed out that the Pauli matrix σ_z in Eq. (7) means that phonons actually couple to the *difference* between the elastic dipole moments in the two wells of the TLS double-well potential. Since the elastic dipole moment in each well should be randomly oriented, the difference should also be random, varying from TLS to TLS. This leads to a broad distribution of couplings γ, in contrast to the standard TLS model where γ is taken to be a constant of order 1 eV. They find a flat distribution for γ^2 with a maximum value $\gamma_{\text{max}} \sim \rho v^2 v_o \sim 5$ eV, which is much larger than 1 eV. (v_o is the volume of a chemical formula unit.) Like YL, Carruzzo and Yu[60] assumed that TLS–TLS interactions dictate the density of states $\bar{P}$, but since γ_{max} is so large, the resulting value of $\bar{P}$ is much lower, and more realistic, than what

YL found. Carruzzo and Yu also pointed out that the TLS coupling to phonons leads to an exponential reduction of the tunneling matrix element due to phonon overlap between the two wells; this is a kind of polaron effect.[61] The idea is that when a TLS tunnels to one well, the neighboring atoms around it relax to accommodate it, making it harder to tunnel to the other well. Unlike the standard TLS model where tunneling depends on the WKB exponent,[10,11] Carruzzo and Yu found integrating out the high frequency phonons produces a tunneling matrix element that depends exponentially on the coupling:

$$\Delta_o = \Delta_{o,\max} e^{-\gamma^2/\gamma_o^2} \tag{16}$$

where the maximum tunneling amplitude $\Delta_{o,\max} \sim 10\,\mathrm{K}$ and γ_o depends on material parameters:

$$\gamma_o = \frac{\sqrt{2}}{3}\sqrt{\rho v^2 v_o \hbar\omega_D} \tag{17}$$

where ω_D is the Debye frequency. Equation (16) implies that the broad distribution γ^2 will lead to a wide range of tunneling amplitudes. The exponential reduction of the tunneling rate with increasing γ means that TLS with large TLS–phonon couplings will have very small tunneling rates and will not be seen, e.g., in internal friction measurements. So only those TLS that tunnel fast enough to respond on an experimental time scale will contribute to measurements. This leads to an experimentally determined effective coupling $\gamma_{\text{eff}} \sim 1\,\mathrm{eV}$. (This energy scale is set by γ_o given in Eq. (17). Plugging typical values into Eq. (17) yields $\gamma_o \sim 0.5\,\mathrm{eV}$.) Using $\gamma_{\text{eff}} \sim 1\,\mathrm{eV}$ in $\gamma^2/\rho v^2$ in Eq. (9), together with the revised value of $\bar{P}$, leads to the correct order of magnitude for C. Since their theory expressed various quantities in terms of material parameters, Carruzzo and Yu were able to make detailed comparisons of their theory with experimental values of C for a variety of glasses as shown in Table 1. Overall the comparisons were very good with a few exceptions, e.g., LAT, which may be due to the inability to properly estimate v_o, the volume of a formula unit.

Table 1. $C_{\rm th}$ for dielectric glasses are from Ref. [60], except polystyrene (PS). Data from Ref. [51] and references therein. $C_{\rm exp} = (C_L + 2C_T)/3$ except for PS where $C_{\rm exp} = C_L$. C_L and C_T are the longitudinal and transverse components of C, respectively. Note that in Ref. [60], $C_{\rm exp} = C_T$. Experimentally, $C_T \sim C_L$.[51]

Glass	ρ[kg/m^3]	v[m/s]	v_o[Å^3]	$T_{\rm D}$[K]	$C_{\rm exp} \cdot 10^4$	$C_{\rm th} \cdot 10^4$
$aSiO_2$	2200	4163	45.3	348	3.0	2.9
BK7	2510	4195	41.8	360	3.1	2.5
SF4	4780	2481	40.7	215	2.6	0.9
SF57	5510	2327	55.2	182	2.7	0.9
SF59	6260	2131	40.2	185	2.6	1.0
V52	4800	2511	61.1	190	4.6	0.8
BALNA	4280	2569	39.9	224	4.5	1.2
LAT	5250	3105	68.2	226	3.7	0.3
Zn glass	4240	2580	45.9	215	3.4	2.0
PS	1050	1670	167	90	3.6	2.4
PMMA	1180	1762	138.4	101	3.1	2.9

2.3. *Exceptions to the rule*

While the values of the internal friction of most amorphous materials at low temperatures lie in the range of 10^{-4} to 10^{-3}, there are exceptions. For example, amorphous silicon films with 1 at. % H produce Q^{-1} over 200 times smaller than the usual glassy range[62] as shown in the following chapter. This may be the result of stress introduced by the hydrogen impurities. This possibility is consistent with the fact that Q^{-1} in high-stress silicon nitride (Si_3N_4) thin films, which show no long-range order in x-ray diffraction or TEM images, is two to three orders of magnitude lower than the universal range.[63] These two experiments imply that stress can greatly reduce dissipation, perhaps by increasing the height of TLS tunnel barriers, or by reducing the TLS–phonon coupling.[64] A more prosaic reason is dissipation dilution in which the stress produces a higher resonant frequency f_o and thus a lower value of $Q^{-1} = \Delta f/f_o$.[64] More recently, the Hellman group has shown that the substrate temperature during the growth of thin films of amorphous silicon can determine the amount of dissipation, with a higher growth temperature producing a denser film and a lower value of Q^{-1} that is presumably due to fewer TLS.[65] Specific heat measurements support this and find that denser amorphous silicon films have substantially fewer TLS.[66] Queen *et al.* suggested that the TLS reside on microvoids which are more numerous in lower density films. Another example is the ultrastable glass, indomethacin. Thin films of indomethacin have virtually no linear term in the specific heat below 1 K, indicating an absence of TLS.[67] Recent simulations do, in fact, find that as the stability of the glass is increased, the density of TLS decreases.[68]

3. Mysteries about Glasses at Low Temperatures: Open Questions for the Future

While there has been much progress on the low temperature properties of glasses over the past 50 years, mysteries still abound. TLS are ubiquitous, but we still do not know what they are microscopically except in a few cases such as $KBr_xKCN_{(1-x)}$, in which the TLS are cyanide molecules.[69,70] It is also not understood why TLS are found in all sorts of materials which have such a wide variety of structures and chemical compositions. Why are there so many active excitations below 1 K and what picks out the 1 K energy scale? One way to answer these questions is with computer simulations, but this has been difficult because the low density of TLS (∼100 ppm) requires systems with a prohibitively large number of atoms in order to obtain good statistics.[71] However, recently, better algorithms and computers have allowed simulations of large systems that find TLS with energy splittings smaller than 1 K.[68] Most of these TLS are very localized and only involve a few atoms, though some involve collective excitations of a couple hundred atoms.[68]

Are there really only two levels? Presumably there can be more energy levels as proposed in the soft potential model[57–59] and the model of localized oscillators coupled to phonons.[55] However, below 1 K, it makes sense to just consider the lowest energy levels. The only condition is that they not be evenly spaced in energy as required by experimental observations of saturation of the ultrasonic attenuation[27,72] and phonon echoes.[73,74] Saturation refers to the attenuation at a given frequency falling to zero at a given frequency as the acoustic intensity increases. The idea is that only a finite number of slowly relaxing TLS can resonantly absorb the sound. As the number of phonons exceeds the number of these TLS, the excess ultrasound will pass through the sample without being attenuated, i.e., the attenuation "saturates" (even though it is really the TLS that are saturated). Phonon echoes are the ultrasonic analog of spin echoes in nuclear magnetic resonance (NMR). Two external ultrasonic phonon pulses separated by a time τ are followed at time 2τ by a spontaneous pulse, i.e., an echo. As Yu and Leggett[25] pointed out, while this implies that there are long-lived states (lifetime $> 2\tau$) with time-reversal, it does not necessarily imply *two*-level systems since systems with $S \geq 1/2$ can also produce phonon echoes.

Leggett has also pointed out that there is no direct experimental evidence that anything is actually tunneling.[25] However, it is difficult to think of another physical mechanism that produces long relaxation times at low temperatures. Evidence for long lifetimes of excited TLS come from experiments such as phonon echoes[73, 74] and the slow time dependence of the specific heat[75, 76] which is predicted to increase logarithmically with time.[77, 78]

The assumptions of the standard model of TLS lead to a flat density of states that are consistent with a linear specific heat and a quadratic thermal conductivity. However, experimentally, the temperature dependence of the specific heat is slightly superlinear[4] and the thermal conductivity is slightly subquadratic.[3, 25] This implies a mild increase with increasing energy in the density of states of the excitations contributing to the specific heat and scattering phonons that are conducting heat. A phenomenological way to describe this would be to invoke a small energy dependent correction factor, e.g., a logarithmic correction of the form[25]

$$n(E) \sim \frac{n_o}{\ln(E_o/E)} \tag{18}$$

Such a density of states would also resolve the discrepancy between the ultrasonic attenuation and the shift in the sound velocity.[27] The TLS model predicts that the ultrasonic attenuation (inverse phonon mean free path) due to resonant scattering is given by

$$\ell^{-1}(\omega, T) = A_\ell \omega \tanh\left(\frac{\hbar\omega}{2k_B T}\right) \tag{19}$$

where A_ℓ is a constant. Performing a Kramers–Kronig transformation of the sound attenuation in Eq. (19) yields the shift in the low frequency sound velocity, as follows:

$$\frac{\Delta v}{v} = \frac{2vA_v}{\pi} \ln\left(\frac{T}{T_o}\right) \tag{20}$$

where A_v is a constant and T_o is a reference temperature. Equations (19)–(20) are consistent with experiment.[1,2,27] However, the TLS model predicts that $A_v = A_\ell$ while the experiment finds that A_v is about 40% larger than A_ℓ.[27] One way to resolve this discrepancy is with a density of TLS states that increases slightly with increasing energy.[25,27] However, this *ad hoc* resolution to the superlinear specific heat, subquadratic thermal conductivity, and discrepancy between the ultrasonic attenuation and the shift in the velocity of sound begs the question as to the underlying reason for such an energy dependence in the density of states. We should point out that an alternative explanation for the superlinear temperature dependence of the specific heat is a time-dependent specific heat that increases slowly with time.[77,79]

Another puzzle involves the shift in the sound velocity.[2,80] At low temperatures where resonant scattering of phonons from TLS dominates, the standard TLS model predicts that the shift in the sound velocity increases logarithmically in temperature and follows Eq. (20), which we can rewrite in terms of C as

$$\frac{\Delta v}{v} = C \ln\left(\frac{T}{T_o}\right) \tag{21}$$

where T_o is a reference temperature. Equation (21) holds in the regime where $\omega\tau_{1m} \gg 1$. Here τ_{1m} is the minimum time for a TLS to relax by emitting a phonon; it corresponds to the case of a symmetric TLS with zero asymmetry energy. At higher temperatures where $\omega\tau_{1m} \ll 1$, relaxation processes can no longer be ignored and produce a velocity shift that decreases logarithmically with temperature as

$$\frac{\Delta v}{v} = -\frac{3}{2} C \ln\left(\frac{T}{T_o}\right) \tag{22}$$

When this is combined with the resonant contribution in Eq. (21), the result is

$$\left.\frac{\Delta v}{v}\right|_{tot} = \left.\frac{\Delta v}{v}\right|_{res} + \left.\frac{\Delta v}{v}\right|_{rel} = -\frac{1}{2} C \ln\left(\frac{T}{T_o}\right) \tag{23}$$

which we also stated in Eq. (12). So the predicted ratio of the slopes between the low and high temperature velocity shifts is 2: (−1). While the observed velocity shifts are logarithmic in temperature with a positive slope at low temperature and a negative slope at high temperature, the experimentally observed ratio of the slopes is 1: (−1).[80,81] The reason for this discrepancy with the TLS model is not understood, though the interactions between TLS may be a factor.[82]

So far we have not distinguished between longitudinal and transverse phonons, though the longitudinal and transverse sound velocities differ, as do the longitudinal and transverse TLS–phonon couplings. One puzzle is that for any given amorphous material,

$$C_L \sim C_T \sim C \tag{24}$$

where C is given by Eq. (9),

$$C_s = \frac{\bar{P}\gamma_s^2}{\rho v_s^2} \tag{25}$$

and the subscript s denotes L (longitudinal) or T (transverse).[51] There is no good explanation for why the longitudinal and transverse values of C_s should be roughly the same. However, this observation leads to[51]

$$\frac{v_L^2}{v_T^2} \sim \frac{\gamma_L^2}{\gamma_T^2} \sim 2.5 \tag{26}$$

The observed ratio of the sound velocities is not so surprising since elasticity theory in an isotropic, homogeneous medium yields[53]

$$\frac{v_L^2}{v_T^2} = 3 \tag{27}$$

However, the equivalence of C_L and C_T as well as the equivalence of the sound velocity ratio to the ratio of TLS–phonon couplings in Eq. (26) is puzzling.

In short, there are still fundamental gaps in our understanding of glasses at low temperatures despite the impressive record of the model of two-level systems (and its modifications) in explaining the observed universalities in the thermal and acoustic properties of amorphous solids.

We started this chapter by stating that, naively, one would expect glasses and crystals to behave similarly at low temperatures. So why are glasses and crystals so different in their thermal and acoustic properties at low temperatures? The answer lies in the types of low-energy excitations. Long length scales are typically associated with low energies, while short length scales are associated with high energies. Thus, in both crystals and glasses, long wavelength phonons have low energies. However, in amorphous materials, there are also low-energy excitations at short length scales, namely tunneling two-level systems.[68] This may be why Vural and Leggett's renormalization group attempt to relate universal low-energy behavior to long length scales has struggled to gain traction:[53] the unique features of glasses are rooted in excitations localized at short length scales.

4. Future Outlook

Our understanding of the low temperature properties of glasses rests on a phenomenological model. The key ingredients are tunneling and randomness in the parameters. Experiments that probe overall response of glassy systems have uncovered discrepancies with the standard TLS model that we discussed in this chapter. These experiments measure quantities averaged over the TLS in the sample. The underlying physics is described by a model which is akin to a black box. Researchers in the field keep tweaking the black box with the hope of resolving this or that issue. This results in more complex models with diminishing explanatory power. More direct access to the behavior of individual TLS is needed. In this respect, we would be remiss not to mention recent qubit spectroscopy experiments that probe individual TLS and their interactions with strain and electric fields as well as interactions between TLS.[83–87] By probing individual TLS, these experiments will help constrain the ingredients entering the TLS model and its derivatives.

More generally, phenomenological models are a great first step in making sense of the data and posing questions. However, we would like to understand how this phenomenology emerges from broader physical considerations. There are a few fairly obvious clues as to what those considerations should be. The first clue is that the TLS model works for so many different amorphous materials, or, stated differently, that such a wide variety of glassy materials exhibit TLS phenomenology. This tells us that the amorphous nature of the system dictates the low-energy excitations, much as the crystalline nature of other systems dictates their low-energy excitations (phonons). The second clue comes from materials like indomethacin.[67] These ultrastable materials, while glassy, deviate markedly from the standard behavior.[67] Thus, not all amorphous structures are equal. This implies that there is at least one additional parameter needed to characterize amorphicity. Hopefully, a single set of concepts will emerge to provide a characterization that depends only on the state, e.g., structure, of the glass. A promising candidate is glass stability,[88] but other concepts may be useful, e.g., fragility.[89] The third clue is provided by the renormalization group approach of Vural and Leggett.[53] Rather than postulate the existence of TLS, they tried to show that TLS phenomenology emerges from one ingredient common to amorphous systems, the presence of non-elastic stresses. A key assumption in their approach is the idea that the low-energy excitations of tiny blocks of glass would, when assembled into bigger and bigger blocks, generate the observed behavior, leading to the conclusion that low-energy physics is all you need. While appealing from a universality standpoint, the limited success of their approach leads us to conclude that the starting point must be at higher energy, a conclusion supported by the fact that the non-elastic stresses originate at the energy scale of the glass transition. Taken together, these clues tell us that we need to start from the processes happening at the glass transition itself. The two energy regimes are intimately connected. Understanding the low temperature physics of amorphous materials will have to start at the glass transition.

Acknowledgments

This research was supported by the U. S. National Science Foundation (NSF) through the University of Wisconsin Materials Research Science and Engineering Center (DMR-1720415).

References

1. W. A. Phillips (ed.). (1981). *Amorphous Solids: Low-Temperature Properties.* Berlin Heidelberg: Springer-Verlag.
2. S. Hunklinger and A. K. Raychaudhuri. (1986). Thermal and elastic anomalies in glasses at low temperatures. *Prog. Low Temp. Phys.* **9**, 265–344.
3. R. C. Zeller and R. O. Pohl. (1971). Thermal conductivity and specific heat of noncrystalline solids. *Phys. Rev. B.* **4**, 2029–2041.
4. J. C. Lasjaunias, A. Ravex, M. Vandorpe, and S. Hunklinger. (1975). The density of low energy states in vitreous silica: Specific heat and thermal conductivity down to 25 mK. *Solid State Comm.* **17**, 1045–1049.
5. A. A. Antoniou and J. A. Morrison. (1965). Low-temperature heat capacity of vitreous Germania. *J. Appl. Phys.* **36**(6), 1873–1877.
6. J. J. DeYoreo, W. Knaak, M. Meissner, and R. O. Pohl. (1986). Low-temperature properties of crystalline $(KBr)_{\{1-x\}}$ $(KCN)_{\{x\}}$: A model glass. *Phys. Rev. B.* **34**(12), 8828–8842.
7. R. B. Stephens. (1973). Low-temperature specific heat and thermal conductivity of noncrystalline dielectric solids. *Phys. Rev. B.* **8**(6), 2896–2905.
8. R. B. Stephens, G. S. Cieloszyk, and G. L. Salinger. (1972). Thermal conductivity and specific heat of non-crystalline solids: Polystyrene and polymethyl methacrylate. *Phys. Lett. A.* **38A**(3), 215–217.

9. C. L. Choy, R. G. Hunt, and G. L. Salinger. (1970). Specific heat of amorphous polymethyl methacrylate and polystyrene below 4 K. *J. Chem. Phys.* **52**(7), 3629–3633.
10. P. W. Anderson, B. I. Halperin, and C. M. Varma. (1972). Anomalous low-temperature thermal properties of glasses and spin glasses. *Phil. Mag.* **25**(1), 1–9.
11. W. A. Phillips. (1972). Tunneling states in amorphous solids. *J. Low Temp. Phys.* **7**(3–4), 351–360.
12. M. P. Zaitlin and A. C. Anderson. (1975). Phonon thermal transport in noncrystalline materials. *Phys. Rev. B.* **12**(10), 4475–4486.
13. J. F. Freeman and A. C. Anderson. (1986). Thermal conductivity of amorphous solids. *Phys. Rev. B.* **34**(8), 5684–5690.
14. M. P. Zaitlin and A. C. Anderson. (1975). An explanation for the plateau in the thermal conductivity of non-crystalline solids. *Phys. Status Solidi B.* **71**(1), 323–327.
15. C. C. Yu and J. J. Freeman. (1987). Thermal conductivity and specific heat of glasses. *Phys. Rev. B.* **36**(14), 7620–7624.
16. R. H. Stolen. (1970). Raman scattering and infrared absorption from low lying modes in vitreous SiO2, GeO2, and B2O3. *Phys. Chem. Glasses.* **11**, 83.
17. U. Buchenau, N. Nucker, and A. J. Dianoux. (1984). Neutron scattering study of the low-frequency vibrations in vitreous silica. *Phys. Rev. Lett.* **53**(24), 2316–2319.
18. U. Buchenau, M. Prager, N. Nucker, A. J. Dianoux, N. Ahmad, and W. A. Phillips. (1986). Low-frequency modes in vitreous silica. *Phys. Rev. B.* **34**(8), 5665–5673.
19. A. K. Raychaudhuri. (1989). Origin of the plateau in the low-temperature thermal conductivity of silica. *Phys. Rev. B.* **39**(3), 1927–1931.
20. A. Marruzzo, W. Schirmacher, A. Fratalocchi, and G. Ruocco. (2013). Heterogeneous shear elasticity of glasses: The origin of the boson peak. *Sci. Rep.* **3**, 1407 and references therein.
21. D. G. Cahill and R. O. Pohl. (1987). Thermal conductivity of amorphous solids above the plateau. *Phys. Rev. B.* **35**(8), 4067–4073.
22. T. Nakayama and R. L. Orbach. (1999). On the increase of thermal conductivity in glasses above the plateau region. *Physica B.* **263–264**, 261–263.
23. J. Joffrin and A. Levelut. (1975). Virtual phonon exchange in glasses. *J. de Physique.* **36**, 811–822.
24. M. W. Klein, B. Fischer, A. C. Anderson, and P. J. Anthony. (1978). Strain interactions and the low-temperature properties of glasses. *Phys. Rev. B.* **18**(10), 5887–5891.

25. C. C. Yu and A. J. Leggett. (1988). Low temperature properties of amorphous materials: Through a glass darkly. *Comments Cond. Matt. Phys.* **14**(4), 231–251.
26. B. Golding, J. E. Graebner, B. I. Halperin, and R. J. Schutz. (1973). Nonlinear phonon propagation in fused silica below 1 K. *Phys. Rev. Lett.* **30**(6), 223–226.
27. B. Golding, J. E. Graebner, and R. J. Schutz. (1976). Intrinsic decay lengths of quasimonochromatic phonons in a glass below 1 K. *Phys. Rev. B.* **14**(4), 1660–1662.
28. J. L. Black and B. I. Halperin. (1977). Spectral diffusion, phonon echoes, and saturation recovery in glasses at low temperatures. *Phys. Rev. B.* **16**(6), 2879–2895.
29. W. Arnold and S. Hunklinger. (1975). Experimental evidence for the direct interaction between two-level systems in glasses at very low temperatures. *Solid State Comm.* **17**, 883–886.
30. C. C. Yu. (1989). Interacting defect model of glasses: Why do phonons go so far? *Phys. Rev. Lett.* **63**(11), 1160–1163.
31. C. C. Yu. (1992). Phase transitions of interacting elastic defects. *Phys. Rev. Lett.* **69**(19), 2787–2790.
32. M. W. Klein. (1990). Microscopic theory for dilute quadrupole glasses. *Phys. Rev. Lett.* **65**(24), 3017–3020.
33. M. W. Klein. (1992). Theory of the dilute interacting quadrupole glass: Four-orientational tunneling quadrupoles. *Phys. Rev. B.* **45**(10), 5209–5216.
34. M. P. Solf and M. W. Klein. (1992). Perturbational approach to glasslike low-energy excitations in interacting tunneling dipoles and quadrupoles. *Phys. Rev. B.* **46**(13), 8147–8153.
35. M. P. Solf and M. W. Klein. (1993). Glasslike low-energy excitations from pairs of interacting tunneling dipoles and quadrupoles: Exact solutions. *Phys. Rev. B.* **47**(17), 11097–11006.
36. M. P. Solf and M. W. Klein. (1994). Two-level tunneling states and the constant density of states in quadrupolar glasses. *Phys. Rev. B.* **49**(18), 12703–12717.
37. E. R. Grannan, M. Randeria, and J. P. Sethna. (1988). Low-temperature properties of a model glass. *Phys. Rev. Lett.* **60**(14), 1402–1405.
38. E. R. Grannan, M. Randeria, and J. P. Sethna. (1990). Low-temperature properties of a model glass. I. Elastic dipole model. *Phys. Rev. B.* **41**(11), 7784–7798.
39. S. V. Maleev. (1988). Intrinsic dynamics of two-level systems in glasses and the virial expansion for the lsing model in a transverse field. *JETP.* **67**(1), 157–162.

40. A. L. Burin. (1991). Universal properties of glasses in a pseudospin model. *JETP Lett.* **54**, 320–323.
41. D. A. Parshin. (1994). Interactions of soft atomic potentials and universality of low-temperature properties of glasses. *Phys. Rev. B.* **49**(14), 9400–9418.
42. S. N. Coppersmith. (1991). Frustrated interactions and tunneling: Two-level systems in glasses. *Phys. Rev. Lett.* **67**(17), 2315–2318.
43. D. J. Salvino, S. Rogge, B. Tigner, and D. D. Osheroff. (1994). Low temperature AC dielectric response of glasses to high DC electric fields. *Phys. Rev. Lett.* **73**(2), 268–271.
44. A. L. Burin, D. Natelson, D. D. Osheroff, and Y. Kagan. (1998). Interactions between tunneling defects in amorphous solids. In P. Esquinazi (ed.), *Tunneling Systems in Amorphous and Crystalline Solids.* Berlin, Heidelberg: Springer, pp. 223–315.
45. H. M. Carruzzo, E. R. Grannan, and C. C. Yu. (1994). Nonequilibrium dielectric behavior in glasses at low temperatures: Evidence for interacting defects. *Phys. Rev. B.* **50**(10), 6685–6695.
46. S. Kirkpatrick and C. M. Varma. (1978). Anomalous specific heat of dipolar spin glasses and ferroelectrics. *Solid State Comm.* **25**(10), 821–823.
47. S. D. Baranovskii, B. I. Shklovskii, and A. L. Efros. (1980). Elementary excitations in disordered systems with localized electrons. *JETP.* **51**(1), 199–207.
48. A. L. Efros and B. I. Shklovskii. (1985). Coulomb interaction in disordered systems with localized electronic states. In A. L. Efros and M. Pollak (eds.), *Electron-Electron Interactions in Disordered Systems. Modern Problems in Condensed Matter Sciences.* Amsterdam: North Holland, pp. 409–482.
49. A. L. Burin. (1995). Dipole gap effects in low energy excitation spectrum of amorphous solids. Theory for dielectric relaxation. *J. Low Temp. Phys.* **100**(3/4), 309–337.
50. A. J. Leggett. (1991). Amorphous materials at low temperatures: Why are they so similar? *Phys. B.* **169**(1), 322–327.
51. J. F. Berret and M. Meissner. (1988). How universal are the low-temperature acoustic properties of glass. *Z. Phys. B Con. Mat.* **70**(1), 65–72.
52. A. L. Burin and Y. Kagan. (1996). On the nature of the universal properties of amorphous solids. *Phys. Lett. A.* **215**(3–4), 191–196.
53. D. C. Vural and A. J. Leggett. (2011). Universal sound absorption in amorphous solids: A theory of elastically coupled generic blocks. *J. Non-Cryst. Solids.* **357**(19–20), 3528–3537.

54. V. Lubchenko and P. G. Wolynes. (2007). The microscopic quantum theory of low temperature amorphous solids. In S. A. Rice (ed.), *Advances in Chemical Physics. Advances in Chemical Physics.* Chichester: John Wiley and Sons, Ltd., pp. 95–206.
55. D. A. Parshin, H. R. Schober, and V. L. Gurevich. (2007). Vibrational instability, two-level systems, and the boson peak in glasses. *Phys. Rev. B.* **76**(6), 064206.
56. M. Schechter, Stamp PCE. (2013). Inversion symmetric two-level systems and the low-temperature universality in disordered solids. *Phys. Rev. B.* **88**(17), 174202.
57. V. G. Karpov, M. I. Klinger, and F. N. Ignat'ev. (1983). Theory of the low-temperature anomalies in the thermal properties of amorphous structures. *Sov. Phys. JETP.* **57**(2), 439–448.
58. M. A. Il'in, V. G. Karpov, and D. A. Parshin. (1987). Parameters of soft atomic potentials in glasses. *Sov. Phys. JETP.* **65**(1), 165–168.
59. Y. M. Galperin, V. G. Karpov, and V. I. Kozub. (1989). Localized states in glasses. *Adv. Phys.* **38**(6), 669–737.
60. H. M. Carruzzo and C. C. Yu. (2020). Why phonon scattering in glasses is universally small at low temperature. *Phys. Rev. Lett.* **124**, 075902-1-5.
61. K. Kassner and R. Silbey. (1989). Interactions of two-level systems in glasses. *J. Phys.-Condens. Mat.* **1**(28), 4599–4610.
62. X. Liu, J. White, R. O. Pohl, E. Iwanizcko, K. M. Jones, A. H. Mahan *et al.* (1997). Amorphous solid without low energy excitations. *Phys. Rev. Lett.* **78**(23), 4418–4421.
63. D. R. Southworth, R. A. Barton, S. S. Verbridge, B. Ilic, A. D. Fefferman, H. G. Craighead, *et al.* (2009). Stress and silicon nitride: A crack in the universal dissipation of glasses. *Phys. Rev. Lett.* **102**(22), 225503.
64. J. Wu and C. C. Yu. (2011). How stress can reduce dissipation in glasses. *Phys. Rev. B.* **84**(17), 174109.
65. X. Liu, D. R. Queen, T. H. Metcalf, J. E. Karel, and F. Hellman. (2014). Hydrogen-free amorphous silicon with no tunneling states. *Phys. Rev. Lett.* **113**(2), 025503.
66. D. R. Queen, X. Liu, J. Karel, T. H. Metcalf, and F. Hellman. (2013). Excess specific heat in evaporated amorphous silicon. *Phys. Rev. Lett.* **110**(13), 135901.
67. T. Perez-Castaneda, C. Rodriguez-Tinoco, J. Rodriguez-Viejo, and M. A. Ramos. (2014). Suppression of tunneling two-level systems in ultrastable glasses of indomethacin. *Proc. Natl. Acad. Sci. USA.* **111**(31), 11275–11280.

68. D. Khomenko, C. Scalliet, L. Berthier, D. R. Reichman and F. Zamponi. (2020). Depletion of two-level systems in ultrastable computer-generated glasses. *Phys. Rev. Lett.* **124**(22), 225901.
69. J. J. De Yoreo, M. Meissner, R. O. Pohl, J. M. Rowe, J. J. Rush, and S. Susman. (1983). Low-energy excitations in KBr{1-x}KCN{x} in the orientational glass state. *Phys. Rev. Lett.* **51**(12), 1050–1053.
70. D. Moy, J. N. Dobbs, and A. C. Anderson. (1984). Low-temperature properties of the orientational glass KBr{1-x} (KCN){x}. *Phys. Rev. B.* **29**(4), 2160–2167.
71. A. Heuer and R. J. Silbey. (1993). Microscopic description of tunneling systems in a structural model glass. *Phys. Rev. Lett.* **70**(25), 3911–3914.
72. J. E. Graebner, L. C. Allen, B. Golding, and A. B. Kane. (1983). Acoustic saturation in a glass at low temperatures. *Phys. Rev. B.* **27**(6), 3697–3708.
73. B. Golding and J. E. Graebner. (1976). Phonon echoes in glass. *Phys. Rev. Lett.* **37**(13), 852–855.
74. J. E. Graebner and B. Golding. (1979). Phonon echoes in a glass at low temperatures. *Phys. Rev. B.* **19**(2), 964–975.
75. M. Meissner and K. Spitzmann. (1981). Experimental evidence on time-dependent specific heat in vitreous silica. *Phys. Rev. Lett.* **46**(4), 265–268.
76. M. T. Loponen, R. C. Dynes, V. Narayanamurti, and J. P. Garno. (1982). The time dependent specific heat of dielectric glasses. *Physica* $B + C$. **109–110**, 1873–1879.
77. J. L. Black. (1978). Relationship between the time-dependent specific heat and the ultrasonic properties of glasses at low temperatures. *Phys. Rev. B.* **17**(6), 2740–2761.
78. A. Nittke, S. Sahling, and P. Esquinazi. (1998). Heat release in solids. In P. Esquinazi (ed.), *Tunneling Systems in Amorphous and Crystalline Solids.* Berlin: Springer, pp. 9–56.
79. J. Zimmermann and G. Weber. (1981). On the excess specific heat of vitreous silica at low temperatures. *Phys. Lett. A.* **86**(1), 32–34.
80. J. Classen, T. Burkert, C. Enss, and S. Hunklinger. (2000). Anomalous frequency dependence of the internal friction of vitreous silica. *Phys. Rev. Lett.* **84**(10), 2176–2179.
81. C. Enss, C. Bechinger, and M. V. Schickfus. (1990). Dynamics of tunneling states in borosilicate glass. In S. Hunklinger, W. Ludwig and G. Weiss (eds.), *Phonons 89: Third International Conference on Phonon Physics and the Sixth International Conference on Phonon Scattering in Condensed Matter; August 21–25, 1989*; Heidelberg. Singapore: World Scientific, pp. 474–476.

82. C. Enss and S. Hunklinger. (2005). *Low-Temperature Physics.* Berlin: Springer-Verlag, p. 573.
83. G. J. Grabovskij, T. Peichl, J. Lisenfeld, G. Weiss, and A. V. Ustinov. (2012). Strain tuning of individual atomic tunneling systems detected by a superconducting qubit. *Science.* **338**(6104), 232–234.
84. B. Sarabi, A. N. Ramanayaka, A. L. Burin, F. C. Wellstood, and K. D. Osborn. (2016). Projected dipole moments of individual two-level defects extracted using circuit quantum electrodynamics. *Phys. Rev. Lett.* **116**(16), 167002.
85. J. Lisenfeld, A. Bilmes, S. Matityahu, S. Zanker, M. Marthaler, M. Schechter, *et al.* (2016). Decoherence spectroscopy with individual two-level tunneling defects. *Sci. Rep.* **6**(1), 23786.
86. J. Lisenfeld, A. Bilmes, A. Megrant, R. Barends, J. Kelly, P. Klimov, *et al.* (2019). Electric field spectroscopy of material defects in transmon qubits. *NPJ Quantum Inf.* **5**(1), 105.
87. C. Muller, J. H. Cole, and J. Lisenfeld. (2019). Towards understanding two-level-systems in amorphous solids: Insights from quantum circuits. *Rep. Prog. Phys.* **82**(12), 124501.
88. M. D. Ediger. (2017). Perspective: Highly stable vapor-deposited glasses. *J. Chem. Phys.* **147**(21), 210901.
89. C. A. Angell. (1995). Formation of glasses from liquids and biopolymers. *Science.* **267**(5206), 1924–1935.

https://doi.org/10.1142/9781800612587_0005

Chapter 5

Exceptions Leading to a New Theory of Universality: Amorphous Solids without Glassy Properties

M. Molina-Ruiz* and F. Hellman†

Department of Physics,
University of California, Berkeley CA, USA
**manelmolinaruiz@gmail.com*
†*fhellman@berkeley.edu*

The anomalous properties of glasses at low temperature and their seeming universality have been a challenge since first observed 50 years ago. The standard tunneling model describes these anomalous properties through the interaction of elastic and electric fields with tunneling two-level systems. The realization of glasses without glassy properties added more questions than answers: What glasses can be made without two-level systems, and why some and not others? Can we connect the ability to make some two-level free amorphous materials to their structure? The reduction of two-level systems at low temperature has been suggested to be related to ultrastability and perhaps ideality — glasses that show enhanced kinetic and thermodynamic stabilities and low enthalpy and entropy. In this chapter, we present experimental results and discuss the properties that may lead to the formation of two-level systems in glasses, and therefore how they can be controlled.

1. Introduction

In 1948, Walter Kauzmann described the dearth of understanding surrounding the nature of disordered materials. This description is a simple paradox that, more than seventy years later, remains unresolved and can be summed up as follows.[1] Creation of a bulk glass requires supercooling a liquid fast enough for the relaxation time of its constituent atoms or molecules to grow faster than they can arrange themselves into a lattice — leaving these particles in a static, disordered array that is akin to a snapshot of the liquid. Such fast supercooling is deemed quenching; quench a liquid too slowly, and the material will find its global free energy minimum, i.e., the crystalline state. The rate of quenching determines the temperature at which the supercooled liquid falls out of equilibrium, called the glass transition temperature, T_g. Faster quenching results in a glass that is more thermodynamically unstable and more disordered (higher enthalpy, configurational entropy, and T_g) than that produced by slower quenching. Kauzmann showed that if one can slow the quench rate sufficiently without dropping into the crystalline state, below the so-called Kauzmann temperature T_K, one would paradoxically create a glass with configurational entropy lower to that of its crystalline counterpart. To overcome this paradox, there must be a phase transition before the entropy of the liquid falls below the crystal one; this state is known as an *ideal glass*. This paradox has incited deep questions about the nature of glasses and the glass transition which Nobel Laureate Anderson deemed "the deepest and most interesting unsolved problem in solid state theory".[2] An intriguing implication of the Kauzmann Paradox is not only that there are different types of disorders but that they are process dependent and are thus tunable.

In 1988, Stillinger reasoned that an ideal glass is only possible for materials such as polymers and covalently bonded atoms that can form arbitrarily large networks.[3] Ediger and Rodríguez-Viejo, independently, show that organic polymer films which lie very low in the energy landscape can be produced by physical vapor deposition methods.[4–8] By growing these films at specific deposition temperatures, typically $0.8\,T_g$, and low deposition rates, typically below 0.5 Å/s, the molecular mobility at the surface of the growing films allowed the molecules to find local energy minima while avoiding crystallization, statements consistent with Hellman's previous work on various amorphous metals and semiconductors.[9,10] Glasses that show higher thermodynamic stability, higher T_g, and lower fictive temperature T_f, and lower entropy than glasses that exhibit an

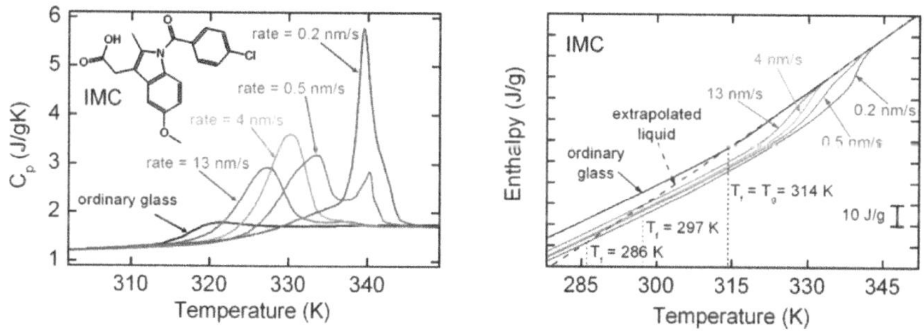

(a) (b)

Fig. 1. From Kearns *et al.*[5] (a) Heat capacity curves observed for vapor-deposited indomethacin (IMC). $T_{\text{substrate}}$ was 265 K (0.84 T_g) for each deposition. Depositions were performed at several rates. Also shown is the ordinary glass C_p curve. (b) Enthalpy curves obtained from integrating the C_p curves in part (a). The temperature at which the extrapolated liquid line (dashed) intersects the enthalpy of the vapor-deposited curves defines T_f as indicated by the dotted vertical lines.

ordinary stability are known as *ultrastable glasses* (Figure 1). The concept of *ideal glass* refers to the *ultrastable glass* with the lowest entropy, or the closest to the crystal entropy. Recently, Samwer demonstrated the first ultrastable metallic glass and showed that the formation of ultrastable glassy materials correlates to the important concept of fragility.[11] The concept of strong and fragile glass formers describes the kinetics on approaching the glass transition from the liquid state. Angell demonstrated a fundamental link between these kinetic properties and the thermodynamic properties of the system, connected through the energy landscape of the structure and the different heat capacities of liquid and solid in fragile and strong liquids.[12–14] Similar data have not yet been taken on inorganic simple materials such as *a*-Si or *a*-Ge.

1.1. *Fragile and strong glasses*

It has been theoretically argued,[15] and observed in attempts to grow ultrastable films of organic materials, that their fragility is a good indicator of the ability to attain ultrastability via vapor deposition techniques. Highly fragile glass formers such as toluene and indomethacin reach high levels of stability, while stronger glass formers, such as alcohols that have high levels of hydrogen bonding and lower surface diffusion, were not successfully grown in an ultrastable state.[16] It is, however, not yet clear whether the difference is the surface mobility or the thermodynamics of the amorphous phase. It is generally true that strong glass formers have T_K much smaller, compared to T_g, than fragile glass formers; indeed, T_K/T_g is related to the thermodynamic fragility. This ratio and the surface mobility argument, whether both are related or not, might be responsible for the difference in vapor deposition success.

2. Anomalous Properties of Glasses at Low Temperatures

The other great long-standing mystery of the glassy state is the nature of the low energy excitations seen in amorphous materials at low temperature, referred to by Nobel Laureate Anthony J. Leggett.[17] Long ago, the low-temperature specific heat of glasses was predicted to converge to the crystalline case; as the phonon wavelength with appreciable non-zero occupation increases with decreasing temperature, the thermodynamic properties should not be affected by a lack of order at the atomic scale. The specific heat should therefore approach the T^3 Debye limit based on the measured sound velocity in these glasses. However, the low-temperature specific heat of amorphous insulating materials over a broad range of compositions displays three components not seen in crystalline materials: a contribution linear in T (c_1), a T^3 contribution in excess of the Debye contribution (c_{ex}), and a bump at relatively low T called the *boson peak*,[18–20] which will be discussed in detail in the following chapters. The linear term is attributed to tunneling or two-level systems (TLSs),[21] which will be further discussed below. The boson peak is thought to arise from non-propagating harmonic modes[22] and is somewhat reminiscent of peaks seen in specific heat of crystalline materials due to the high density of non-propagating zone boundary phonons. The soft-potential model (discussed in detail in Chapter 8) was proposed to account for the low-energy tail of the boson peak, and it therefore provides satisfactory results in several systems with no need for an excess cubic term. In other systems, the soft-potential model does not provide a satisfactory description of the data and an excess cubic term is still needed to reconcile theory and experiment.[23]

Zeller and Pohl in 1971 showed their remarkable plots of "universal" thermodynamic properties in a wide variety of amorphous systems, properties that were strikingly similar to each other and at the time inexplicably different compared to those of their crystalline counterparts.[18] A set of universal properties is seen in specific heat, thermal conductivity, and various spectroscopic studies, despite widely different compositions.[24,25] Fifty years later, we have greatly increased our understanding of these materials and their properties, but we have yet to conclusively explain this "universality", such that in 1991, one of the "great questions" in Condensed Matter and Materials Physics according to Nobel Laureate A. Leggett was as follows: "Why are all amorphous materials at low temperatures so similar?"[17] This "universal" behavior includes an excess specific heat which we will refer to as low energy excitations (LEE); LEE include a linear term (unexpected in an insulating material) and a cubic term that is well above the Debye phonon term calculated from elastic properties. It also includes mechanical losses that lead directly to thermal noise in the mirror coatings used for the detection of gravitational waves, and dielectric losses that are deleterious to systems such as qubits. These properties are typically explained by the standard tunneling model (STM) that are caused by atoms or groups of atoms tunneling between adjacent configurations in the energy landscape.[26,27] By this definition, the *ideal glass* would not have any TLSs since it has no adjacent configurations. The nature of the tunneling entities and the near universal density of TLSs in traditional glasses remain poorly understood, given enormous variations in chemical constituents. Models beyond the STM have been developed, including a many-body strain-mediated model,[28] a thermal activation of TLS model,[29] and a soft-potential model,[30,31] but no proof yet exists that these interpretations are broadly applicable.

Previous work on a-Si shows that the reduction in TLS density correlates with atomic density as it gets closer to the crystal one,[32–34] suggesting a connection between ultrastability and TLS density as also shown for organic glassy films of indomethacin (IMC).[35]

2.1. *Random first-order transition theory*

The random first-order transition (RFOT) theory of the glass transition, put forward by Kirkpatrick, Thirumalai, and Wolynes[36–40] and developed extensively by them and others over the past thirty years,[41,42] provides a compelling argument for the universal behavior of the TLS density (in liquid quenched glasses). The RFOT theory pictures a glass as a mosaic of frustrated domain walls, frozen at the glass transition, that separate dynamically correlated less energetically frustrated regions of material. The size of these regions is nearly universal because it depends on the value of the Lindemann ratio and the RMS fluctuation in atomic position divided by the nearest-neighbor distance, which is nearly universal across many different chemical compositions at a fixed ratio of temperature relative to the glass transition temperature. It has been suggested that the low-temperature TLSs arise from these domain walls, and then the TLS density scales as the inverse of the size of the regions separated by these domain walls.[43] This picture also provides an elegant explanation for why ultrastable glasses would show low TLS density[32–35]: A liquid-quenched ultrastable glass would have reached lower temperature before undergoing the glass transition, implying an increase in the correlation length and larger dynamically correlated regions. Furthermore, it argues for the effectiveness of vapor deposition as a means of creating ultrastable glasses, because the correlation length is greatly enhanced on the surface of the growing film, given high enough substrate temperature.[44–46]

To date, most experimental results showing low-T universal behavior are based on glasses quenched from melt. However, with the advent of thin film growth techniques, amorphous films are made from materials that cannot be quenched from the liquid state, including such technologically important materials as a-Si and a-Si:H. There is evidence, described below, that vapor deposition leads, under certain growth conditions, to amorphous materials with vanishingly small TLSs, and that these materials are ultrastable and likely nearly ideal glasses (to date, only demonstrated directly for indomethacin). This is posited to be because high surface mobility at growth temperatures near $0.8\,T_g$, where bulk mobility is nonexistent, enables the material to reach this low-lying state. It has therefore been hypothesized that "universal" low-temperature glass properties are a consequence of the universal nature of the liquid quenching process through T_g rather than intrinsic to the non-crystalline structure.

2.2. *The standard tunneling model*

Amorphous materials are extensively and widely used for research and technological applications. Nevertheless, the underlying physics that gives rise to the anomalous properties of disordered solids at low temperatures is not completely understood. The development of new technologies that operate at low temperature, such as quantum computing, demands a better understanding of the processes behind these anomalous properties. The STM, developed and improved since 1972,[26,27] presents a theoretical and mathematical framework that explains the observed anomalous properties at low temperature, caused when tunneling two-level systems interact with propagating phonons throughout the solid.

In the STM, the TLS density $\bar{P}$ is derived from the TLS distribution function $f(\Delta, \lambda)$, which is only a function of the asymmetry between states Δ and the tunneling parameter λ. The main assumptions of the STM are that Δ and λ are independent of each other and uniformly distributed, therefore $f(\Delta, \lambda) \approx \bar{P}$, where $\bar{P}$ is an energy-independent parameter. The STM provides tools to quantify $\bar{P}$, which can be obtained by acoustic and specific heat measurements. Acoustic measurements provide insight on both the imaginary and real parts of the solid elastic properties, i.e., mechanical loss (attenuation) and sound velocity (dispersion). The STM predicts that $\bar{P}$ is proportional to the mechanical loss when coupled to phonons via elastic fields.[21] At acoustic frequencies in the kHz frequency range, relaxation dominates over resonant interactions above ~70 mK. When the minimum TLS relaxation time $\tau_{\min} \ll \omega^{-1}$, all TLSs equilibrate within the timescale of an oscillation, and the mechanical loss $Q^{-1}(T)$ is temperature-independent, typically at temperatures from 0.1 to 10 K, and produces a plateau in $Q^{-1}(T)$. The STM predicts that the mechanical loss at this plateau, Q_0^{-1}, is given by

$$Q_0^{-1} = \frac{\pi}{2} \frac{\bar{P}\gamma_\alpha^2}{\rho v_\alpha^2} \tag{1}$$

where α indicates the wave polarization (longitudinal or transverse), γ_α is the coupling constant between TLS and phonons, ρ the mass density, and v_α the sound velocity. The interaction between a phonon and a TLS is given by the deformation potential, $\gamma_i = \partial\Delta_i/2\partial u$, where u is the strain tensor. γ_i, thus, depends on the specific TLS. The STM, however, assumes that all TLSs in a sample have the same deformation potential value, the coupling constant γ_α. For traditional glasses, Q_0^{-1} (empirically) ranges from 10^{-4} to 10^{-3} and varies little with differing chemical compositions.[25] Recently, Carruzzo and Yu proposed a model that considers self-interacting TLSs and a wide distribution of deformation potentials (see Chapter 4).[47] This model replaces γ by an effective quantity, γ_{eff}, due to the reduction in tunneling caused by phonon overlap between TLS wells, and explains why Q_0^{-1} falls in a universal range of small values known as the "glassy range". Measurements obtained using superconducting qubits are more consistent with the predictions of this model than with those made by the STM.[48]

In disordered solids and typically below 10 K, other contributions to $Q^{-1}(T)$ are negligible at this temperature regime, giving us access to a direct measurement of the TLS density $\bar{P}$ through $Q^{-1}(T)$ (Eq. (1)), assuming γ_α and v_α are known. γ_α, however, is obtained from acoustic properties measured at very low temperatures and has been determined for only a few disordered solids.[49]

The specific heat $C_P(T)$ linear term c_1T is associated with the density of TLS n_0 of the TLSs that equilibrate with the phonon bath on the measurement time scale. Therefore, n_0 is proportional to $\bar{P}$. The TLS density derived from specific heat measurements n_0

first proposed by Black and Halperin,[50,51] and later experimentally verified,[52,53]

$$\mathrm{n}_0 = \frac{1}{2} ln \left(\frac{4t}{\tau_{\min}} \right) \bar{P} \tag{2}$$

where t is the experimental measurement time and $\tau_{\min}$ is the TLS minimum relaxation time that in the dominant phonon approximation[51] can be expressed as

$$\tau_{\min}^{-1} = \frac{8\mathrm{k}_\mathrm{B}^2}{\pi \rho \hbar^4} \left(\frac{\gamma_l^2}{v_l^5} + \frac{2\gamma_t^2}{v_t^5} \right) T^3 \tag{3}$$

and

$$\mathrm{n}_0 = \frac{6}{\pi^2 k_B^2} \frac{n_{at}}{N_A} c_1 \tag{4}$$

where T is the temperature at which γ_α and v_α are measured, n_{at} is the atomic density, k_B Boltzmann's constant, and N_A Avogadro's number. Typically, Eq. (2) can be written as $n_0 \sim 10\bar{P}$.[54,55]

The STM also predicts that the dielectric loss $\tan \delta_0$ is proportional to $\bar{P}$, where TLSs couple to electric fields via electric dipole moments.[21] Dielectric loss is frequently measured using superconducting resonators or qubits at GHz frequencies and mK temperatures to avoid TLS saturation[56–58]; at ~5 GHz and below ~120 mK, resonant interactions are the dominant loss mechanisms. In the resonant regime, and when $\hbar\omega = k_B T$, the dielectric loss is temperature-independent and proportional to $\bar{P}$ as

$$\tan \delta_0 = \pi \frac{\bar{P} p_0^2}{3\epsilon} \tag{5}$$

where p_0 is the electric dipole moment of the TLSs, and the permittivity $\epsilon = \epsilon_0 \epsilon_r$. Note the similarity of Eqs. (1) and (5), and the importance of the parameters γ and p_0, which are not well known and not easily measured. Note further that while both are commonly taken as a single value, they are each in fact properties of a specific TLS, and hence likely to be distributions.

While $\bar{P}$ appears in the expressions of all three measurements (specific heat, mechanical loss, and dielectric loss), as well as in other standard measures of the low-temperature glassy properties (thermal conductivity), there are few to no measurements of these different measures of TLS on the same samples to permit a quantitative comparison to be made.[50,51,59]

The STM describes how each TLS interacts with phonons through the deformation potential, γ_i, or dipole moment, $p_{0,i}$. Mathematically, it then assumes that their distributions, $f(\gamma_i)$ and $f(p_{0,i})$, are Dirac delta functions that yield a unique value for all TLSs, in this case either γ or p_0. This leads to the prediction in Eqs. (1) and (5), which is then often viewed as a phenomenological model for the mechanical and dielectric losses within the STM, where the individual distribution values cannot be accessed. These assumptions, if not accurate, may lead to wrong interpretations of the STM predictions.

3. Amorphous Solids without Glassy Properties

The nature of defects in the amorphous state is a thermodynamically simple concept, even if they are not structurally obvious: They are local or semi-local arrangements of atoms or molecules that, because of the many places they could occur and arrangements they could take, add energy and entropy to the ideal glass. That the ideal glass, an intrinsically high-T high-energy scale description related to the glass transition thermodynamics, should be directly connected with low-T TLSs, an intrinsically low-energy scale description, is a remarkable hypothesis, and far outside the purview of models for either set of "universal" glass properties. The connection between thermodynamic and quantum properties, i.e., that ultrastable glasses show a depletion of TLSs, is not yet clear nor established. Amorphous silicon is the first and to date only inorganic dielectric that has shown a drastic reduction (two orders of magnitude) in the TLS density as a function of growth conditions,[33] as shown in Figure 2. The temperature-independent mechanical loss Q_0^{-1} of a-Si shows a similar behavior with growth temperature (Figure 3). Lower TLS densities correlate extremely well with higher atomic densities. It has not yet been proven thermodynamically that a-Si is an ultrastable

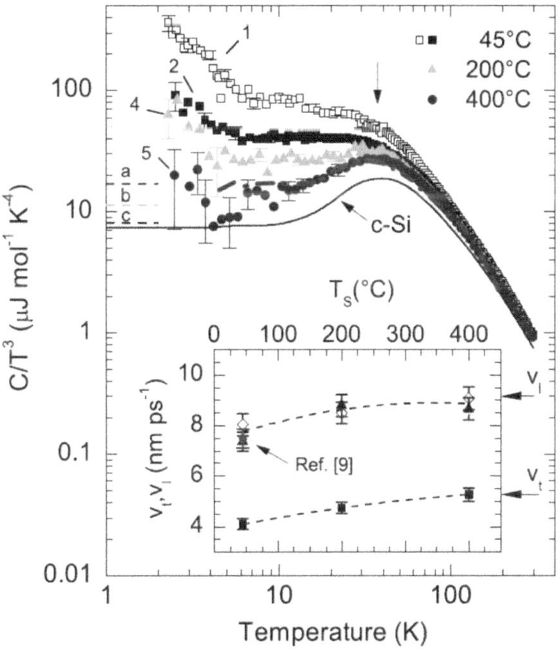

Fig. 2. From Queen *et al.*[33] Specific heat of a-Si plotted as C/T^3 versus T for several growth temperatures and thicknesses. (Samples identified by number.) Horizontal dashed lines show low T Debye specific heat for Debye temperature θ_D = (a) 479, (b) 560, and (c) 611 K (corresponding or near to samples 1–5). Data for crystalline silicon (solid line) with $\theta_D = 645$ K. Vertical arrow identifies the location of what appears to be a boson peak for a-Si; notably close to the temperature of the well-known van Hove singularity of c-Si found at 35 K. Inset: v_t (solid squares) for thick and v_l for thick (solid triangles) and thin (open diamonds) films vs growth temperature. Arrows indicate polycrystalline average: $v_l(v_t)$ = 9.14(5.28) nm/ps.

glass, since its glass transition is intrinsically difficult to access. While a variety of vapor-deposited organic glasses have proven to be grown in an ultrastable state, a vanishing TLS density (low-temperature glassy properties) has only been measured on such material (IMC), where a correlation between stability and TLS density was seen.[35]

Hellman has proposed that TLS density is reduced in low-enthalpy and low-entropy glasses because there are no low-lying nearby energy states; a state with similarly low energy would require a major rearrangement of the structure, and hence no

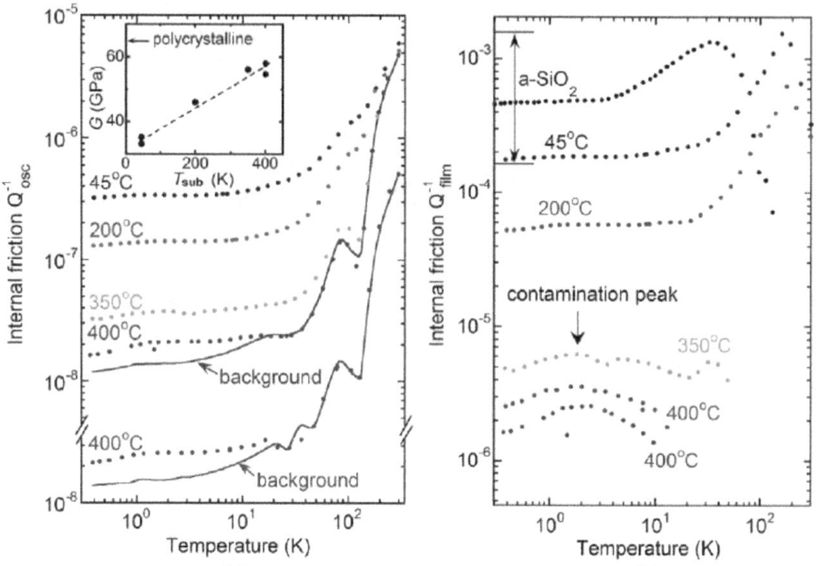

Fig. 3. From Liu *et al.*[34] (a) Mechanical loss of the oscillator $Q^{-1}_{\rm osc}$ with the a-Si thin film. The inset shows the film shear modulus ($G_{\rm film}$) vs growth temperature ($T_{\rm sub}$). The dashed line is a guide to the eye. (b) Sample mechanical loss $Q^{-1}_{\rm film}$, and a 107-nm-thick dry thermal oxide for comparison. The label "contamination peak" denotes that a large portion of $Q^{-1}_{\rm film}$ for films with $T_{\rm sub} = 350°$C and 400°C includes contamination induced loss in the substrate, not related to the a-Si films. The double arrow denotes the "glassy range".

tunneling between such states would be possible.[60–64] This suggests a connection between TLS formation and the liquid quenching process, specifically the RFOT theory of the development of ordered domain formation with disordered boundaries. It is this connection that IMC and a-Si films evince, and if shown to be true, would profoundly alter our description of the glassy state.

A remarkable additional fact was first noted in Ref. [31] for which we now have even more clear-cut evidence: There is not only a strong dependence of atomic density and TLS density on growth temperature and growth rate but also a strong dependence of these densities on film thickness. Moreover, it is not simply that thicker films have lower TLS density and higher atomic density than thinner films, they have lower TLSs on an absolute scale than a thin film.

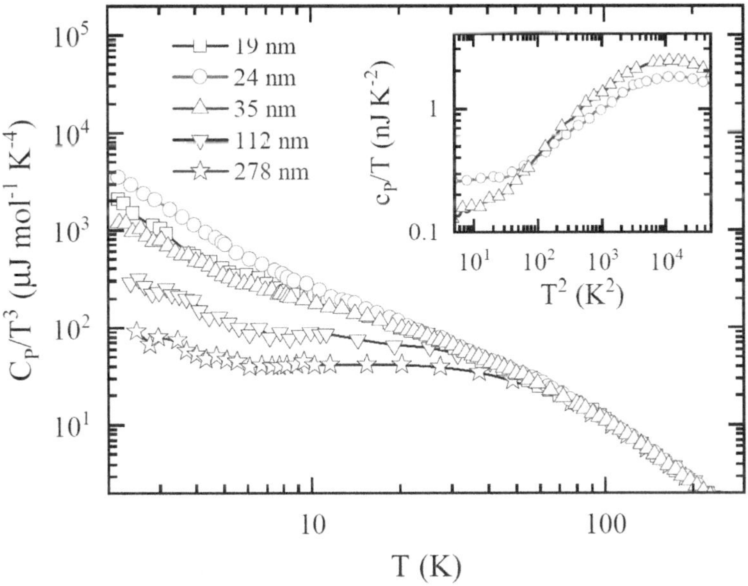

Fig. 4. Specific heat scaled as C_p/T^3 versus T of a-Si thin films grown at room-temperature with thicknesses of 19 nm (squares), 24 nm (circles), 35 nm (upward-pointing triangles), 112 nm (downward-pointing triangles), and 278 nm (stars). Data for 112-nm- and 278-nm-thick a-Si films is are reported in Ref. [34]. Inset: Heat capacity C_p the 24-nm- and 35-nm-thin a-Si films shown as C_p/T vs T^2. C_p for the 24-nm film (light grey circles) is higher than for the 35-nm film (darker grey triangles) at low temperature, where C_p is dominated by TLSs, whereas it is lower at high temperature, where C_p is dominated by vibrational states.

Evidence for this is shown in Figure 4. The low-T heat capacity of an ~100-nm-thick film was shown to be greater than that of an ~300-nm film of the same area (in absolute numbers, not per volume or thickness).[33] These results indicate that the 300-nm film cannot be simply the 100-nm film plus another 200 nm of a-Si. This surprising result has been reproduced in more heat capacity data of thinner a-Si films (Figure 4). In Figure 4 inset, a thinner film (24 nm) shows a larger linear heat capacity term, proportional to the TLS density, than a thicker film (35 nm), which confirms that the TLSs are suppressed as films grow thicker.

It has also been shown in a-Si that sound velocity increases with increasing growth temperature (see Figure 2 inset), while its dependence on thickness is much less, within error bars of no dependence.[33,65,66] These observations lead to the conclusion that elastic properties do not directly affect the mechanisms that remove TLSs from a-Si.

These data offer an opportunity to identify the structures that promote or eliminate the formation of TLSs in a-Si. Organic molecules, such as IMC, are difficult to structurally characterize due to their structural complexity. Simpler inorganic systems, such as amorphous silica (a-SiO_2), have been extensively studied.[67–71] However, the structural complexity of even these diatomic systems has limited identification of the structural features responsible for TLSs. Silicon is a simpler system; it is monoatomic, covalently bonded, and tetrahedral. The first evidence of the strong relationship between structure and anomalous properties in a-Si was provided by Hellman and coworkers who established a strong correlation between atomic density and TLS density for films grown at different temperatures. We therefore turn now to a detailed set of structural measurements on a-Si.

3.1. *Connection between TLS and structure*

The structure of disordered solids requires a variety of perspectives: (i) the arrangement of the atoms, which is described by short- and medium-range order metrics, including nearest-neighbor distances, bond angles, and ring structures, and (ii) defects, such as nanovoids and broken bonds or unpaired electrons, i.e., floppy bonds or dangling bonds. The atomic density of electron-beam evaporated a-Si depends on both, although dominated by defects.[66] Figure 5 shows that atomic density of a-Si increases systematically and monotonically with thickness and growth temperature and reduces with growth rate. This plot was critical to understanding the dependence of TLSs on both growth temperature and thickness.

The position and width of the Raman transverse-optic peak provides information of the local strain $tr(\epsilon)$ and the average bond angle deviation $\Delta\theta$, i.e., the atomic disorder of the amorphous structure compared to its crystalline counterpart.[72,73] Dangling bond density was determined by electron paramagnetic resonance

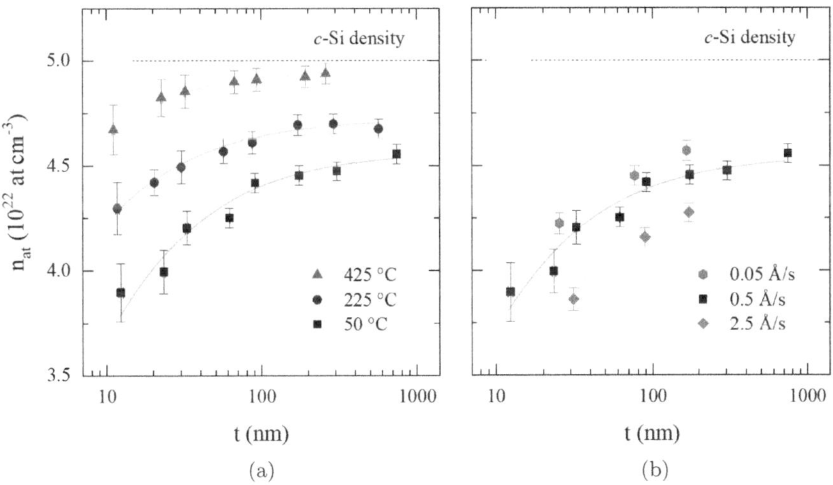

Fig. 5. From Jacks *et al.*[66] Atomic density n_{at} of a-Si thin films as a function of thickness t for (a) growth temperatures of 50°C (squares), 225°C (circles), and 425°C (triangles), and (b) growth rates of 0.05 Å/s (hexagons), 0.5 Å/s (squares), and 2.5 Å/s (diamonds). In (a), all films are grown at 0.5 Å/s, and in (b), all films are grown at 50°C. In both plots, the density of c-Si is given for reference (dashed horizontal line), and solid lines are guides to the eye.

measurements, which show a clear Landé g-factor of 2.0055, evidence of the classic dangling bond in silicon. Figure 6 shows that bond angle deviation, local strain, and dangling bond density decrease with increasing growth temperature, whereas they change less, and non-monotonically, with thickness, unlike the atomic density shown in Figure 5. Figures 6(a) and 6(b) suggest that a relaxation process occurs between 30 and 50 nm. Detailed high- and low-resolution transmission electron microscopy, fluctuation electron microscopy, and plasmon peak shift measurement studies on 30-nm-thick films show no difference in bond lengths or local structure with growth temperature, hence not explaining the atomic density differences seen in Figure 5; they do show differences in medium-range order, but these differences are not large enough to explain Figure 5 data on atomic density.[66] Dangling bond density has also been measured for these films and found to be $\sim 10^{18}$ cm^{-3}, as expected for a-Si grown without hydrogen. This number depends on growth

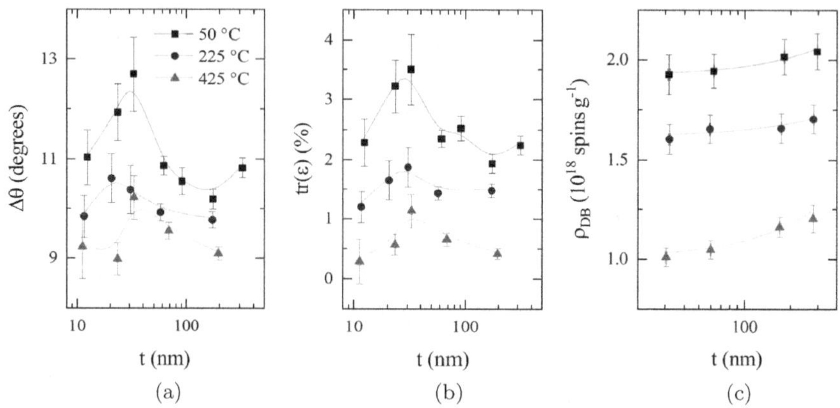

Fig. 6. From Jacks *et al.*[66] (a) Bond angle deviation $\Delta\theta$, (b) local strain $tr(\epsilon)$, and (c) dangling bond density ρ_{DB} as a function of thickness t of a-Si films grown at 50°C (squares), 225°C (circles), and 425°C (triangles). Lines are guides to the eye. Films grown at higher temperature, for all thicknesses, have less bond angle disorder, less local strain, and lower dangling bond density. There is additionally a dependence of these parameters on film thickness, with a peak seen ~30–50 nm that suggests a structural reorganization occurs at that thickness.

temperature somewhat strongly (factors of two to three, but not orders of magnitude), and depends only weakly on film thickness, as shown in Figure 6(c).

The low atomic density values measured for the a-Si films grown at room temperature, particularly those of the thinnest films (Figure 5), cannot solely be explained by atomic disorder or local strain (Figure 6). Nor can this be explained by missing bonds, measured by EPR characterization of dangling bonds, which are of order $10^{18}\,\mathrm{cm}^{-3}$.[66] The low atomic density is instead found to correlate with the presence of nanovoids, a common defect in amorphous materials, even though not commonly measured. Nanovoids, i.e., empty volume, in evaporated a-Si films were characterized by Doppler broadening positron annihilation spectroscopy.[66] These observations in combination with the previously discussed structural characterization and specific heat results, and therefore the strong correlation between TLSs and atomic density, suggest that nanovoids, most likely their surfaces, are the source of TLSs in a-Si.

The structural changes observed when films' thickness increases, summarized in the plots of atomic density vs thickness in Figure 5 and seen to some extent in Figure 6, raise the question of whether these changes occur continuously through the thickness of the film as it grows, or if the film reconstructs throughout its thickness as it grows. To put this in simple terms, a film that is 100 nm thick grown at room temperature has a density of 4.4 g/cm^3, while a film that is 30 nm thick has a density of 4.2 g/cm^3; does the thicker film consist of a low-density layer at the bottom, which gradually becomes denser through its thickness, or is the thicker film denser throughout? While the former model may seem more intuitively obvious, it is not what is experimentally found in etched thicker films. Several thick films, 300 nm thick, were etched using an isotropic KOH solution and measured at progressively thinner thicknesses. If the films' morphology and atomic distribution are constant through the film, then the etched film would have the same density and other properties as the as-deposited thicker film, which would NOT be the same as the as-grown thin film and would indicate that the bottom layer of the thick film undergoes a structural *reorganization* as the film grows thicker.

Atomic density and bond angle deviation of several etched a-Si films are compared to as-deposited films (Figure 7). Atomic density and bond angle deviation (Figures 7(a) and 7(b), respectively) show some degree of reorganization of the early layers of the films. These results confirm that thicker a-Si films are not just more ordered because disorder decreases as the film grows thicker, but because disordered layers near the substrate reorganize during growth.

This reorganization was first suggested based solely on heat capacity data,[33] but the results discussed in Figure 7 give structural evidence that it occurs, supporting the heat capacity data. Notably, Floro *et al.*[74] showed that evaporated a-Si grows under compressive stress up to a given thickness, which depends on growth temperature (e.g., 50 nm at 100°C), above which the film relaxes. While the work by Floro *et al.* suggests a critical thickness and does not speak to reorganization of the bottom layers, we note that in epitaxial films, stress-induced reconstruction is common where defects that permit

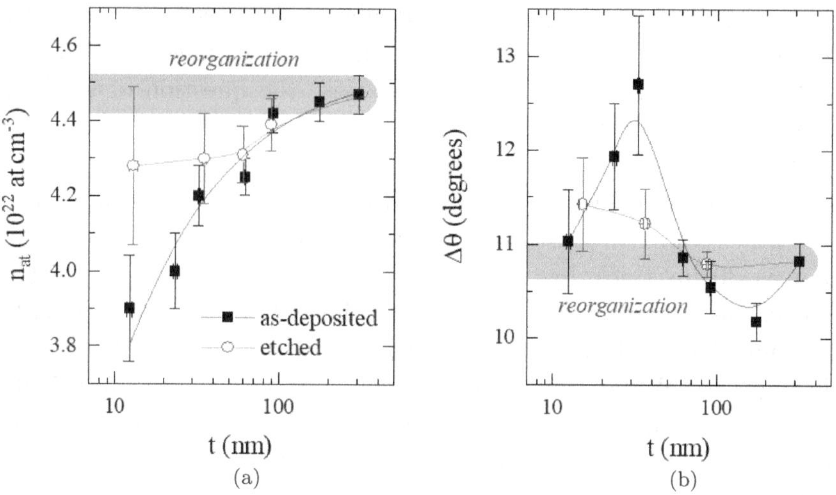

Fig. 7. From Jacks *et al.*[66] (a) Atomic density n_{at} and (b) bond angle deviation $\Delta\theta$ as a function of thickness t for as-deposited (closed squares) and etched (open circles) samples. All films were grown at 50°C and 0.5 Å/s; films etched were ~300 nm thick. Lines are guides to the eye. Gray areas labeled *reorganization* show the extrapolated atomic density for low thicknesses found for the ~300-nm films prior to etching.

the relaxation usually form at the film/substrate interface, causing a thick film to have a uniform structure that is different from the thin film it started as. A similar effect to the epitaxial reconstruction may be expected as well in amorphous films.

These results suggest the explanation for why TLSs in amorphous silicon are suppressed in thicker films, as seen in Figure 4; films relax and reorganize during growth, which reduces the density of nanovoids and in turn leads to the low TLSs of thick films. That sound velocity depends on growth temperature and far less on thickness (and hence does not correlate directly with atomic density) suggests that sound velocity depends on bond distances and bond angles and is not strongly affected by nanovoids. The overconstrained nature of the a-Si network likely precludes the existence of TLSs *within* the network; instead, TLSs exist in more loosely bonded regions, specifically, in the surfaces of nanovoids and their surroundings.

4. Universal and Non-universal Values of TLSs

Results presented in Section 3 show that the number of TLSs in amorphous silicon films may be varied systematically over several orders of magnitude, between vanishingly small and large, simply by changing growth parameters. The observed reduction of nanovoids as films grow thicker, at higher temperatures or slower growth rate, leads to the reduction of TLSs in a-Si films. Two additional questions arise at this point that can be tested by comparing a-Si grown under different conditions: How accurate are the assumptions and consequently the predictions of the STM? And how do the TLSs interact with different applied fields (elastic and electric)? In this Section, results that help to understand the nature of TLSs and their interaction mechanisms with the environment are discussed.

4.1. *Interaction of TLSs with elastic fields: The coupling constant*

During the years, amorphous silicon prepared by different techniques, including both physical and chemical vapor deposition techniques, and in different ways (as-deposited, annealed, hydrogenated, ion-implanted, etc.) has been studied by either specific heat[23,33] or mechanical loss[34,75,76] measurements. Because growth techniques[76] and preparation conditions[75] affect the measured TLS density of amorphous silicon, electron-beam evaporated films grown under ultra-high vacuum conditions with carefully controlled growth parameters have been measured by specific heat and mechanical loss and used to demonstrate trends that are in agreement with the predictions made by the STM.[65] Specifically, in that early work, the TLSs as measured by specific heat and mechanical loss are proportional to each other, albeit analyzed with the coupling constant of germanium (0.3 eV), which later was shown to be three times smaller than the experimental value of the coupling constant of silicon (~1 eV).[77] This correction modified the original reported ratio of $n_0/\bar{P} \sim 10$, derived from Eq. (2), to $n_0/\bar{P} \sim 100$, showing that, despite the proportionality, a-Si deviates strongly from this prediction of the STM.

The study of hydrogenated amorphous silicon (a-Si:H) has further questioned the predictions made by the STM, and therefore the understanding of the low-temperature physics of disordered solids. Films of a-Si:H grown by the hot-wire chemical vapor deposition (HWCVD) technique showed very low mechanical loss values, of order 10^{-6}, compatible with low TLS densities,[75] whereas specific heat measurements on similar a-Si:H films reported large TLS densities.[23] The value of $n_0/\bar{P}$ calculated from the ratio of these measurements yields a value of $>10^3$, even further outside plausible ratios expected from the STM.

This apparent contradiction has been further studied by comparing mechanical loss and specific heat measurements of a-Si:H films grown by HWCVD before and after annealing.[78] A film of a-Si:H grown at 370°C with 7 at.% H was simultaneously grown on a double-paddle oscillator, for mechanical loss measurements, and on a nanocalorimeter, for specific heat measurements. Both devices were measured, then annealed at 200°C (well below growth temperature, where H is known to diffuse but does not leave the film) and measured again in the annealed state. The mechanical loss of a-Si:H, which is proportional to the TLS density (see Eq. (1)), shows a very small (8%) increase between the as-prepared and the annealed states (see Figure 8). The specific heat, however, shows a very large reduction of the linear term at low temperatures (see Figure 9), which is proportional to the density of TLSs that equilibrates within the measurement timescale n_0 (see Eq. (4)), therefore, proportional to the TLS density $\bar{P}$ (see Eq. (2)). Mechanical loss results thus suggest that $\bar{P}$ does not change upon annealing, whereas specific heat results suggest otherwise. What is the origin of this apparent contradiction?

The TLS density $\bar{P}$ of a-Si:H films can be derived from the data presented in Figures 8 and 9, specifically from the specific heat values of n_0 using Eq. (2), along with the experimental value of $t = 10\,\mathrm{ms}$, and $\tau_{\min} \approx 1.25\,\mathrm{ns}$, obtained from Eq. (3) with $T = 2\,\mathrm{K}$, the lowest temperature of the specific heat measurements. This approach yields $n_0/\bar{P} \approx 9$, and therefore $\bar{P} \approx 2.4 \times 10^{47}\,\mathrm{J}^{-1}\,\mathrm{m}^{-3}$ before annealing, and $\bar{P} \approx 3.4 \times 10^{46}\,\mathrm{J}^{-1}\,\mathrm{m}^{-3}$ after. We note again that the logarithm

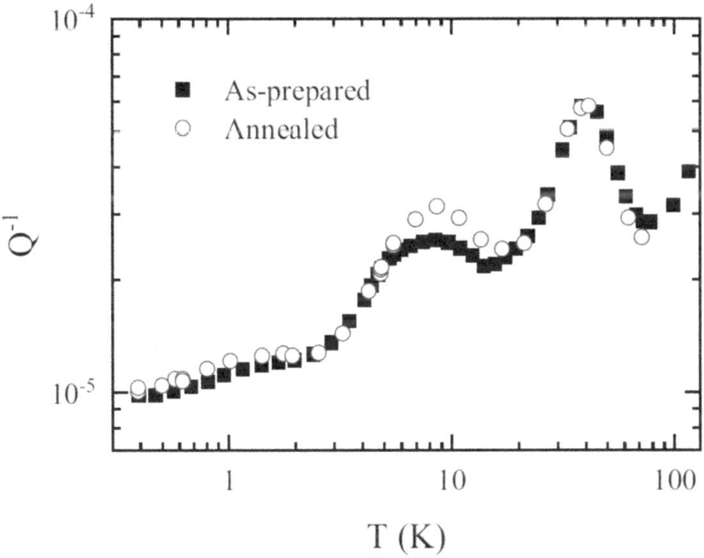

Fig. 8. From Molina-Ruiz *et al.*[78] Mechanical loss Q^{-1} as a function of temperature T for an a-Si:H film grown at $T_S = 370°C$ (7 at. % H) in its as-prepared (solid squares) and annealed (open circles) states. The loss difference between the bare DPO and the DPO with the film is of almost one order of magnitude, which ensures the accuracy of the measurement.

in Eq. (2) makes the choice of $\tau_{\min}$ not important (three orders of magnitude change in $\tau_{\min}$ modifies $n_0/\bar{P}$ less than twofold).

The experimental values of Q_0^{-1} along with the values of $\bar{P}$ calculated above, and combined in Eq. (1), yield $\gamma = 7\,\text{meV}$ and $19\,\text{meV}$ for the as-prepared and annealed states, respectively. These results, however, lead to the non-intuitive conclusion that both $\bar{P}$ and γ must change upon annealing in such a way that Q_0^{-1} barely changes. The usual approach of assuming a value for γ, e.g., $1\,\text{eV}$, leads to a value of $\bar{P}$ from Q_0^{-1} that is orders of magnitude less than that found from $C_P(T)$ measurements, i.e., to a decoupling between propagating acoustic waves from TLSs, and thus not explaining why the large $\bar{P}$ found in specific heat does not cause losses (an equivalent statement to the low γ derived on comparing the two measurements, and to the quite unphysical values of $n_0/\bar{P} = 10^3$ and 10^4 in the as-prepared and annealed states, respectively).

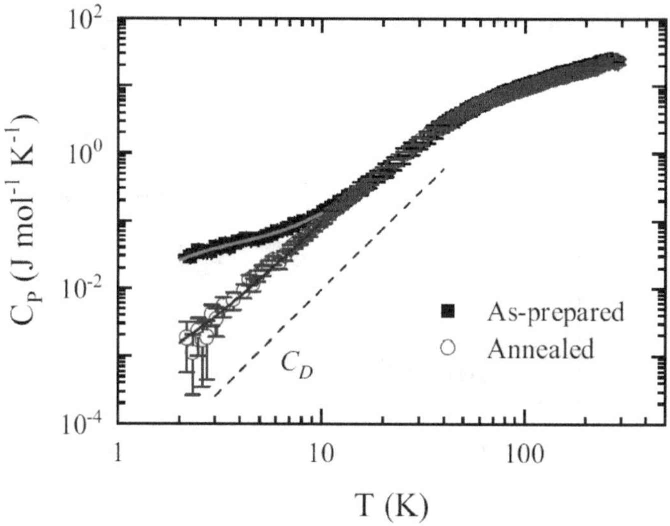

Fig. 9. From Molina-Ruiz *et al.*[78] Specific heat of a-Si:H film grown at T_S = 370°C (7 at. % H) in the as-prepared (closed squares) and annealed (open circles) states. The Debye specific heat (dashed line), $C_D = C_D T^3$, is calculated from the measured sound velocity v. The solid lines show the fitting to the as-prepared and annealed states (see Ref. [78] for fitting details).

Very few studies have been made that estimate γ,[49, 54, 77, 79–81] which typically ranges from 0.1 to 1.5 eV and has not been measured for a-Si:H. In the literature, γ is considered to be a material constant. While the values of γ reported above are unexpectedly low, there is nothing in the STM that prevents such low values. These low γ values are on the order of those reported in H-Nb systems,[80, 81] suggesting that H atoms might be responsible for the tunneling states in a-Si:H. The increased value of γ, obtained for the annealed state, and the fact that the longitudinal sound velocity does not change upon annealing imply that upon annealing the minimum relaxation time decreases significantly (7-fold) since $\tau_{\min} \propto v^5/\gamma^2$ (see Eq. (3)). It is remarkable that Q_0^{-1}, proportional to $\bar{P}\gamma^2$, increases very slightly after annealing, whereas n_0 and thus $\bar{P}$ decrease noticeably. This apparent contradiction is mathematically solved if γ increases after annealing. The fact that sound velocity does not change upon annealing, however, suggests that the silicon structure is not altered, so it is difficult to then see why γ would change. However, if the structural rearrangement of hydrogen atoms leads, as measured by $C_P(T)$, to a reduction of TLSs that turn out to be those with the lowest γ_i values, then γ should increase. In Section 2.2, we suggested that the coupling constant γ could be the average value of the distribution of coupling constants $f(\gamma_i)$. This hypothesis explains why a reduction in $\bar{P}$ upon annealing would correlate with an increase in γ, or vice versa.

Previous work has found that large variations in specific heat are often not associated with changes in thermal conductivity.[24,82] A similar inconsistency is found in $C_P(T)$ and $Q^{-1}(T)$ measurements of vapor deposited amorphous silicon nitride, notably a material that contains of order 1–2 at.% H.[83,84] An alternative approach to the decoupling between specific heat and acoustic properties results was proposed by Black,[51] in which he posited that samples may contain "anomalous" TLSs, besides "standard" TLSs. Black argued that anomalous TLSs relax fast enough to be measured by specific heat techniques, but too slow to be detected by acoustic properties techniques. In this approach, Black considered an additional TLS density for the anomalous states that would only contribute to the specific heat measurements. Despite the addition of anomalous TLSs, which to some extent improved the agreement between the STM and the experimental results, Black could not satisfactorily reconcile theory and results, particularly after combining results from acoustic and short-time (~0.1 ms time scale) specific heat measurements. We suggest that these apparent discrepancies are due to a misunderstanding of the coupling mechanisms between TLSs and propagating waves, i.e., the deformation potential and, particularly, that it varies from TLS to TLS, and hence affects the value of the coupling constant to be used within the STM. We *speculate* that if γ_i is not the same for all TLSs within the system, then the average value of such a distribution, γ, may depend on the relaxation times probed. We note that the $C_P(T)$ and $Q^{-1}(T)$ measurements presented in this Section allow one to access TLSs whose relaxation time τ ranges from $\tau_{\min}$ to ~10 ms for $C_P(T)$ and is ~30 μs for the 5 kHz $Q^{-1}(T)$ results.

4.2. *Interaction of TLSs with elastic and electric fields*

Low-temperature dielectric losses are connected to a density of TLSs in much the same way as low-temperature mechanical losses, but with a different coupling constant: The relevant parameter is the dipole moment of a TLS $p_{0,i}$, which, like the deformation potential γ_i, is different for each TLS. Notably, in the monoatomic material a-Si, it is not clear what the dipole moment of a TLS is attributable to since there is no ionic bond. Previous work studied the interaction of TLSs with both elastic and electric fields in borosilicate glass at low temperatures.[59] In these experiments, a fraction of the mechanical loss due to TLSs was resonantly saturated using electric fields, and vice versa, suggesting the existence of TLSs with both elastic and electric response. Several studies on a-SiO_2 showed that mechanical loss is mostly independent of the concentration of OH impurities, whereas dielectric loss was highly sensitive to them.[25,85] These works suggested that not all TLSs that couple to elastic fields also couple to electromagnetic fields; their estimate was that only half of the states measured by mechanical loss are also detected by dielectric loss.[86] It is not clear whether this response is caused by the same, or different, species of TLSs, whether there is a relationship between the two types of loss, or whether there are predictors for the expected dissipation rates of a material. In a simple STM, the same TLSs and hence the same $\bar{P}$ would be relevant, with some fraction of these possessing a dipole moment.

To disentangle whether the loss mismatch of a-SiO_2 is a particularity of OH impurities, or caused by the nature of the interaction between TLSs and the different applied external fields, mechanical, and dielectric losses of a-Si films was measured and studied using superconducting microwave loss measurements.[87] Mechanical and dielectric loss measurements are obtained in different regimes, from the relaxation regime at low frequency and the resonant interaction regime at high frequency, respectively. Nonetheless, within the STM, the density of TLS $\bar{P}$ can be determined from each type of loss. However, the coupling constant γ and the dipole moment

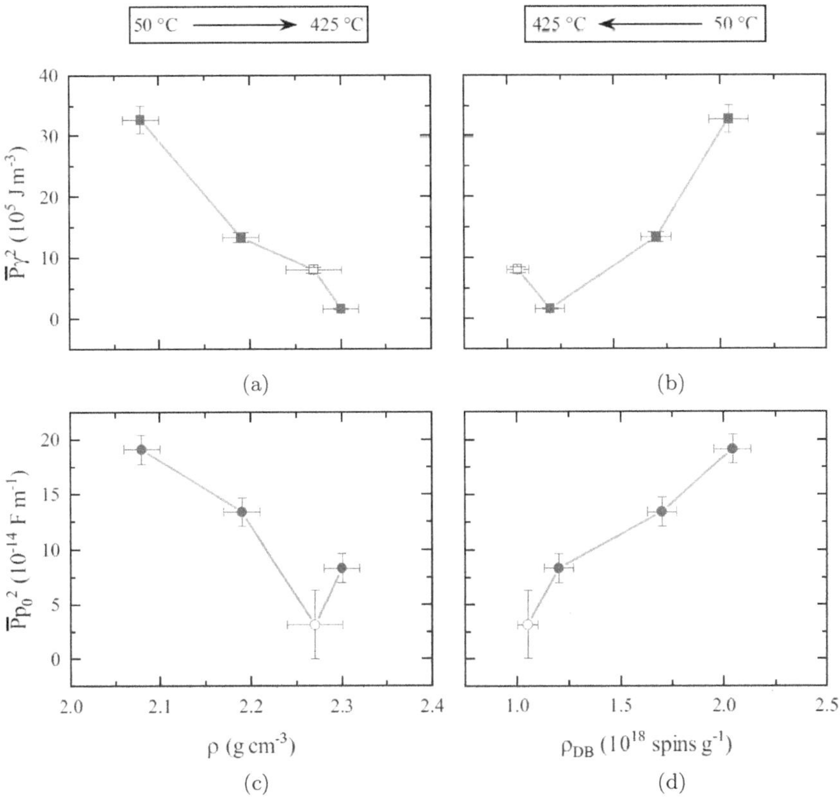

Fig. 10. From Molina-Ruiz *et al.*[87] Mechanical loss-derived quantity $\bar{P}\gamma^2$ (squares) and dielectric loss-derived quantity $\bar{P}p_0^2$ (circles) as a function of density ρ (left column) and dangling bonds density ρ_{DB} (right column). Both quantities are, respectively, derived from Eqs. (1) and (5). Closed symbols are thick films (180 or 300 nm), and open symbols are thin films (60 nm). Lines connecting data points are a guide to the eye. Black arrows indicate the growth temperature increase for the samples reported in these plots.

p_0 that appear in Eqs. (1) and (5) are not well understood or measured for many materials, and as previously noted, are generally poorly understood average values for the TLSs. For this reason, Figures 10(a) and 10(c) show the products $\bar{P}\gamma^2$ and $\bar{P}p_0^2$ (which come directly from experimental measurements) as a function of the mass density ρ, while Figures 10(b) and 10(d) show the same data as a function of the dangling bond density ρ_{DB}. Both products

$\bar{P}\gamma^2$ and $\bar{P}p_0^2$ decrease as the growth temperature increases. However, their dependence on ρ and ρ_{DB} is different. Moreover, they are not proportional to each other, meaning that $(\gamma/p_0)^2$ is not constant for a-Si grown under different conditions.

Further examination of Figures 10(a) and 10(c) reveals that $\bar{P}\gamma^2$ shows a correlation with mass density ρ, while $\bar{P}p_0^2$ does not; the 60-nm sample grown at 425°C shows the lowest dielectric loss value despite an intermediate mass density. Similarly, Figures 10(d) and 10(b) show that dielectric loss $\bar{P}p_0^2$ is monotonic with the density of dangling bonds ρ_{DB}, while the mechanical loss-derived $\bar{P}\gamma^2$ is not; the 300-nm sample grown at 425°C shows the lowest mechanical loss value despite reporting an intermediate dangling bond density. Thus, mechanical loss Q_0^{-1} correlates with the density of atoms, while dielectric loss $\tan\delta_0$ appears to be related to the density of dangling bonds. However, a reduction in ρ_{DB} by a factor of 2 corresponds with a factor of 6 reduction in $\tan\delta_0$, which suggests that the dielectric loss is not simply proportional to the number of dangling bonds.

Comparing only the thick samples, the factor of 34 reduction of mechanical loss with increased growth temperature is in stark contrast to the factor of 2.3 reduction of dielectric loss. This difference can be explained by two different hypotheses: (1) The density of TLS $\bar{P}$ is a universal quantity for a given sample, and differences between mechanical and dielectric loss are due to different behaviors of the coupling constant and dipole moment with growth temperature, or (2) there are different species of TLSs yielding independent values of $\bar{P}$ for mechanical and dielectric loss.

In the framework of the first hypothesis, $\bar{P}$ is a universal number, relevant to both types of measurements. The 15 times relative difference between Q_0^{-1} and $\tan\delta_0$ with growth temperature (see Ref. [87]) is then explained by a different dependence of γ and p_0 with growth temperature. Atomic density and sound velocity are easy to measure, but the coupling constant γ, however, is more difficult. The relationship between coupling constant and elastic properties, $\gamma^2/(\rho v^2)$, is a constant for a wide variety of glasses,[49] which implies that $\bar{P}$ is reduced by a factor of 34 with increasing growth temperature from 50 to 425°C (see Eq. (1)). If $\bar{P}$ decreases by a factor of 34, then p_0^2/ϵ must increase 15 times (Eq. (5)) to account for the smaller 2.3 times decrease in $\tan\delta_0$. To reconcile these results with the predictions made by the STM, the average TLS dipole moment p_0 would therefore need to increase almost four times with increasing growth temperature, to allow for a TLS density reduction of 34 times, an unlikely scenario since neither the charge nor separation of atoms participating in the tunneling is likely to increase fourfold. Hypothesis (1) therefore seems quite unlikely.

The second hypothesis concerns the nature of the structural defects that give rise to TLSs. Two different subsets of defects, one that strongly couples to elastic fields and another that strongly couples to electric fields, could yield two different densities of TLSs: $\bar{P}_m$ and $\bar{P}_e$, respectively. In this case, the coupling constant and dipole moment could remain almost independent of growth parameters. The second hypothesis is therefore the most likely scenario.

Since mechanical loss correlates with atomic density, while dielectric loss correlates with dangling bond density, considering different $\bar{P}_m$ and $\bar{P}_e$ is a natural and intuitive concept. It has been suggested that TLSs that yield $\bar{P}_m$ are spatially correlated with nanovoids or low-density regions,[34] and it has also been shown that nanovoids are the structural defects associated with the observed reduced atomic density in a-Si films.[66,88] By analogy, TLSs that yield $\bar{P}_e$ are likely connected with dangling bonds. In a monatomic material such as a-Si, dangling bonds are a natural source of electric dipole moments, although how those states are related to tunneling two-level systems is unclear.

The mechanical and dielectrics loss results presented in this Section, and their analysis, obtained for two quite different frequency ranges (kHz and GHz), could alternatively support the idea that the STM is incomplete, specifically that the TLS density $\bar{P}$ is not energy independent, which would be the case if the TLS parameters, Δ and λ, are not independent or are not uniformly distributed in energy (see Section 2.2). To disentangle whether these results provide evidence of different types of TLSs, or alternatively evidence that $\bar{P}$ is not energy independent, acoustic measurements in the GHz frequency range would be valuable. A frequency dependence of the distribution of states would significantly modify the STM; such a model, which predicts dephasing and noise in superconducting microresonators due to TLSs in amorphous materials, has been explored.[89]

5. Conclusions and Perspectives

The TLS density in a-Si films obtained from specific heat spans more than three orders of magnitude. High-density films, close to the crystal silicon value, show a vanishing TLS density,[33] whereas low-density films show the largest TLS density reported to date for any material. Vapor-deposited films, specifically fragile glasses, have been shown to be able to explore a larger energy landscape and range of structural configurations than traditionally quenched glasses.[4–6] That high and low TLS densities in a-Si align with the conditions that produce high enthalpy and entropy and low enthalpy and entropy glasses, respectively, suggests a new approach to suppressing TLSs and hence low loss materials, which is important for a wide range of applications such as detection of gravitational waves and quantum computing. Two conditions appear to be essential to produce glasses with low TLS density: that growth occurs at a temperature at which the material is in a low entropy and low enthalpy state, yet still has enough surface atomic mobility to reach equilibrium at the surface during growth. These conditions suggest that fragile glasses are required, where the Kauzmann temperature T_K is not very far below the glass transition temperature. An alternative hypothesis may require that the atomic bonding be constrained, as it is in a-Si (tetrahedral bonding), as originally suggested by Phillips.[27] The bonding in indomethacin is however van der Waals, which is not an overconstrained system. It therefore remains to be determined if fragility is a sufficient condition or if details of bonding type are also crucial.

The comparison of the specific heat to the mechanical loss results suggests that TLSs in amorphous silicon and amorphous hydrogenated silicon are uncharacteristically decoupled from acoustic waves. The ratio obtained by assuming a standard value for γ (1 eV) leads to $n_0/\bar{P} \sim 10^2$ for silicon, and $n_0/\bar{P} \sim 10^4$ for hydrogenated silicon, which is orders of magnitude larger than predicted by the standard tunneling model. Alternatively, the coupling constant of a-Si:H, derived from the STM prediction of $n_0/\bar{P} \approx 10$, yields $\gamma = 7\,\text{meV}$, which increases to $19\,\text{meV}$ after annealing, suggesting that γ is not a constant material quantity, but instead strongly depends on the microstructure. The low values of the coupling constant suggest that the TLSs that couple to applied elastic fields in amorphous hydrogenated silicon are likely related to the presence of hydrogen within the amorphous silicon network. This view of the STM, in which the coupling constant of a given material changes with its microstructure, explains the apparent decoupling between n_0 and Q_0^{-1} observed for several glasses.

Mechanical loss Q_0^{-1} of a-Si films is inversely proportional to atomic density, while dielectric loss $\tan\delta_0$ is proportional to dangling bond density.[87] The variations of mechanical and dielectric losses suggest different underlying structural origins, the former attributed to nanovoid surfaces and the latter to the structure near a dangling bond.

To date, only indomethacin and silicon-, germanium-, and carbon-based glasses have reported TLS densities and/or mechanical losses well below the "glassy range", which suggests that besides the hypothesis, made long ago by Phillips, that only tetrahedrally bonded materials can overcome the structural limitations that give rise to TLSs, additional mechanisms play a role preventing their formation. These results suggest that the ultrastable glass may be the key to low TLS density. On the other side of the same coin, TLS densities from specific heat, which lie well above the TLS densities derived from the universal "glassy range", are found in higher enthalpy and higher entropy glasses, and TLS densities well below the universal "glassy range" are found in low enthalpy, low entropy, ultrastable, near-ideal glasses. This observed relation between quantum (TLSs) and thermodynamic (T_g and T_K) properties suggests a connection between the structural features that enable each one of these seemingly very different — from an energetic point of view — properties.

A better understanding of the coupling between TLSs and external fields (elastic and electric) in disordered solids, their relationship with structure, and how the defects that give rise to TLSs interact through different dissipation mechanisms is still needed to disentangle the underlying physics of the anomalous properties of amorphous materials at low temperatures.

References

1. W. Kauzmann. (1948). The nature of the glassy state and the behavior of liquids at low temperatures. *Chem. Rev.* **43**, 219.
2. P. W. Anderson. (1995). Through the glass lightly. *Science.* **267**, 1615.
3. F. H. Stillinger. (1988). Supercooled liquids, glass transitions, and the Kauzmann paradox. *J. Chem. Phys.* **88**, 7818.
4. S. F. Swallen, K. L. Kearns, M. K. Mapes, Y. S. Kim, R. J. McMahon, M. D. Ediger, T. Wu, L. Yu, and S. Satija. (2007). Organic glasses with exceptional thermodynamic and kinetic stability. *Science.* **315**, 353.
5. K. L. Kearns, S. F. Swallen, M. D. Ediger, T. Wu, Y. Sun, and L. Yu. (2008). Hiking down the energy landscape: Progress toward the Kauzmann temperature via vapor deposition. *J. Phys. Chem. B.* **112**, 4934.
6. K. J. Dawson, K. L. Kearns, L. Yu, W. Steffen, and M. D. Ediger. (2009). Physical vapor deposition as a route to hidden amorphous states. *Proc. Natl. Acad. Sci. U.S.A.* **106**, 15165.
7. E. León-Gutiérrez, G. Garcia, A. F. Lopeandía, M. T. Clavaguera-Mora, and J. Rodríguez-Viejo. (2010). Size effects and extraordinary stability of ultrathin vapor deposited glassy films of toluene. *J. Phys. Chem. Lett.* **1**, 341.
8. E. León-Gutiérrez, A. Sepúlveda, G. Garcia, M. T. Clavaguera-Mora, and J. Rodríguez-Viejo. (2010). Stability of thin film glasses of toluene and ethylbenzene formed by vapor deposition: An in situ nanocalorimetric study. *Phys. Chem. Chem. Phys.* **12**, 14693.
9. F. Hellman. (1994). Surface-induced ordering: A model for vapor-deposition growth of amorphous materials. *Appl. Phys. Lett.* **64**, 1947.
10. D. Querlioz, E. Helgren, D. R. Queen, F. Hellman, R. Islam, and D. J. Smith. (2005). Beneficial effects of annealing on amorphous Nb–Si thin-film thermometers. *Appl. Phys. Lett.* **87**, 221901.
11. H.-B. Yu, Y. Luo, and K. Samwer. (2013). Ultrastable metallic glass. *Adv. Mater.* **25**, 5904.
12. C. A. Angell. (1991). Relaxation on liquid, polymers and plastic crystal — strong/fragile patterns and problems. *J. Non-Cryst. Solids.* **131–133**, 13.

13. L.-M. Martinez and C. A. Angell. (2001). A thermodynamic connection to the fragility of glass-forming liquids. *Nature.* **410**, 663.
14. V. Molinero, S. Sastry, and C. A. Angell. (2006). Tuning of tetrahedrality in a silicon potential yields a series of monatomic (Metal-like) glass formers of very high fragility. *Phys. Rev. Lett.* **97**, 075701.
15. C. A. Angell. (2015). On the uncertain distinction between fast landscape exploration and second amorphous phase (Ideal Glass) interpretations of the ultrastable glass phenomenon. *J. Non-Cryst. Solids* **407**, 246.
16. M. Tylinski, Y. Z. Chua, M. S. Beasley, C. Schick, and M. D. Ediger. (2016). Vapor-deposited alcohol glasses reveal a wide range of kinetic stability. *J. Chem. Phys.* **145**, 174506.
17. A. J. Leggett. (1991). Amorphous materials at low temperatures. Why are they so similar? *Phys. B Condens. Matter.* **169**, 322.
18. R. C. Zeller and R. O. Pohl. (1971). Thermal conductivity and specific heat of noncrystalline solids. *Phys. Rev. B.* **4**, 2029.
19. D. A. Parshin. (1993). Soft potential model and universal properties of glasses. *Phys. Scr.* **T49A**, 180.
20. D. A. Parshin. (1994). Interactions of soft atomic potentials and universality of low-temperature properties of glasses. *Phys. Rev. B.* **49**, 9400.
21. W. A. Phillips. (1987). Two-level states in glasses. *Rep. Prog. Phys.* **50**, 1657.
22. R. B. Stephens. (1973). Low-temperature specific heat and thermal conductivity of noncrystalline dielectric solids. *Phys. Rev. B.* **8**, 2896.
23. M. Molina-Ruiz, H. C. Jacks, D. R. Queen, Q. Wang, R. S. Crandall, and F. Hellman. (2020). Two-level systems and growth-induced metastability in hydrogenated amorphous silicon. *Mater. Res. Express.* **7**, 095201.
24. R. B. Stephens. (1976). Intrinsic low-temperature thermal properties of glasses. *Phys. Rev. B.* **13**, 852.
25. R. O. Pohl, X. Liu, and E. Thompson. (2002). Low-temperature thermal conductivity and acoustic attenuation in amorphous solids. *Rev. Mod. Phys.* **74**, 991.
26. P. W. Anderson, B. I. Halperin, and C. M. Varma. (1972). Anomalous low-temperature thermal properties of glasses and spin glasses. *Philos. Mag.* **25**, 1.
27. W. A. Phillips. (1972). Tunneling states in amorphous solids. *J. Low Temp. Phys.* **7**, 351.
28. C. C. Yu and A. J. Leggett. (1988). Low temperature properties of amorphous materials: Through a glass darkly. *Comments Cond. Mat. Phys.* **14**, 231.

29. C. M. Varma, R. C. Dynes, and J. R. Banavar. (1982). Thermally created tunnelling states in glasses. *J. Phys. C Solid State Phys.* **15**, L1221.
30. V. G. Karpov, M. I. Klinger, and F. N. Ignat'ev. (1983). Theory of the low-temperature anomalies in the thermal properties of amorphous structures. *Sov. Phys. JETP.* **57**, 439.
31. M. A. Il'in, V. G. Karpov, and D. A. Parshin. (1987). Parameters of soft atomic potentials in glasses. *Zh. Eksp. Teor. Fiz.* **92**, 291.
32. B. L. Zink, R. Pietri, and F. Hellman. (2006). Thermal conductivity and specific heat of thin-film amorphous silicon. *Phys. Rev. Lett.* **96**, 055902.
33. D. R. Queen, X. Liu, J. Karel, T. H. Metcalf, and F. Hellman. (2013). Excess specific heat in evaporated amorphous silicon. *Phys. Rev. Lett.* **110**, 135901.
34. X. Liu, D. R. Queen, T. H. Metcalf, J. E. Karel, and F. Hellman. (2014). Hydrogen-free amorphous silicon with no tunneling states. *Phys. Rev. Lett.* **113**, 025503.
35. T. Pérez-Castañeda, C. Rodríguez-Tinoco, J. Rodríguez-Viejo, and M. A. Ramos. (2014). Suppression of tunneling two-level systems in ultrastable glasses of indomethacin. *Proc. Natl. Acad. Sci. U.S.A.* **111**, 11275.
36. T. R. Kirkpatrick and D. Thirumalai. (1987). Dynamics of the structural glass transition and the P-spin — interaction spin-glass model. *Phys. Rev. Lett.* **58**, 2091.
37. T. R. Kirkpatrick and D. Thirumalai. (1987). P-spin-interaction spin-glass models: Connections with the structural glass problem. *Phys. Rev. B.* **36**, 5388.
38. T. R. Kirkpatrick and P. G. Wolynes. (1987). Connections between some kinetic and equilibrium theories of the glass transition. *Phys. Rev. A.* **35**, 3072.
39. T. R. Kirkpatrick, D. Thirumalai, and P. G. Wolynes. (1989). Scaling concepts for the dynamics of viscous liquids near an ideal glassy state. *Phys. Rev. A.* **40**, 1045.
40. T. R. Kirkpatrick and D. Thirumalai. (1989). Random solutions from a regular density functional hamiltonian: A static and dynamical theory for the structural glass transition. *J. Phys. A. Math. Gen.* **22**, L149.
41. G. Biroli and J.-P. Bouchaud. (2012). The random first-order transition theory of glasses: A critical assessment. In P. G. Wolynes and V. Lubchenko (eds.), *Structural Glasses and Supercooled Liquids: Theory, Experiment, and Applications*, pp. 31–113. Wiley, Hoboken, New Jersey.

42. V. Lubchenko. (2015). Theory of the structural glass transition: A pedagogical review. *Adv. Phys.* **64**, 283.
43. V. Lubchenko and P. G. Wolynes. (2001). Intrinsic quantum excitations of low temperature glasses. *Phys. Rev. Lett.* **87**, 195901.
44. J. D. Stevenson and P. G. Wolynes. (2008). On the surface of glasses. *J. Chem. Phys.* **129**, 234514.
45. M. A. Ramos, T. Pérez-Castañeda, R. J. Jiménez-Riobóo, C. Rodríguez-Tinoco, and J. Rodríguez-Viejo. (2015). Do tunneling states and boson peak persist or disappear in extremely stabilized glasses? *Low Temp. Phys.* **41**, 412.
46. Y. Zhang, E. C. Glor, M. Li, T. Liu, K. Wahid, W. Zhang, R. A. Riggleman, and Z. Fakhraai. (2016). Long-range correlated dynamics in ultra-thin molecular glass films. *J. Chem. Phys.* **145**, 114502.
47. H. M. Carruzzo and C. C. Yu. (2020). Why phonon scattering in glasses is universally small at low temperatures. *Phys. Rev. Lett.* **124**, 075902.
48. H. M. Carruzzo, A. Bilmes, J. Lisenfeld, Z. Yu, B. Wang, Z. Wan, J. R. Schmidt, and C. C. Yu. (2021). Distribution of two-level system couplings to strain and electric fields in glasses at low temperatures. *Phys. Rev. B.* **104**, 134203.
49. J. F. Berret and M. Meissner. (1988). How universal are the low temperature acoustic properties of glasses? *Z. Phys. B.* **70**, 65.
50. J. L. Black and B. I. Halperin. (1977). Spectral diffusion, phonon echoes, and saturation recovery in glasses at low temperatures. *Phys. Rev. B.* **16**, 2879.
51. J. L. Black. (1978). Relationship between the time-dependent specific heat and the ultrasonic properties of glasses at low temperatures. *Phys. Rev. B.* **17**, 2740.
52. M. Meissner and K. Spitzmann. (1981). Experimental evidence on time-dependent specific heat in vitreous silica. *Phys. Rev. Lett.* **46**, 265.
53. M. T. Loponen, R. C. Dynes, V. Narayanamurti, and J. P. Garno. (1982). Measurements of the time-dependent specific heat of amorphous materials. *Phys. Rev. B.* **25**, 1161.
54. J. Y. Duquesne and G. Bellessa. (1985). Ultrasonic study of tunnelling defects in amorphous Se, Ge and Se-Ge compounds. *Philos. Mag. B.* **52**, 821.
55. J. M. Grace and A. C. Anderson. (1989). Low-temperature specific heat and thermal conductivity of a glassy polymer under applied pressure. *Phys. Rev. B.* **40**, 1901.
56. B. Sarabi, A. N. Ramanayaka, A. L. Burin, F. C. Wellstood, and K. D. Osborn. (2016). Projected dipole moments of individual two-level defects extracted using circuit quantum electrodynamics. *Phys. Rev. Lett.* **116**, 167002.

57. J. D. Brehm, A. Bilmes, G. Weiss, A. V. Ustinov, and J. Lisenfeld. (2017). Transmission-line resonators for the study of individual two-level tunneling systems. *Appl. Phys. Lett.* **111**, 112601.
58. D. Niepce, J. J. Burnett, M. G. Latorre, and J. Bylander. (2020). Geometric scaling of two-level-system loss in superconducting resonators. *Supercond. Sci. Technol.* **33**, 025013.
59. P. Doussineau, A. Levelut, and T.-T. Ta. (1977). Interaction between electromagnetic and elastic waves in a borosilicate glass at low temperatures. *J. Phys. Lett.* **38**, 37.
60. F. Hellman. *Ideality and Tunneling Two-Level Systems in Covalently Bonded Thin Film Glasses*, in *Unifying Concepts in Glass Physics VI* (Aspen, 2015).
61. F. Hellman. (2017). Ideality and Tunneling Level Systems (TLS) in amorphous Si. In *American Physical Society Meeting.* New Orleans.
62. F. Hellman. (2017). Tunneling Level Systems (TLS) in amorphous silicon films — are these ideal glasses? In *Fundamental Problems in Quantum Condensed Matter.* Riverside.
63. F. Hellman. (2019). Ideality and Tunneling Level Systems (TLS) in amorphous silicon films. In *Simons Collaboration on Cracking the Glass Problem Annual Meeting.* New York.
64. F. Hellman. (2019). Ideality and Tunneling Two-Level Systems (TLS) in amorphous silicon films. In *ENS Workshop: Two-Level Systems in Glasses.* Paris.
65. D. R. Queen, X. Liu, J. Karel, H. C. Jacks, T. H. Metcalf, and F. Hellman. (2015). Two-level systems in evaporated amorphous silicon. *J. Non-Cryst. Solids.* **426**, 19.
66. H. C. Jacks, M. Molina-Ruiz, M. H. Weber, J. J. Maldonis, P. M. Voyles, T. H. Metcalf, X. Liu, and F. Hellman. (2022). Structural tunability and origin of two-level systems in amorphous silicon. *Phys. Rev. Materials* 6, 045604.
67. J. Zimmermann and G. Weber. (1981). On the excess specific heat of vitreous silica at low temperatures. *Phys. Lett.* **86A**, 32.
68. H. V. Löhneysen, H. Rüsing, and W. Sander. (1985). Structural relaxation effects on the low-temperature properties of vitreous silica. *Z. Phys. B Condens. Matter.* **60**, 323.
69. U. Buchenau, H. M. Zhou, N. Nucker, K. S. Gilroy, and W. A. Phillips. (1988). Structural relaxation in vitreous silica. *Phys. Rev. Lett.* **60**, 1318.
70. D. Tielburger, R. Merz, R. Ehrenfels, and S. Hunklinger. (1992). Thermally activated relaxation processes in vitreous silica: An investigation by brillouin scattering at high pressures. *Phys. Rev. B.* **45**, 2750.

71. A. D. Fefferman, R. O. Pohl, A. T. Zehnder, and J. M. Parpia. (2008). Acoustic properties of amorphous silica between 1 and 500 MK. *Phys. Rev. Lett.* **100**, 195501.
72. D. A. Strubbe, E. C. Johlin, T. R. Kirkpatrick, T. Buonassisi, and J. C. Grossman. (2015). Stress effects on the raman spectrum of an amorphous material: Theory and experiment on a-Si:H. *Phys. Rev. B.* **92**, 241202(R).
73. D. Beeman, R. Tsu, and M. F. Thorpe. (1985). Structural information from the Raman spectrum of amorphous silicon. *Phys. Rev. B.* **32**, 874.
74. J. A. Floro, P. G. Kotula, S. C. Seel, and D. J. Srolovitz. (2003). Origins of growth stresses in amorphous semiconductor thin films. *Phys. Rev. Lett.* **91**, 096101.
75. X. Liu and R. O. Pohl. (1998). Low-energy excitations in amorphous films of silicon and germanium. *Phys. Rev. B.* **58**, 9067.
76. X. Liu, M. R. Abernathy, T. H. Metcalf, B. Jugdersuren, J. C. Culbertson, M. Molina-Ruiz, and F. Hellman. (2021). Comparing amorphous silicon prepared by electron-beam evaporation and sputtering toward eliminating atomic tunneling states. *J. Alloys Compd.* **855**, 157431.
77. A. Fefferman, A. Maldonado, E. Collin, X. Liu, T. Metcalf, and G. Jernigan. (2017). Elastic measurements of amorphous silicon films at MK temperatures. *J. Low Temp. Phys.* **187**, 654.
78. M. Molina-Ruiz, H. C. Jacks, D. R. Queen, T. H. Metcalf, X. Liu, and F. Hellman. (2021). Decoupling between propagating acoustic waves and two-level systems in hydrogenated amorphous silicon. *Phys. Rev. B.* **104**, 024204.
79. J. Y. Duquesne and G. Bellessa. (1983). Rayleigh waves in amorphous germanium at low temperature: Evidence for the existence of tunnelling defects. *J. Phys. C Solid State Phys.* **16**, L65.
80. G. Bellessa. (1985). Low-temperature ultrasonic study of trapped hydrogen in niobium. *Phys. Rev. B.* **32**, 5481.
81. W. Morr, A. Muller, G. Weiss, H. Wipf, and B. Golding. (1989). Isotope dependence of hydrogen tunneling in niobium. *Phys. Rev. Lett.* **63**, 2084.
82. S. Hunklinger, L. Piche, J. C. Lasjaunias, and K. Dransfeld. (1975). On the nature of low-energy excitations in amorphous materials. *J. Phys. C Solid State Phys.* **8**, L423.
83. B. L. Zink and F. Hellman. (2004). Specific heat and thermal conductivity of low-stress amorphous Si-N membranes. *Solid State Commun.* **129**, 199.
84. X. Liu, T. H. Metcalf, Q. Wang, and D. M. Photiadis. (2007). Elastic properties of several silicon nitride films. *Mater. Res. Soc. Symp. Proc.* **989**, A22.

85. M. von Schickfus and S. Hunklinger. (1976). The dielectric coupling of low-energy excitations in vitreous silica to electromagnetic waves. *J. Phys. C Solid State Phys.* **9**, L439.
86. W. Arnold, C. Laermans, and S. Hunklinger. (1976). *Saturation of the Acoustic Attenuation in Borosilicate Glass by Microwaves at Low Temperatures*, in *1976 Ultrasonics Symposium*, Vol. 49 (IEEE, 1976), pp. 659–661.
87. M. Molina-Ruiz, Y. J. Rosen, H. C. Jacks, M. R. Abernathy, T. H. Metcalf, X. Liu, J. L. DuBois, and F. Hellman. (2021). Origin of mechanical and dielectric losses from two-level systems in amorphous silicon. *Phys. Rev. Mater.* **5**, 035601.
88. H. C. Jacks. (2018). *Identifying Structures Responsible for Two-Level Systems in Amorphous Silicon*, University of California Berkeley.
89. L. Faoro and L. B. Ioffe. (2015). Interacting tunneling model for two-level systems in amorphous materials and its predictions for their dephasing and noise in superconducting microresonators. *Phys. Rev. B.* **91**, 014201.

https://doi.org/10.1142/9781800612587_0006

Chapter 6

Vibrational Dynamics of Non-Crystalline Solids

Giacomo Baldi*, Aldo Fontana and Giulio Monaco

Department of Physics, University of Trento,
Via Sommarive 14, Povo (Tn) 38123, Italy
**giacomo.baldi@unitn.it*

The boson peak (BP) is an excess of vibrational states over the Debye law appearing at terahertz frequencies. It is found in all glasses and marks the crossover between the long-wavelength behavior, where the solid can be considered as an isotropic continuum, and the region where the wavelength of the sound wave starts to experience the microscopic details of the structure. This chapter is devoted to a review of the main experimental observations regarding the vibrational dynamics of amorphous solids, as detected by neutron and X-ray scattering techniques. A first part of the chapter is devoted to the measurements of the BP and its evolution as a function of external parameters, such as temperature, pressure, or density. The second part of the chapter reviews the wavevector evolution of the dynamic structure factor, which provides evidence of a pseudo-acoustic propagating mode up to frequencies comparable to those of the BP. This longitudinal mode has a sound attenuation which follows the Rayleigh scattering law and a negative dispersion that can partially explain the deviation from the Debye law. At higher frequencies, the inelastic spectrum presents a complex pattern of vibrations, with evidences of two peaks in various systems. To conclude, we will highlight the information that can be gained on the nature of the glass vibrational modes from a comparison with the dynamics of the corresponding polycrystal.

1. Introduction

At the macroscopic scale, a glass is well described as an isotropic continuum. Classical elasticity theory[1] correctly predicts the propagation of elastic waves in the long-wavelength limit, albeit the theory ignores the presence of internal stresses, natively stored in the material during its production. However, the density of vibrational states (DOS), measured by means of Raman or inelastic neutron scattering, deviates from the Debye prediction, whose validity would be expected for the sound waves (see Chapter 1). At frequencies of a few terahertz the ratio between the density of states and the Debye law shows a peak, usually termed "boson peak" for reasons that will be clear in the next section. The nature of the vibrational modes associated with this peak is matter of active research. The typical frequency of the BP corresponds to wavelengths of tens of interatomic spacings, meaning that the microscopic structure of the solid cannot be neglected in the description of the vibrational states. The determination of the vibrational modes of a disordered solid is significantly more complex than for a single crystal because the absence of translational periodicity precludes the use of concepts such as the reciprocal space and the Bloch theorem. For this reason, many models have been proposed but a microscopic description of the terahertz vibrations of glasses is still missing.

Without attempting to present a comprehensive account of all the various theoretical approaches, we will here mention a few of them. The soft potential model[2] is based on an extension of the concept of defects with two-level states.[3] It assumes the presence of soft potentials and of quasi-localized excitations in addition to the sound waves (see Chapter 8). The idea of the presence of soft spots in the glass, regions of lower elastic modulus with respect to the surroundings, is common also to another approach that considers the BP as vibrations of nanometer sized regions.[4] A different approach, initially based on numerical calculations[5] and more recently formalized in a theoretical framework,[6] is the fluctuating elasticity theory. The model is based on the idea that the microscopic elastic modulus of glasses has random spatial fluctuations and predicts the emergence of a boson peak and the Rayleigh scattering of the sound waves[7–10] (see Chapter 9). This model assumes the validity of the harmonic approximation, an assumption common also to the random matrix theory.[11] Another family of models describes the boson peak within the jamming scenario,[12] originally developed for a-thermal glasses, such as granular materials and colloids. Finally, we can mention a model where the BP is considered as a smeared first van Hove singularity, pushed down in frequency with respect to the crystal by hybridization effects.[13]

Quite recently, numerical simulation studies (see Chapter 10) have succeeded in the determination of the nature of the vibrational modes in the frequency range of the BP and below, employing systems composed of millions of atoms.[14–19] These studies suggest the presence of a non-negligible fraction of quasi localized excitations that coexist with the sound waves. However, these simulations are performed on model systems, based on quite simple interatomic potentials. Moreover, it has been shown that the density of quasi local modes decreases significantly in more stable glasses, which resemble more closely those obtained experimentally.[19] Consequently, the relevance of these findings for real glasses is still to be demonstrated.

This chapter presents a review of the main experimental observations on the vibrational dynamics of amorphous solids. We restrict our attention mainly to scattering experiments performed by means of neutron and X-rays. Section 2 is devoted to the results obtained by measurements of the density of vibrational states. These studies include the dependence of the boson peak on the thermodynamic parameters, such as temperature and pressure, and on the variation of macroscopic quantities, such as density and elastic moduli. The section is closed by a paragraph where the DOS of glasses is compared to the one of the corresponding crystals. Section 3 summarizes the main experimental observations on the wavevector dependence of the dynamic structure factor. These include the dispersion and broadening of the inelastic peaks, the Rayleigh scattering of the sound waves, evidences of transverse dynamics and the comparison with the dynamics of polycrystals.

2. Vibrational Density of States

The vibrational density of states, $g(\omega)$, is a well-defined quantity both in crystals and in amorphous materials. It was recognized long ago,[20] by means of neutron inelastic scattering, that the DOS is similar in glasses and in the corresponding crystalline form, as shown in Figure 1 for the case of vitreous silica (upper panel) and of α-quartz (lower panel). At sufficiently low frequencies the DOS of crystals follows the Debye law, $g_D(\omega) = 3\omega^2/\omega_D^3$, where ω_D is the Debye frequency. However, the DOS of glasses differs from that of crystals in this low frequency range, below a few terahertz.

The corresponding vibrations are responsible for the well-known anomalous low temperature thermal properties of amorphous solids (see Chapters 1 and 2 for details). The glass DOS does not extrapolate to the expected Debye behavior and the deviation is better appreciated by plotting the "reduced" DOS, $g(\omega)/\omega^2$. For the case of the canonical glass of silica this quantity extrapolates to a value which is more than twice the one expected from the Debye model (see Figure 2(b)) from Ref. [21]).

Moreover, the reduced DOS presents a peak, the "boson peak". In most glasses the BP is located at frequencies close to one terahertz and corresponds to a peak in the specific heat over temperature cube, C_p/T^3, at a few degrees Kelvin. The example of the silica glass is shown in Figure 2, where Raman (panel a) and inelastic neutron scattering data (panel b) are compared. The name "boson peak" takes its origin from the fact that the peak in $g(\omega)/\omega^2$ is temperature independent at sufficiently low temperatures, as shown in Figure 2(c) for silica. Consequently, in scattering experiments, both with electromagnetic waves and with neutrons, the intensity at the frequency of the BP follows the Bose-Einstein statistics, at least at sufficiently low temperatures. At higher temperatures the peak shifts, typically towards lower energies. The situation in silica is quite peculiar because this system becomes harder with increasing temperature and the BP correspondingly shifts to higher energies.

The BP in the reduced Raman scattering spectra (Figures 2(a)) and 2(c) is shifted to higher frequency with respect to its position in the neutron spectra because the light-vibrations coupling constant,[23]

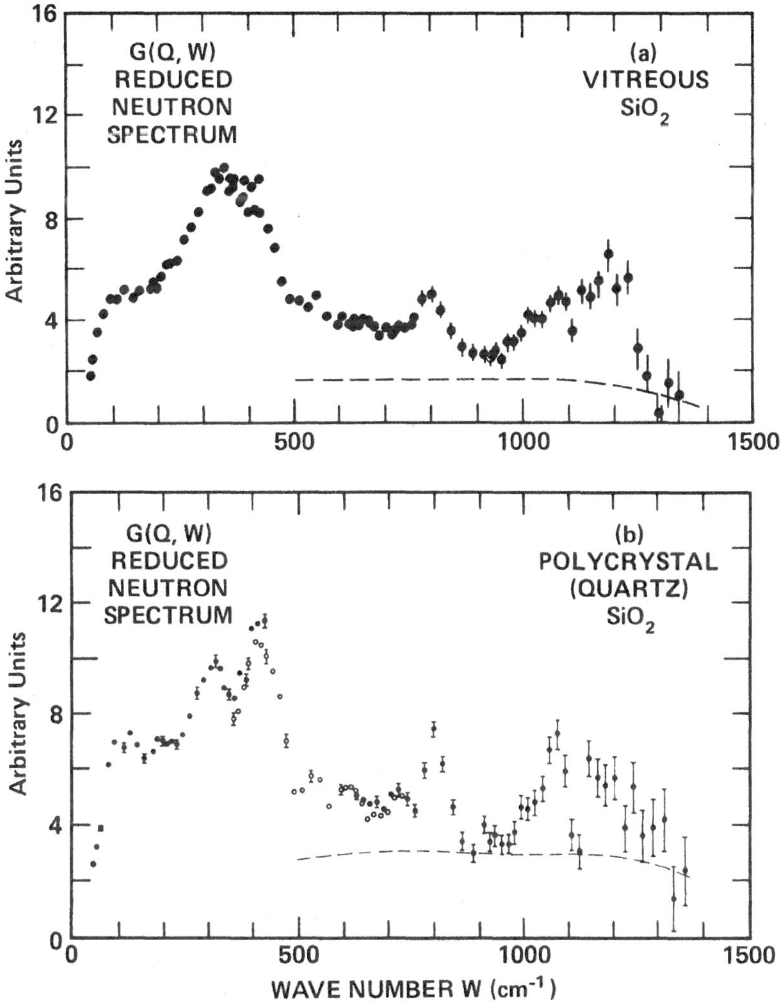

Fig. 1. DOS, $g(\omega)$, measured by neutron inelastic scattering in vitreous silica (upper panel) and in crystalline quartz (lower panel). Reprinted with permission from Ref. [20]. Copyright 1983 by the American Physical Society.

$C(\omega)$, is approximately linear in ω around the BP.[21,24] It is worth mentioning that, already in their first seminal contribution,[25] Buchenau and coworkers observed that the BP modes of vitreous silica were dominated by coupled rotations of corner sharing SiO_4 tetrahedra.

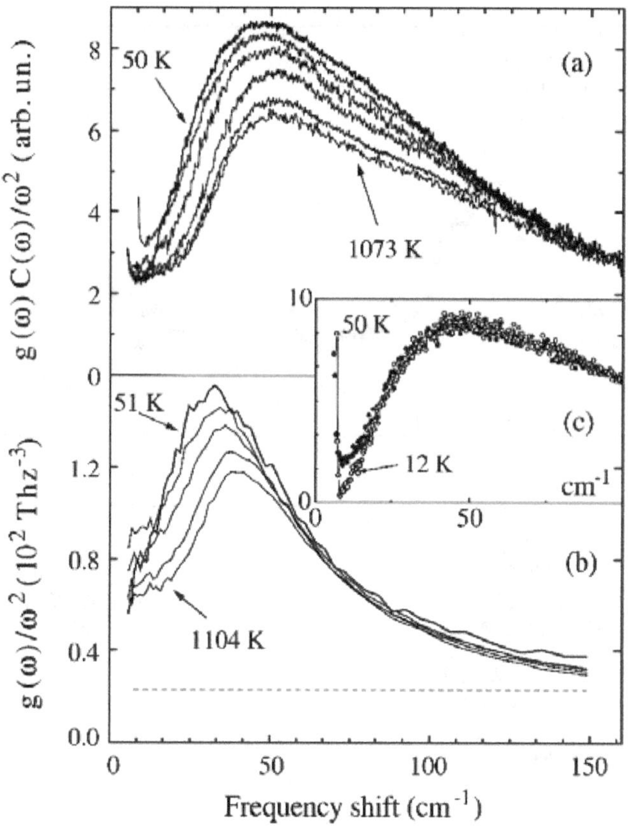

Fig. 2. Reduced density of vibrational states, $g(\omega)/\omega^2$, of vitreous silica as a function of temperature from Ref. [21] Panel (a) Raman data (temperatures from top to bottom: 45, 323, 423, 523, 873, and 1073 K). Panel (b) inelastic neutron scattering (from top to bottom: 51, 318, 523, 873, and 1104 K). The dashed line is the Debye level computed from the Brillouin sound velocities at 50 K. Panel (c) Reduced Raman spectra at 12 and 50 K. All the temperatures are below the glass transition, which in SiO_2 is approximately 1450 K.[22] Reproduced with permission from Ref. [21].

2.1. *Boson peak shape and intensity*

Malinovsky and coworkers showed that the BP shape is quite similar in different kind of solids, from network forming to metallic glasses, and suggested a universal shape for the excess DOS over the Debye level in the form of a logarithmic-normal function.[26,27]

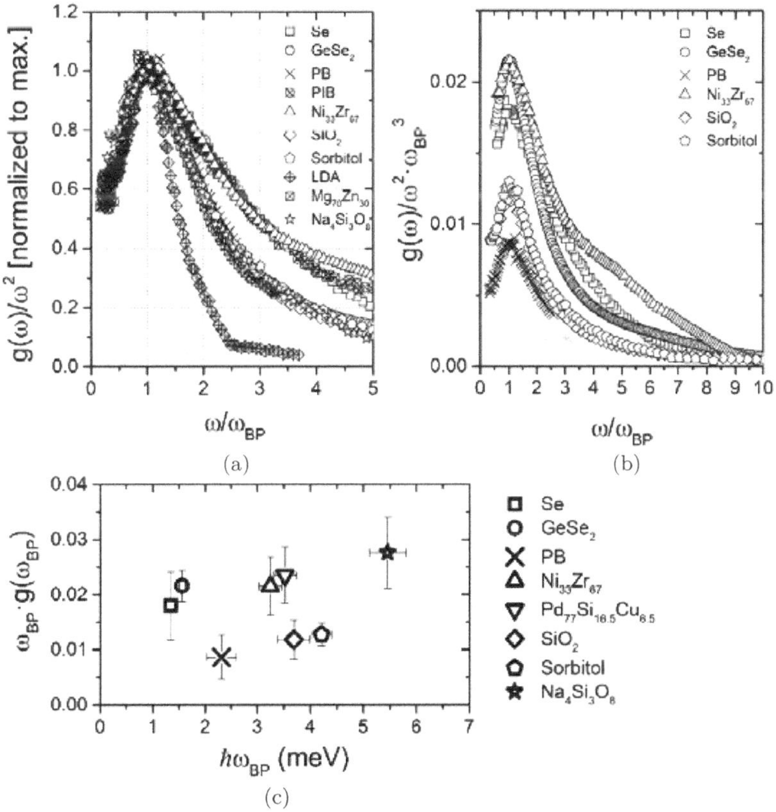

Fig. 3. Boson peak of a selection of glasses determined by neutron scattering: Se (squares),[28] $GeSe_2$ (circles),[31] polybutadiene (PB, crosses),[32] poly(isobutylene) (PIB, crossed circles),[29] $Ni_{33}Zr_{67}$ (up triangles),[30] $Pd_{77}Si_{16.5}Cu_{6.5}$ (down triangles),[33] SiO_2 (diamonds),[34] sorbitol (pentagons),[35] low density amorphous ice (LDA, crossed diamonds),[36] $Mg_{70}Zn_{30}$ (crossed squares),[26] $3(SiO_2)$-$2(Na_2O)$ (stars).[37] Panel (a) reduced DOS plotted as a function of ω/ω_{BP} and normalized to the maximum. Panel (b) reduced DOS scaled according to the mode conservation, Eq. (1). Panel (c) Maximum of the curves like those in panel (b) for eight glasses as a function of the BP position.

We have compared the reduced DOS of a set of ten amorphous solids, by plotting the different curves, all measured by neutron scattering, normalized to their maximum value and as a function of $\tilde{\omega} = \omega/\omega_{BP}$. The result is shown in Figure 3(a) and suggests that the BP shape is not universal, although it is similar for a set of

6 out of the 10 glasses considered. It should be mentioned, however, that this observation could be affected by the way the data are treated and by the fact that the spectra are collected from different neutron spectrometers. Keeping this in mind, we observe that Se,[28] poly(isobutylene) (PIB)[29] and a metallic glass ($Ni_{33}Zr_{67}$)[30] show a similar normalized reduced DOS in the chosen $\tilde{\omega}$ range, with an excess wing at high frequencies with respect to the other glasses.

On the contrary, low density amorphous ice[36] (LDA) has a reduced number of states at frequencies above the BP. The shape of the reduced DOS below ω_{BP} is instead quite similar in the ten systems considered. Since $g(\omega)$ counts the number of states, it was later recognized[38] that a rescaling of the ordinate axis to $\tilde{\omega} = \omega/\omega_{BP}$ should be accompanied by a corresponding renormalization of the DOS to:

$$\tilde{g}(\tilde{\omega}) = g(\omega) \cdot \frac{d\omega}{d\tilde{\omega}} = g(\omega) \cdot \omega_{BP} \tag{1}$$

Despite the fact that the maximum of the reduced DOS has strong variations between the considered glasses, those differences almost disappear once the spectra are scaled by the BP position and the mode conservation of Eq. (1), which implies a multiplication by ω_{BP}^3, is taken into account. The result is shown in Figure 3(b) for a subset of the glasses considered in Figure 3(a). The position of the maximum has been determined by fitting with a logarithmic-normal distribution a small portion around the peak. The BP intensity scaled in this way lies in the range 0.01 to 0.03 for all these materials. The precision of the scaling is affected by the uncertainty in the determination of the peak position and intensity. This is highlighted in Figure 3(c) where we report the maximum of the rescaled reduced DOS as a function of the BP position. The maximum of $\tilde{g}(\tilde{\omega})/\tilde{\omega}^2$ is simply the product of the BP frequency times the density of states at the BP, a quantity related to the integral of the DOS from zero to ω_{BP}. This quantity appears to be almost independent of the BP position and with variations between different systems which are often within the uncertainty. Consequently, it seems that the number of vibrational states at the BP does not vary appreciably with the peak position and is unrelated to the Debye level, which is considerably different between the selected glasses.

Sokolov and coworkers observed that the ratio between the specific heat and the Debye expectation at the temperature of the BP maximum correlates well with the fragility of the glass, with a higher value for stronger glasses.[39] Because of the direct relationship between specific heat and DOS[40] this correlation implies a higher value of $g(\omega_{BP})/g_D(\omega_{BP})$ for a strong glass like silica (fragility index ~20) and a lower one for a fragile glass like sorbitol (fragility index ~100). The result of Figure 3(c)) suggest that the number of vibrational states is not strongly correlated with the glass fragility, which is instead probably correlated with the Debye level, at least for the systems considered in Ref. [39].

We can conclude that there is some evidence that the total number of vibrational modes at low frequencies, below the BP region, is not markedly different in different glasses. There is, on the contrary, a stronger variation in the BP position which can range from 1 to 10 meV.[41] However, at present there is no clear microscopic explanation of the value of ω_{BP}, we only know that it is typically higher in systems with a higher value of the elastic modulus and there is some indication that the local structure plays a role.[42,43]

2.2. *Boson peak dependence on external parameters*

Various studies have been performed with the aim to elucidate the dependence of the low frequency DOS on the variation of some external parameter, like pressure,[29,44] temperature,[38,43] sample density for permanently densified glasses,[45–51] quenching rate,[52] aging[53,54] or sample preparation by chemical vitrification.[55] All these studies indicate that the BP shifts to higher frequencies following an increase of the macroscopic elastic moduli. When the variation is small, below a few percent, the change of the DOS can be described by the elastic medium transformation, taking only into account the variation of the Debye frequency,[52] as shown in Figure 4(a) for the case of a silicate glass enriched in iron.

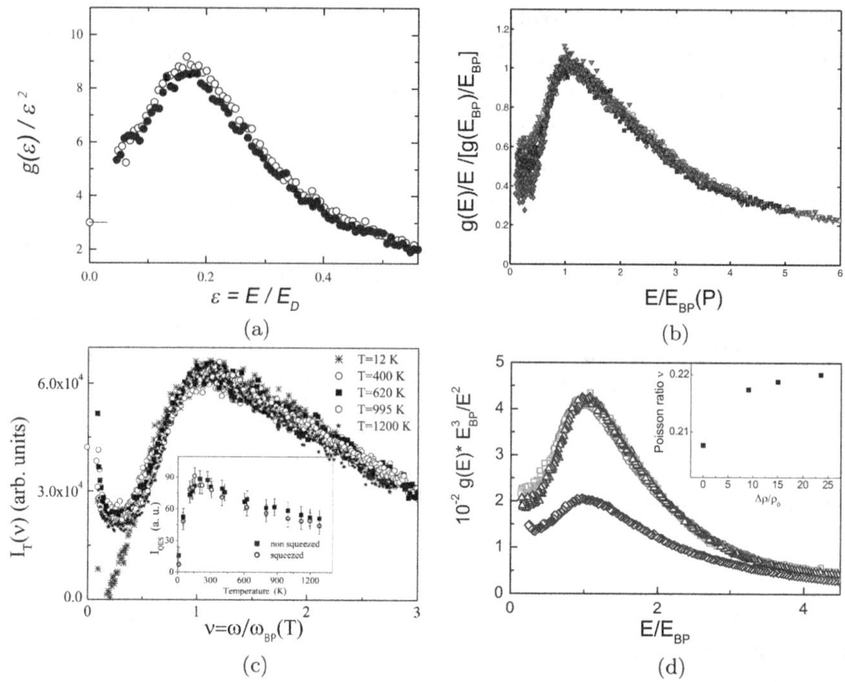

Fig. 4. Panel (a): The difference in the BP of an hyperquenced (open symbols) and an annealed (solid symbols) silicate glass is well described by the elastic medium transformation, i.e., by the variation of the Debye frequency. Reprinted with permission from [Ref. [52]. Copyright 2006 by the American Physical Society. Panel (b): The shape of the reduced DOS of PIB is not affected by pressure, although the BP position changes by more than a factor of two. Data obtained by neutron scattering at: atmospheric pressure (dots), 0.4 (triangles), 0.8 (squares) and 1.4 GPa (diamonds). Reprinted with permission from Ref. [29]. Copyright 2007 by the American Physical Society. Panel (c): Reduced Raman intensity of vitreous silica at the indicated temperatures rescaled following Eq. (1) and the hypothesis that the Raman coupling function is linear in frequency. The inset shows the Raman quasielastic intensity at 9.5 cm^{-1}. Reprinted with permission from Ref. [38]. Copyright 2007 by the American Physical Society. Panel (d): Neutron scattering data of permanently densified GeO_2 glasses.[49] Normal glass (diamonds) and glasses permanently densified at the following pressures: 2 GPa (circles), 4 GPa (triangles) and 6 GPa (squares). The reduced DOS is rescaled following the mode conservation, Eq. (1). The inset shows the Poisson ratio as a function of the relative density variation. Reproduced with permission from Ref. [49] with the permission of AIP Publishing.

In the case of larger variations of ω_{BP}, which are typically obtained by the application of pressure or by permanent densification, the modification of the DOS cannot be described by the macroscopic elastic medium transformation alone. Niss and coworkers, comparing neutron scattering and Brillouin measurements of poly(isobutylene) (PIB) as a function of applied pressure, observed a variation of ω_{BP} twice larger than that expected from the sound velocity values.[29] They showed that the shape of the reduced DOS was not affected by the application of pressure (see Figure 4(b)), even if ω_{BP} at the highest applied pressure doubled its value at ambient conditions. Permanent densification of a silicate glass enriched in iron[48] indicated that the elastic medium transformation can account for the modification of the reduced DOS only if the local structure of the glass is not affected by the densification. In this material the first step of permanent densification induced some degree of modification of the local environment of the atoms, as suggested by Mossbauer spectroscopy, with the consequence that the variation of the BP shape from the normal density sample to the permanently densified ones could not be described by a simple elastic medium transformation. A subsequent work on permanently densified germanium dioxide[49] showed that the BP variation with density was higher than expected from the variation of the Debye energy. The reduced DOS of the permanently densified samples varied according to the mode conservation law of Eq. (1), as shown in Figure 4(d). However, as in the previous example, the first step of densification induced some severe modification of the local structure which induced a strong change of the DOS. Equation (1) accounts well for the modification of the DOS in the case of the temperature evolution of the BP in vitreous silica (Figure 3(c)), where it shifts from approximately 4 meV at low temperatures to 6 meV at the glass transition temperature.

2.3. *Boson peak as a broadened van-Hove singularity*

In recent years, some inelastic scattering and specific heat measurements have revealed a significant connection between the boson peak of glasses and the first van Hove singularity present in the corresponding crystal. The most studied glass was silica[51] but other systems show a similar phenomenology, like the glassy mineral aegirine[42] and v-GeO_2[56] The absence of periodicity characteristic of disordered systems such as glasses makes it difficult to describe their microscopic dynamics and the differences with respect to crystals pose several intriguing issues. In a crystal, in the long wavelength limit, the experiments agree with the Debye prediction while in glasses, as discussed in the previous paragraphs, this does not happen. The connection between the macroscopic elasticity and the microscopic vibrational dynamics of crystals is well understood. At the scale where the wavelength becomes comparable with the interatomic distances, the vibrational excitations gradually abandon their long-wavelength character, and this is marked by a bending-over of the acoustic branches, giving rise to the van-Hove singularities. This causes in crystals a deviation of the DOS from the low frequency Debye dependence. The boson peak in a glass and the lowest van-Hove singularity of the corresponding crystal are both expected to appear in the low frequency region where the sound waves experience a transition from the macroscopic continuum limit to the microscopic regime. However, it must be considered that usually the vitreous system and the corresponding crystal can have a notably different mass density. In fact, the lower density of glasses with respect to the corresponding crystals is a well-known consequence of the way glasses are prepared from the melt. The density dependence of the boson peak and its tendency towards the first van Hove singularity of the corresponding crystal has been demonstrated by nuclear resonant analysis of inelastic X-ray scattering[57] on the aegirine glass,[42] by means of in-situ high pressure measurements.

This observation has been later confirmed[51] by comparing the DOS of permanently densified silica with the same density of crystalline α-quartz and normal density silica with cristobalite. The experimental results are shown in Figure 5 where the DOS, $g(E)$, and the reduced DOS, $g(E)/E^2$, of α-cristobalite, $\rho = 2.29$ g/cm^3, and α-quartz, $\rho = 2.65\,\mathrm{g/cm^3}$, are compared with those of v-SiO_2, $\rho = 2.20\,\mathrm{g/cm^3}$, and permanently densified v-SiO_2, $\rho = 2.65\,\mathrm{g/cm^3}$, respectively. The $g(E)$ of glasses and of the corresponding crystals are

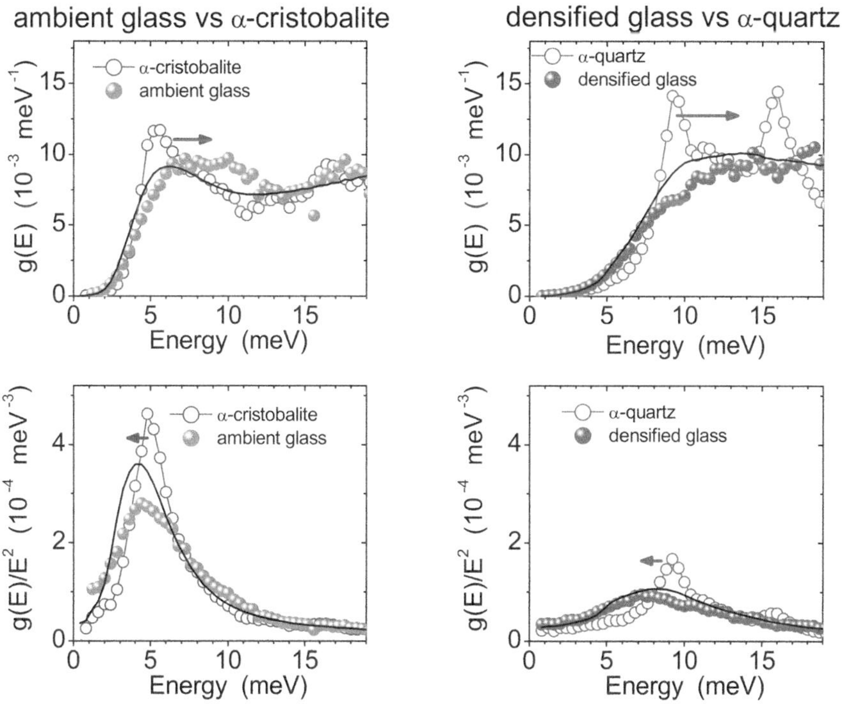

Fig. 5. Left: Comparison of the DOS (upper panel) and reduced DOS (lower panel) of ambient SiO_2 glass, full symbols, and of crystalline α-crystobalite, open symbols. Right: DOS (upper panel) and reduced DOS (lower panel) of permanently densified SiO_2 glass, full symbols, and of crystalline α-quartz, open symbols. The lines are obtained from the dispersion curves of the crystals including the effect of the smearing-out of the pseudo-Brillouin zone boundary given by the width of the first sharp diffraction peak of the glass. The arrows indicate the peak shift associated to this mechanism. Reprinted with permission from Ref. [51]. Copyright by the American Physical Society.

very similar and differ mainly because glasses present a smoother and slightly broader DOS, with a stringent qualitative and quantitative correspondence. This similarity is even more evident if we look at the BP. In fact, the boson peak of v-SiO_2 is located nearly at the same energy of that of the van Hove singularity of α-cristobalite. Concerning the comparison between the densified silica and α-quartz, the energies are slightly different, but these small differences can be attributed to the distribution of the sizes of the local structures of the glass. The pseudo-Brillouin zone of glasses has a finite width that, once included in the crystal dispersion curves, can quantitatively account for the shift in the peak position as shown by the continuous curves in the figure and by the arrows. It is worth highlighting one detail, which can be appreciated in the bottom right panel, where the BP of densified silica is compared with the reduced DOS of α-quartz. At low frequency the reduced DOS of the glass tends to a constant value, very close to that of the crystal and to the Debye level. Consequently, the densified silica is an example of a glass without appreciable excess of modes with respect to the Debye expectation. Other evidences are found also in boron oxide[58] and in glycerol.[59,60]

A similar observation, of a relation between the BP and the lowest energy van Hove singularity of the corresponding crystal, was done by Taraskin and coauthors in molecular dynamics (MD) simulations.[13] However, they interpreted the difference in energy between the BP of the glass and the first van Hove singularity of the crystal not in terms of a difference in mass density but as an effect of disorder that should shift, at the boundaries of the vibrational spectrum, the modes to lower energies.

3. Dispersion and Scattering of Vibrational Modes

Even though the density of states provides many important information on the dynamics of disordered systems, as documented in the previous paragraphs, it cannot give conclusive results on the nature of the excitations. A richest wealth of information can be gained by studying the wave-vector resolved dynamical structure factor, $S(q, \omega)$. A variety of experimental techniques has been developed to probe this quantity in different regions of the $q - \omega$ space, but neutron or X-ray inelastic scattering are better suited to probe the terahertz frequency range of interest to our discussion.

Neutron scattering is the technique of choice for the study of the vibrational dynamics of crystals, where the kinematical restrictions imposed by the limited neutron energy can be avoided by probing the sample dynamics outside of the first Brillouin zone. Brillouin neutron scattering can be performed also in the first zone but only in samples with a sound velocity smaller than that of the incoming neutron.

This limitation restricts the technique mainly to liquid samples, while in solid amorphous materials and in polycrystals the longitudinal vibrational modes in the first Brillouin zone is most often outside of the kinematical range, apart from specific glasses with low sound velocity.[31,61,62] X-ray inelastic scattering is a powerful method to study the dynamics of systems that do not possess translational periodicity, such as glasses, liquids, or polycrystalline materials, and as such it is complementary to neutron inelastic scattering.

We will provide in this section an overview of the main experimental results obtained on the investigation of the vibrational dynamics of glasses by probing the wave-vector resolved spectral distribution, $S(q, \omega)$, with these experimental methods.

3.1. *Dynamic structure factor*

Assuming for simplicity to deal with a monoatomic solid, the coherent cross section for both neutrons and X-ray scattering is proportional to the dynamical structure factor[63]:

$$S(\boldsymbol{q}, \omega) = \frac{1}{2\pi N} \int_{-\infty}^{+\infty} dt e^{i\omega t} \sum_{l,l'} \langle e^{-i\boldsymbol{q}\cdot\hat{\boldsymbol{r}}_l(t)} e^{i\boldsymbol{q}\cdot\hat{\boldsymbol{r}}_{l'}(0)} \rangle, \tag{2}$$

where $\hat{\boldsymbol{r}}_l(t)$ is the position operator of atom l at time t and N is the number of atoms of the system. In a solid composed of more than one atomic species, the scattering cross section is proportional to a weighted average of the partial dynamical structure factors, where the weights are the appropriate scattering lengths for neutrons or the atomic form factors for the Thompson scattering of X-rays.

The dynamic structure factor is the sum of an elastic, $S^{(0)}(\boldsymbol{q}, \omega)$, and of an inelastic term, $S^{(in)}(\boldsymbol{q}, \omega)$. Assuming the atomic displacement of atom l, $\hat{\boldsymbol{u}}_l(t)$, around the equilibrium position $\hat{\boldsymbol{x}}_l$, to be sufficiently small we can write the elastic term as:

$$S^{(0)}(\boldsymbol{q}, \omega) = \frac{1}{N} \left| \sum_l e^{-i\boldsymbol{q}\cdot\hat{\boldsymbol{x}}_l} e^{-\frac{W_l(\boldsymbol{q})}{2}} \right|^2 \delta(\omega), \tag{3}$$

where $W_l(\boldsymbol{q}) = \langle (\boldsymbol{q} \cdot \hat{\boldsymbol{u}}_l)^2 \rangle$ is the Debye-Waller factor for atom l. In a perfect crystal $S^{(0)}$ is different from zero only when the Laue condition is satisfied and $\boldsymbol{q}$ equals a reciprocal lattice vector. In a glass the structural disorder implies the presence of a non-negligible elastic component at all values of q. The inelastic term $S^{(in)}$ in the harmonic approximation is dominated by the one-phonon contribution:

$$S^{(1)}(\boldsymbol{q},\omega) = \frac{\hbar}{Nm}\sum_{\lambda}\left|\sum_{l} e^{-i\boldsymbol{q}\cdot\hat{\boldsymbol{x}}_l} e^{-\frac{W_l(q)}{2}}\boldsymbol{q}\cdot\boldsymbol{e}_l(\lambda)\right|^2 \boldsymbol{x}$$
$$\frac{1}{2\omega_\lambda}\left\{\left[n(\omega_\lambda,T)+1\right]\delta(\omega-\omega_\lambda)+n(\omega_\lambda,T)\,\delta(\omega+\omega_\lambda)\right\}. \tag{4}$$

In this expression m is the mass of the atom, $\boldsymbol{e}_l(\lambda)$ is the λth eigenvector at atom l of eigenfrequency ω_λ and $n(\omega,T)$ is the Bose population factor for vibrations of frequency ω at temperature T. Positive frequencies in Eq. (4) correspond to the Stokes side where the peak intensity of the ground state ($T = 0$) is non-zero.

For an ideal crystal the $S^{(1)}$ is peaked at frequencies corresponding to the dispersion curves of phonons with a wave-vector $\boldsymbol{k}$ satisfying the relation $\boldsymbol{q} = \boldsymbol{k} + \boldsymbol{G}$, where $\boldsymbol{G}$ is a reciprocal lattice vector. Consequently, the dynamic structure factor of a crystal is composed of a set of sharp peaks, which are broadened by defects and anharmonicities. In cases where the measurement is confined within the first Brillouin zone, the one-phonon term includes contributions coming only from the longitudinal component of the eigenmodes. The nature of the eigenmodes, which cannot be associated to a well-defined $\boldsymbol{k}$ vector, of a disordered solid implies $S^{(1)}$ to be composed of broad peaks even in the harmonic approximation. The broadening comes from the sum over the eigenvalues λ in Eq. (4).

The experimental inelastic X-ray scattering (IXS) spectra are typically analyzed in terms of one or more damped harmonic oscillation (DHO) functions that describe the inelastic part of the spectrum. The single DHO model for the dynamic structure factor is[64]:

$$\mathrm{S}(q,\omega) = S(q)\left\{f_q\delta(\omega) + (1-f_q)\frac{\hbar\omega}{k_BT}\left[n(\omega,T)+1\right]\right.$$
$$\left.\times\frac{1}{\pi}\frac{\Omega^2(q)\,\Gamma(q)}{\left[\omega^2-\Omega^2(q)\right]^2+\omega^2\Gamma^2(q)}\right\}, \tag{5}$$

where $S(q)$ is the static structure factor. The parameters of the model are: the non-ergodicity parameter, f_q, defined as the ratio of elastic to the total intensity, the position of the peak in the longitudinal current, $\Omega(q)$, and the parameter $\Gamma(q)$ that estimates the full width at half maximum of the inelastic peak.

This simple model is useful to analyse the experimental spectra, which are often measured at a fixed q as a function of frequency. However, the model itself finds a proper justification only in the hydrodynamic regime of simple liquids. In the case of solids or highly viscous systems it is conveniently used to determine the positions and widths of the inelastic features, but it has clear limitations, in particular at high q values. This will be shown in the next paragraph.

3.2. *Single damped harmonic oscillator analysis*

3.2.1. *Dispersion of the inelastic peak position*

An example where the single DHO model works well is reported in Figure 6 for the case of glassy glycerol. The spectrum presents a marked elastic line whose width is fixed by the instrumental response function, since possible relaxation processes contributing to the width of the central line are undetectable with the available energy resolution of approximately 1.5 meV.

The inelastic features are composed of a Stokes and an anti-Stokes peak and present some dispersion as a function of q, as highlighted in the right panel of the figure. Here the Stokes side is plotted after subtraction of the elastic line, whose intensity is determined by a fit of the entire spectrum. The validity of the DHO model as a proper description of the glass spectra has been debated in the literature.[66–68] However, it is now commonly accepted as a useful fitting model, better suited to describe the spectra in the low q region, far from the position of the first sharp diffraction peak (FSDP) in the static structure factor. As the wavevector is increased, the DHO model has the tendency to overestimate the inelastic intensity at the centre of the spectrum.[69] The dispersion of the Brillouin peak has been documented in a variety of systems, from oxides to organic

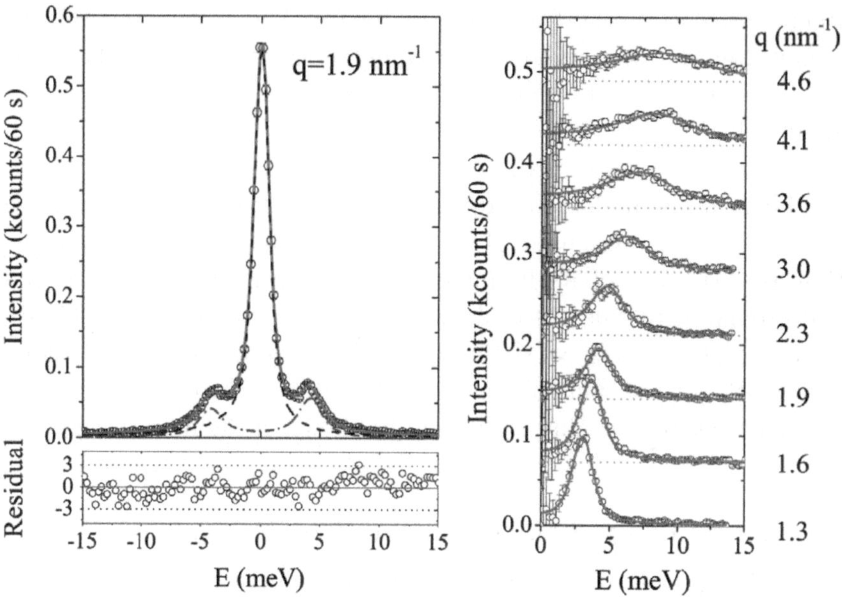

Fig. 6. Left column: Inelastic X-ray scattering spectrum of glassy glycerol at the indicated exchanged wavevector together with the best fitting function (continuous line), modelled as an elastic line (dashed) plus a damped harmonic oscillator describing the inelastic component (dot-dashed). See text for details on the data analysis. The bottom panel shows the residues between the model and the data. Right column: inelastic intensity on the Stokes side at selected qs, together with the best fitting functions. The elastic line is subtracted after the fit is performed on the entire spectrum. Reproduced with permission from Ref. [65].

and metallic glasses.[31,33,35,37,41,61,62,65,66,68–84] A few examples are reported in Figure 7.

The dispersion of the peak position as a function of q in a metallic glass[85] is shown in the upper left panel of the figure, together with the apparent sound velocity shown in the inset. At low wave-vectors the dispersion is linear and follows the sound velocity of the longitudinal acoustic mode. At higher qs the dispersion resembles that of a crystalline material, with a sinusoidal behaviour suggesting the presence of a pseudo-Brillouin zone even in a disordered structure. The zone boundary, Q_P in the figure, has a value which is approximately one half of the FSDP position. The presence of a certain degree of periodicity of the dispersion curve has been sometimes interpreted in

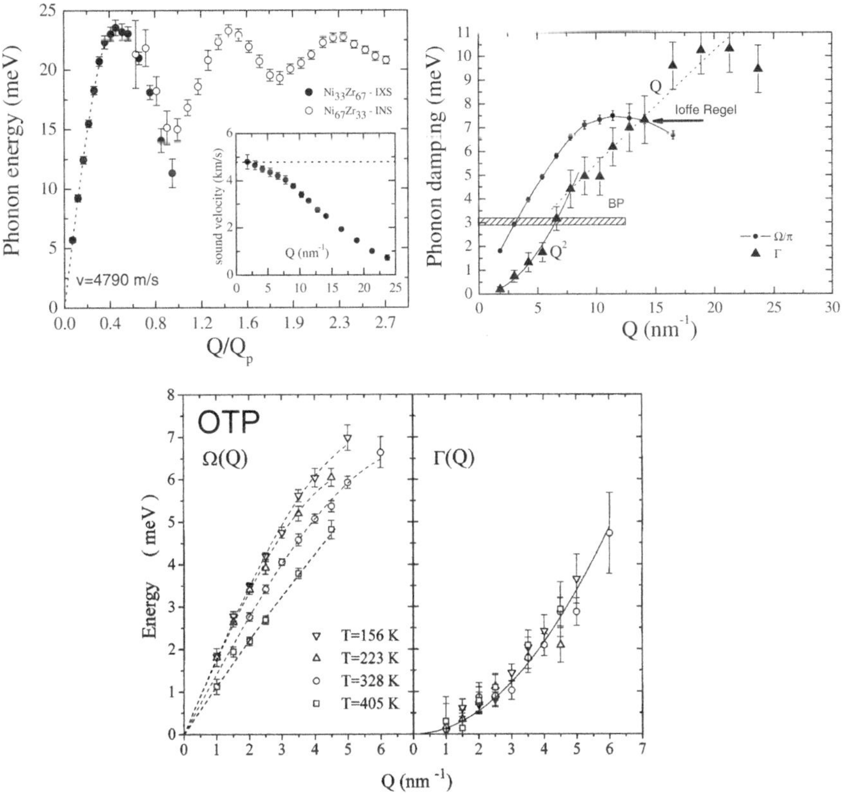

Fig. 7. Top: Dispersion (left) and damping (right) of the inelastic peak of a metallic glass. The left panel compares data for $Ni_{33}Zr_{67}$ measured by IXS[85] with those of $Ni_{67}Zr_{33}$ investigated by neutron inelastic scattering (INS),[86] scaling the exchanged wavevector by the first sharp diffraction peak position, Q_P. ($Q_P^{Ni_{33}Zr_{67}} = 25.7\,\mathrm{nm}^{-1}$ and $Q_P^{Ni_{67}Zr_{33}} = 29\,\mathrm{nm}^{-1}$). Inset: Wavevector dependence of the apparent sound velocity of $Ni_{33}Zr_{67}$ probed by IXS. The phonon damping of $Ni_{33}Zr_{67}$ is shown in the right panel (triangles) together with the dispersion curve plotted as Ω/π (dots). The intersection between the two data sets defines the Ioffe-Regel limit, here observed in the second pseudo-Brillouin zone, much above the boson peak frequency (horizontal bar). Reprinted with permission from Ref. [85]. Copyright 2006 by the American Physical Society. Bottom: Dispersion (left) and damping (right) of OTP as a function of temperature across the glass transition ($T_g = 243\,\mathrm{K}$). Reprinted with permission from Ref. [71]. Copyright 1998 by the American Physical Society.

terms of Umklapp scattering.[87] Metallic glasses are characterized by a FSDP located at relatively high wavevectors and typically sharper than that of other glasses, a consequence of the dominantly radial metallic bonding. The clear signature of a pseudo-Brillouin zone and a periodicity which persists even in the fourth zone are most probably consequences of the somehow larger residual order in their disordered structure with respect to network forming, organic or polymeric glasses.

Clear evidences of a dispersion curve are observed in almost all the studied glasses, at least within the first pseudo-Brillouin zone. An example is the orto-terphenyl glass (OTP)[71] shown in the bottom panel of Figure 7.

However, the width of the inelastic peak, Γ, of OTP becomes quickly comparable to the peak position as the border of the pseudo-Brillouin zone is approached (bottom part of the figure, right panel). When this happens, the inelastic feature cannot be considered as a damped vibrational mode but more properly as the spectral distribution of vibrational modes at the selected q. A common observation is the reduction of the sound velocity as the temperature is increased, reflecting the solid to liquid transition. On the contrary, the peak broadening is typically found to be temperature independent within experimental uncertainty,[88] suggesting the predominantly harmonic nature of the vibrational modes, although evidence of an-harmonic damping has been reported in silicate glasses.[76]

3.2.2. *Peak broadening and the Ioffe-Regel limit*

In many glasses the broadening of the inelastic peak is found to grow quadratically with the exchanged wavevector. This is the case in both systems of Figure 7. The metallic glass shows $\Gamma \sim q^2$ in the low q region, followed by an approximately linear dependence at higher qs. This linear regime is a peculiarity of metallic glasses, reminiscent of the behavior of liquid metals.[89] The OTP glass shows a quadratic increase of Γ with q in the entire range shown in the bottom panel of Figure 7. The precise nature of this quadratic dependence is still not clear, despite its observation goes back more than twenty years. For the specific case of vitreous silica, we have a quite robust interpretation, which stems from the comparison with the corresponding crystalline structure.[90] We will discuss this point in more detail in the following paragraphs. A quadratic increase of the peak broadening with frequency can also be a consequence of an-harmonicities, as we observed in a sodium silicate glass.[76]

It is commonly accepted that the inelastic peak can be considered as a damped longitudinal acoustic phonon if the broadening is much smaller than the peak position. A qualitative criterion to assess whether the peak can be treated as a single eigenmode is the Ioffe-Regel (IR) limit,[91] defined by the condition:

$$l \sim \lambda. \tag{6}$$

The IR limit is reached at the frequency where the amplitude mean free path, l, equals a wavelength, λ. The condition can be rewritten in terms of the full width at half maximum, Γ, and of the position, Ω, of the inelastic peak as:

$$\Gamma \sim {}^{\Omega}\!/_{\pi}. \tag{7}$$

It was initially believed, in analogy with the theory of electron conduction, that the IR limit marked the transition from propagating to localized excitations. That this is not the case became evident from MD simulations, showing that the vibrational modes above the IR limit have a high participation ratio, characteristic of extended vibrational modes.[92] The localization threshold is found at much higher frequency, towards the end of the spectrum (for instance, in silica at approximately 30 THz, or 1,000 cm^{-1}, see Figure 1).

The IR limit in many glasses, including oxides, polymers and organic ones, has been found at a frequency comparable to that of the BP.[41] This coincidence between ω_{IR} and ω_{BP} has been debated in the literature.[93–95] In particular, it is not observed in metallic glasses.[33,85] An example is the NiZr of Figure 7, where the IR frequency is almost an order of magnitude higher than the BP frequency, as shown in the upper right panel of the figure. This observation further indicates how the dispersion curves of metallic glasses are well defined, with sharp inelastic features persisting at q values significantly higher than in other materials.

3.2.3. *Rayleigh-like scattering and negative dispersion*

Improvement in the energy resolution, luminosity and available flux of the IXS beamlines has allowed the detection of Rayleigh-like scattering of elastic waves in glasses.[35,41,65,69,76,78,96,97] In analogy with the scattering of electromagnetic radiation by small particles, localized impurities in a solid can cause a sound attenuation which increases as the fourth power of the frequency.[98,99] Recent theories for the vibrations in disordered solids suggest that Rayleigh scattering should be present also in glasses.[10] An increase of the sound damping as the fourth power of the wavevector has been indeed measured by means of IXS in a few systems, including vitreous silica,[69] sodium silicate glasses,[76] alkali borates,[41] glycerol[65] and an organic glass.[35] In all these systems the Rayleigh scattering regime is measured at low frequencies, below the IR limit and is accompanied by a negative dispersion of the apparent sound velocity.

The data for silica at a temperature slightly above the glass transition are reported in Figure 8. The sound velocity, top panel, has an initial decrease from the macroscopic value of the longitudinal sound wave followed by an increase at frequencies exceeding the boson peak position. The negative dispersion typically accompanies the Rayleigh scattering, while the positive dispersion is not always observed. The BP frequency at this temperature is highlighted by the vertical dashed line and marks the upper frequency for the Rayleigh scattering, as shown in the middle panel. Also shown is the coincidence between ω_{IR} and ω_{BP}. At higher frequencies, the width of the inelastic peak, estimated by a DHO model, follows a quadratic frequency dependence, like those seen in Figure 7. The bottom panel of the figure shows the reduced DOS as a function of temperature, highlighting the shift of the BP position as the temperature is increased, a peculiarity of vitreous silica whose elastic modulus increases with temperature (DOS from neutron scattering as in Figure 2, bottom panel, but in a wider T range).

The negative dispersion evidenced by the reduction of the sound velocity in the low frequency part of the figure (upper panel) indicates that the Debye approximation is no longer valid at these frequencies.

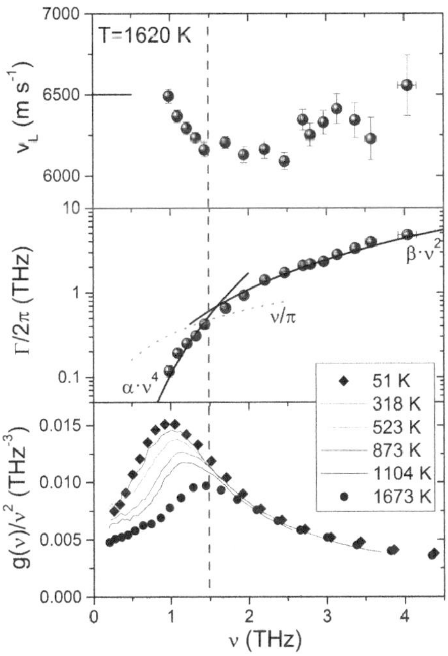

Fig. 8. Inelastic X-ray (the two upper panels) and neutron scattering (lower panel) data of v-SiO_2. The IXS data are collected at 1620 K, above the glass transition temperature. The upper panel shows the apparent sound velocity as determined from the ratio of the inelastic peak position and the exchanged wavevector. The peak broadening increases as the fourth power of the frequency (Rayleigh scattering) below the boson peak position and quadratically above. The two power-law dependences are indicated by the continuous lines. The vertical dashed line indicates the BP position determined by INS at a similar temperature (bottom panel).[21,100] The Ioffe-Regel frequency perfectly matches the BP frequency in this system, as indicated by the dashed line in the middle panel showing the quantity ν/π. Reprinted with permission from Ref. [69]. Copyright 2010 by the American Physical Society.

A deviation from the linear dispersion of long wavelengths sound waves has a direct effect on the density of states, giving rise to a deviation from the Debye law. However, a quantitative evaluation of the contribution of the observed negative dispersion to the DOS requires an estimate of the dispersion of the transverse acoustic modes, which are not directly accessible experimentally, at least in this low frequency regime. Assuming that the same negative dispersion affects also the transverse branch one can estimate the reduced DOS, which is in reasonable agreement with the boson peak position and intensity both in glycerol and in sorbitol glasses.[35,65] A more precise estimate is obtained by including the broadening of the longitudinal and transverse branches. However, the small departure from a linear dispersion is not sufficient to account for the BP intensity in the case of silica and silicate glasses.[69,76] In silica the agreement between the dispersion curve and the DOS can be recovered by considering the bending of the acoustic-like branch towards the border of the pseudo-Brillouin zone,[51] in analogy with the van-Hove singularity of the corresponding crystal, as discussed in Section 2.3.

This bending is not observed in the analysis reported in Figure 8, where the inelastic features are described in terms of a single excitation model. Evidences of a more complex phenomenology, with at least two separate inelastic peaks, is found in the high q range and will be discussed in Section 3.3.

3.2.4. *Sound attenuation in glasses*

Rayleigh scattering persists up to the boson peak frequency, above which the peak width follows a different law, growing with the square of the frequency in most glasses (see Figures 7 and 8). The microscopic origin of the Rayleigh scattering in glasses is debated. One interpretation is found within the soft potential model.[101] The model (see more in Chapter 8) assumes the existence of quasi-localized vibrational modes, which should coexist with the sound waves in the low frequency range, below the BP. Interaction between the sound waves and the quasi localized modes is predicted to give rise to the strong scattering, with the damping growing as the fourth power of frequency.[101] This view has received some support very recently, because MD simulations on model systems composed of a huge number of particles allow one to probe with accuracy the low frequency part of the DOS, revealing that indeed a considerable fraction of the vibrational modes are quasi localized in this frequency range.[16,18] However, it has also been shown that the density of quasi localized modes depends on the way the glass is prepared.

Specifically, more stable glasses tend to have a much lower density of quasi-localized, non-phononic, excitations with respect to glasses obtained by a fast quench.[19,102] Since the experiments are typically performed on stable glasses, it is difficult to judge at present if this mechanism is indeed the one responsible for the Rayleigh scattering.

A different view considers Rayleigh scattering in glasses to arise from local elastic modulus fluctuations[5,7] (see more in Chapter 9). These elastic heterogeneities would strongly scatter the sound waves, giving rise to the $\Gamma \sim q^4$ law. The strong scattering regime is observed when the wavevector is sufficiently small that the condition: $qa \ll 1$ is satisfied, where a is the mean size of the elastic modulus heterogeneities. In this view the BP would correspond to that frequency where the wavelength satisfies $\lambda \sim 2\pi a$ and the sound wave starts to feel the microscopic heterogeneity of the glass.

The IXS technique can probe the Rayleigh scattering only in a quite restricted frequency window, mainly because of limitations in the available energy resolution. A more detailed investigation of the sound attenuation in glasses requires the use of multiple experimental probes, including ultrasonic techniques, Brillouin light (BLS)[103,104] and UV (BUVS) scattering,[105,106] pico-second optical methods (POT)[107] and the tunneling junction (TJ) technique.[108] A summary of some of the data available for vitreous silica is reported in Figure 9. The frequency range between 100 GHz and 1 THz is the most difficult to probe. This region can be investigated, at present, only with sophisticated ultrasonic techniques based on the generation of acoustic pulses with optical methods.[109–111]

Another possible approach, which has been recently demonstrated, is the transient grating spectroscopy with deep UV radiation produced by a free electron laser.[112] However, both approaches can probe the sound attenuation of longitudinal waves only in thin films of a thickness of a few tens of nanometers, contrary to the bulk samples that can be studied with Brillouin scattering of light or X-rays.

The continuous line in Figure 9 is an extrapolation of the Rayleigh scattering observed by IXS towards lower frequencies, suggesting that the sound damping probed by the lower frequency

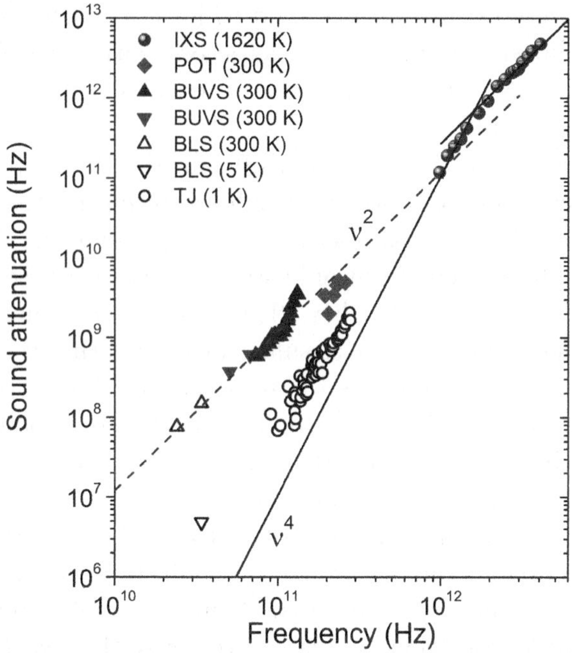

Fig. 9. Sound attenuation in vitreous silica as a function frequency at the temperatures indicated in the legend. Collection of data obtained with various techniques: inelastic X-ray scattering (IXS)[69] above T_g; picosecond optical technique (POT),[107] Brillouin ultraviolet scattering (BUVS) with synchrotron radiation[106] (upper full triangles) and with a laser source[105] (lower full triangles). Brillouin light scattering (BLS)[103,104] and tunneling junction method (TJ).[108] Reprinted with permission from Ref. [69]. Copyright 2010 by the American Physical Society.

methods is dominated by other attenuation mechanisms. Specifically, the room temperature data follow a quadratic frequency dependence, typical of an-harmonic and relaxational contributions. More recent data.[109–111] confirm this trend up to approximately 300 GHz. However, also the data at lower temperatures remain higher than the expected Rayleigh term. Room temperature IXS data, not shown in the figure, give some evidence that this discrepancy could be associated with the variation of the BP position with temperature.[113] The Rayleigh scattering seems to be slightly shifted to lower frequencies at room temperature, following the shift of the boson peak, in qualitative agreement with the TJ data at 5 K. Nevertheless, the room temperature IXS data are quite noisy and it is difficult to draw a definite conclusion from the experiments.

Recently, a series of numerical works has observed the possible existence of a logarithmic correction to the Rayleigh law, suggesting that the discrepancy between the TJ and the IXS data in silica is indeed related to this correction.[114] Those numerical works indicate the origin of the logarithmic correction in a long-range correlation of the elastic modulus fluctuations.[115] Other MD works associate this anomalous Rayleigh scattering to the presence of the previously mentioned quasi localized vibrational modes.[102]

3.3. *Evidences of two excitations*

The existence of acoustic-like excitations propagating up to terahertz frequencies in all the amorphous solids studied so far has stimulated the search for evidences of transverse acoustic modes.[31, 33, 61, 62, 82, 116–120] Numerical simulations.[82, 95, 121–123] suggest the boson peak to have a predominantly transverse character, stimulating research in this direction. In a purely acoustic picture, the weight of the transverse modes in the DOS far exceeds that of the longitudinal ones because of their lower sound velocity, as in the Debye model. An X-ray or neutron scattering experiment is, however, blind to the transverse vibrational modes of an isotropic solid, being sensitive only to vibrations with a component parallel to the exchanged wavevector q, as discussed in Section 3.1. Indeed, only a single inelastic peak is measured, at least for wavevectors below one-half of the pseudo-Brillouin zone boundary. At shorter wavelengths — higher qs — a second inelastic feature appears in the spectra and becomes progressively more and more evident as q increases, as reported by Ruzicka *et al.*[116] in a first X-ray scattering study on silica.

The inelastic scattering intensity together with the corresponding current correlation functions of v-SiO_2 are shown in Figure 10 at selected q values, as measured in a later experiment.[119] The inelastic spectra show two evident broad features. The lower-frequency inelastic component is better evidenced in the plots of

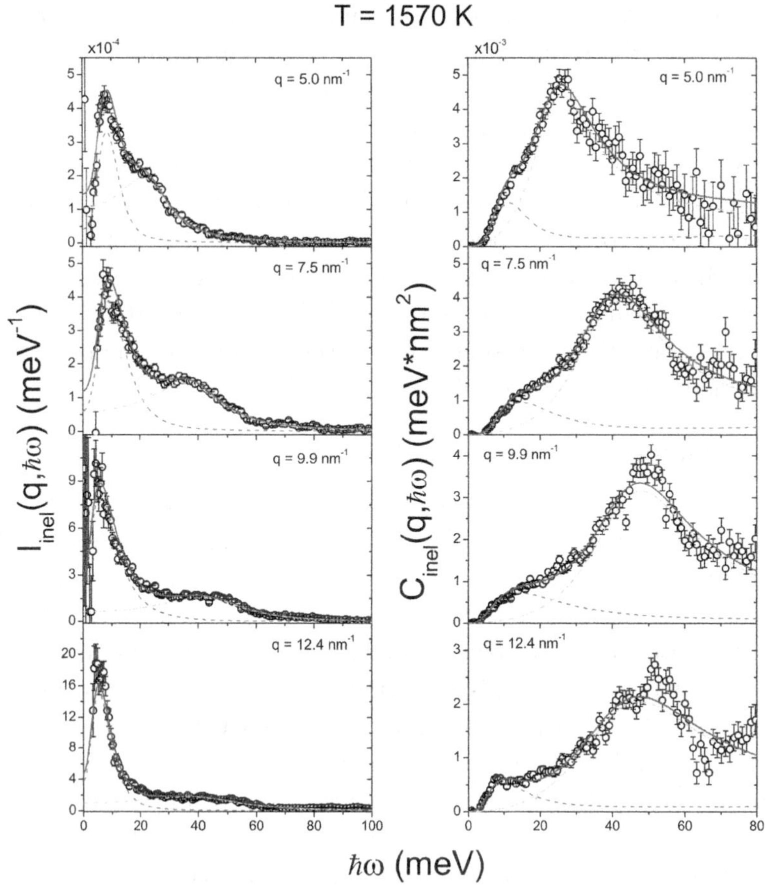

Fig. 10. Inelastic part of the IXS spectra measured at T=1570 K on v-SiO_2 at selected exchanged wave vectors; for the details of data treatment see Ref. [119]. In the left panel the spectra are plotted as $I_{inel}(q\omega)$. The continuous line is the best-fitting function obtained by means of two DHO functions (dashed and dashed-dotted lines) convoluted with the instrumental resolution function. The right panel shows the same spectra plotted as longitudinal inelastic currents; same symbols as in the left panel. Reprinted with permission from Ref. [119]. Copyright 2008 by the American Physical Society.

the inelastic intensity, $I_{inel}(q,\omega)$, left column of the figure, while the higher frequency one is enhanced in the current correlation function, $C_{inel}(q,\omega)$, where $C_{inel} = \omega^2/q^2 \cdot I_{inel}$, by the ω^2 factor. Evidences of two inelastic features appearing within the first pseudo-Brillouin zone have been reported for systems of different structure and composition both with IXS and INS. Examples include other oxide glasses,[82] metallic glasses,[33,80] chalcogenides,[31,61] organic materials[62,117,120] and even water.[118] The case of the metallic glass $Pd_{77}Si_{16.5}Cu_{6.5}$ is included in Figure 11.

The two inelastic features observed in v-SiO_2 present a dispersion with q, as shown in the upper panel of Figure 11, left side, together with points determined in previous investigations and including neutron data for the lower frequency peak. It is worth noting that data previously shown in the upper panels of Figure 8 span the wavevector range from approximately 1 to $4.5\,nm^{-1}$. The macroscopic longitudinal and transverse sound velocities measured by Brillouin light scattering[124] are used to calculate the linear dispersion curves shown in the figure. Two inelastic features are present in the spectra when q exceeds some $4\,nm^{-1}$. The first few points of the lower frequency excitation are in good agreement with the linear dispersion of the macroscopic transverse sound wave, suggesting that the second excitation which appears in the spectrum is indeed the transverse acoustic mode. More precisely, this mode corresponds to the projection of the transverse acoustic mode in the longitudinal current, which is not negligible because of the high value of the wavevector. The phenomenon is sometimes called "mixing" and was initially revealed in numerical simulation studies where it appeared clear that the wave polarization was progressively lost as the wavevector increased towards the pseudo-Brillouin zone boundary.[122,125]

In fact, the comparison of the transverse and longitudinal current correlation functions, calculated from MD simulations, indicated that, on increasing q, the transverse peak appeared also in the longitudinal current and, vice-versa, the longitudinal peak appeared in the transverse one. It should also be kept in mind that the two inelastic features are broad peaks since they appear at a wavevector well above the Ioffe-Regel limit. Consequently, the two inelastic features cannot be considered as two excitations but more properly as a distribution of many eigenstates. It is also of interest to consider the energies of the peaks of the two inelastic features.

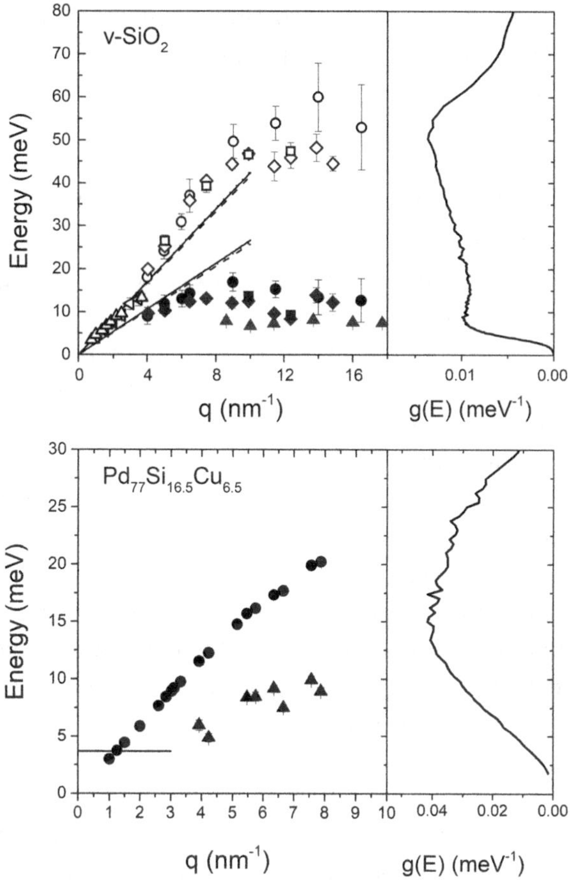

Fig. 11. Top. Left side: dispersion relation of v-SiO_2 from the IXS and INS data. Open (longitudinal branch) and filled (transverse branch) diamonds: T = 1570 K, open and filled squares: T = 920 K, from Ref. [119]; open and filled circles: Ref. [116], T = 1270 K; up triangles: from Ref. [126], T = 1375 K; down triangles: Ref. [66], T = 1050 K. The position of the peak in the INS spectra is reported as filled triangles.[119] The lines correspond to the macroscopic sound velocity measured by Brillouin light scattering[124] for the longitudinal and transverse sound waves: $v_L = 6320$ m/s and $v_T = 3910$ m/s at T = 920 K (dashed), $v_L = 6440$ m/s and $v_T = 4040$ m/s at T = 1270 K (continuous), $v_L = 6500$ m/s and $v_T = 3980$ m/s at T = 1570 K (dotted). Reprinted with permission from Ref. [119]. Copyright 2008 by the American Physical Society. Right side: $g(\omega)$ measured by INS, revised data from Ref. [34]. Bottom. Left side: dispersion relation of the $Pd_{77}Si_{16.5}Cu_{6.5}$ metallic glass from IXS measurements. Longitudinal branch (full circles) and transverse branch (full triangles). The straight line indicates the position of the boson peak. Right side: $g(\omega)$ measured by means of INS. Reprinted with permission from Ref. [33] with the permission of AIP Publishing.

The acoustic modes of quartz and cristobalite are restricted to below some 20 meV. Above this energy the vibrations of the crystalline polymorphs of silica are optic modes. This observation suggests that the lower frequency inelastic peak of vitreous silica has a predominantly acoustic character, while the higher frequency one is mostly equivalent to the optic modes of the corresponding crystal. The dispersion of the higher frequency peak displays, in fact, a strong deviation from the linear dispersion of the longitudinal branch just above the point where the second excitation appears in the spectrum. This kind of "positive dispersion" is possibly related to a partial mixing of acoustic and optical branches, whose presence in the glass is confirmed by Raman, neutron scattering and infrared absorption measurements.[20,127] Both inelastic features lose any appreciable dispersion above 6–7 nm^{-1}. This q value is close to the pseudo-Brillouin zone boundary estimated as one-half the position of the first sharp diffraction peak in the static structure factor.

The dispersion curves of the metallic glass $Pd_{77}Si_{16.5}Cu_{6.5}$ are shown in the lower panel of Figure 11. A second, lower frequency excitation, appears also in this system when q exceeds approximately one-half of the pseudo-Brillouin zone boundary. The DOS, $g(\omega)$, obtained by inelastic neutron scattering at room temperature on v-SiO_2 is reported in the right side of the top panel of Figure 11, to compare it with the dispersion curves. The comparison between the DOS and the dispersion curves shows that the dynamic structure factor evolves towards the density of vibrational states as q is increased, as expected from the incoherent approximation, whose validity is assumed in the calculation of $g(\omega)$ from the neutron spectra.

In the case of vitreous silica, the figure indicates that the lower frequency inelastic feature evolves towards the lower peak of $g(\omega)$, located at approximately 10 meV. When seen as reduced DOS, $g(\omega)/\omega^2$, this first peak is the boson peak and is located at 4 meV. The higher frequency inelastic feature of the silica glass evolves towards a second peak, appearing in the DOS at about 50 meV. The convergence of the lower frequency peak towards the transverse acoustic dispersion at low qs confirms that at least a part of the BP intensity has a transverse nature. This discussion does not exclude the presence of additional modes in the low frequency part of the spectrum, such as the quasi localized excitations observed in recent numerical simulations.[16–19] However, the presence of those modes is not revealed in scattering experiments as the main contribution to the Brillouin scattering, even at terahertz frequencies, comes from the acoustic-like excitations discussed above.

3.4. *Dynamics of glasses and corresponding polycrystals*

Scattering experiments carried out within the first pseudo-Brillouin zone indicate that the dynamic structure factor evolves as a function of q and displays oscillations reminiscent of crystals dispersion curves. However, in most systems, with the notable exception of metallic glasses, the Ioffe-Regel limit is reached at a relatively low frequency, comparable to the position of the boson peak. At least in these glasses we cannot thus talk of vibrational branches, where well-defined excitations follow a dispersion curve. On the contrary, the experiments suggest the presence of a distribution of vibrational modes whose amplitude has a non-trivial q-dependence, gradually evolving towards that expected from the density of states. Consequently, scattering experiments do not tell us much about the nature of the glasses vibrational modes at frequencies exceeding the Ioffe-Regel limit. At lower frequencies, on the contrary, the Brillouin spectra clearly indicate the presence of a longitudinal acoustic mode with a sound velocity compatible with the one determined from macroscopic measurements.

A line of research has tried to compare the dynamics of the glass with that of the corresponding crystal, to obtain some additional information on the nature of the terahertz vibrations of glasses, in the same spirit of the approaches discussed in Section 2.3.

The idea is to compare a glass with the crystalline polymorph with similar macroscopic properties, such as sound velocities and mass density. Unfortunately, it is not always possible to perform this comparison either because of difficulties in preparing the crystal or because of the presence of more than one crystalline structure preventing the identification of the one that corresponds more closely to the glass, in terms of macroscopic elastic properties. Consequently, the number of studies is still quite limited and includes investigations performed on ethanol,[74] amorphous water,[72] vitreous silica,[128] GeO_2,[129] BeF_2[129] and permanently densified silica.[90]

We will here focus our attention to the case of the permanently densified silica with a mass density close to that of the crystal of α-quartz. The study has been performed by means of IXS and has investigated in detail the low q region, including the range extending from the typical macroscopic response up to the border of the pseudo-Brillouin zone.[90] The Stokes side of the inelastic spectra of the glass and of α-quartz polycrystal are compared in Figure 12 at selected q values together with the spectrum of the polycrystal determined by an ab-initio calculation. A striking result is the observation of two regimes. A low q regime where the inelastic spectra of the polycrystal and of the glass are different and a high q region where the spectra of the two samples are very similar. The transition between the two takes place at a crossover wavevector $Q_c \sim 2\,\text{nm}^{-1}$. In the low q region the spectrum of the polycrystal is broader than that of the glass, while above the crossover wavevector the inelastic spectra of the two become almost indistinguishable. The polycrystal spectrum presents sharper features with respect to that of the glass at all the probed qs, with peaks that correspond to the single crystal branches averaged over all the possible directions in reciprocal space. In the limit of high qs the incoherent approximation applies, and the spectrum resembles the density of states (panel f of Figure 12).

The distinction between the two regimes, above and below Q_c, is better highlighted by fitting the spectrum with a single excitation model, a simple damped harmonic oscillator. The DHO model allows one to give a rough estimate of the width of the spectrum, even when it is composed by a distribution of normal modes and not by a single excitation. The result of this analysis is shown in Figure 13. Below Q_c the glass spectrum is well described by the single DHO model, as shown in Figure 12, panels (a) and (b). The mode can be identified as a longitudinal acoustic vibration because its dispersion follows the longitudinal sound velocity.

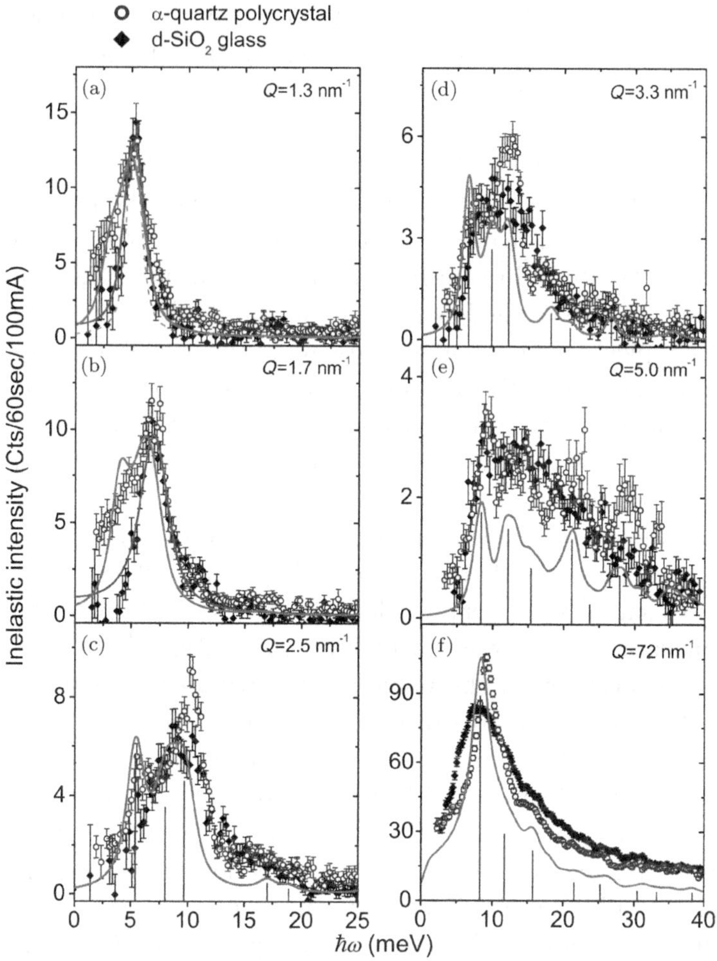

Fig. 12. Comparison of the inelastic spectra, Stokes side, of an α-quartz polycrystal (open circles) and of permanently densified SiO_2 glass (full diamonds) with matching mass densities. The spectra are measured by means of IXS and shown at selected wave-vectors. The continuous line is the IXS spectrum of a polycrystal of α-quartz determined with an ab-initio calculation. The vertical lines indicate the average positions of the different branches of the crystal. The three lines of lower frequency derive from the acoustic branches while the higher frequency ones correspond to the optic branches of the single crystal. It is worth noting that the spectra are reported in absolute units, without any rescaling. Panels (a) and (b) include also the fit of the glass spectrum with a DHO model. The instrumental resolution function is reported as a dashed line in panel (a). Reprinted with permission from Ref. [90]. Copyright 2013 by the American Physical Society.

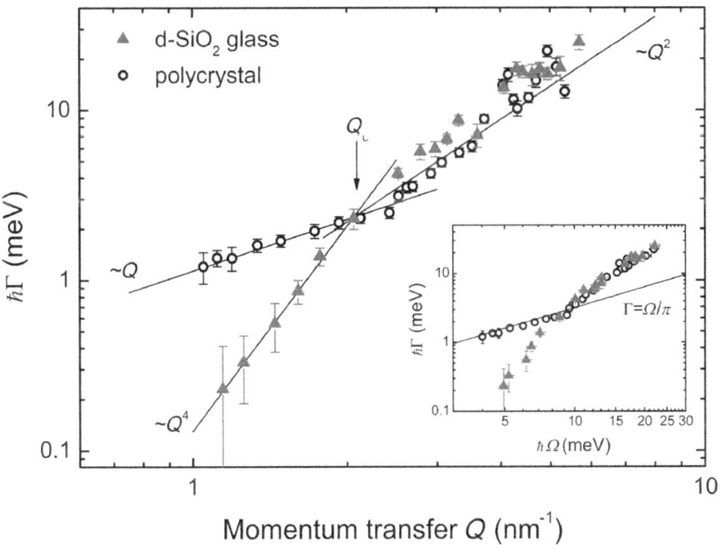

Fig. 13. Full width at half maximum of the DHO model used to fit the spectra of a permanently densified silica glass (full triangles) and of α-quartz polycrystal (open circles). The lines indicate different power law behaviors. Q_c is the crossover wavevector above which the spectra of the glass and the polycrystal become very similar. Inset: same data as a function of the frequency of the inelastic peak (the maximum of the DHO, which corresponds to the maximum of the current correlation function). The line indicates the Ioffe-Regel condition. Reprinted with permission from Ref. [90]. Copyright 2013 by the American Physical Society.

The sound attenuation follows the Rayleigh scattering law and increases rapidly with q until the spectrum becomes comparable to that of the polycrystal at Q_c. In this low q region the spectrum of the polycrystal is broader than that of the glass and its width follows a linear dependence on q. The ab-initio calculation well reproduces the experimental spectrum and shows that this linear increase of the linewidth is a consequence of the contribution of the quasi-transverse branches, whose polarization is not purely transverse because of the anisotropy of quartz. At wavevectors above Q_c the peak broadening for both the glass and the polycrystal follows a quadratic dependence on q. This growth for the polycrystal is due to the quadratic increase with q of the optic mode intensity, whose effect is a corresponding spectral broadening.

It is worth noting that the Ioffe-Regel condition coincides within experimental uncertainty with the linear broadening of the polycrystal spectrum, as shown in the inset of Figure 13. This apparent coincidence can be understood as follows. The mode dispersion in this frequency range is approximately linear and the ab-initio calculation suggests the contribution of all three acoustic branches. If, for simplicity, we assume to have only one quasi-transverse and one quasi-longitudinal branch with a similar weight, the peak broadening is expected to be:

$$\Delta\omega \sim (\mathrm{v}_L - \mathrm{v}_T)\, q \sim \left(1 - \frac{\mathrm{v}_T}{\mathrm{v}_L}\right) \omega \sim \frac{\omega}{3}. \tag{8}$$

Here, v_L and v_T are the macroscopic longitudinal and transverse sound velocities, with $\mathrm{v}_T \sim 2/3 \cdot \mathrm{v}_L$ for silica. Condition (8) is very similar to the Ioffe-Regel limit, Eq. (7). Moreover, the transition between the macroscopic regime for the glass, where a single acoustic mode has a damping which follows the Rayleigh law, and the microscopic range, where the spectra of the polycrystal are very similar to those of the glass, takes place at the boson peak energy, at about 9 meV (Figure 12(f)). Since the boson peak has approximately the same position of the first van-Hove singularity of the crystal, the proximity of the Ioffe-Regel and the BP frequency is possibly a coincidence, related to the shape of the dispersion of the quasi-transverse branch.

The comparison of the dynamics in glasses and in the corresponding polycrystals can help to shed some light on the nature of the glass vibrational modes in this low frequency regime, at least for the case of densified silica. One can imagine the glass elasticity to be heterogeneous at the nanometer or sub-nanometer scale, the typical length being set by $1/Q_c$. For wavelengths exceeding $2\pi/Q_c$ the glass behaves as an isotropic continuum and the fluctuations of the local elastic modulus give rise to the Rayleigh scattering. At small wavelengths, $\lambda < 2\pi/Q_c$, the elastic wave feels the local atomic structure and the dynamics is similar to the one of the corresponding polycrystal. However, because of the high anisotropy of quartz, it is difficult to make predictions on the nature of the vibrational modes of the polycrystal, because the crystal symmetry is lost by the average in reciprocal space and many eigenmodes contribute to the spectrum. Nevertheless, the similarity between the polycrystal and the glass spectrum suggests that the modes around the boson peak frequency are predominantly extended and not localized. This is one of the main conclusions of the studies here summarized, together with the estimate of the average size of the local elastic modulus fluctuations, $\sim 1/Q_c \sim 0.5\,\text{nm}$.

Extensions of this approach to other glasses are not trivial, both because of limitations in the wavevector range which is experimentally accessible and because of the difficulty of finding the proper corresponding polycrystal of a given glass. At the same time there are indications of counterexamples where the dynamics of the glass and of the corresponding crystal do not match. This discrepancy has been reported in a two-dimensional model system[130] and in numerical simulations of a Lennard-Jones glass.[14]

4. Conclusions

We have reviewed the main experimental observations of the vibrational dynamics of amorphous solids performed by means of scattering methods. The measurement of the density of vibrational states indicates, in all the studied glasses, the presence of a boson peak, whose position spans a range of approximately a decade in frequency, centered around one terahertz. The shape of this feature is similar in very different glasses, particularly at frequencies below the peak. At present, the connection between the frequency of the BP and other parameters of the glass, such as its fragility or its elasticity, remains elusive. When the glass is subjected to variations of external parameters, such as pressure, density, temperature or annealing time, the BP has a shift and a corresponding variation of intensity, that normally follows a simple counting law, related to the normalization of the density of states. Applying the same counting law to a series of different glasses and scaling all the density of states by the BP position, we observe that the peak intensity has fluctuations to within a factor two or three but with considerable uncertainties.

A few studies have highlighted the similarity of the BP with the first van-Hove singularity of the corresponding crystal with comparable mass density and elastic properties. This equivalence is quite convincing in strong network forming glasses. Another interesting result is that the density of states of a few systems, densified silica,[51] boron oxide[58] and glycerol,[59] converges towards the Debye limit at low frequencies. These findings pose new questions on the role of the structural disorder as origin of the boson peak. Nevertheless, their generality should be verified by new scattering measurements on a wider selection of glasses.

Evidences of pseudo-acoustic modes propagating at terahertz frequencies in glasses have been provided by inelastic X-ray scattering experiments. The technique detects, in many glasses, an acoustic mode that follows, at low frequencies, the macroscopic longitudinal sound velocity. In those systems where the measurements can be performed at frequencies lower than the BP position, this acoustic mode has an attenuation that follows the Rayleigh scattering law. In the same frequency range the mode shows a negative dispersion, which can explain quantitatively the BP intensity in the cases of glycerol and sorbitol glasses.[35,65] In silicate glasses the observed negative dispersion alone is not sufficient to justify the deviation from the Debye law observed at the BP.[76,113] A quantitative comparison of the DOS of the glass with that of the corresponding crystal of similar density suggests that the BP is equivalent to the first van-Hove singularity.[51,90] However, the acoustic mode becomes ill-defined above the Ioffe-Regel limit, so that the nature of the glass vibrational modes around the BP frequency remains unclear.

When the exchanged wavevector exceeds the value where the inelastic spectrum is peaked at the BP frequency, the spectrum starts to display a complex phenomenology. We have reviewed the evidences of transverse dynamics mixed to the longitudinal branch and the appearance of two inelastic features. In a few cases, the wavevector evolution of the inelastic spectrum can be compared with that of the corresponding polycrystal. The dynamics is different below the BP frequency, where the glass shows a single damped pseudo-acoustic mode as expected for an isotropic solid. Above the BP the spectra of the glass and of the corresponding polycrystal become very similar, indicating the contribution of a multitude of eigenvectors. In the polycrystal these modes are associated with the various branches of the single crystal averaged over the directions in reciprocal space.

Recent numerical simulations of model systems composed of millions of atoms have highlighted the coexistence of phonon like modes with quasi localized excitations in glasses in the proximity of the BP frequency.[16, 17, 19, 102] Although the applicability of these results to real systems is still under scrutiny, they motivate the realization of new experiments searching for evidences of these excitations. Another interesting issue regards the origin of the Rayleigh scattering and whether it is related to those quasi local modes or whether it originates from the elastic modulus fluctuations present in glasses at the nanometer scale. Recent theories for the elasticity in glasses suggest these fluctuations to extend to the long range, with the consequence that the Rayleigh law should be modified by a logarithmic correction.[114, 115, 131, 132] Finally, the nice agreement between the spectra of the glass and of the corresponding polycrystal observed in a few systems seems questionable for the Lennard Jones glass, leaving the connection between the BP and the first van Hove singularity of the corresponding crystal an open question.

References

1. L. D. Landau and E. M. Lifshitz. (1986). *Theory of Elasticity.* 3rd ed. Oxford: Elsevier.
2. Y. M. Galperin, V. G. Karpov, and V. I. Kozub. (1989). Localized states in glasses. *Adv. Phys.* **38**, 669–737.
3. W. A. Phillips. (1987). Two-level states in glasses. *Rep. Prog. Phys.* **50**, 1657–1708.
4. E. Duval, A. Mermet, and L. Saviot. (2007). Boson peak and hybridization of acoustic modes with vibrations of nanometric heterogeneities in glasses. *Phys. Rev. B.* **75**, 024201.
5. W. Schirmacher, G. Diezemann, and C. Ganter. (1998). Harmonic vibrational excitations in disordered solids and the "Boson Peak." *Phys. Rev. Lett.* **81**, 136–139.
6. W. Schirmacher. (2006). Thermal conductivity of glassy materials and the "boson peak". *Europhys. Lett.* **73**, 892–898.

7. W. Schirmacher, G. Ruocco, and T. Scopigno. (2007). Acoustic attenuation in glasses and its relation with the Boson peak. *Phys. Rev. Lett.* **98**, 025501.
8. W. Schirmacher, B. Schmid, C. Tomaras, G. Viliani, G. Baldi, G. Ruocco, and T. Scopigno. (2008). Vibrational excitations in systems with correlated disorder. *Phys. Status Solidi.* **5**, 862–866.
9. W. Schirmacher, C. Tomaras, B. Schmid, G. Baldi, G. Viliani, G. Ruocco, and T. Scopigno. (2010). Sound attenuation and anharmonic damping in solids with correlated disorder. *Condens. Matter Phys.* **13**, 23606.
10. C. Ganter and W. Schirmacher. (2010). Rayleigh scattering, long-time tails, and the harmonic spectrum of topologically disordered systems. *Phys. Rev. B.* **82**, 094205.
11. T. S. Grigera, V. Martín-Mayor, G. Parisi, and P. Verrocchio. (2001). Vibrational spectrum of topologically disordered systems. *Phys. Rev. Lett.* **87**, 085502.
12. L. E. Silbert, A. J. Liu, and S. R. Nagel. (2005). Vibrations and diverging length scales near the unjamming transition. *Phys. Rev. Lett.* **95**, 098301.
13. S. N. Taraskin, Y. L. Loh, G. Natarajan, and S. R. Elliott. (2001). Origin of the Boson peak in systems with lattice disorder. *Phys. Rev. Lett.* **86**, 1255–1258.
14. G. Monaco and S. Mossa. (2009). Anomalous properties of the acoustic excitations in glasses on the mesoscopic length scale. *Proc. Natl. Acad. Sci.* **106**, 16907–16912.
15. H. Mizuno, S. Mossa, and J.-L. Barrat. (2014). Acoustic excitations and elastic heterogeneities in disordered solids. *Proc. Natl. Acad. Sci. U. S. A.* **111**, 11949–11954.
16. H. Mizuno, H. Shiba, and A. Ikeda. (2017). Continuum limit of the vibrational properties of amorphous solids. *Proc. Natl. Acad. Sci.* **114**, E9767–E9774.
17. E. Lerner and E. Bouchbinder. (2018). Frustration-induced internal stresses are responsible for quasilocalized modes in structural glasses. *Phys. Rev. E.* **97**, 032140.
18. L. Angelani, M. Paoluzzi, G. Parisi, and G. Ruocco. (2018). Probing the non-Debye low-frequency excitations in glasses through random pinning. *Proc. Natl. Acad. Sci.* **115**, 8700–8704.
19. L. Wang, A. Ninarello, P. Guan, L. Berthier, G. Szamel, and E. Flenner. (2019). Low-frequency vibrational modes of stable glasses. *Nat. Commun.* **10**, 26.

20. F. L. Galeener, A. J. Leadbetter, and M. W. Stringfellow. (1983). Comparison of the neutron, Raman, and infrared vibrational spectra of vitreous SiO_2, GeO_2, and BeF_2. *Phys. Rev. B.* **27**, 1052–1078.
21. A. Fontana, R. Dell'Anna, M. Montagna, F. Rossi, G. Viliani, G. Ruocco, M. Sampoli, U. Buchenau, and A. Wischnewski. (1999). The Raman coupling function in amorphous silica and the nature of the long-wavelength excitations in disordered systems. *Europhys. Lett.* **47**, 56–62.
22. R. H. Doremus. (2002). Viscosity of silica. *J. Appl. Phys.* **92**, 7619–7629.
23. F. L. Galeener and P. N. Sen. (1978). Theory for the first-order vibrational spectra of disordered solids. *Phys. Rev. B.* **17**, 1928–1933.
24. A. Fontana, F. Rossi, G. Viliani, S. Caponi, E. Fabiani, G. Baldi, G. Ruocco, and R. D. Maschio. (2007). The Raman coupling function in disordered solids: A light and neutron scattering study on glasses of different fragility. *J. Phys. Condens. Matter.* **19**, 205145.
25. U. Buchenau, N. Nücker, and A. J. Dianoux. (1984). Neutron scattering study of the low-frequency vibrations in vitreous silica. *Phys. Rev. Lett.* **53**, 2316–2319.
26. V. K. Malinovsky, V. N. Novikov, P. P. Parshin, A. P. Sokolov, and M. G. Zemlyanov. (1990). Universal form of the low-energy (2 to 10 meV) vibrational spectrum of glasses. *Europhys. Lett.* **11**, 43–47.
27. V. K. Malinovsky, V. N. Novikov, and A. P. Sokolov. (1991). Log-normal spectrum of low-energy vibrational excitations in glasses. *Phys. Lett. A.* **153**, 63–66.
28. W. A. Phillips, U. Buchenau, N. Nücker, A.-J. Dianoux, and W. Petry. (1989). Dynamics of glassy and liquid selenium. *Phys. Rev. Lett.* **63**, 2381–2384.
29. K. Niss, B. Begen, B. Frick, J. Ollivier, A. Beraud, A. Sokolov, V. N. Novikov, and C. Alba-Simionesco. (2007). Influence of pressure on the Boson peak: Stronger than elastic medium transformation. *Phys. Rev. Lett.* **99**, 055502.
30. J.-B. Suck. (1996). Temperature dependence of the total dynamic structure factor of the metallic glass $Ni_{33}Zr_{67}$. *J. Non. Cryst. Solids.* **205–207**, 592–596.
31. L. Orsingher, G. Baldi, A. Fontana, L. E. Bove, T. Unruh, A. Orecchini, C. Petrillo, N. Violini, and F. Sacchetti. (2010). High-frequency dynamics of vitreous $GeSe_2$. *Phys. Rev. B.* **82**, 115201.
32. B. Frick, U. Buchenau, L. Comez, A. Sokolov, C. Masciovecchio, A. Mermet, G. Ruocco, F. Sette, and L. Willner. (1999). High-frequency dynamics of glass-forming polybutadiene. *Phys. Rev. E — Stat. Physics, Plasmas, Fluids, Relat. Interdiscip. Top.* **59**, 4470–4475.

33. P. Bruna, G. Baldi, E. Pineda, J. Serrano, J.-B. Suck, D. Crespo, and G. Monaco. (2011). Communication: Are metallic glasses different from other glasses? A closer look at their high frequency dynamics. *J. Chem. Phys.* **135**, 101101.
34. E. Fabiani, A. Fontana, and U. Buchenau. (2008). Neutron scattering study of the vibrations in vitreous silica and germania. *J. Chem. Phys.* **128**, 244507.
35. B. Ruta, G. Baldi, V. M. Giordano, L. Orsingher, S. Rols, F. Scarponi, and G. Monaco. (2010). Communication: High-frequency acoustic excitations and boson peak in glasses: A study of their temperature dependence. *J. Chem. Phys.* **133**, 041101.
36. M. M. Koza. (2008). Vibrational dynamics of amorphous ice structures studied by high-resolution neutron spectroscopy. *Phys. Rev. B.* **78**, 064303.
37. G. Baldi, A. Fontana, G. Monaco, L. Orsingher, S. Rols, F. Rossi, and B. Ruta. (2009) Connection between Boson peak and elastic properties in silicate glasses. *Phys. Rev. Lett.* **102**, 195502.
38. S. Caponi, A. Fontana, F. Rossi, G. Baldi, and E. Fabiani. (2007). Effect of temperature on the vibrational density of states in vitreous SiO_2: A Raman study. *Phys. Rev. B.* **76**, 092201.
39. A. P. Sokolov, R. Calemczuk, B. Salce, A. Kisliuk, D. Quitmann, and E. Duval. (1997). Low-temperature anomalies in strong and fragile glass formers. *Phys. Rev. Lett.* **78**, 2405–2408.
40. U. Buchenau, M. Prager, N. Nücker, A. J. Dianoux, N. Ahmad, and W. A. Phillips. (1986). Low-frequency modes in vitreous silica. *Phys. Rev. B.* **34**, 5665–5673.
41. B. Rufflé, G. Guimbretière, E. Courtens, R. Vacher, and G. Monaco. (2006). Glass-specific behavior in the damping of acousticlike vibrations. *Phys. Rev. Lett.* **96**, 045502.
42. A. I. Chumakov, G. Monaco, A. Monaco, W. A. Crichton, A. Bosak, R. Rüffer, A. Meyer, F. Kargl, L. Comez, D. Fioretto, H. Giefers, S. Roitsch, G. Wortmann, M. H. Manghnani, A. Hushur, Q. Williams, J. Balogh, K. Parliński, P. Jochym, and P. Piekarz. (2011). Equivalence of the Boson peak in glasses to the transverse acoustic van hove singularity in crystals. *Phys. Rev. Lett.* **106**, 225501.
43. M. Zanatta, G. Baldi, S. Caponi, A. Fontana, C. Petrillo, F. Rossi, and F. Sacchetti. (2011). Debye to non-Debye scaling of the Boson peak dynamics: Critical behavior and local disorder in vitreous germania. *J. Chem. Phys.* **135**, 174506.
44. L. Hong, B. Begen, A. Kisliuk, C. Alba-Simionesco, V. N. Novikov, and A. P. Sokolov. (2008). Pressure and density dependence of the boson peak in polymers. *Phys. Rev. B.* **78**, 134201.

45. Y. Inamura, M. Arai, O. Yamamuro, A. Inaba, N. Kitamura, T. Otomo, T. Matsuo, S. Bennington, and A. Hannon. (1999). Peculiar suppression of the specific heat and boson peak intensity of densified SiO_2 glass. *Phys. B Condens. Matter.* **263–264**, 299–302.
46. Y. Inamura, M. Arai, T. Otomo, N. Kitamura, and U. Buchenau. (2000). Density dependence of the boson peak of vitreous silica. *Phys. B Condens. Matter.* **284–288**, 1157–1158.
47. Y. Inamura, M. Arai, M. Nakamura, T. Otomo, N. Kitamura, S. M. Bennington, A. C. Hannon, and U. Buchenau. (2001). Intermediate range structure and low-energy dynamics of densified vitreous silica. *J. Non. Cryst. Solids.* **293–295**, 389–393.
48. A. Monaco, A. I. Chumakov, G. Monaco, W. A. Crichton, A. Meyer, L. Comez, D. Fioretto, J. Korecki, and R. Rüffer. (2006). Effect of densification on the density of vibrational states of glasses. *Phys. Rev. Lett.* **97**, 135501.
49. L. Orsingher, A. Fontana, E. Gilioli, G. Carini, G. Carini, G. Tripodo, T. Unruh, and U. Buchenau. (2010). Vibrational dynamics of permanently densified GeO_2 glasses: Densification-induced changes in the boson peak. *J. Chem. Phys.* **132**, 124508.
50. M. Zanatta, G. Baldi, S. Caponi, A. Fontana, E. Gilioli, M. Krish, C. Masciovecchio, G. Monaco, L. Orsingher, F. Rossi, G. Ruocco, and R. Verbeni. (2010). Elastic properties of permanently densified silica: A Raman, Brillouin light, and X-ray scattering study. *Phys. Rev. B.* **81**, 212201.
51. A. I. Chumakov, G. Monaco, A. Fontana, A. Bosak, R. P. Hermann, D. Bessas, B. Wehinger, W. A. Crichton, M. Krisch, R. Rüffer, G. Baldi, G. Carini Jr., G. Carini, G. D'Angelo, E. Gilioli, G. Tripodo, M. Zanatta, B. Winkler, V. Milman, K. Refson, M. T. Dove, N. Dubrovinskaia, L. Dubrovinsky, R. Keding, and Y. Z. Yue. (2014). Role of disorder in the thermodynamics and atomic dynamics of glasses. *Phys. Rev. Lett.* **112**, 025502.
52. A. Monaco, A. I. Chumakov, Y.-Z. Yue, G. Monaco, L. Comez, D. Fioretto, W. A. Crichton, and R. Rüffer. (2006). Density of vibrational states of a hyperquenched glass. *Phys. Rev. Lett.* **96**, 205502.
53. E. Duval, L. Saviot, L. David, S. Etienne, and J. F. Jal. (2003). Effect of physical aging on the low-frequency vibrational density of states of a glassy polymer. *Europhys. Lett.* **63**, 778–784.
54. E. A. A. Pogna, A. I. Chumakov, C. Ferrante, M. A. Ramos, and T. Scopigno. Tracking the connection between disorder and energy landscape in glasses using geologically hyperaged amber. *J. Phys. Chem. Lett.* **10**, 427–432.

55. S. Caponi, S. Corezzi, D. Fioretto, A. Fontana, G. Monaco, and F. Rossi. (2019). Raman-scattering measurements of the vibrational density of states of a reactive mixture during polymerization: Effect on the Boson peak. *Phys. Rev. Lett.* **102**, 027402 (2009).
56. G. Baldi, G. Carini, G. Carini, A. Chumakov, R. Dal Maschio, G. D'Angelo, A. Fontana, E. Gilioli, G. Monaco, L. Orsingher, B. Rossi, and M. Zanatta. (2016). New insights on the specific heat of glasses. *Philos. Mag.* **96**, 754–760.
57. A. I. Chumakov, A. Q. R. Baron, R. Rüffer, H. Grünsteudel, H. F. Grünsteudel, and A. Meyer. (1996). Nuclear resonance energy analysis of inelastic X-ray scattering. *Phys. Rev. Lett.* **76**, 4258–4261.
58. N. V. Surovtsev, A. P. Shebanin, and M. A. Ramos. (2003). Density of states and light-vibration coupling coefficient in B_2O_3 glasses with different thermal history. *Phys. Rev. B.* **67**, 024203.
59. J. Wuttke, W. Petry, G. Coddens, and F. Fujara. (1995). Fast dynamics of glass-forming glycerol. *Phys. Rev. E.* **52**, 4026–4034.
60. U. Buchenau. (2019). Neutron scattering investigations of the boson peak. *Condens. Matter Phys.* **22**, 43601.
61. M. Zanatta, A. Fontana, A. Orecchini, C. Petrillo, and F. Sacchetti. (2013). Inelastic neutron scattering investigation in glassy $SiSe_2$: Complex dynamics at the atomic scale. *J. Phys. Chem. Lett.* **4**, 1143–1147.
62. N. Violini, A. Orecchini, A. Paciaroni, C. Petrillo, and F. Sacchetti. (2012). Neutron scattering investigation of high-frequency dynamics in glassy glucose. *Phys. Rev. B.* **85**, 134204.
63. S. W. Lovesey. (1984). *Theory of Neutron Scattering from Condensed Matter.* Oxford: Oxford University Press. https://global.oup.com/academic/product/theory-of-neutron-scattering-from-condensed-matter-9780198520283?lang=en&cc=it.
64. J. P. Boon and S. Yip. (1991). *Molecular Hydrodynamics.* New York: Dover Publications, Inc..
65. G. Monaco and V. M. Giordano. (2009). Breakdown of the Debye approximation for the acoustic modes with nanometric wavelengths in glasses. *Proc. Natl. Acad. Sci.* **106**, 3659–3663.
66. P. Benassi, M. Krisch, C. Masciovecchio, V. Mazzacurati, G. Monaco, G. Ruocco, F. Sette, and R. Verbeni. (1996). Evidence of high frequency propagating modes in vitreous silica. *Phys. Rev. Lett.* **77**, 3835–3838.
67. O. Pilla, A. Cunsolo, A. Fontana, C. Masciovecchio, G. Monaco, M. Montagna, G. Ruocco, T. Scopigno, and F. Sette. (2000). Nature of the short wavelength excitations in vitreous silica: An X-ray Brillouin scattering study. *Phys. Rev. Lett.* **85**, 2136–2139.

68. M. Foret, E. Courtens, R. Vacher, and J.-B. Suck. (1996). Scattering investigation of acoustic localization in fused silica. *Phys. Rev. Lett.* **77**, 3831–3834.
69. G. Baldi, V. M. Giordano, G. Monaco, and B. Ruta. (2010). Sound attenuation at terahertz frequencies and the boson peak of vitreous silica. *Phys. Rev. Lett.* **104**, 195501.
70. F. Sette, M. H. Krisch, C. Masciovecchio, G. Ruocco, and G. Monaco. (1998). Dynamics of glasses and glass-forming liquids studied by inelastic X-ray scattering. *Science (80-.).* **280**, 1550–1555.
71. G. Monaco, C. Masciovecchio, G. Ruocco, and F. Sette. (1998). Determination of the infinite frequency sound velocity in the glass former o-terphenyl. *Phys. Rev. Lett.* **80**, 2161–2164.
72. H. Schober, M. M. Koza, A. Tölle, C. Masciovecchio, F. Sette, and F. Fujara. (2000). Crystal-like high frequency phonons in the amorphous phases of solid water. *Phys. Rev. Lett.* **85**, 4100–4103.
73. A. Matic, L. Börjesson, G. Ruocco, C. Masciovecchio, A. Mermet, F. Sette, and R. Verbeni. (2001). Contrasting behaviour of acoustic modes in network and non-network glasses. *Europhys. Lett.* **54**, 77–83.
74. A. Matic, C. Masciovecchio, D. Engberg, G. Monaco, L. Börjesson, S. C. Santucci, and R. Verbeni. (2004). Crystal-like nature of acoustic excitations in glassy ethanol. *Phys. Rev. Lett.* **93**, 145502.
75. T. Scopigno, R. Di Leonardo, G. Ruocco, A. Q. R. Baron, S. Tsutsui, F. Bossard, and S. N. Yannopoulos. (2004). High frequency dynamics in a monatomic glass. *Phys. Rev. Lett.* **92**, 025503.
76. G. Baldi, V. M. Giordano, B. Ruta, R. Dal Maschio, A. Fontana, and G. Monaco. (2014). Anharmonic damping of terahertz acoustic waves in a network glass and its effect on the density of vibrational states. *Phys. Rev. Lett.* **112**, 125502.
77. P. Benassi, M. Nardone, A. Giugni, G. Baldi, and A. Fontana. (2015). Collective excitations in liquid and glassy 3-methylpentane. *Phys. Rev. B.* **92**, 104203.
78. B. Ruta, G. Baldi, F. Scarponi, D. Fioretto, V. M. Giordano, and G. Monaco. (2012). Acoustic excitations in glassy sorbitol and their relation with the fragility and the boson peak. *J. Chem. Phys.* **137**, 214502.
79. T. Ichitsubo, S. Hosokawa, K. Matsuda, E. Matsubara, N. Nishiyama, S. Tsutsui, and A. Q. R. Baron. (2007). Nanoscale elastic inhomogeneity of a Pd-based metallic glass: Sound velocity from ultrasonic and inelastic X-ray scattering experiments. *Phys. Rev. B.* **76**, 140201.
80. T. Ichitsubo, E. Matsubara, K. Miyagi, W. Itaka, K. Tanaka, and S. Hosokawa. (2008). Low-temperature elastic moduli of a Pd-based metallic glass showing positive phonon dispersion. *Phys. Rev. B.* **78**, 052202.

81. T. Ichitsubo, W. Itaka, E. Matsubara, H. Kato, S. Biwa, S. Hosokawa, K. Matsuda, J. Saida, O. Haruyama, Y. Yokoyama, H. Uchiyama, and A. Q. R. Baron. (2010). Elastic inhomogeneity and acoustic phonons in Pd-, Pt-, and Zr-based metallic glasses. *Phys. Rev. B.* **81**, 172201.
82. L. E. Bove, E. Fabiani, A. Fontana, F. Paoletti, C. Petrillo, O. Pilla, and I. C. V. Bento. (2005). Brillouin neutron scattering of v-GeO2. *Europhys. Lett.* **71**, 563–569.
83. M. D. Ruiz-Martín, M. Jiménez-Ruiz, F. J. Bermejo, and R. Fernández-Perea. (2006). High-frequency collective excitations in molten and glassy Te studied by inelastic neutron scattering. *Phys. Rev. B.* **73**, 094201.
84. A. Matic, D. Engberg, C. Masciovecchio, and L. Börjesson. (2001). Sound wave scattering in network glasses. *Phys. Rev. Lett.* **86**, 3803–3806.
85. T. Scopigno, J.-B. Suck, R. Angelini, F. Albergamo, and G. Ruocco. (2006). High-frequency dynamics in metallic glasses. *Phys. Rev. Lett.* **96**, 135501.
86. T. Otomo, M. Arai, Y. Inamura, J.-B. Suck, S. M. Bennington, and K. Suzuki. (1998). Observation of collective excitations in the amorphous alloy $Ni_{67}Zr_{33}$. *J. Non. Cryst. Solids.* **232–234**, 613–618.
87. T. Scopigno, M. D'Astuto, M. Krisch, G. Ruocco, and F. Sette. (2001). Observation of Umklapp processes in noncrystalline materials. *Phys. Rev. B.* **64**, 012301.
88. G. Ruocco, F. Sette, R. Di Leonardo, D. Fioretto, M. Krisch, M. Lorenzen, C. Masciovecchio, G. Monaco, F. Pignon, and T. Scopigno. (1999). Nondynamic origin of the high-frequency acoustic attenuation in glasses. *Phys. Rev. Lett.* **83**, 5583–5586.
89. M. Zanatta, F. Sacchetti, E. Guarini, A. Orecchini, A. Paciaroni, L. Sani, and C. Petrillo. (2015). Collective ion dynamics in liquid zinc: Evidence for complex dynamics in a non-free-electron liquid metal. *Phys. Rev. Lett.* **114**, 187801.
90. G. Baldi, M. Zanatta, E. Gilioli, V. Milman, K. Refson, B. Wehinger, B. Winkler, A. Fontana, and G. Monaco. (2013). Emergence of crystal-like atomic dynamics in glasses at the nanometer scale. *Phys. Rev. Lett.* **110**, 185503.
91. A. F. Ioffe and A. R. Regel. (1960). *Progress in Semiconductors*, A. F. Gibson (Ed.). Vol. 4, pp. 237–291. New York: Wiley.
92. S. Taraskin and S. Elliott. (2000). Ioffe-Regel crossover for plane-wave vibrational excitations in vitreous silica. *Phys. Rev. B. Condens. Matter Mater. Phys.* **61**, 12031–12037.

93. G. Ruocco, A. Matic, T. Scopigno, and S. N. Yannopoulos. (2007). Comment on "Glass-Specific Behavior in the Damping of Acousticlike Vibrations." *Phys. Rev. Lett.* **98**, 079601.
94. B. Rufflé, G. Guimbretière, E. Courtens, R. Vacher, G. Monaco, Rufflé *et al.* (2007). Reply. *Phys. Rev. Lett.* **98**, 079602.
95. H. Shintani and H. Tanaka. (2008). Universal link between the boson peak and transverse phonons in glass. *Nat. Mater.* **7**, 870–877.
96. G. Baldi, V. M. Giordano, G. Monaco, and B. Ruta. (2011). High frequency acoustic attenuation of vitreous silica: New insight from inelastic X-ray scattering. *J. Non. Cryst. Solids.* **357**, 538–541.
97. B. Rufflé, M. Foret, E. Courtens, R. Vacher, and G. Monaco. (2003). Observation of the onset of strong scattering on high frequency acoustic phonons in densified silica glass. *Phys. Rev. Lett.* **90**, 095502.
98. J. W. S. Rayleigh. (1896). *The Theory of Sound.* London: Mac Millan.
99. P. G. Klemens. (1955). The scattering of low-frequency lattice waves by static imperfections. *Proc. Phys. Soc. Sect. A.* **68**, 1113–1128.
100. A. Wischnewski, U. Buchenau, and A. Dianoux. (1998). Sound-wave scattering in silica. *Phys. Rev. B - Condens. Matter Mater. Phys.* **57**, 2663–2666.
101. U. Buchenau, Y. M. Galperin, V. L. Gurevich, D. A. Parshin, M. A. Ramos, and H. R. Schober. (1992). Interaction of soft modes and sound waves in glasses. *Phys. Rev. B.* **46**, 2798–2808.
102. A. Moriel, G. Kapteijns, C. Rainone, J. Zylberg, E. Lerner, and E. Bouchbinder. (2019). Wave attenuation in glasses: Rayleigh and generalized-Rayleigh scattering scaling. *J. Chem. Phys.* **151**, 104503.
103. R. Vacher and J. Pelous. (1976). Behavior of thermal phonons in amorphous media from 4 to 300 K. *Phys. Rev. B.* **14**, 823–828.
104. R. Vacher, H. Sussner, and S. Hunklinger. (1980). Brillouin scattering in vitreous silica below 1 K. *Phys. Rev. B.* **21**, 5850–5853.
105. P. Benassi, S. Caponi, R. Eramo, A. Fontana, A. Giugni, M. Nardone, M. Sampoli, and G. Viliani. (2005). Sound attenuation in a unexplored frequency region: Brillouin ultraviolet light scattering measurements in v-SiO_2. *Phys. Rev. B.* **71**, 172201.
106. C. Masciovecchio, G. Baldi, S. Caponi, L. Comez, S. Di Fonzo, D. Fioretto, A. Fontana, A. Gessini, S. C. Santucci, F. Sette, G. Viliani, P. Vilmercati, and G. Ruocco. (2006). Evidence for a crossover in the frequency dependence of the acoustic attenuation in vitreous silica. *Phys. Rev. Lett.* **97**, 035501.
107. A. Devos, M. Foret, S. Ayrinhac, P. Emery, and B. Rufflé. (2008). Hypersound damping in vitreous silica measured by picosecond acoustics. *Phys. Rev. B.* **77**, 100201.

108. W. Dietsche and H. Kinder. (1979). Spectroscopy of phonon scattering in glass. *Phys. Rev. Lett.* **43**, 1413–1417.
109. S. Ayrinhac, M. Foret, A. Devos, B. Rufflé, E. Courtens, and R. Vacher. (2011). Subterahertz hypersound attenuation in silica glass studied via picosecond acoustics. *Phys. Rev. B.* **83**, 014204.
110. C. Klieber, E. Peronne, K. Katayama, J. Choi, M. Yamaguchi, T. Pezeril, and K. A. Nelson. (2011). Narrow-band acoustic attenuation measurements in vitreous silica at frequencies between 20 and 400 GHz. *Appl. Phys. Lett.* **98**, 211908.
111. A. Huynh, E. Péronne, C. Gingreau, X. Lafosse, A. Lemaître, B. Perrin, R. Vacher, B. Rufflé, and M. Foret. (2017). Temperature dependence of hypersound attenuation in silica films via picosecond acoustics. *Phys. Rev. B.* **96**, 174206.
112. F. Bencivenga, R. Mincigrucci, F. Capotondi, L. Foglia, D. Naumenko, A. A. Maznev, E. Pedersoli, A. Simoncig, F. Caporaletti, V. Chiloyan, R. Cucini, F. Dallari, R. A. Duncan, T. D. Frazer, G. Gaio, A. Gessini, L. Giannessi, S. Huberman, H. Kapteyn, J. Knobloch, G. Kurdi, N. Mahne, M. Manfredda, A. Martinelli, M. Murnane, E. Principi, L. Raimondi, S. Spampinati, C. Spezzani, M. Trovò, M. Zangrando, G. Chen, G. Monaco, K. A. Nelson, and C. Masciovecchio. (2019). Nanoscale transient gratings excited and probed by extreme ultraviolet femtosecond pulses. *Sci. Adv.* **5**, eaaw5805.
113. G. Baldi, V. M. Giordano, and G. Monaco. (2011). Elastic anomalies at terahertz frequencies and excess density of vibrational states in silica glass. *Phys. Rev. B.* **83**, 174203.
114. S. Gelin, H. Tanaka, and A. Lemaître. (2016). Anomalous phonon scattering and elastic correlations in amorphous solids. *Nat. Mater.* **15**, 1177–1181.
115. A. Lemaître. (2018). Stress correlations in glasses. *J. Chem. Phys.* **149**, 104107.
116. B. Ruzicka, T. Scopigno, S. Caponi, A. Fontana, O. Pilla, P. Giura, G. Monaco, E. Pontecorvo, G. Ruocco, and F. Sette. (2004). Evidence of anomalous dispersion of the generalized sound velocity in glasses. *Phys. Rev. B.* **69**, 100201.
117. T. Scopigno, E. Pontecorvo, R. Di Leonardo, M. Krisch, G. Monaco, G. Ruocco, B. Ruzicka, and F. Sette. (2003). High-frequency transverse dynamics in glasses. *J. Phys. Condens. Matter.* **15**, S1269–S1278.
118. E. Pontecorvo, M. Krisch, A. Cunsolo, G. Monaco, A. Mermet, R. Verbeni, F. Sette, and G. Ruocco. (2005). High-frequency longitudinal and transverse dynamics in water. *Phys. Rev. E.* **71**, 011501.

119. G. Baldi, V. M. M. Giordano, G. Monaco, F. Sette, E. Fabiani, A. Fontana, and G. Ruocco. (2008). Thermal conductivity and terahertz vibrational dynamics of vitreous silica. *Phys. Rev. B.* **77**, 214309.
120. G. Baldi, P. Benassi, A. Fontana, A. Giugni, G. Monaco, M. Nardone, and F. Rossi. (2017). Damping of vibrational excitations in glasses at terahertz frequency: The case of 3-methylpentane. *J. Chem. Phys.* **147**, 164501.
121. S. N. Taraskin and S. R. Elliott. (1998). Nature of the low-frequency vibrational excitations in vitreous silica. *J. Non. Cryst. Solids.* **232–234**, 205–211.
122. O. Pilla, S. Caponi, A. Fontana, J. R. Gonçalves, M. Montagna, F. Rossi, G. Viliani, L. Angelani, G. Ruocco, G. Monaco, and F. Sette. (2004). The low energy excess of vibrational states in v-SiO_2: The role of transverse dynamics. *J. Phys. Condens. Matter.* **16**, 8519–8530.
123. O. Pilla, L. Angelani, A. Fontana, J. R. Goncalves, and G. Ruocco. (2003). Structural and dynamical consequences of density variation in vitreous silica. *J. Phys. Condens. Matter.* **15**, S995–S1005.
124. A. Polian, D. Vo-Thanh, and P. Richet. (2002). Elastic properties of a-SiO_2 up to 2300 K from Brillouin scattering measurements. *Europhys. Lett.* **57**, 375–381.
125. M. Sampoli, G. Ruocco, and F. Sette. (1997). Mixing of longitudinal and transverse dynamics in liquid water. *Phys. Rev. Lett.* **79**, 1678–1681.
126. C. Masciovecchio, V. Mazzacurati, G. Monaco, G. Ruocco, T. Scopigno, F. Sette, P. Benassi, A. Cunsolo, A. Fontana, M. Krisch, A. Mermet, M. Montagna, F. Rossi, M. Sampoli, G. Signorelli, and R. Verbeni. (1999). Acoustic nature of the boson peak in vitreous silica. *Philos. Mag. B.* **79**, 2013–2020.
127. A. J. Leadbetter. (1969). Inelastic cold neutron scattering from different forms of silica. *J. Chem. Phys.* **51**, 779–786.
128. G. Baldi, V. M. Giordano, B. Ruta, and G. Monaco. (2016). On the nontrivial wave-vector dependence of the elastic modulus of glasses. *Phys. Rev. B.* **93**, 144204.
129. A. J. Leadbetter. (1973). *Chemical Applications of Thermal Neutron Scattering*, B. T. M. Willis (Ed.). London: Oxford University Press, pp. 146–172.
130. Y. Wang, L. Hong, Y. Wang, W. Schirmacher, and J. Zhang. (2018). Disentangling boson peaks and Van Hove singularities in a model glass. *Phys. Rev. B.* **98**, 174207.

131. A. Lemaître. (2017). Inherent stress correlations in a quiescent two-dimensional liquid: Static analysis including finite-size effects. *Phys. Rev. E.* **96**, 052101.
132. C. Caroli and A. Lemaître. (2019). Fluctuating elasticity fails to capture anomalous sound scattering in amorphous solids. *Phys. Rev. Lett.* **123**, 055501.

https://doi.org/10.1142/9781800612587_0007

Chapter 7

Low-frequency Vibrational Spectroscopy of Glasses

Benoit Rufflé*, Marie Foret and Bernard Hehlen

Laboratoire Charles Coulomb (L2C), University of Montpellier, CNRS, Montpellier, France
**benoit.ruffle@umontpellier.fr*

Atomic vibrations in perfect, slightly defective, or mixed crystals are to a large extent well understood since many decades. Theoretical descriptions are thus in excellent agreement with the experiments. As a consequence, phonon-related properties like specific heat, thermal conductivity, or sound attenuation are also well explained in these solids. This is not yet the case in glasses where the lack of periodicity generates enormous difficulties in theoretical treatments as well as in experiments or in numerical simulations. Thanks to recent developments along all these lines, comprehensive studies have emerged in the last decades and several decisive advances have been made. This chapter is thus devoted to a discussion of the nature of the vibrational properties in glasses with particular emphasis on the low-frequency part of the vibrational density of states, including the acoustic excitations, and of the experimental techniques used in their study.

1. Introduction

1.1. *Thermal properties*

The lack of periodicity in glasses implies that no reciprocal lattice can be defined. As a consequence, the wavevector $\mathbf{q}$ is no longer a good quantum number for the description of the vibrational excitations, and the phonon states cannot be characterized by dispersion curves $\omega(q)$. The quantity which can be safely used is the vibrational density of states $g(\omega)$ (vDOS), which is the distribution of the number of vibrational modes as a function of the frequency ω. This quantity is of paramount importance as it also makes the connection between the microscopic and macroscopic aspects of atomic vibrations. The isochoric heat capacity $C_V(T)$ is, for example, directly derived from $g(\omega)$.

The second macroscopic quantity of great interest is the thermal conductivity which relates heat transport and propagative elastic waves in dielectric crystals. Early measurements of these two thermal properties rapidly showed evidence of significant differences when compared to the crystalline counterpart. It is then natural to start the chapter with a very short review of what is usually called the low-temperature thermal anomalies of glasses (see Chapters 2 and 3 for an extended discussion of the latter).

1.1.1. *Heat capacity*

The temperature dependence of the heat capacity of crystalline dielectrics is well understood. Well below the Debye temperature θ_D, the isochoric heat capacity C_V follows the Debye law, $C_V = C_D T^3$, where $C_D \propto \theta_D^{-3}$ is given by the sound velocities and the atomic density. At higher T, C_V shows a slower increase depending on the details of the atomic structure through the phonon dispersion relation and then flattens eventually to the classical Dulong–Petit limit at high temperatures.

Figure 1(a) shows the reduced heat capacity C_P/T^3 at low temperature of three SiO_2 polymorphs: vitreous silica, α-quartz, and α-cristobalite. Below 5 K, C_P/T^3 of α-quartz follows the expected constant Debye value, whereas the bump at higher T is fully described by the curvature of the acoustic phonon branches. According to their comparable densities and structure, it is, however, more appropriate to compare silica glass with α-cristobalite. For that crystal, the constant Debye value of C_P/T^3 holds to about 2 K while the large increase at higher T, giving rise to a large bump around 13 K, originates essentially from the flattening of the low-velocity transverse acoustic branch in the <110> direction.[1,2] Hence, both crystalline polymorphs nicely illustrate the general trends mentioned

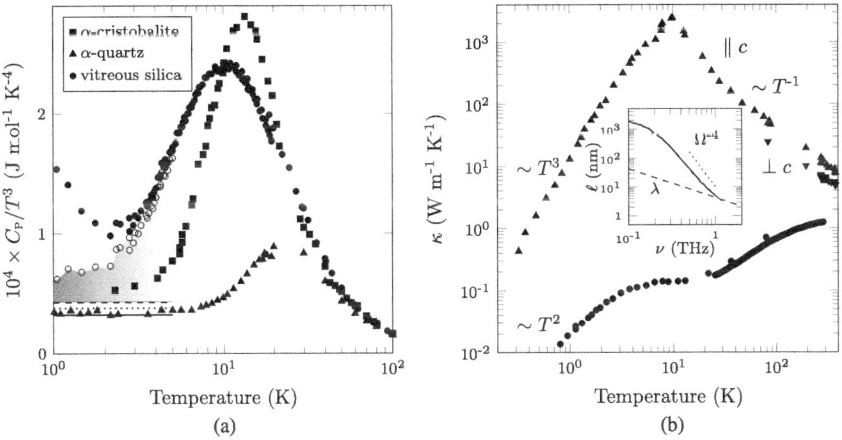

Fig. 1. (a) Low-temperature heat capacity of vitreous silica (dots[4–9]), α-cristobalite (squares[1,4,5]), and α-quartz (triangles[1,4,8]) plotted as C_P/T^3 against T. The horizontal lines are the corresponding constant Debye values (dashed, dotted, and solid line, respectively). The circles show the heat capacity of vitreous silica lessened by the two-level systems (TLS) contribution.[10] (b) Temperature dependence of the thermal conductivity of vitreous silica (dots[8,11,12]) and α-quartz (up and down triangles for the two principal directions[8,13–15]). The square is a room T measurement on α-cristobalite.[16] Inset: Frequency dependence of the mean free path of vitreous silica in the dominant phonon approximation.

earlier with characteristic temperatures related to their local order and atomic packing. It is also worth recalling that optic modes contribute to C_P. The Raman active zone center mode around $50\,\text{cm}^{-1}$[3] of α-cristobalite is, for example, expected to strengthen C_P just above the maximum of C_P/T^3.

Conversely, the heat capacity of vitreous silica is much higher than the Debye expectation at 2 K and further never reaches it at low T, indicating low-frequency extra modes. At the lowest T, most of this excess originates from tunneling states giving rise to an almost linear contribution $C_P \propto T$; see Chapter 2. Subtracting this contribution to C_P still leaves an excess over the Debye prediction as illustrated by the circles and the dashed region in Figure 1(a). At higher T, C_P/T^3 of vitreous silica displays a peak around 10 K analogous to that of α-cristobalite in agreement with their similar structure and atomic density.

Such common features indicated very early[1] that a peak in C_P/T^3 at low T is not a peculiarity of glasses[17,18] although it does not preclude that part of the vibrational spectrum responsible for this peak in glasses may be very different in nature from that in the corresponding crystal. Opposite to the case of α-cristobalite, it is, for example, not possible to reconcile the low-T part of the broad hump in C_P/T^3 of vitreous silica with the known linear dispersion of sound waves up to at least 440 GHz,[19] the latter giving a constant Debye value up to ~5.5 K. This maximum above the Debye expectation in the C_P/T^3 of glasses is directly related to the peak that shows up in the *reduced* vibrational density of states $g(\omega)/\omega^2$ (mostly not in $g(\omega)$ itself) and originally referred to as the *boson peak* in Raman spectroscopy; see Section 4.2.

1.1.2. *Thermal conductivity*

The temperature dependence of the thermal conductivity κ of crystalline dielectrics is also well understood. Using the standard gas kinetic equation, κ is approximately described in these solids by $\kappa = \frac{1}{3}C_V v\ell$, where C_V is the specific heat per volume of the acoustic phonons providing the thermal transport, v is their velocity of propagation, and ℓ is their mean free path. Starting from ambient temperature, κ increases with decreasing temperatures as ℓ increases: phonons are scattered by anharmonic Umklapp processes which become less frequent as there are less and less phonons. At a certain point, ℓ reaches sample dimensions and κ decreases rapidly with the T^3 dependence of C_V. Figure 1(b) illustrates such an ideal behavior in the case of α-quartz in the c-axis direction. It also shows a room temperature measurement of the thermal conductivity of α-cristobalite. Interestingly, the value $\sim 6\,\mathrm{W\,m^{-1}\,K^{-1}}$ is almost identical to that obtained perpendicular to the c-axis in α-quartz, suggesting a similar behavior.

In striking contrast, dielectric glassy materials are known to exhibit much lower thermal conductivities with markedly different temperature dependence. As an example, κ of vitreous silica is plotted in Figure 1(b). It follows an approximate T^2 law at very low T instead of the T^3 dependence expected from the

Debye approximation. This initial T^2 rise originates from interactions between phonons and tunneling states, reducing ℓ in glasses.[10,20,21] At higher T, κ displays a remarkable plateau around 10 K, a temperature close to the hump in C_P/T^3. It was earlier recognized that a very efficient phonon scattering mechanism was needed to generate such a plateau, at least with $\ell \propto \omega^{-4}$,[22] suggesting a Rayleigh-type scattering mechanism arising from the glass disorder.

Based on a dominant phonon approximation $\hbar\omega \sim 3.8\ k_B T$, the frequency dependence of the mean free path can be estimated from the above-mentioned kinetic equation of κ. The resulting $\ell(\omega)$ for vitreous silica is plotted in the inset in Figure 1(b) as a line. The mean free path exceeds 1 μm below 100 GHz, a value much larger than the acoustic wavelength λ_a illustrating the propagative character of these acoustic phonons. In this frequency range, resonant relaxation by tunneling states dominates the acoustic attenuation at the corresponding temperatures, i.e., below 1 K as seen on the top x-scale, and leads to $\ell \propto \omega^{-1}$. At higher frequencies, ℓ drops rapidly and then follows an approximate ω^{-4} trend. Around 1 THz, ℓ is comparable to λ_a, indicating that the Ioffe–Regel criterion $\ell = \lambda_a/2$ is fulfilled in this frequency region (here, ℓ is the energy mean free path). It corresponds to the propagation of the sound wave over a distance of less than half of its wavelength. Above the corresponding frequency ω_{IR}, sound waves do not propagate and cannot transfer energy anymore. The wavevector is no longer a good quantum number and the notion of phonon becomes ill defined. At higher temperatures, κ rises again and eventually saturates at values around $\sim$1 W m^{-1} K^{-1} following approximately $C_P(T)$. As the heat transport cannot be mediated anymore by propagating sound waves, it is generally admitted that κ is governed diffusion mechanisms.[23,24]

1.2. *Vibrations in glasses*

In many aspects, the vibrational spectrum of glasses is similar to that of their crystalline counterparts. However, the anomalous thermal properties displayed by the disordered solids indicate that some fundamental differences must exist. Well-defined low-frequency sound waves also exist in glasses as their wavelengths are much larger than the disorder length scale. At higher frequencies, the rapid decrease of the acoustic phonon mean free path related to the plateau in κ around 10 K seems, however, specific to disordered solids. Whether and how this peculiar scattering of the acoustic phonons occurring at high frequencies is related to the excess above the Debye expectation in the low-temperature specific heat at temperatures below the maximum in C_P/T^3 remains a central and still unsettled question. It is further interesting to note that while similar humps do exist in the low-temperature reduced specific heat both for vitreous silica and α-cristobalite, indicating similar vibrational density of states in that region, thermal transport mechanisms at the nanometer scale seem very different in the two polymorphs.

Section 2 briefly describes the (ω, q) range which must be addressed by the experimental techniques as well the quantities probed using scattering techniques. Section 3 is devoted to the spectroscopy of acoustic phonons, whereas Section 4 discusses the different experimental techniques aiming at measuring the low-frequency part of the vibrational density of states, in particular, in the boson peak region.

2. Inelastic Spectroscopy

2.1. *Dispersion diagram and experimental techniques*

Figure 2 displays a (ω, q) map of the main experimental techniques that can probe vibrational excitations in disordered solids. The diagonal region represents the linear dispersion relation for sound waves of velocity between 1 km s^{-1} and 6 km s^{-1}. Brillouin light scattering (BLS), inelastic scattering with ultra-violet radiation (IUVS), x-rays (IXS), or neutron (INS) probes an increasing well-defined q range associated with an increasing ω range according to the linear dispersion law. Conversely, ultrasonics (US) is not a scattering experiment but covers specific frequency ranges. Accordingly, no scattering vector $\mathbf{q}$ is really defined for this technique. Picosecond ultrasonics (PU) forms actually a family of optical techniques using pulsed lasers and belongs to the two categories depending on the

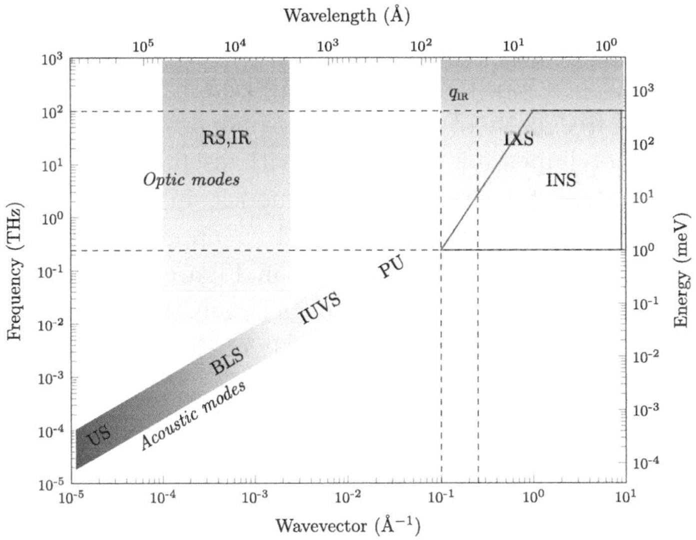

Fig. 2. Dispersion diagram and (ω, q) windows of experimental techniques probing vibrational excitations in disordered materials. Ultrasonics (US), Brillouin light scattering (BLS), Raman scattering (RS), infrared absorption (IR), inelastic scattering with ultra-violet radiation (IUVS), picosecond acoustics (PU), inelastic neutron (INS) and x-ray scattering (IXS). The horizontal dashed lines mark the domain of network optic vibrations in glasses while the diagonal band indicates approximately the domain of acoustic excitations, merging with the optic ones at high q.

setup. Concerning the optic modes, the (ω, q) ranges covered by scattering techniques that can probe these excitations and measure the vibrational density of states are shown approximately as vertical regions in Figure 2. In increasing q range order, one finds Raman scattering (RS), IXS, and INS. Due to kinematics constraints, the (ω, q) range covered by INS is somewhat limited as compared to the one covered by IXS as shown by the approximate trapezoidal shape in Figure 2. Additionally, there also exists non-scattering techniques like infrared (IR) spectroscopy or nuclear inelastic scattering (NIS) which give information on the vibrational density of states in a q region related to the incoming radiation.

Even if much of the dispersion diagram seems nowadays accessible for studying acoustic excitations from the macroscopic length scale down to the atomic one, it is far from being the case for most glasses. Moreover, it is often feasible for the longitudinal acoustic mode only. As mentioned in the preceding section, of peculiar interest is the mean free path of the acoustic phonons, a quantity often difficult, if not impossible, to obtain with useful accuracy. This is dramatically true for the transverse modes, in particular, in the high q range above about $5 \times 10^{-3}\,\text{Å}^{-1}$, approaching the Ioffe–Regel crossover region, defined by q_{IR} in Figure 2. This is all the more important as the transverse modes largely dominate the Debye density of states. Consequently, the study of the longitudinal modes will give only incomplete information.

A crucial point revealed by Figure 2 is that the acoustic linear dispersion curves merge with the low-frequency optic vibrations in the Ioffe–Regel crossover region, where stands the boson peak feature around 1 THz. This explains why, concomitantly to the study of acoustic excitations, a very large effort of the scientific community has been devoted to the understanding of the low-frequency part of the vibrational density of states of disordered materials.

2.2. *Inelastic scattering intensity*

The incident radiation impinging on a material is inelastically scattered with the intensity $I(\mathbf{q}, \omega)$ corresponding to the space and time Fourier transforms of the autocorrelation function of the physical quantity A which couples to the incoming radiation:

$$I(\mathbf{q}, \omega) \propto FT\left\{\int_V \int_t A(\mathbf{r}, t)\, A(\mathbf{r} + \mathbf{r}', t + t')\, \mathrm{d}t\mathrm{d}\mathbf{r}\right\} = C(A')S(\mathbf{q}, \omega), \tag{1}$$

where A' is the derivative of A over the atomic displacements U of the mode, while the correlation function $C(A')$ contains the selection rules. The latter depends on the strength of the coupling of the incident radiation to vibration and, therefore, modulates the intensity of its spectral shape $S(\mathbf{q}, \omega)$. The physical quantity A is the electronic density ρ_e for X-ray scattering, whereas it stands for the coherence length b for neutron scattering and the dielectric susceptibility χ for light scattering. In the latter case, the dynamical dielectric susceptibility χ' is usually expressed in terms of the polarizability tensor $\bar{\bar{\alpha}}'$ (Raman scattering), the hyper-polarizability tensor $\bar{\bar{\bar{\beta}}}'$ (hyper-Raman scattering), and the photo-elastic tensor $\bar{\bar{\bar{p}}}'$ (Brillouin scattering). All the atomic vibrations can be detected using neutron scattering, enabling the phonon dispersion curves to be measured over a large $(\mathbf{q}, \omega)$ range, however, limited by the kinematic conditions and the instrumental resolution. The normalized integral over all $\mathbf{q}$ values leads to the number of vibrations per unit frequency also called the neutron vibrational density of states $g(\omega)$ (vDOS). The latter is, however, modulated by the coherent length of the atoms and remains an approximated quantity, in particular, for polyatomic systems.

As mentioned earlier, the vibrations are very well defined in crystals so are their spectral response functions. In that case, the group theory provides the mathematical support to compute the selection rules contained in $C(A')$ for all scattering experiments. The situation is again much more complicated in glasses owing to the structural disorder. Each elementary unit of the network is connected to neighboring units, inducing strong structural local distortions. The modes of a given symmetry may also involve different numbers of neighboring units. These effects modulate $C(A')$ and $S(\mathbf{q}, \omega)$ from site to site inhibiting up to now the development of a general theory describing the vibrational spectra in glasses.

2.3. *Coherent and incoherent scattering*

The relative spatial localization of extended waves in disordered media induces a loss of coherence of the scattering process due to the partly pointless vibration wavevector $\mathbf{Q}$. Acoustic or optic phonons in perfect crystals arc plane waves with atomic displacements defined by

$$U(\mathbf{r}, t) = u_0 e^{\pm i(\mathbf{Q}\mathbf{r} - \Omega t)}. \tag{2}$$

They are extended modes characterized by a spatial extension larger than the probed wavelength. The scattered radiation results from the interference integral over all the modes in the scattering volume V, which in case of light scattering reads

$$\mathbf{E}_s = \int_V \delta\mathbf{P} e^{-i\mathbf{k}_s\mathbf{r}} dV, \tag{3}$$

where $\delta\mathbf{P} = \epsilon_0\chi' U(\mathbf{r},t)\mathbf{E}_i$ is the modulation of the electrical polarization induced by the incident field $\mathbf{E}_i = \mathbf{E}_0 e^{i(\mathbf{k}_i\mathbf{r}-\omega t)}$. Putting Eq. (2) in Eq. (3) leads to

$$\mathbf{E}_s \propto \mathbf{E}_0 \int_V \chi' u_0 e^{i((\mathbf{k}_i-\mathbf{k}_s\pm\mathbf{Q})\mathbf{r}-(\omega_i\mp\Omega)t)} d\mathbf{r}, \tag{4}$$

where $\mathbf{r}$ integrates over all orientations so that $\mathbf{E}_s$ is zero unless $\mathbf{k}_i - \mathbf{k}_s \pm \mathbf{Q} = 0$, i.e., for momentum conservation. For a given scattering geometry, only the vibrations whose wavevector $\mathbf{Q}$ matches one of the experiments $\mathbf{Q} = \pm(\mathbf{k}_i - \mathbf{k}_s) = \mathbf{q}$ will be active, denoting a coherent scattering process.

On the extreme opposite side, the atomic displacements of fully localized atomic motions are defined by $U(t) = u_0 e^{\pm i\Omega t}$ as $\mathbf{Q}$ is not a relevant quantum number. $\mathbf{k}_s\mathbf{r}$ is constant over the very small spatial extension of the mode, and if χ' is also supposed to be constant over the volume V (gases and non-viscous liquids), Eq. (4) transforms into

$$\mathbf{E}_s \propto \chi' u_0 \mathbf{E}_0 e^{-i(\omega_i\mp\Omega)t}, \tag{5}$$

showing that the vibration scatters at all $\mathbf{q}$ values, indicating incoherent scattering.

For intermediate situations, i.e., damped plane waves or quasi-localized waves in disordered solids, the atomic displacements of these vibrations can be modeled by $U(\mathbf{r},t) = u_0 e^{-r/\ell_c} e^{\pm i(\mathbf{Q}\mathbf{r}-\Omega t)}$, where ℓ_c is the coherence length of the vibrational mode. For sufficiently small values of l_c, the spatial localization of the mode induces a wavevector spectral broadening ΔQ. In that case, the contribution at the spectral frequency ω is the sum over all modes of frequency $\Omega = \omega$ having a wavevector spectral component $\mathbf{Q}$ matching the scattering wavevector $\mathbf{q}$ of the experiment. The above conclusion holds in glasses for optic modes as well as for short wavelength acoustic waves, since the latter progressively transform into quasi-localized vibrations at THz frequencies.

Alternatively, high-frequency acoustic modes in amorphous solids can be treated by considering plane waves propagating in an inhomogeneous elastic medium[25] where χ' is position-dependent. It leads to a spatial modulation of the photoelastic tensor which now reads $\bar{\bar{\bar{p}}}'(\mathbf{r})$. The product $\bar{\bar{\bar{p}}}'(\mathbf{r})U(\mathbf{r},t)$ in $\delta\mathbf{P}$ yields to distorted acoustic waves whose coherence length ℓ_c is limited by the local elastic inhomogeneities. The full treatment shows that the light scattering spectrum consists of the usual Brillouin peaks, i.e., coherent scattering, and a ω^2 incoherent background induced by these local heterogeneities.

3. Spectroscopy of Acoustic Phonons

3.1. *Attenuation and sound velocity in dielectric crystals*

In perfect dielectric crystals, it is well known that the dominant mechanism for sound attenuation originates from the interaction between the acoustic wave and high-frequency phonons constituting the thermal bath.[26] This interaction also leads to a significant variation of the sound velocity. Within the original Akhiezer relaxation model,[27] it is assumed that the temperature is sufficiently high for the bath to be appreciably populated. The thermal phonon mode frequencies Ω_i in equilibrium are then perturbed by a sound wave of frequency Ω by $\gamma_i = \partial \ln \Omega_i / \partial e$, where e is the strain developed by the sound wave. As a rough estimate, the γ_i can be rationalized by using a mean-square average Grüneisen parameter $\gamma^2 = \overline{\gamma_i^2}$. The perturbed thermal bath relaxes towards equilibrium with a characteristic thermal time $\tau_{\rm th}$ via anharmonic interactions and dissipates the energy of the sound wave with an internal friction

$$Q^{-1}(\Omega, T) = \frac{A\Omega\tau_{\rm th}}{1+\Omega^2\tau_{\rm th}^2}, \tag{6}$$

where $A = \gamma^2 C_{\rm V} T v / 2\rho v_{\rm D}^3$.[26] In the latter equation, $C_{\rm V}$ is the specific heat per unit volume, ρ is the mass density, and $v_{\rm D}$ is the Debye velocity.

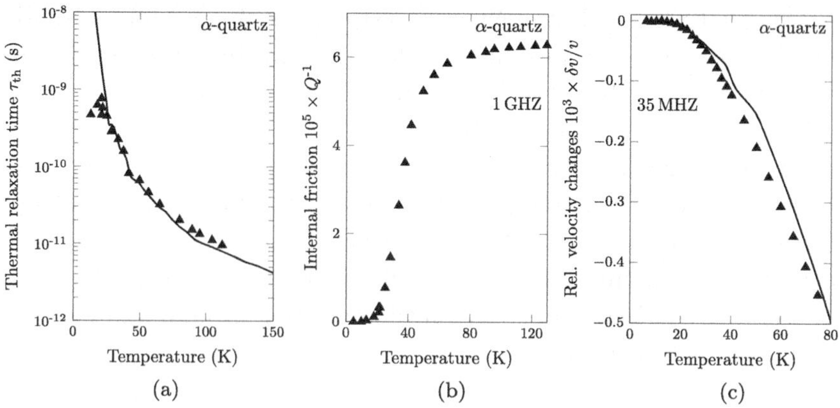

Fig. 3. (a) Temperature dependence of the thermal relaxation time $\tau_{\rm th}$ in α-quartz. The line is calculated from the kinetic equation $\kappa = \frac{1}{3}C_{\rm V}v_{\rm D}^2\tau_{\rm th}$, while the symbols are obtained from Q^{-1} of panel (b) using Eq. (6) and $\gamma^2 \simeq 1$.[29] (b) Internal friction of 1 GHz longitudinal waves in Z-cut α-quartz.[28] (c) Fractional velocity changes of 35 MHz shear waves in BC-cut α-quartz.[30] The line is obtained using $\tau_{\rm th}(T)$ from the kinetic equation and $\gamma^2 \simeq 2.6$.

The temperature dependence of the thermal phonon lifetime $\tau_{\rm th}$ can be roughly estimated from κ using the kinetic equation $\kappa = \frac{1}{3}C_{\rm V}v_{\rm D}^2\tau_{\rm th}$. The corresponding values for crystal quartz are shown in Figure 3(a) as a solid line. As $\tau_{\rm th}$ is typically short and further decreases rapidly with T, a significant damping of the sound wave is expected at high frequencies only. In the case of quartz, $(2\pi\tau_{\rm th})^{-1} \simeq$ 1 GHz around 35 K. The internal friction of 1 GHz longitudinal waves in a Z-cut crystal quartz[28] is plotted in Figure 3(b). At very low temperatures, $\Omega\tau_{\rm th} \gg 1$ and the internal friction is very small. The rapid increase of Q^{-1} with increasing T is essentially driven by the temperature dependence of $C_{\rm V}$, leading to $A \propto T^4$. At higher temperatures, both A and $\tau_{\rm th}$ evolve moderately so does Q^{-1}. In the latter region, $\Omega\tau_{\rm th} \ll 1$ and $Q^{-1} \propto \Omega$. The acoustic damping $\alpha \propto \Omega^2$ is thus best observed using high-frequency techniques. Another estimate of $\tau_{\rm th}$ can be extracted from the measured Q^{-1} and Eq. (6) using a temperature-independent Grüneisen parameter $\gamma^2 \simeq 1$ as demonstrated by the symbols in Figure 3(a). The agreement between the two sets of values emphasizes the physical significance of $\tau_{\rm th}$ as a mean lifetime of the vibrational excitations in the thermal bath.

The frequency-dependent damping induces a frequency-dependent correction to the sound velocity *via* the Kramers–Kronig transform

$$\frac{\delta v}{v}(\Omega, T) = -\frac{1}{2}\frac{A}{1+\Omega^2\tau_{\text{th}}^2}, \tag{7}$$

where $\delta v = v - v_\infty$. In a crystal, one expects that v_∞, the high-frequency limit of v, is constant in T. For $\Omega\tau_{\text{th}} \ll 1$, which is the case in ultrasonics for crystal quartz, one simply has $\delta v/v = -A/2$. Such a decrease in the sound velocity with T is illustrated in Figure 3(b) for 35 MHz shear waves in BC-cut quartz.[30] It compares well with the predicted temperature dependence of $\delta v/v$ computed from Eq. (7) and τ_{th} from κ using $\gamma^2 \simeq 2.6$ for the shear acoustic mode. All in all, this approach gives a simple and coherent picture of how anharmonicity affects sound waves in dielectric crystals.

3.2. *Attenuation and sound velocity in glasses*

As mentioned earlier, sound waves are key properties to understand the low-temperature thermodynamic anomalies of disordered systems. In spite of its long history, the subject remains of much actuality and activity, particularly at high frequencies in relation with nanoscale thermal transport. Other additional mechanisms do affect sound velocity and attenuation in glasses. The direction of this paragraph is to review progress during the last decades in the understanding of the frequency and temperature dependencies of acoustic properties in glasses and which experimental technique can be used for that purpose.

3.2.1. *Ultrasonics and mechanical techniques*

At sonic and ultrasonic frequencies, there are several well-established methods to measure the sound velocity $v(\Omega, T)$ and the energy sound damping constant $\alpha(\Omega, T)$ related to the energy mean free path $\ell = \alpha^{-1}$ or to the internal friction $Q^{-1} = \alpha v/\Omega$. Figure 4 shows examples of internal friction and velocity changes in dielectric glasses measured at ultrasonic frequencies as a function of temperature measured in the late 1960's.[31] Opposite to the case of crystalline

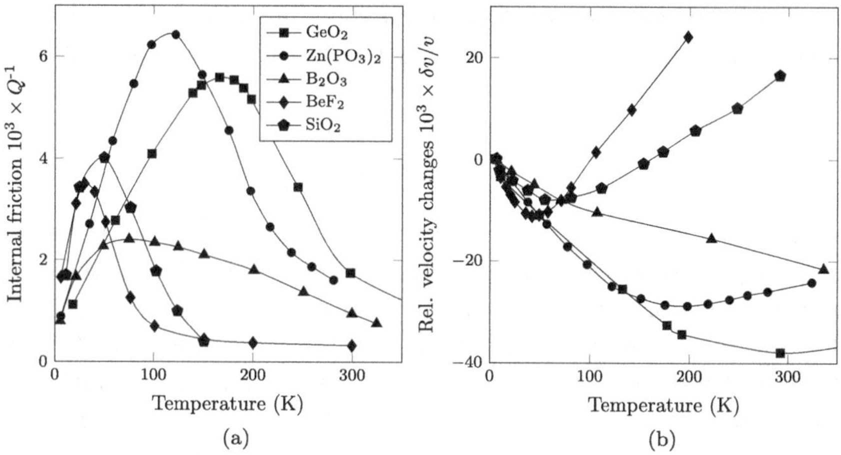

Fig. 4. (a) Internal friction and b) fractional velocity changes of 20 MHz longitudinal waves in different glass formers.[31]

quartz plotted in Figure 3(b), the internal friction in these glasses increases rapidly at low temperatures, $Q^{-1} \propto T$, evidencing a clear maximum which depends on the material. Moreover, the values of Q^{-1} observed here are about two orders of magnitude larger than in quartz, while the frequency is 50 times lower. It is clear that the internal friction shown in Figure 4(a) must originate from another damping mechanism rather than anharmonic interactions.

There are also marked differences in the low-T decay of the relative changes in the sound velocity illustrated in Figure 4(b) as compared to Figure 3(c) for quartz. According to the Kramers–Kronig relation with the internal friction, the onset in $\delta v/v$ is much slower in crystalline quartz than in these glasses. Furthermore, the velocity changes with T show a clear minimum for four of the five glasses. This anomalous behavior is specific to tetrahedrally coordinated glasses[31] and is related to progressive structural changes with increasing T. In contrast, the velocity changes in B_2O_3 at sufficiently high T are similar to those observed in quartz. It was indeed soon identified that anharmonicity was governing the velocity changes in glasses in this temperature range[32,33] except for the few that are tetrahedrally coordinated.

The ultrasonic absorption peaks observed in glasses at low T are usually described using a thermally activated structural relaxation process (TAR) represented by a broad distribution of asymmetric double-well potentials.[20,21,34–36] Owing to structural disorder, several atoms or groups of atoms can occupy two or more equilibrium positions. In that picture, their displacements correspond to these double-well potentials. From site to site, the heights V of the potential barrier and their asymmetries Δ are distributed according to a probability $P(\Delta, V)$. The sound wave couples to the defect site via a deformation potential $\gamma = \frac{1}{2}\frac{\partial\Delta}{\partial e}$, breaking the thermal equilibrium population which relaxes within the characteristic time

$$\tau = \tau_0 \exp\left(\frac{V}{T}\right) \operatorname{sech}\left(\frac{\Delta}{2T}\right), \tag{8}$$

where $1/\tau_0$ is the attempt frequency. Integrating over all the thermally activated relaxations, the internal friction can be calculated following a similar approach used to described anharmonicity leading to[29,37]

$$Q^{-1}_{\mathrm{TAR}} = \frac{\gamma^2}{\rho v^2 T}\int_{-\infty}^{\infty} \mathrm{d}\Delta \int_0^{\infty} \mathrm{d}V P(\Delta, V)\operatorname{sech}^2\frac{\Delta}{2T}\left(\frac{\Omega\tau}{1+\Omega^2\tau^2}\right). \tag{9}$$

The most advanced analysis so far has been carried out on vitreous silica in the early 2000s.[29] This is illustrated in Figure 5(a) which shows the temperature dependence of the internal friction at several representative sonic and ultrasonic frequencies. The peak position depends logarithmically on T,[38] supporting thermally activated relaxation as the main source of the internal friction. All the datasets are simultaneously adjusted with Eq. (9) using for Δ and V independent Gaussian distributions with cutoff Δ_c and V_0, respectively. The same set of parameters is also used to describe the low-T part of the sound velocity variations using the following equation:

$$\left(\frac{\delta v}{v}\right)_{\mathrm{TAR}} = -\frac{1}{2}\frac{\gamma^2}{\rho v^2 T}\int_{-\infty}^{\infty} \mathrm{d}\Delta \int_0^{\infty} \mathrm{d}V P(\Delta, V)\operatorname{sech}^2\frac{\Delta}{2T}\left(\frac{1}{1+\Omega^2\tau^2}\right), \tag{10}$$

which is the Kramers–Kronig transform of Eq. (9). Interestingly, the three datasets covering nearly five decades, corrected from this relaxational effect, collapse on a single curve at low temperature revealing that the dip around 50 K originates from dynamical effects. Oppositely, the anomalous continuous increase of the sound velocity at high temperatures is specific to the tetrahedral networks and thus is of structural origin.

The parameters used to draw the lines in Figure 5 are $V_0 \simeq 660\,\mathrm{K}$, $\Delta_c \simeq 80\,\mathrm{K}$, $\tau_0 \simeq 0.6\,\mathrm{ps}$, and a dimensionless constant strength $C = 1.4 \times 10^{-3}$ related to γ^2 and the density of relaxing defects,[29] the latter being estimated at $N \simeq 0.1\,\mathrm{nm}^{-3}$. This phenomenological description of internal friction at sonic and ultrasonic frequencies recently found a strong support from atomic-scale modeling.[41] In that work, statistical properties of the two-level systems responsible for sound absorption above 10 K are obtained for a model of vitreous silica. The agreement with the above analysis is very good, both qualitatively and quantitatively. Additionally, the microscopic nature

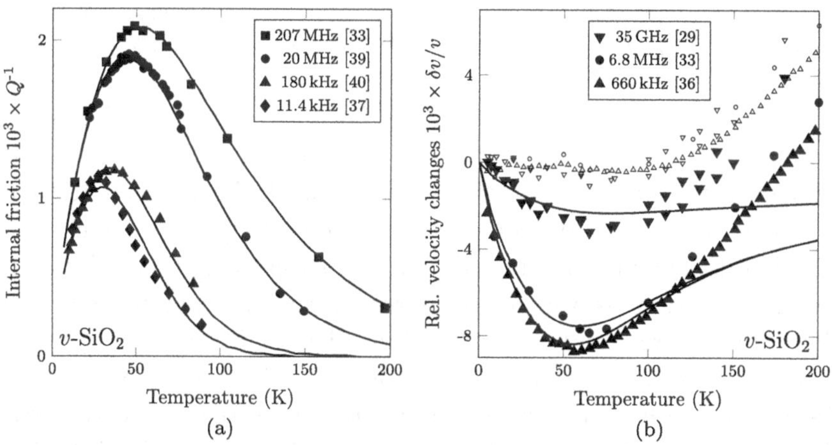

Fig. 5. (a) Internal friction of v-SiO_2 at sonic and ultrasonic frequencies as a function of temperature taken from the literature.[33,37,39,40] The lines show the adjustment of the data to Eq. (9). (b) Fractional velocity changes of v-SiO_2 at sonic, ultrasonic, and hypersonic frequencies[29,33,36] (solid symbols). The lines show the adjustment of low-T part of the data to Eq. (10), whereas the smaller hollow symbols are the experimental $\delta v/v$ corrected from the $(\delta v/v)_{\mathrm{TAR}}$ effect.

of the relaxing defects is revealed. It involves the cooperative rotation of Si–O–Si bonds; see Figure 6 in Ref. [41]. Similar results have been obtained later for some other glasses like GeO_2[42] and B_2O_3,[43,44] where the boroxol rings were identified as the units subjected to thermally activated relaxations or binary borate glasses.[45,46]

3.2.2. *Brillouin light scattering*

Increasing frequency up to the hypersonic regime, it was soon anticipated that anharmonicity should contribute significantly to the sound attenuation in glasses as is the case in dielectric crystals (see Section 3.1).[33] In this frequency domain, typically in the range of several tens of GHz, Brillouin light scattering (BLS) is a method of choice for sufficiently transparent glasses. BLS deals with light inelastically scattered by the density fluctuations exiting in the probed material which are related to the thermally activated acoustic modes. The scattering vector $\mathbf{q}$ is fixed by the geometry, the laser wavelength λ and the refractive index n of the material, $q = 4\pi n \sin(\theta)/\lambda$, where θ is the angle between the incident and the scattered light directions. It also corresponds to the wavevector $\mathbf{Q}$ of the observed acoustic mode $Q = 2\pi/\lambda_a$, where λ_a is the acoustic wavelength. BLS technique measures a Brillouin shift frequency, which is actually the frequency Ω of the probed sound wave. Ω relates to the acoustic velocity v through $\Omega = v/q$. The line width Γ of the Brillouin spectrum quantifies the acoustic energy attenuation α at the frequency Ω or the inverse energy mean free path ℓ^{-1} by $\alpha = \ell^{-1} = \Gamma/v$, where Γ is expressed in angular frequencies just like Ω. The internal friction is actually a direct output of a BLS experiment as $Q^{-1} = \Gamma/\Omega$. The ratio Ω/Γ is thus the quality factor of the damped harmonic oscillator (DHO) used to adjust the Brillouin spectrum.

Typical BLS experiments are nowadays carried out using a tandem multi-pass plane Fabry–Perot interferometer (T-FPI) developed 40 years ago by J. R. Sandercock.[47] This spectrometer is characterized by high contrast, better than 10^{10} thanks to the six Fabry–Perot (FP) etalons in series, and high flexibility due to the tandem arrangement. It gives a frequency resolving power of about

2×10^6 corresponding to a spectral bandwidth of 250 MHz (HWHM) for a Brillouin frequency shift of 30 MHz. Compared to the very abundant ultrasonic results, there exist relatively few damping data at frequencies above 1 GHz in glasses. This is essentially due to the fact that the frequency resolving power of the T-FPI hardly meets the demanding needs for proper sound attenuation measurements in glasses well below the glass transition temperature.

This experimental difficulty can be overcome by using a Brillouin spectrometer consisting of a tandem arrangement of a multi-pass plane FP etalon followed by a confocal one.[48] The plane FP acts here as a prefilter effectively eliminating other spectral components and reducing considerably the elastic signal strength. It is maintained at a fixed spacing which is dynamically adjusted to one of the Brillouin lines with the help of an electro-optically modulated signal at the Brillouin frequency. The confocal FP is piezoelectrically scanned for the spectra analysis. Different confocal FPs can be used to adjust the spectrometer resolution which is ranging from 10 MHz to 100 MHz. In this way, a high contrast (better than 10^{10}) and a high-frequency resolving power ($10^7 - 10^8$) is obtained.

As far as sound damping is concerned, most BLS experiments relate actually to the longitudinal acoustic (LA) mode. This is due to the fact that experiments are generally performed in the backscattering geometry offering several advantages: (i) higher Brillouin frequency shift; (ii) lower uncertainty of velocity measurements; (iii) lower geometrical broadening of the Brillouin line due to the finite aperture. Selection rules for Brillouin scattering indicate, however, that transverse acoustic modes are not active in the backscattering geometry for isotropic materials. One concludes that high-resolution BLS gives accurate and absolute measurements of Q^{-1} as both the Brillouin frequency shift Ω and the Brillouin line width Γ can be obtained simultaneously.

We pursue the case of vitreous silica which is by far the most studied system. Figure 6(a) shows the internal friction of v-SiO_2 at hypersonic frequencies as a function of temperature. A maximum of the damping coefficient is observed around 150 K similar to what is shown in Figure 5(a). However, Q^{-1} does not decrease rapidly to

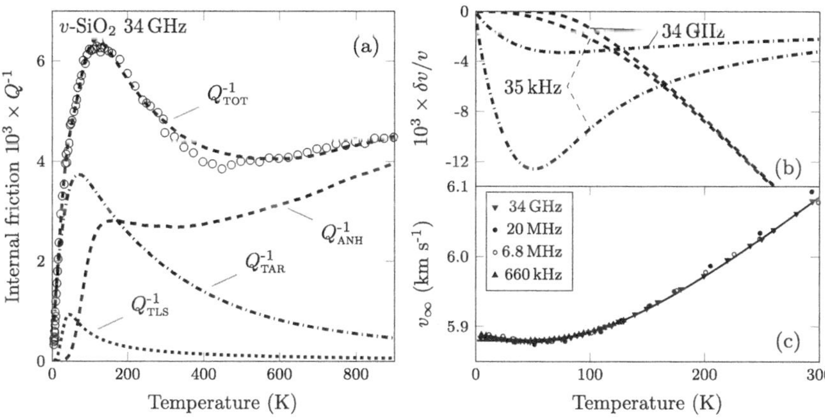

Fig. 6. (a) Internal friction of v-SiO_2 at hypersonic frequencies as a function of temperature.[29, 33, 49] The solid line shows the adjustment of the curve to the sum of the three damping processes $Q^{-1}_{TOT} = Q^{-1}_{TLS} + Q^{-1}_{TAR} + Q^{-1}_{ANH}$. (b) Calculated fractional velocity changes at sonic and hypersonic frequencies resulting from TAR and anharmonic processes. (c) Unrelaxed velocities v_∞ after correction for TAR and anharmonic processes at four selected frequencies. The solid line is a Wachtman's adjustment of the data.

zero above that maximum but remains at a high level up to the highest temperatures, indicating the appearance of another damping mechanism in this frequency range. From Eq. (9), we can anticipate that Q^{-1}_{TAR} cannot account for the entire Q^{-1}. At the peak position corresponding to $\omega\tau \simeq 1$, Q^{-1}_{TAR} is indeed nearly independent of Ω. Second, anharmonicity is still in the regime $\Omega\tau_{th} \ll 1$ in that temperature/frequency range, i.e., $Q^{-1}_{ANH} \propto \Omega$. This is illustrated in Figure 6(a) where the dash-dotted line is the Q^{-1}_{TAR} values calculated with the parameters given in the preceding section for the TAR damping mechanism in v-SiO_2 to which a small contribution from the incoherent scattering of the two-level systems is added (Q^{-1}_{TLS} dotted line; see Chapter 4). The difference, Q^{-1}_{ANH}, is the contribution due to the anharmonic coupling of the LA waves with the thermally excited vibrational modes. It clearly appears that at a sufficiently high temperature and frequency, Q^{-1}_{ANH} takes over Q^{-1}_{TAR}. The analysis of the anharmonic damping has been done using Eq. (6) with $A(T) = \gamma^2 C_V T v/2\rho v_D^3$.[29] The amplitude of Q^{-1}_{TAR} determines the value of the Grüneisen parameter $\gamma^2 = 3.8$, the unique unknown parameter entering $A(T)$, while the temperature dependence of the thermal time τ_{th} is extracted from the shape of $Q^{-1}_{\text{TAR}}(T)$. The values of τ_{th} are proportional to T^{-2} in the approximate range 100 K to 300 K, evolving towards a T^{-1} behavior at higher temperatures in a similar way to those of α-quartz plotted in Figure 3 but three times shorter. It is important to stress again at that point that contrarily to the case of α-quartz, it is not possible to obtain $\tau_{\text{th}}(T)$ in that temperature range from the thermal conductivity. The kinetic equation $\kappa = \frac{1}{3}C_v v \ell$ is indeed valid only for propagating modes, a condition that is presumably no more fulfilled beyond the plateau in $\kappa(T)$ located around 10 K in v-SiO_2. Finally, the validity of the analysis of the internal friction solely based on the TAR contribution at ultrasonic frequencies illustrated in Figure 5 is verified *a posteriori*.

The effect of anharmonicity on the velocity dispersion can be calculated following Eq. (7). This is illustrated in Figure 6(b) where the relative sound velocity variations $\delta v/v$ induced by anharmonicity are plotted for two different frequencies and compared to those due to the thermally activated relaxation processes. While $(\delta v/v)_{\text{TAR}}$ strongly depends on the frequency below 100 K, $(\delta v/v)_{\text{ANH}}$ remains essentially unaffected by changing ν, in particular, above 100 K. At higher temperatures, the magnitude of $(\delta v/v)_{\text{ANH}}$ increases almost linearly with temperature. Linear decrease of the sound velocity has indeed been observed in a large number of glasses.[31,32,44,50] Finally, Figure 6(c) shows the unrelaxed values v_∞ of the frequency-dependent measured sound velocity v that are corrected for the two damping processes, $v_\infty = v - [\delta v_{\text{TAR}} + \delta v_{\text{ANH}}]$. One observes that the four series of data covering nearly five decades in frequency nicely collapse on a single curve $v_\infty(T)$. The latter is not a constant but rather increases rapidly with T and can be adjusted to a Wachtman's equation.[51] This is a peculiarity of tetrahedrally coordinated glasses and associated with a progressive structural change occurring in the network when the temperature is raised.[52]

It is expected that the above-mentioned competing damping processes are not unique to vitreous silica. Even if accurate sound attenuation data are scarce at hypersonic frequencies, some attempts

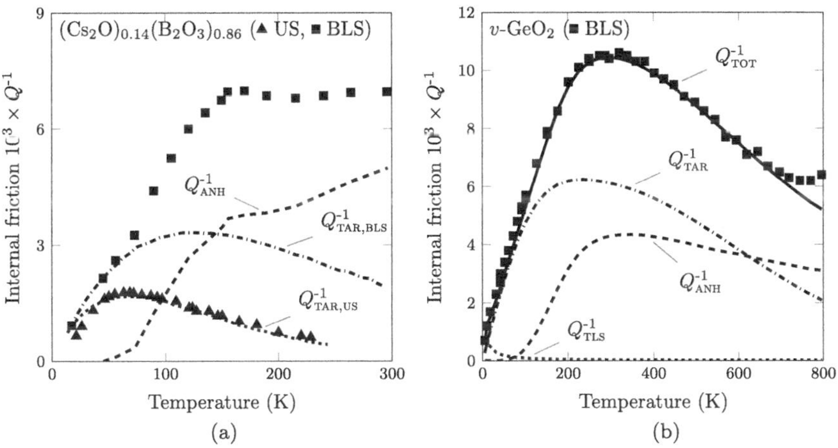

Fig. 7. (a) Internal friction of $(Cs_2O)_{0.14}(B_2O_3)_{0.86}$ at ultrasonic[53] and hypersonic[45] frequencies. The dotted line is the extrapolation at hypersonic frequencies of the Q^{-1}_{TAR} solid line adjusted to the ultrasonic data. The dash-dotted line is the anharmonic contribution to the damping resulting from the subtraction of Q^{-1}_{TAR} from the BLS data.[45] (b) Internal friction of v-GeO_2 at hypersonic frequencies. The solid line shows the adjustment of the BLS data to the sum of the three damping processes $Q^{-1}_{\mathrm{TOT}} = Q^{-1}_{\mathrm{TLS}} + Q^{-1}_{\mathrm{TAR}} + Q^{-1}_{\mathrm{ANH}}$.[42]

have been made in that direction. Figure 7 illustrates the case for two other systems. In Figure 7(a), the internal friction of a cesium borate glass at ultrasonic frequencies had been adjusted to TAR alone following Eq. (9).[53] Its extrapolation at hypersonic frequencies, shown by the dotted line, largely fails to reproduce the Brillouin light scattering data. Clearly, the BLS damping data must contain an appreciable contribution from anharmonicity, shown by the dash-dotted line. The latter has been analyzed using Eq. (6)[45] leading to the determination of τ_{th} and of the mean Grüneisen parameter for that glass as well. Figure 7(b) shows the case of vitreous germania, v-GeO_2, for which several sound absorption experiments exist at low frequencies ensuring a fair estimate of Q^{-1}_{TAR}. In a nutshell, an analysis similar to that done for v-SiO_2 leads to comparable conclusions.[42]

This does not seem to be restricted to oxide glasses as illustrated by the internal friction of poly(methyl methacrylate) (PMMA) obtained at ultrasonic[54,55] and hypersonic[48,56] frequencies and plotted in Figure 8(a). It has been shown that the TAR model alone

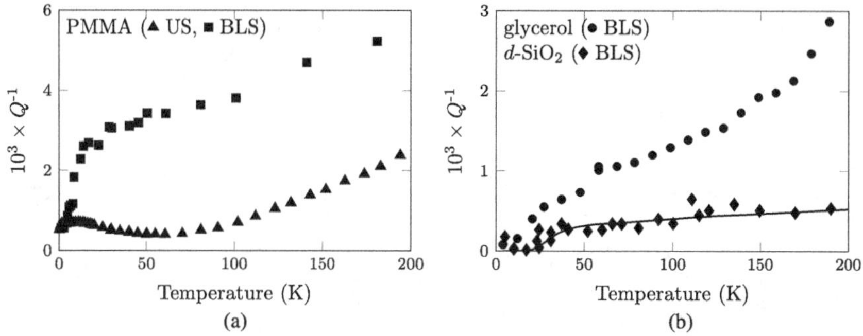

Fig. 8. (a) Internal friction of PMMA at ultrasonic[54,55] and hypersonic[48,56] frequencies. (b) Internal friction obtained using BLS in densified silica[57] and in glycerol.[58,59] Note the different y-scale of the plots compared to Figure 7.

fails to explain consistently the internal friction in the HZ–GHz frequency domain above 10 K.[60] The authors concluded that an additional relaxational mechanism seems to appear at high frequency whose strength increases with T. They invoked a constant loss spectrum ($Q^{-1}_{\text{LOSS}} \propto \omega$) whose exact nature is not clear. Alternatively, anharmonicity could explain the missing contribution to the sound damping at a high frequency, a possibility actually suggested by the authors themselves. Permanently densified glasses seem to show a much lower contribution from the TAR, likely related to the loss of the free volume.[61] It offers unique possibilities to study the sound attenuation originating from anharmonicity at hypersonic frequencies. Figure 8(b) illustrates this point with densified silica glass having the density of crystal quartz. In that case, the TAR is almost completely suppressed.[61] The internal friction measured at GHz frequencies is indeed quite small as compared to that of normal silica.[57] Hence, adjusting the measured Q^{-1} with Eq. (6) gives the solid line in Figure 8(b). The suppressing of Q^{-1}_{TAR} with densification seems also to happen in vitreous boron oxide but accurate high-resolution BLS data are still missing to properly estimate the anharmonicity in d-B_2O_3.[44] Finally, Figure 8(b) also shows internal friction data for glycerol obtained using BLS. The damping coefficient is surprisingly low for a molecular glass and no well-defined peak, characteristic of TAR processes is observed. No attempt has been made yet to described these data using Eq. (6).

3.2.3. *Inelastic scattering of UV radiation*

Brillouin scattering at a shorter wavelength can be used to increase the Brillouin shift frequency. This can be done using excitation with UV radiation[62,63] as long as the samples are sufficiently transparent.[64] This has been attempted, for example, in glycerol, using the second harmonic of a visible laser operating at 488 nm and the IUVS spectrometer located at the Elettra synchrotron light source in Trieste, Italy.[62] In this operation mode, the total resolving power is 1.1×10^6. This setup allows an increase of Q by a factor of 2. Figure 9(a) shows the Brillouin line width at $T = 150\,\mathrm{K}$ ($T_g = 187\,\mathrm{K}$) extracted from the inelastic spectrum obtained with IUVS at $\simeq 43\,\mathrm{GHz}$. It is compared to the values obtained at lower frequencies using stimulated Brillouin gain spectroscopy (SBG)[65]

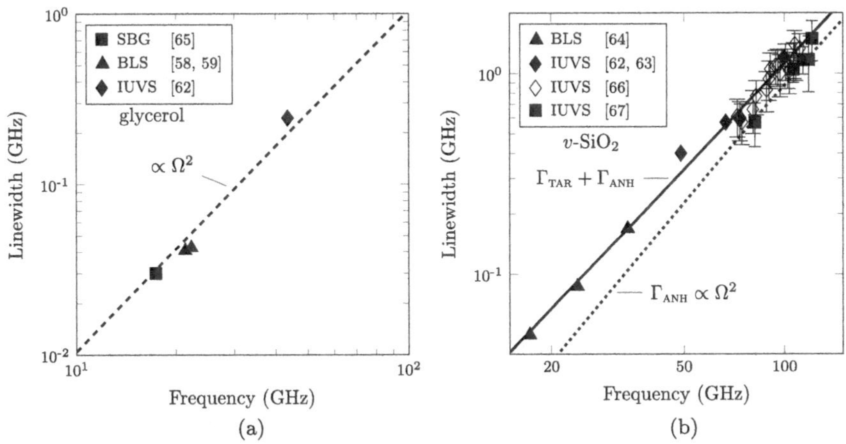

Fig. 9. (a) Frequency dependence of the Brillouin line width in glycerol using SBG,[65] BLS,[58,59] and IUVS[62] at $T = 150\,\mathrm{K}$. The dashed line is the $\Gamma \propto \Omega^2$ adjustment of the data. (b) Frequency dependence of the Brillouin line width in vitreous silica using BLS[64] and IUVS[62,63,66,67] at $T = 300\,\mathrm{K}$. The line is the theoretical expectation based on Eqs. (6) and (9) used to describe the internal friction at lower frequency in v-SiO_2 as explained in the preceding paragraphs. The dotted line is Γ_{ANH} only, showing a quadratic behavior.

and BLS.[58,59] The data show that, in this frequency range, the sound attenuation coefficient closely follows a Ω^2 dependence, as expected for anharmonicity, and in agreement with the temperature dependence discussed earlier.

Vitreous silica is known to have a rather high-absorption edge at $E_i \approx 8\,\mathrm{eV}$ allowing the opportunity to access Brillouin frequencies above 100 GHz and a momentum transfer region of $0.14\,\mathrm{nm}^{-1}$. In that case, the IUVS spectrometer uses the UV beam produced by the synchrotron, and the total resolving power is 5×10^5. Figure 9(b) shows part of the first IUVS results obtained at two wavelengths[62] or at two angles[63] plotted as solid symbols. The graph also shows high-resolution BLS values obtained with visible light at three scattering angles.[64] The IUVS data remarkably align with the expected damping values calculated from Eqs. (6) and (9) with the parameters used to described internal friction of vitreous silica at lower frequencies. One observes that the power-law frequency exponent of $\Gamma(\Omega)$ is about 1.7 at 20 GHz, still slightly smaller than the quadratic behavior of the growing anharmonic contribution Γ_{ANH}.

The understanding of the sound attenuation in vitreous silica and more generally in glasses was challenged by deep UV Brillouin scattering experiments, i.e., close to the absorption gap, claiming an unexpected dramatic increase of the line width beyond 110 GHz in v-SiO_2.[66] Below this frequency, these new data agreed very well with the preceding IUVS ones. At higher frequencies, a much steeper increase was clearly observed that was interpreted as a crossover associated with elastic disorder. The latter was, however, not detected by a second experiment carried out using a high-grade superpolished silica sample up to about 125 GHz.[67] As we will see in the next section devoted to picosecond ultrasonics, several other experiments covering this frequency range also disproved the deep UV results. As a consequence, only the deep UV Brillouin data which truly relate to acoustic damping are plotted in Figure 9(b) as hollow symbols. Brillouin scattering close to the absorption gap is clearly a delicate experiment. Approaching the gap from below, the collected Brillouin signal considerably decreases, owing to rapidly increasing sample absorption of both the incident and the scattered light. At first sight, the unexpected increase of the Brillouin line width observed in the deep UV data above 110 GHz could be related to finite-size effects induced by the dramatic increase of the imaginary part of the refractive index approaching the absorption gap.[64,68] However, this effect is still negligible in the investigated region.[67] If not an artifact, the anomalous increase in damping could originate from growing refractive index fluctuations close to the gap producing an enhanced uncertainty broadening in Brillouin scattering.[69]

3.2.4. *Picosecond ultrasonics*

There is, then, a limit set by the optical absorption edge to the proper determination of acoustic attenuation in dielectric samples using Brillouin light scattering. It corresponds to acoustic waves around 100 GHz in the favorable case of vitreous silica. The samples then remain opaque to electromagnetic radiation up to the X-ray region where they again become transparent. Brillouin spectroscopy known as inelastic x-ray scattering then allows the observation of sound properties in the THz frequency range; see Section 3.2.5.

In the crucial frequency region between 50 GHz and 1 THz, there exists but single spectroscopy known as picosecond ultrasonics (PU). This delicate but very promising technique is essentially based on the generation and the detection of GHz – THz acoustic waves propagating through a thin glass film by ultrashort optical pulses typically shorter than 1 ps. This optical technique has been pioneered in the 1980s,[70] but first attenuation measurements on a glass have been reported in the 1990s only.[71] Generally, an ultrashort light pulse is focused onto a small area of a thin metallic film deposited on the dielectric glass sample, which is itself often a film deposited on a substrate. This light pulse, the pump, is absorbed by the metallic layer which expands generating a longitudinal strain pulse into the glass. Typically, for a 12 nm thick aluminum film on vitreous silica, the launched acoustic pulse is a plane wave packet of about 20 nm spatial extent whose Fourier spectrum peaks around 250 GHz.[72] The propagation of the acoustic pulse in the media and its reflection at the different interfaces can be monitored using a second time-delayed light pulse focused on the same spot. The main part of this probe light is reflected at the fixed interfaces, whereas a weaker component is reflected by the propagating strain pulse. All these reflections interfere leading to oscillations of the measured reflectivity with the time delay due to the strain pulse bouncing back and forth in the sample.

In the early work by Zhu *et al.*,[71] the frequency dependence of the sound attenuation in vitreous silica layers was determined from the decay of successive echoes reaching the Al transducer after a round trip in the glass. A single-crystal tungsten substrate was used to produce numerous and strong echoes thanks to the large acoustic mismatch between v-SiO_2 and tungsten. Based on the modification of the shape of the acoustic pulse reflecting the frequency-dependent attenuation coefficient of its spectral components, this method allows accessing frequencies well above the Brillouin frequency. These early data obtained at room temperature are plotted as stars in Figure 10 showing an approximate quadratic behavior. It is interesting to note that extrapolating this Ω^2 behavior (dash-dotted line in Figure 10) to the THz region leads to values comparable to those reported using

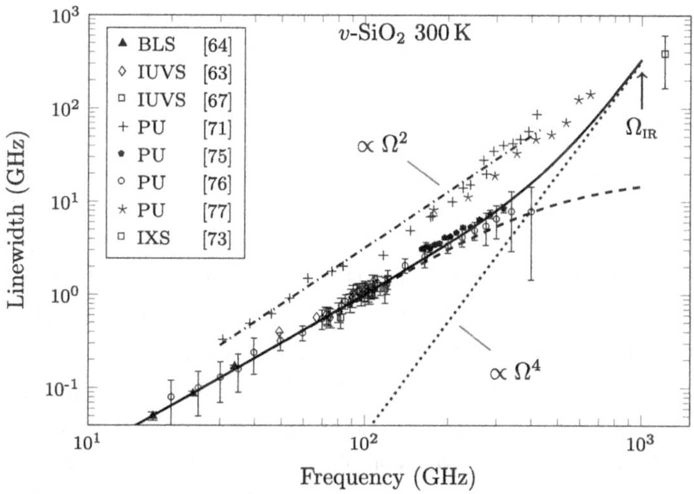

Fig. 10. Frequency dependence of the Brillouin line width in vitreous silica at room temperature using BLS,[64] IUVS,[62,63,67] different PU schemes,[71,74–77] and IXS.[73] The solid line is the sum of $\Gamma_{\text{TAR}} + \Gamma_{\text{ANH}}$ discussed in the preceding section plus the expected Ω^4 law leading to the Ioffe–Regel crossover at Ω_{IR}.

IXS (hollow square in Figure 10).[73] We will come back to this point in the next section. Up to about 200 GHz, the picosecond ultrasonic data roughly follow the expected frequency dependence shown by the dashed line discussed in the preceding section as the sum of $\Gamma_{TAR} + \Gamma_{ANH}$. However, it became clear with time that the sound attenuation was systematically overestimated by a factor between 2 and 3. It is now well admitted that incomplete corrections for losses at the interfaces were the likely source of errors.

Those difficulties could be alleviated with a different scheme using the oscillations produced by the interference of the probe partly reflected at the sample surface with its reflection by the moving acoustic pulse. In that case, the probe light interacts with one of the Fourier components of the acoustic pulse according to momentum conservation. This *modus operandi* is actually the time-domain analog of the usual Brillouin light scattering and as such suffers, in principle, from the same limits in the accessible frequency domain.[78] If the dielectric sample film is deposited on a substrate with significantly higher refractive index and acoustic velocity like crystalline silicon, the oscillations detected in the substrate correspond then to much higher-frequency acoustic waves. A study of the amplitude of these higher-frequency acoustic oscillations in the silicon substrate as a function of the glass film thickness allows then determining hypersound attenuation in the amorphous sample at a frequency beyond the usual BLS limit. This scheme has been used for silica films on crystalline silicon[74,75] and the more recent data covering the frequency range 165 GHz to 320 GHz are plotted in Figure 10 as solid pentagons. They nicely superimpose with the theoretical expectations (dashed line) based on Eqs. (6) and (9) up to about 250 GHz in a domain where the anharmonic contribution $\Gamma_{ANH} \propto \Omega^2$ becomes dominant as described in Figure 9(b). At higher frequencies, the latter tends to saturate, while the measured sound attenuation still increases. This is in fair agreement with the solid line which takes into account the expected Ω^4 law leading to the Ioffe–Regel crossover at Ω_{IR}. The latter is located around 1 THz in vitreous silica as shown in Figure 10 together with the lowest point obtained using IXS.[73] This was not yet a clear observation of the

Ω^4 law in that frequency range but thanks to the very high quality of the data, the onset seems to have been detected here.

Similar results have been obtained later on using a narrowband ultrafast photoacoustic approach that allows covering more than a decade in frequency.[76] The attenuation coefficients, converted to line width, are shown in Figure 10 as hollow disks. As expected, for a multiband technique, the most precise values are obtained in the mid-range from 50 to 270 GHz while data in the frequency wings are rapidly approximate. It is particularly true for the data above 300 GHz which cannot allow the observation of the anticipated Ω^4 law. However, the PU technique is continuously improving and other works reported ultra-broadband PU data reaching 650 GHz in v-SiO_2[77] shown by stars in Figure 10. In that case, high-frequency Fourier components in the acoustic pulse are obtained replacing the metallic transducer with a piezoelectric nanolayer one. Unfortunately, the reported sound attenuation extracted from the decaying echoes is still clearly overestimated as in Ref. [71]. Nevertheless, with a proper analysis of the interface contribution, one can envision in the near future to bridge the gap between BLS and IXS so as to firmly detect the onset of the expected Ω^4 law up to the IR crossover at THz frequencies.

One must note, however, that most of the PU data focus on vitreous silica due to the tremendous requirements for making decent glass films having properties comparable to bulk samples. Nevertheless, similar measurements for other amorphous materials have been sparsely collected to investigate the universality of the attenuation at high frequency.[79] In the latter work, a general $\Gamma \sim \Omega^2$ law is evidenced up to about 320 GHz for the amorphous polymers PMMA, Polyethyl methacrylate (PEMA), and polystyrene (PS) as well as for the metallic glass TiNi, all at room temperature. Concerning PMMA, the attenuation coefficient measured at low frequencies using BLS nicely extrapolates the obtained Ω^2 law at ambient T as shown in Figure 11(a). This does not seem to be the case on the high-frequency side where a single measurement at about 600 GHz using IXS shows an attenuation coefficient about 70% larger,[80] leaving room for the observation of a Ω^4 law in PMMA.

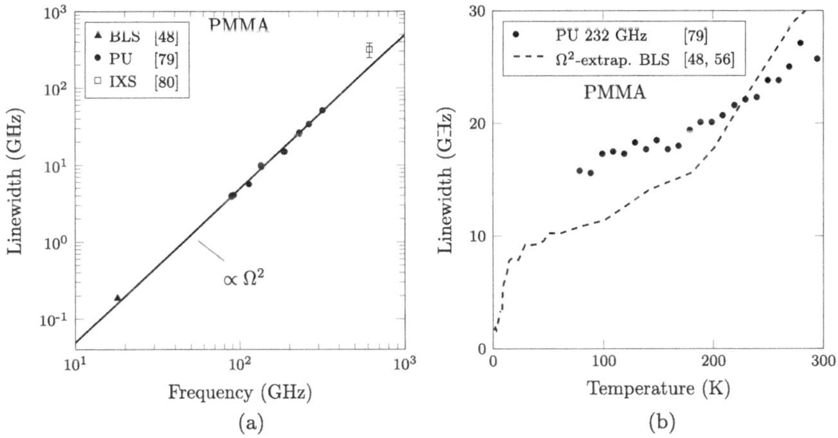

Fig. 11. (a) Frequency dependence of the Brillouin line width in PMMA at room temperature using BLS,[48] PU,[79] and IXS.[80] The solid line is the $\Gamma \propto \Omega^2$ best fit of the PU data. (b) Temperature dependence of the Brillouin line width in PMMA at 232 GHz[79] compared to Ω^2-extrapolated 18 GHz BLS data (dashed line).

Besides, the temperature dependence of the sound attenuation can be studied using PU as well. Both the transient and steady-state heating of the sample by the laser pump must be controlled, in particular, at low T where heat capacity is rapidly decreasing. A typical result is displayed in Figure 11(b) for PMMA at 232 GHz .[79] The phonon line width is continuously decreasing with decreasing temperature, in agreement with possibly anharmonicity-dominated sound absorption. The temperature dependence seems, however, different from that obtained using BLS around 18 GHz as shown in Figure 11(b) where the dashed line is the measured BLS line width[48] extrapolated to 232 GHz with a $\Gamma \propto \Omega^2$ law. This probably means that several competing sound damping mechanisms, each with its own frequency and temperature dependence, are present in PMMA as well. More recently, the temperature dependence of the sound absorption up to about 200 GHz has been studied in vitreous silica showing fair agreement with the theoretical expectations based on Eqs. (6) and (9).[81]

An attractive perspective given by the rapidly developing picosecond ultrasonic technique is finally the possibility to observe the predicted Ω^4 law for sound attenuation in a (ω, T) region where this process would dominate over relaxation mechanisms. The first direction is to push the technique towards higher frequencies using more complex transducers as acoustic pulse generator/detector. The second path is to sufficiently reduce the temperature to suppress the relaxations leaving the disorder-induced process to dominate the acoustic attenuation.

3.2.5. *Inelastic neutron and X-ray scattering*

Glass samples become again transparent in the soft X-ray region. The latter then allows the observation of sound properties in the THz range using a Brillouin scattering technique known as inelastic X-ray scattering (IXS). Due to the large decrease of the wavelength compared to visible light, the experiment must be done at relatively small angles to achieve the lowest momentum transfer.[82] Hence, to observe Brillouin scattering of THz acoustic excitations in amorphous materials, IXS experiments are made in forward scattering in transmission, while conventional BLS uses backscattering to increase the sensitivity. The instrumental resolution of the IXS spectrometer is now around $1\,\text{meV} \simeq 0.25\,\text{THz}$, while the lowest achievable q value is, in practice, about $1\,\text{nm}^{-1}$. On the other hand, inelastic neutron scattering (INS) is not a technique of choice for such experiments at least in amorphous materials with rather high sound velocities. Close to $q = 0$, it is indeed not possible to measure sound waves whose velocities are larger than those of the incident neutrons due to kinematic conditions. Amorphous selenium is one of the few examples where LA phonons are sufficiently slow to be accessible by INS.[83] Both techniques, however, share the same drawback which is their inability in probing transverse acoustic modes at low q.

(a) Two competing scenarios

Measuring the acoustic modes in glasses in the (ω, q) region of the expected IR crossover together with the search for a damping $\Gamma \propto \Omega^4$ has been a strongly debated issue emerging in the late nineties when IXS experiments became operational. The very first IXS experiments in glasses were reported on glycerol and LiCl:6H_2O in the q region

2–8 nm^{-1}.[82] The measured spectra, which are proportional to the dynamical structure factor $S(q,\omega)$, consist of a strong elastic peak plus a weaker LA Brillouin doublet modeled as a damped harmonic oscillator (DHO) both convoluted with the instrumental resolution. In the classical approximation and neglecting Debye–Waller effects, $S(q,\omega)$ can be written as

$$S(q,\omega) = f_q S(q)\delta(\omega) + (1-f_q)S(q)\frac{1}{\pi}\frac{\Omega_q^2\Gamma_q}{\left[\omega^2-\Omega_q^2\right]^2+\left[\omega\Gamma_q\right]^2}, \quad (11)$$

where $S(q)$ is the structure factor and f_q is the elastic fraction of the scattering at this q. Ω_q is the Brillouin frequency shift, related to the phase velocity of sound waves $v_q = \Omega_q/q$, while Γ_q is the full width at half maximum of the Brillouin peak, associated with the sound attenuation. Parameters with the index q depend on the magnitude of the scattering vector q but not on the frequency ω.

The pioneering values obtained in glassy glycerol at $T = 145\,\mathrm{K}$ are displayed in Figures 12(a) and 12(b) showing an almost linear dependence of Ω_q with q and an approximate quadratic q dependence of Γ_q. The linear dispersion was corresponding to a sound velocity $v_{\mathrm{LA}} \simeq 3350\,\mathrm{ms}^{-1}$, lower but close to the known longitudinal sound velocity measured at a much larger acoustic wavelength. Very similar results were obtained for the second glass, LiCl:6H_2O, leading the authors to conclude that propagating sound waves were existing at THz frequencies in those two glasses. This picture was reinforced when first IXS data measured in the archetypal glass v-SiO_2 also revealed linear dispersion with a longitudinal sound velocity comparable to the macroscopic one together with a quadratic q dependence of the line widths. The latter behavior was further shown to be reasonably consistent with the sound attenuation coefficient known at lower frequencies at that time[71,73] as illustrated in Figure 12(c) where the solid line is the adjustment of the sole IXS data to a $\Gamma_q \propto q^{\alpha}$ law with $\alpha \simeq 2$.

Interestingly, a distinct combined analysis of room temperature INS and IXS data in vitreous silica proposing a very different scenario appeared simultaneously.[84] In that work, the inelastic part of the

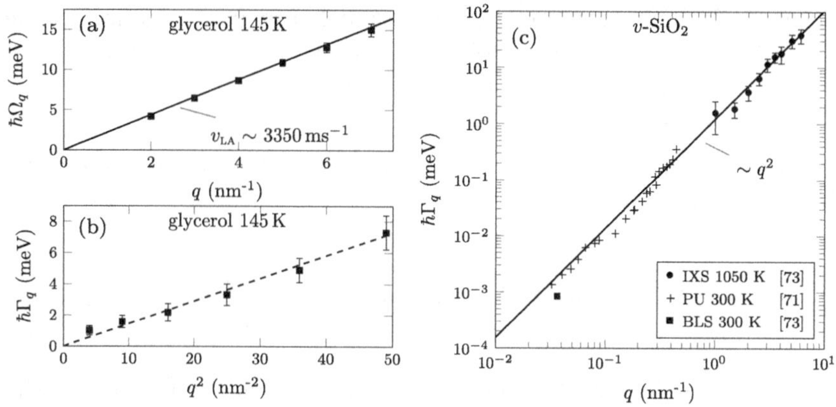

Fig. 12. Wavenumber dependence of (a) the Brillouin frequency shift Ω_q and (b) the Brillouin line width Γ_q in glycerol at 145 K measured using IXS.[82] The solid line in (a) is the $\Omega_q = v_{LA}q$ best fit giving a sound velocity $v_{LA} \simeq 3350\,\mathrm{ms}^{-1}$, while the dashed line in (b) is the $\Gamma_q \propto q^2$ best fit.[82] (c) Wavenumber dependence of the Brillouin line width Γ_q in vitreous silica at 1050 K measured using IXS,[73] PU,[71] and BLS[73] at 300 K. The solid line is the $\Gamma_q \propto q^\alpha$ best fit of the IXS data giving $\alpha \simeq 1.95$.[73]

spectra was analyzed with a $S_{\mathrm{in}}(q,\omega)$ function derived from an effective-medium approximation (EMA) for the force constants of a percolating network that reads[85]

$$S_{\mathrm{in}_{\mathrm{EMA}}}(q,\omega) = \frac{k_B T q^2}{\pi m \omega} \frac{q^2 W''(\omega)}{\left[\omega^2 - q^2 W'(\omega)\right]^2 + \left[q^2 W''(\omega)\right]^2}. \tag{12}$$

Equation (12) is a generalization of the second term in Eq. (11), appearing clearly when the second moment rule giving $(1 - f_q)S(q)\Omega_q^2 = k_B T q^2/m$ is used in the later:[86]

$$S_{\mathrm{in}_{\mathrm{DHO}}}(q,\omega) = \frac{k_B T q^2}{\pi m \omega} \frac{\omega \Gamma_q}{\left[\omega^2 - \Omega_q^2\right]^2 + \omega^2 \Gamma_q^2}. \tag{13}$$

In the EMA, the disorder-induced inhomogeneous force constants are replaced by a complex, frequency-dependent, uniform function $W = W' - iW''$. $W'(\omega)$ and $W''(\omega)$ result in effective sound velocity

$c_{\mathrm{EMA}}(\omega)$ and scattering length ℓ_{EMA} which depend on the frequency[85]

$$c_{\mathrm{EMA}}(\omega) = \frac{|W(\omega)|}{\mathrm{Re}\,\sqrt{W(\omega)}}, \qquad \ell_{\mathrm{EMA}}(\omega) = \frac{1}{\omega}\frac{|W(\omega)|}{\mathrm{Im}\,\sqrt{W(\omega)}}. \qquad (14)$$

The physical picture is that, as ω increases towards ω_{IR}, the definition in q of the vibrations becomes less and less precise due to the disorder. Thus, vibrations of different frequencies have an appreciable Fourier component at the scattering vector q. This leads to dependence in frequency (rather than in wavevector, which is not well defined for the vibrations) of the parameters entering $S_{\mathrm{in}}(q,\omega)$. Although by themselves measurements of $S(q,\omega)$ cannot demonstrate phonon propagation or localization, it seemed relevant to investigate the applicability of the two models.

To illustrate the difference between Eqs. (12) and (13), typical spectral shapes are shown in Figure 13. The frequency scale is normalized by the hypothetical Ioffe–Regel frequency ω_{IR} occurring

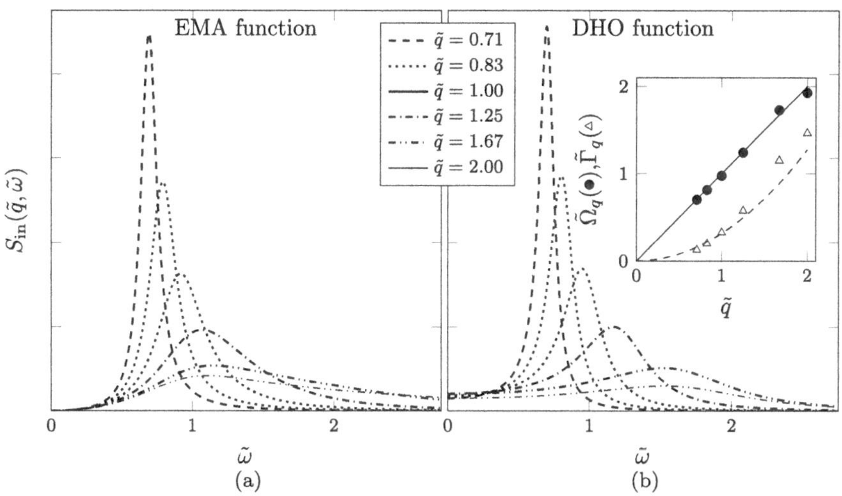

Fig. 13. Spectral shapes of (a) the EMA function (Eq. (12)) at several q values around q_{IR}[84] and (b) the DHO function (Eq. (13)) obtained using the best Ω_q and Γ_q values that fit the EMA profiles. Inset: q-dependence of Ω_q and Γ_q. The line is $\Omega_q = c_0 q$, whereas the dashed line is $\Gamma_q \propto q^2$. The scales are normalized by ω_{IR} or q_{IR}.

when the line width Γ_q reaches Ω_q/π at the wavenumber $q_{\rm IR} = \omega_{\rm IR}/c_0$, where c_0 is the constant low-frequency sound velocity. Figure 13(a) shows EMA spectral shapes at several q values around $q_{\rm IR}$ obtained from a phonon–fracton crossover model[87] which was also used to describe IXS data in v-SiO_2.[84] In that case, the effective sound velocity $c_{\rm EMA}(\omega)$ is c_0 well below $\omega_{\rm IR}$ and increases as $\sqrt{\omega}$ above, while the sound damping coefficient $\Gamma_{\rm EMA}(\omega) = c_{\rm EMA}(\omega)/\ell_{\rm EMA}(\omega)$ rapidly increases as ω^4 below $\omega_{\rm IR}$ and then flattens to a linear frequency dependence above. Figure 13(b) displays the corresponding DHO spectral shapes obtained using the best Ω_q and Γ_q values that adjust the EMA profiles. At small q values, below $q_{\rm IR}$, the spectral shapes in Figures 13(a) and 13(b) are quite similar showing a sharp Brillouin peak. The main difference arises at larger q values when the line width is large. The homogeneous DHO spectrum shows an increasing high intensity for ω tending to 0 compared to the peak intensity. Conversely, the EMA spectrum, which is intrinsically inhomogeneous, shows an intensity going to zero together with ω. Furthermore, the EMA spectrum for $q \gg q_{\rm IR}$ becomes practically q independent with a peak around $\omega_{\rm IR}$.

The inset in Figure 13(b) shows the q-dependence of the normalized Ω_q and Γ_q values. An almost linear dispersion of the former is observed even for $q > q_{\rm IR}$, while the sound damping fairly compares to a quadratic behavior. These schematic graphs illustrate how contradictory conclusions could be drawn from similar IXS experiments: one supporting the existence of propagating sound waves up to high-q values[73,82] characterized by linear dispersion and quadratic sound damping and the other one implying a strong disorder-induced scattering of these excitations leading to a Ioffe–Regel crossover.[84] However, the limited statistics and energy resolution of the early IXS experiments clearly prevented any definite conclusions at that time.

(b) Search for a quartic attenuation law

The two competing views have been extensively discussed for many years, thereby including better experiments, progresses in numerical modeling, and theoretical advances. On the experimental side, the linear dispersion of the longitudinal acoustic excitation and the quadratic trend of the line width using IXS were reported for many different amorphous systems.[88–98] However, spectroscopic evidence for the alternative scenario were slowly showing up.[83, 99–101] The latter were concomitantly strengthened by numerical simulations, in particular, of v-SiO_2 models. IR crossovers for acoustic excitations were observed both for LA and TA modes at ~1 THz[102, 103] in very good agreement with the value obtained from the analysis of INS data[84] and ideas developed in Refs. [83, 99–101]. Molecular dynamics further demonstrated that the IR crossover was actually characterizing a continuous transition from propagative acoustic waves to diffusive excitations and not localized vibrations as suggested by the phonon–fracton approach.[103] Localized modes exist only at much higher frequencies. The disorder-induced quartic behavior of the acoustic damping was nevertheless not observed due to the low-q box-size limitation of the simulation.

Similarly, direct observation of the ω^4 law using x-rays was still a very demanding experiment. From the thermal conductivity, it was indeed anticipated that the expected ω_{IR} hardly exceed 1–2 THz in most materials. Combined with the typical sound velocities found in amorphous solids, it leads to crossover wavevectors q_{IR} in the range 1–2 nm^{-1}. This happened to be the lowest accessible limit in IXS. Furthermore, with a Lorentzian-like instrumental resolution profile of about 1.7 meV (FWHM), features below 1 THz were easily buried into the wings of the intense elastic signal. However, studying selected materials with IR crossovers expected at high values appeared to be the right strategy for further progresses.

The first example is densified silica which is about 20% denser than normal vitreous silica. The expected crossover is then pushed to higher values, $\omega_{\mathrm{IR}} \sim 2.2$ THz and $q_{\mathrm{IR}} \sim 2.2\,\mathrm{nm}^{-1}$,[99, 101] as opposed to the estimates $\omega_{\mathrm{IR}} \sim 1$ THz and $q_{\mathrm{IR}} \sim 1\,\mathrm{nm}^{-1}$ for v-SiO_2.[84] d-SiO_2 is also more homogeneous so that the elastic intensity is significantly reduced. For these reasons, it was anticipated that the crucial region below and near ω_{IR} might be accessible to spectroscopy in d-SiO_2 using existing IXS capabilities.[99, 101] Figure 14 shows typical inelastic spectra obtained at $T = 565$ K.[104] The measured spectra were fitted

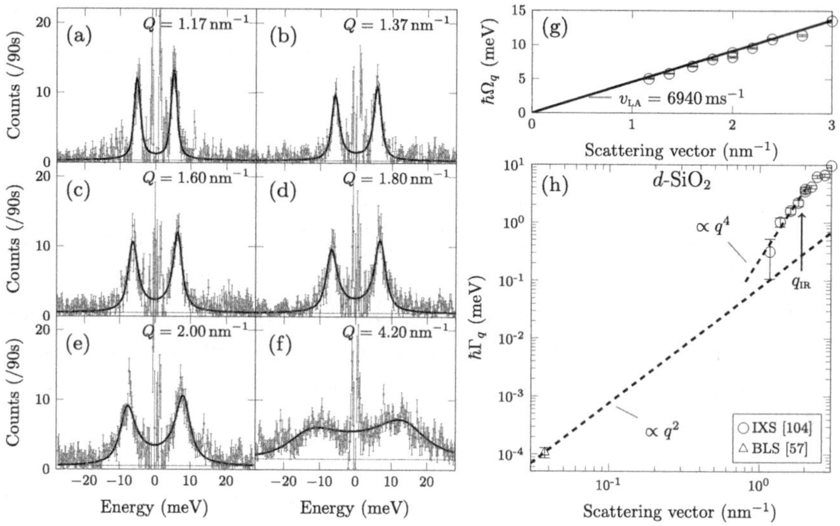

Fig. 14. (a–f) IXS spectra of d-SiO$_2$ at several q values and their DHO fits of the inelastic part after subtraction of a central peak freely adjusted in the fitting process.[104] The background shown by the thin baselines is fixed to the detector noise, measured independently. (g) Evolution of Ω_q with q where the line extrapolates the BLS result. (h) Evolution of Γ_q with q where the dashed line extrapolates the quadratic behavior of the sound attenuation from BLS results.[57] The solid line is a $\Gamma_q \propto q^4$ law interpolating the five lowest points.[104]

to a DHO plus a delta function for the elastic part, convoluted with the Lorentzian instrumental function. In Figure 14(a), the width of the Brillouin peak is mostly instrumental, while in Figure 14(e), it is due in large part to a real broadening of the Brillouin signal.

The values obtained for Ω_q are reported in Figure 14(g) and show an almost linear q-dependence as in other materials and in agreement with the discussion of Figure 13. The q dependence of the resulting Γ_q values is shown in Figure 14(h). Γ_q increases very rapidly with q and is significantly above the values that can be extrapolated from the precise BLS measurement. As discussed before, for d-SiO$_2$, see Figure 8, the sound attenuation at long wavelength originates from anharmonicity in this temperature range and it seems reasonable to assume $\Gamma_q \propto q^2$ in the sub-THz range shown as the dashed line in Figure 14(h). This homogeneous broadening contributes only to

10% of the line width measured using IXS at $q = 1.37\,\text{nm}^{-1}$. Below about $2\,\text{nm}^{-1}$, the four points can be well described with a $\Gamma_q \propto q^4$ expression, giving the solid line in Figure 14(h). At larger q values, Γ_q extracted from the DHO fits seems to increase as q^2, in agreement with observations in other materials. It was strongly suggesting that a *new damping mechanism* should exist in between. Clearly, the latter must lead to a rapid increase of the line width, faster than q^2, as two different q^2 laws could not even crossover.

In addition, the DHO parameters fulfill the Ioffe–Regel criterion ($\Gamma_q = \Omega_q/\pi$) at $q \simeq 2\,\text{nm}^{-1}$ corresponding approximately to $\hbar\omega_{\text{IR}} \simeq \hbar\omega_{\text{BP}}$. Finally, another piece of the puzzle came from the shape of the inelastic spectra. It seems clear from Figure 14 that with increasing q, the DHO profile demands a larger and larger signal at small ω which is not measured as already noted in the preceding experiments in d-SiO_2.[99,101] This is particularly clear in Figure 14(f) at $q \simeq 2q_{\text{IR}}$, an averaged spectrum obtained after a week of measurement. All these spectra were also adjusted to an EMA function using Eq. (12), giving similar results at low-q values but much better fits approaching q_{IR} and above as expected. The sole use of the DHO profile was then to evidence the existence of a new sound damping mechanism for the LA waves which approximately behaves as q^4 or ω^4 up to an IR crossover.[104]

(c) Fate of the high-frequency acoustic waves

It was crucial to examine to what extent the observations reported for d-SiO_2 were common to glasses. To that effect, a second glass, lithium diborate $Li_2O{-}2B_2O_3$ or LB2, was carefully studied using IXS upon approaching the expected IR crossover at two temperatures. Again, the feeling that the IR frequency should be related to the boson peak guided the choice of a material with a boson peak located at a high energy to facilitate the spectroscopy. Furthermore, some preliminary results already suggested that such an IR crossover should exist in LB2.[94] The same procedure as the one previously described for analyzing d-SiO_2 IXS data was applied to LB2.

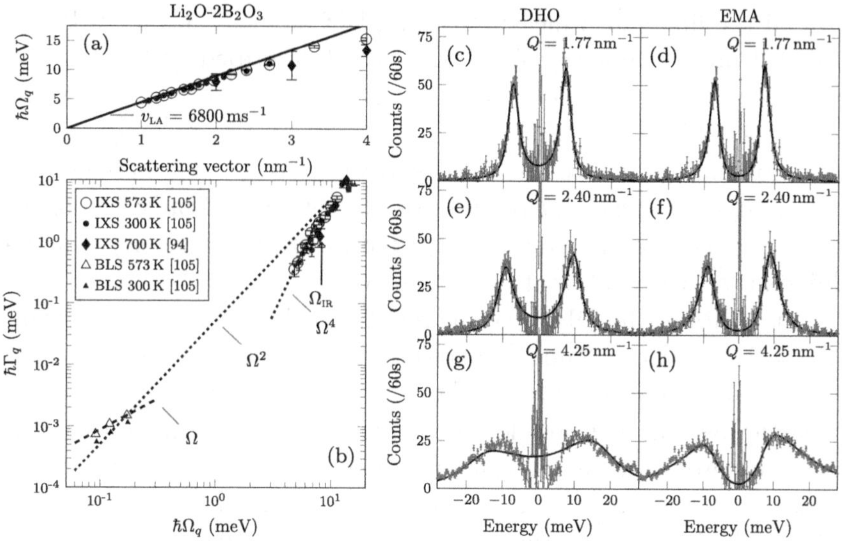

Fig. 15. (a) Evolution of Ω_q with q where the line extrapolates the BLS result for LB2 glass. (b) Evolution of Γ_q with Ω_q where the dashed line at low energy extrapolates the linear behavior of the sound attenuation from BLS results.[106] The solid line is a $\Gamma_q \propto q^4$ law interpolating the lowest q points of IXS data.[105] The dotted line is a quadratic behavior extrapolating IXS data obtained above the IR crossover. (c–h) Comparison between IXS spectra adjusted to a DHO lineshape or to an EMA lineshape showing the failure of the former approaching q_{IR}.[105]

The parameters extracted from the DHO fits of the IXS data are shown in Figures 15(a) and 15(b).[105] They exhibit very little dependence on T, if any. Below about $2\,nm^{-1}$, there is only a slight departure from linear dispersion as seen in Figure 15(a) where the slope of the line is the BLS sound velocity at 573 K .[106] There is, however, a very rapid increase of the full width Γ_q in this q region, as shown in Figure 15(b). The latter is definitely faster than a q^2 behavior and is consistent with a q^4 law as can be inferred from the solid and the dotted lines. Above about $2\,nm^{-1}$, Γ_q increases slower with an exponent close to 2 in agreement with numerous reports. All these results are fully consistent with those which have been obtained previously for densified silica including the occurrence of the Ioffe–Regel crossover q_{IR} below $2\,nm^{-1}$ which corresponds to an IR frequency ω_{IR} close to the frequency of the maximum of the boson peak ω_{BP}. The two series of points reported in Figure 15(b) at lower energy have been obtained using BLS analyzed with a high-resolution tandem spectrometer[106] for three different scattering angles. They show that contrarily to the d-SiO_2 case the sound attenuation depends linearly on the frequency in the hypersonic range pointing to TAR processes as the dominant broadening mechanism in this (ω, T) region. The anharmonic contribution should take over at higher frequencies.

As discussed before, the experimental evidence for a disorder-induced q^4 sound attenuation mechanism leading to a Ioffe–Regel crossover indicates that the DHO cannot be the adequate representation of the spectral lineshape approaching q_{IR}. This is illustrated in Figures 15(c)–15(h) which compare inelastic IXS spectra just below and above q_{IR} fitted to the DHO lineshape on the left and to an EMA lineshape on the right.[105, 107] Well below q_{IR}, the experimentally selected scattering vector q matches acoustic modes of well-defined Q number and the line width Γ_q truly reflects an inverse lifetime. In this case, the DHO model gives an excellent representation of the spectral shape. In the opposite limit, where Q is ill defined at a given mode frequency Ω, a spectrum $S(q, \omega)$ at constant $\Omega = \omega$ just reflects an average spatial profile of the excitation packets at ω.[108] Conversely, a spectrum $S(q, \omega)$ at constant q is then the sum of contributions of all the modes with $\Omega = \omega$ at this particular scattering q value. This is inhomogeneous broadening. There is no basis for using the DHO in this case. The failure of the DHO above q_{IR} is clearly illustrated in Figures 15(e), 15(g) and can be observed in Figures 14(e) and 14(f) as well.

The question of why such a rapid increase of the Brillouin line width was not detected in all the other glasses investigated so far was elucidated by considering the values of the ratio Γ_q/Ω_q obtained in these IXS experiments. Figure 16 shows values of that ratio vs Ω_q for ten different glasses. For each glass, the points were arbitrarily interpolated with a straight line in the IR crossover region. Its intercept with the dashed horizontal line $\Gamma_q/\Omega_q = 1/\pi$ gave an estimate for $\omega_{IR,LA}$. One observes that only for LB2 and d-SiO_2 are high-quality data available below the crossover. In the other cases,

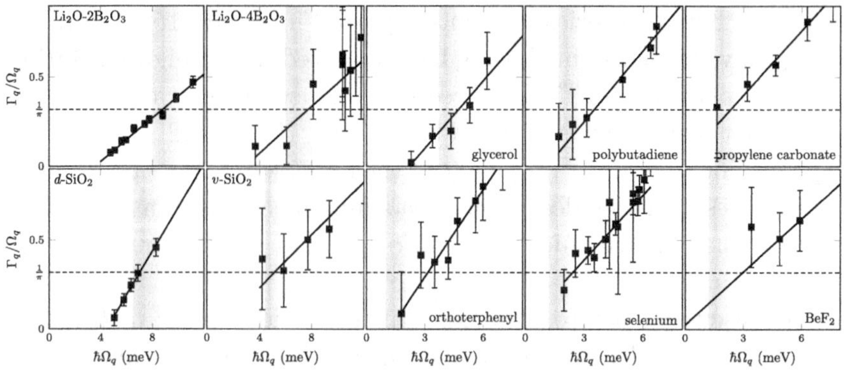

Fig. 16. Determination of the Ioffe–Regel energy $\hbar\omega_{\mathrm{IR}}$ (intercept with the dashed line) for different glasses and comparison with the energy region of the maximum of the boson peak $\hbar\omega_{\mathrm{BP}}$ (hatched region). References for the experimental points can be found in Ref. [105]. The positions and ranges of the BPs are discussed in Refs. [107, 109].

this region where the q^4 was expected was simply not investigated at that time. For all these glasses, the estimated $\omega_{\mathrm{IR,LA}}$ values were clearly in the same frequency range than the maximum of the boson peak ω_{BP}, marked by the hatched region, and not a decade above.[110] All in all, after a decade of vivid scientific discussions, it appeared that a large part of the picture originally proposed in Ref. [84] was experimentally verified. It must be noted that the latter observation made on the LA modes gave an incomplete view. The low-frequency region of the vibrational density of states is indeed largely dominated by the transverse acoustic excitations, i.e., as $2c_{\mathrm{L}}^3/c_{\mathrm{T}}^3$. Even if no experimental evidence exists yet, it is largely admitted that ω_{BP} is closed to $\omega_{\mathrm{IR,TA}}$ for all glasses, while the ratio $\omega_{\mathrm{IR,LA}}/\omega_{\mathrm{IR,TA}}$ is system-dependent. The case of orthoterphenyl in Figure 16 already illustrated this point with $\omega_{\mathrm{IR,LA}}/\omega_{\mathrm{BP}} > 2$, whereas computer simulation of a soft sphere glass even showed a larger different behavior of the two branches with well-defined LA modes up to the pseudo zone boundary and IR crossover of the TA modes at the boson peak frequency.[111] Similarly, well-defined LA modes have been observed later at frequencies well above ω_{BP} in metallic glasses.[112–114]

(d) Interaction of theory and experiment

Concomitant with the latter experimental evidence, theoretical approaches were developed to describe vibrations in glasses. They can be grossly separated into two main categories. The first one, the so-called Soft Potential Model (SPM), assumes that there exists in glasses, non-acoustic, soft quasi-local vibrations (QLVs) at low frequencies.[115] The latter form an excess g_{LOC} of the vibrational density of states above the Debye contribution, $g_{\mathrm{LOC}}(\omega) \propto \omega^4$.[116] Their mutual interactions, mediated by the acoustic phonons, lead to the universal shape of the boson peak.[117] In return, the acoustic phonons are strongly scattered by the QLVs leading to $\Gamma \propto \omega^4$ up to an IR crossover at $\omega_{\mathrm{IR}} \simeq \omega_{\mathrm{BP}}$.[10,118,119] As the SPM is an extension of the standard tunneling model,[20,21] it offers a unified approach and a comprehensive understanding of the thermo-mechanical anomalies observed in glasses and discussed in Section 1.1. For more details, the reader is referred to Chapter 8.

The second category considers harmonic elastic disorder. A self-consistent effective medium approach based on fluctuating elastic constants has been worked out in great detail and is now often termed Fluctuating Elasticity Theory (FET).[120,121] The elastic disorder induces an excess of the reduced vibrational density of states over the Debye theory, which is interpreted as the boson peak located at a fraction of the Debye frequency. Similar to the SPM results, the acoustic waves are strongly scattered in the boson peak frequency range.[23] In this region, a crossover from acoustic damping proportional to q^4 at low frequency to q^2 at high frequency is obtained as observed experimentally.[110] For more details, the reader is referred to Chapter 9.

Both theories give very similar predictions, for which increasing the disorder leads to a downshift of the boson peak, eventually driving the system to instability. The boson peak anomaly marks the crossover from plane wave-like vibrational states to disorder-dominated states with increasing frequency. The inelastic structure factor related to the scattering of the acoustic modes takes the EMA form given by Eq. (12) with distinct self-energy functions $W(\omega)$ for the two theoretical treatments,[23,122] permitting quantitative comparisons with experiments and computational simulations. It must be stressed than conversely to the FET approach which leads to the vDOS of the eigenmodes of the disordered system; the SPM starts from the coexistence at a low frequency of pure propagating acoustic modes and boson peak modes with $g_{\mathrm{EXC}}(\omega)$ which are not the eigenmodes.[105, 107, 109, 123]

Numerical simulation has become an increasingly powerful tool to shed light on the vibrational properties of glasses. It can provide very relevant physical quantities that are not yet accessible to experiments and thus are certainly useful for testing theories. On one hand, elastic heterogeneity at a typical length scale $\xi \simeq 10 - 15\sigma$, where σ is the atomic scale, is well documented in glasses. It leads to the breakdown of the continuum elastic description and thus of the Debye approximation at ξ which corresponds to the acoustic wavelength of frequencies close to ω_{IR}.[124–128] On the other hand, there is also substantial evidence from simulation that QLVs exist in disordered systems.[24, 111, 129–131] More recently, large-scaled simulations gave new insights into the low-frequency continuum limit, evidencing the ω^4 law for the QLVs thus supporting the SPM approach.[132–139]

As far as the experiment is concerned, first comparisons using the IXS Brillouin line widths of the LA modes in several glasses showed a much better agreement of the experimental data with the SPM expectations.[140] The strength of the q^4 behavior derived from FET fell systematically well below the observed widths. This can be partly attributed to the fact that the model used was not able to produce sufficiently intense boson peaks above the Debye level. This was particularly true for vitreous silica.[140] It was later shown that, in principle, the agreement between experiments and FET calculations could be improved. Introducing spatial correlations in the elastic disorder,[141] or replacing the original Gaussian distribution of the elastic constants with some non-Gaussian ones,[142] leads indeed to a significant enhancement of the boson peak close to the instability. High-quality new IXS data were later obtained on the H-bonded glass glycerol which, as guessed from Figure 16, was the next system to study. IXS spectroscopy has been performed in great details across the expected crossover $q_{\mathrm{IR}} \simeq 2$–$2.5\,\mathrm{meV}$. All the features previously discussed for d-SiO_2 and LB2 were confirmed, extending the phenomenology to molecular glasses.[143] Additionally, a marked softening of the sound velocity was evidenced below the IR crossover, thus exactly in the region of the rapid increase of the acoustic damping. A negative dispersion is actually the Kramers–Kronig counterpart of the q^4 behavior of the sound attenuation and as such was further evidence of its existence. Both the SPM and FET approaches produce a softening of the apparent sound velocity for frequency below the boson peak.[122,144] Similar IXS results were obtained afterward in silica melt,[145] glassy sorbitol,[146,147] and $2Na_2O{-}3SiO_2$.[148] Further comparisons between the original FET model[23,110] and IXS data for the case of silica melt and glassy sorbitol concluded that FET were able to reproduce qualitatively the experimental data but not quantitatively all the features.[147,149] For more details, the reader is referred to Chapter 7.

The relationship between the boson peak spectrum and the Debye density of states, the characteristic of the continuous elastic medium was also studied. A successful scaling was supposed to promote the idea that vibrational modes forming the boson peak were only acoustic as developed by the FET. Small changes of the boson peak spectrum following the evolution of the Debye vDOS were indeed observed in some glasses,[150–152] while the Debye scaling clearly failed in others.[49,153–155] In the vitreous silica case, it was found that the temperature dependence of the boson peak spectrum rather scales to the sound velocity based on the unrelaxed bulk modulus with scaling exponents in line with the SPM. All in all, the Debye scaling approach was not really successful to discriminate between both theories.

The importance of coordination number or fluctuations of the particle contact number and applied stress on elastic properties of disordered systems has been underlined recently.[156] An effective medium theory based on these two quantities has been developed which reproduced all the features obtained using the FET approach including Eq. (12). Interestingly, this work emphasizes the relevance of the effective sound velocity $c_{\mathrm{EMA}}(\omega)$ and scattering length $\ell_{\mathrm{EMA}}(\omega)$ of Eq. (14) as the physical quantities characterizing the total response of the system. Above the IR crossover, these quantities differ significantly from those extracted from a simple analysis of the IXS data using Eq. (13) which leads to

$$c_q(\omega) = \sqrt{\mathrm{Re}\, W(\omega)}, \quad \ell_q(\omega) = \frac{1}{\omega} \frac{\mathrm{Re}\, W(\omega)}{\mathrm{Im}\, \sqrt{W(\omega)}}, \tag{15}$$

with $c_q = \Omega_q/q$ and $\ell_q = c_q/\Gamma_q$. In addition, analysis in terms of constant-energy scans using Eq. (12) instead of usual constant-wavenumber scans using Eq. (13) provides the opportunity to easily include the diffuse Umklapp scattering contribution as illustrated by U. Buchenau.[86] Depending on the system, this contribution can also greatly affect the obtained parameter values. Hence, a promising way to test the several theories is to directly use Eq. (12) on constant-energy spectra to find the frequency dependence of the complex self-energy $W(\omega)$ taking into account the quadratic Umklapp scattering contribution.

4. Spectroscopy of the Low-frequency Region of the vDOS

This section aims at describing the nature of the modes constituting the low-frequency region of the vibrational density of states and to what extent this information can be obtained from the experiments. For example, inelastic neutron scattering and inelastic x-ray scattering are sensitive to all vibrations but the q-dependence of the scattering allows separating different types of excitations. The vibrational contrast in light experiments is more pronounced even though the well-established crystalline or molecular selection rules are lifted out by the structural disorder. In glasses, those are gathered in a coupling-to-light function, $C(\omega)$, whose expression (including $C(\omega) = 0$ for inactive vibrations) depends on the mode symmetry (atomic displacements) as well as on the experimental technique. Comparing the vibrational responses of infrared spectroscopy, Raman scattering, and hyper-Raman scattering can, therefore, be very informative. It is particularly true in simple glasses where selection rules are more pronounced.

4.1. *Inelastic neutron scattering*

One of the first experimental breakthroughs in the understanding of the boson peak modes occurred in the 80's using inelastic neutron scattering (INS). Atomic vibrations are all actives in INS and hence it is a technique of choice to access the density of vibrational states $g(\omega)$ of a material (vDOS). One drawback, however, is that it is impossible (except by doing a series of experiments with atomic isotopic substitutions) to normalize the experimental data by the scattering length of each type of atom constituting the material. This possibly distorts the experimental result when one wants to compare it, e.g., with atomistic simulations. Within the Debye model, the boson peak is an excess of modes over the ω^2 dependence of $g(\omega)$ at low frequency and hence it is observed in a $g(\omega)/\omega^2$ plot; see Figure 17(a). Despite the above-mentioned limitations, at the frequencies of the boson peak, the specific heat $C_P(T)$ calculated from the density of vibrational states compares well with the experimental data, at least in vitreous silica.[9,157]

In the pioneering work of Buchenau *et al.*,[9] a quantitative analysis of the low-frequency vDOS of v-SiO_2 is performed by decomposing the inelastic signal into two components: one proportional to $q^2S(q)$ and another one to q^2e^{-2W}, where $S(q)$ is the elastic structure factor and e^{-2W} is the Debye–Waller factor. This implies two kinds of low-frequency vibrations: a *coherent* contribution corresponding to *in-phase* motions of neighboring atoms or molecular units whose inelastic structure factor reproduces $S(q)$ and hence the first sharp diffraction peak (FSDP), and a *quasi-incoherent* part corresponding to uncorrelated atomic motions with its own oscillations as a function of the scattering wavevector q. Long wavelength acoustic modes are collective excitations and belong to the first family, while local vibrations (at least to the extent of a few structural units) or optic-like modes belong to the second. These two contributions add to construct the total vDOS. In vitreous silica, the quasi-incoherent excitations were interpreted in terms of coupled librations of SiO_4 units. A simplified schematic representation is given in Figure 17(a). Interestingly, these motions also define the soft mode of the β-α transition in quartz[160] and are known to be unstable

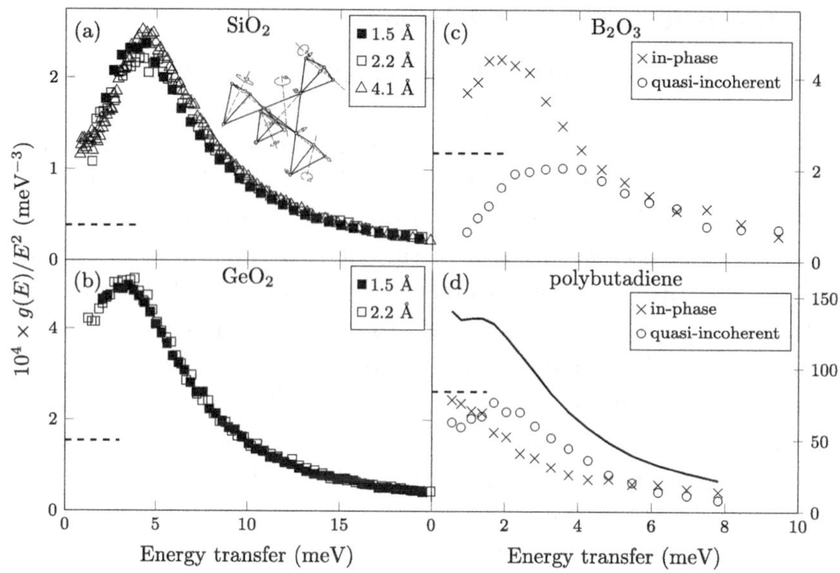

Fig. 17. Boson peak of (a) SiO_2 and (b) GeO_2 at different wavevectors[157] and schematic representation of coupled SiO_4 librations in an undistorted structural model containing five units.[9] Boson peak of (c) B_2O_3[158] and (d) polybutadiene[159] decomposed into in-phase and quasi-incoherent excitations. The dashed lines indicate the Debye level.

in quartz and in cristobalite. As such, they naturally vibrate at low frequency.[2] Similar decompositions were made in v-B_2O_3[158] and polybutadiene,[159] and the two contributions are well separated as shown in Figures 17(c) and 17(d). Differently, they superimpose in silica, suggesting a strong hybridization of the modes. The possibility that librations of coupled BO_3 triangles and B_3O_6 rings participate in the boson peak of v-B_2O_3 was proposed later,[161] providing thereby a possible explanation for the quasi-incoherent contribution in boron oxide as well.

The second campaign of INS experiments was performed about 10 years after using the high flux of the reactor at the Institut Laue–Langevin at Grenoble. The high quality of the data allowed improving the conclusions of the first campaign as well as to compare the situation in v-SiO_2 and v-GeO_2 (see Figure 17).[157] The spectroscopy allowed defining at each frequency ω, a function $S_\omega(q)$

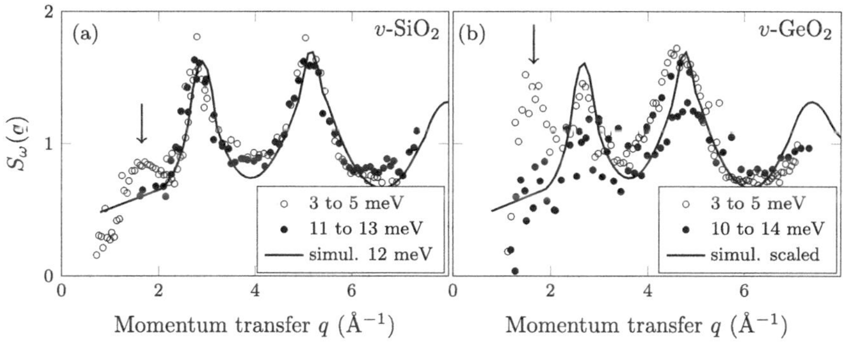

Fig. 18. Oscillation function of the inelastic structure factor integrated over two frequency domains for (a) v-SiO_2 and (b) v-GeO_2. The arrows mark the position of the first sharp diffraction peak.[157]

corresponding to the oscillations of the inelastic structure factor over a purely incoherent scattering (Figure 18). The appearance of an additional oscillation at the position of the FSDP ($\sim$1.6 Å^{-1}), solely in the low-frequency response (3–5 meV), shows that in-phase atomic motions contribute mostly at a frequency close to or below the BP maximum. For frequency above 10 meV, typically, $S_\omega(q)$ only reproduces the oscillations around 3 and 5 Å^{-1} characteristic of coupled librations of SiO_4 or GeO_4 tetrahedra (solid lines). The full analysis reveals that in vitreous silica, about 25% of in-phase motions and 75% of quasi-incoherent excitations contribute to the BP below its maximum (40% and 60%, respectively, in vitreous germania). Above the maximum, the in-phase part decreases so do the long wavelength excitations.

Among all these excitations, all but the long wavelength acoustic modes can be identified as boson peak modes, the former defining the Debye level (dashed lines in Figure 17). In silica, germania, and boron oxide, coupled librations of rigid elementary units (SiO_4 and GeO_4 tetrahedra, BO_3 triangles, and boroxol rings) contribute partly to the excess of modes, but it is important to notice that the sole contribution of *in-phase* motions also exceeds the Debye level (see, e.g., Figure 17(c)). This shows that in addition to acoustic *plane waves*, the in-phase contribution contains modes that also built up the reservoir of boson peak modes. When considering the connected networks of silica, germania, and boron oxide, it is natural that local librations of elementary structural units (SiO_4 tetrahedra, BO_3 triangles, and B_3O_6 rings) induce slightly delocalized atomic translations, providing thereby in-phase-type contributions to the boson peak.

Attempts to measure the dispersion curves of long wavelength acoustic phonons in glasses with neutrons have also been made. The main issue in using INS is that in addition to working in near forward scattering ($\theta \simeq 1$–2°) with good instrumental resolution, the experiment must also fulfill drastic kinematic conditions (conservation of momentum and energy). The latter basically require that the probed sound velocities must be slower than the speed of the incident neutrons. Fulfilling all these conditions at once is very challenging, but still experiments using the thermal neutrons of a triple axis spectrometer have been performed in vitreous selenium, a glass with a low longitudinal sound velocity, $v_{\mathrm{LA}} \simeq 1800\,\mathrm{ms}^{-1}$.[83] Spectra were recorded at wavevectors as low as $0.2\,\text{Å}^{-1}$, as shown in Figure 19(a). The cutoff at high frequency is due to the kinematic conditions and limits the analysis. However, one clearly observes the narrow lineshapes of the longitudinal phonons at low wavevector which rapidly broaden with increasing q. Above about $1\,\text{Å}^{-1}$, the measured spectral shape starts mirroring the BP measured by time-of-flight experiment; see Figure 19(b). This qualitatively demonstrates the Ioffe–Regel crossover of LA phonons as discussed in Section 3.2.5.

An analysis of the q-dependence of the signal, similar to that performed in Refs. [9, 157], highlighted three spectral components: the Brillouin LA response at low q (dashed line in Figure 19(a)), the quasi-incoherent or local-mode component (dot-dashed line in Figures 19(a) and 19(b)), and the in-phase component developing at high q (gray region in Figure 19(b)). The latter was called acoustic Umklapp scattering in Ref. [83] as it corresponds to q-conserving inelastic processes occurring by exchanging part of the neutron momentum with the peaks of the elastic structure factor. It is worth noting that transverse modes dominate in this scattering process

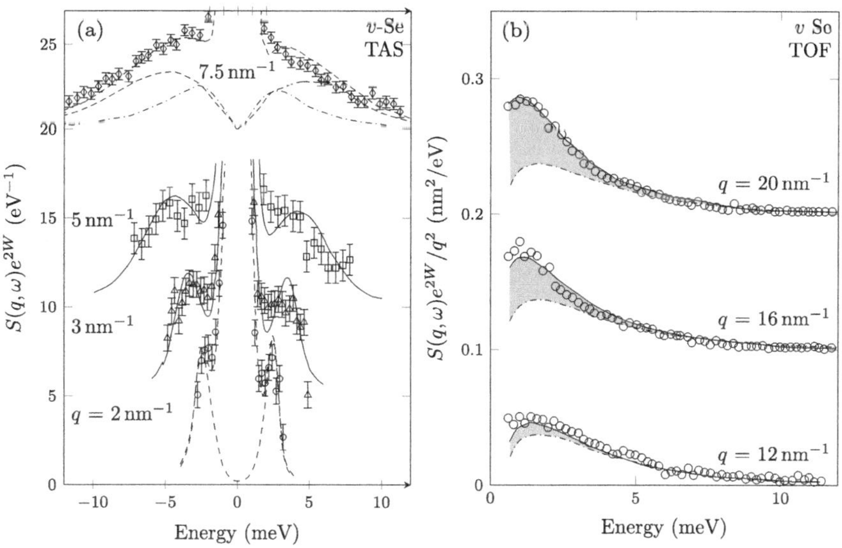

Fig. 19. INS triple axis experiment in vitreous selenium. LA plane waves are seen at the smallest q values. Dashed lines in (a) and (d) illustrate the Brillouin component alone. The solid lines represent the overall fits (see Ref. [83] for details). Note the unusual abscissa scale, ω/q, in meV.nm.

and the analysis therefore provides a unique way for obtaining high-frequency spectroscopic information on TA modes in glasses. Similar to INS in silica and germania, its structure factor mimics that of $S(q)$, i.e., $\propto q^2S(q)$, and we recall that for the quasi-incoherent part, one simply has a signal $\propto q^2$. LA and quasi-incoherent contributions as well as the total scattering response have been fitted with EMA functions derived from the effective-medium approximation (Eq. (12)), and the in-phase component corresponds to the difference signal. The analysis shows that as q increases, the quasi-incoherent contribution progressively overwhelms that of acoustic phonons. At high q, the in-phase component (Umklapp scattering) adds to these local modes to construct the boson peak. One also observes in Figure 19(b) that the in-phase contribution disappears around 3 to 5 meV revealing that TA phonons cease to propagate at very low frequency in v-Se.

4.2. *Inelastic spectroscopies of light*

In a light scattering experiment, the dipole $\mathbf{p}$ induced by an incident electric field $\mathbf{E}$ can be expanded in terms of $\mathbf{E}$,

$$\mathbf{p} = \alpha.\mathbf{E} + \frac{1}{2}\beta : \mathbf{EE} + \cdots , \tag{16}$$

where α and β are the polarizability and hyper-polarizability tensors, respectively. The first-order term gives the Raman signal whose spectrum $I(\omega)$ for a mode σ corresponds to the space and time Fourier transform of the modulation of α by the mode amplitude W_σ, $\alpha'_\sigma = (\partial\alpha/\partial W_\sigma)$ (see Section 2.2). Contrary to INS where all modes participate to the scattering, a vibration will be active or not in light spectroscopy depending on the symmetry properties of the atomic displacements W_σ (eigenvectors). In Raman, the density of vibrational states $g_\sigma(\omega)$ of the mode σ is observed via the coupling-to-light coefficient C_σ^{RS} so that $I(\omega)$ reads

$$I(\omega) = C_\sigma^{\mathrm{RS}} g_\sigma(\omega)\frac{n(\omega)+1}{\omega}, \tag{17}$$

where $n(\omega)$ is the Bose population factor and a sum over all the modes σ is implicit. As the boson peak appears in a $g(\omega)/\omega^2$ plot, it is worthwhile defining the "reduced intensity" $I_{\mathrm{red}}(\omega)$ which can also be related to the imaginary part of the Raman susceptibility χ"

$$I_{\mathrm{red}}(\omega) = \frac{I(\omega)}{\omega(n(\omega)+1)} = C_\sigma^{\mathrm{RS}}\frac{g_\sigma(\omega)}{\omega^2} \propto \frac{\chi''(\omega)}{\omega}. \tag{18}$$

C_σ^{RS} is known for every space groups and normal modes in crystals and molecules. These selection rules are lifted in amorphous materials owing to the structural disorder and C_σ^{RS} transforms into a frequency-dependent function $C_\sigma^{\mathrm{RS}}(\omega)$. However, for simple network glasses with elementary structural units similar to these of the crystalline polymorphs, the vibrational spectrum resembles a smeared-out version of the crystal one. For example, the similarities between the Raman spectrum of vitreous silica and that of cristobalite or α-quartz are striking as illustrated in Figure 20.[162] One major difference, however, is the broad boson peak response appearing at low frequency in all vitreous materials. The knowledge of $C_{\mathrm{BP}}^{\mathrm{RS}}(\omega)$ (we will call it $C^{\mathrm{RS}}(\omega)$

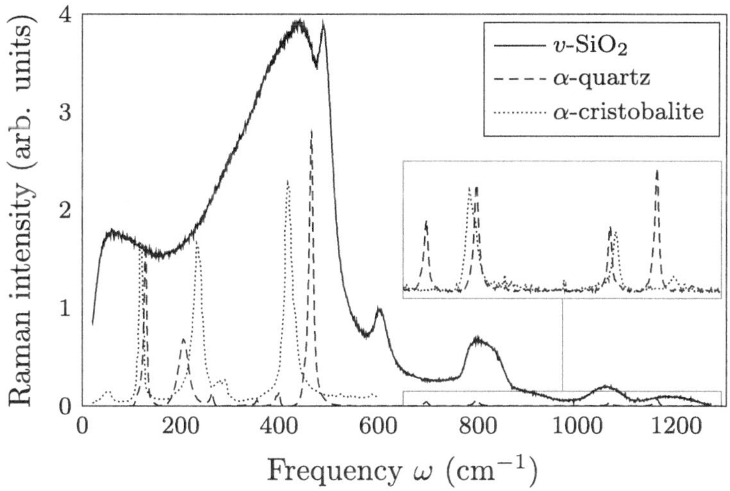

Fig. 20. Raman spectra of vitreous silica, α-quartz, and α-cristobalite.[162]

for ease of writing) and its frequency dependence are, therefore, of crucial importance for understanding the nature of the vibrations at BP frequencies.

Within classical expectations, $C^{\mathrm{RS}}(\omega)$ is a constant for local or quasi-local vibrations.[163] For acoustic waves propagating in a mechanically inhomogeneous medium, Martin and Brenig[25] predicted a response composed by the usual sharp Brillouin peaks at GHz frequencies and a background rising as ω^2 originating from the incoherent contribution of the local heterogeneities so that $C^{\mathrm{RS}}(\omega) \propto \omega^2$. This points out a fundamental difference between Raman scattering in crystals and glasses at low frequency. In the former, the signal is null, while in the latter, the disorder opens up a new scattering channel for short wavelength acoustics phonons based on incoherent processes (see Section 2.3). This density of states of acoustic phonons builds the low-frequency part of the boson peak (typically up to ω_{BP}) to which one may add a coupling with low-lying optic modes.

These predictions have been the subject of extensive experimental and theoretical investigations to unravel the nature of the boson peak modes. The comparison between the boson peak position ω_{BP}

Table 1. Frequency (in cm^{-1}) of the boson peak maximum ω_{BP} in a Raman reduced spectra and in a $g(\omega)/\omega^2$ plot in INS or specific heat measurement compiled from Refs. [164, 166] and additional data (when indicated).

	Glass	ω_{max}^{RS}	ω_{max}^{INS}		Glass	ω_{max}^{RS}	ω_{max}^{INS}
1	PMMA	16.5	12.5	9	Se	$23^{(2)}$	$16^{(3)}$
2	SiO_2	51.5	33.5	10	PC	12–15	11.0
3	$Na_2O-18GeO_2$	48.5		11	PS	14–18	11.5
4	SF1	38.5		12	CKN		20.5
5	F1	43.5		13	$GeSe_2$	10.0	
6	BaSF64	83.5		14	GeO_2	$41^{(4)}$	27.0
7	LaSF9	64.0		15	$Ag_2O-6B_2O_3$	22.5	
8	B_2O_3	$23^{(1)}$	18.0	16	As_2S_3		16.5

Note: [1]from Ref. [167], [2]from Ref. [168], [3]from Ref. [83], [4]from Ref. [169].

in Raman and the spectral broadening and position of the first sharp diffraction peak measured by x-ray diffraction has highlighted the close link between this vibrational feature and the medium range order[164] (Figures 21(a) and 21(b)). Based on these observations, it was proposed that the position of the BP relates to a characteristic length scale in the glass, typically in the nanometer range. The microscopic origin of the nanometric heterogeneities, either elastic, structural, or dynamical, is still controversial, but it is likely that all of these effects coexist with a glass of specific weight. An early qualitative model developed by Duval *et al.*[165] proposed that the frequency of the Raman BP is inversely proportional to the size of the nanodomains, $\xi \simeq v/\omega_{BP}$, where v is the transverse or longitudinal acoustic velocity for torsional or spherical modes, respectively. For most glasses, ξ, v, and ω_{BP} vary by a factor of about two only, thus giving similar numbers which limit the predictive power of the model.

Since each spectroscopy is sensitive to particular atomic motions, another way to probe the boson peak modes is to extract the frequency dependence of its coupling-to-light coefficient. According to Eq. (18), $C^{RS}(\omega)$ is extracted from the comparison between the Raman intensity with $g(\omega)$ obtained using either INS or specific heat. The results reveal that below $\sim 2\omega_{BP}$, the coupling coefficient exhibits

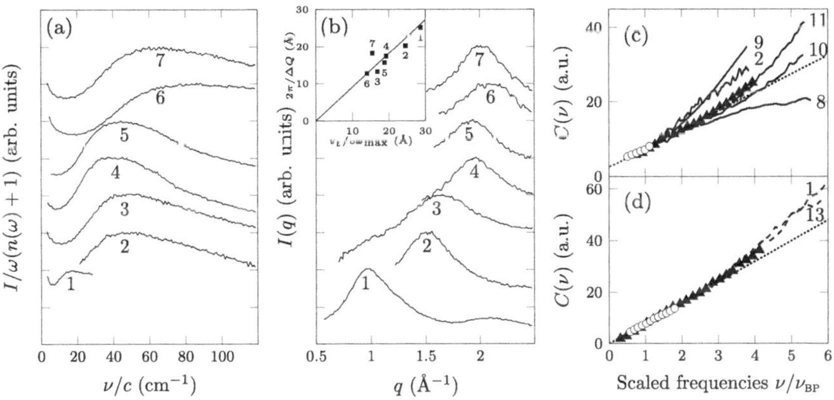

Fig. 21. (a) Boson peak and (b) FSDP of different glasses (numbers 1 to 7 in Table 1). The inset in (b) shows the correlation between characteristic lengths derived from the glass structure and vibrations.[164] (c) Frequency dependence of the coupling coefficient $C^{RS}(\omega)$ for glasses 2, 8–11 in Table 1, CKN (circles), and $Ag_2O{-}6B_2O_3$ (triangles) vs ν/ν_{BP}. The dotted line in (c) is $C^{RS}(\nu) \propto \nu/\nu_{BP} + 0.5$. (d) Same as in (c) for glasses 1 and 13 in Table 1, As_2O_3 (triangles), and GeO_2 (circles). The dotted line is $C^{RS}(\nu) \propto \nu/\nu_{BP}$.[166]

a universal behavior $C^{RS}(\omega) = A + B\omega/\omega_{BP}$[166] with $A = 0$ or 0.5 depending on the glass (Figures 21(c) and 21(d)). For frequencies above $\sim 2\omega_{BP}$, no universal character is found and the BP shape becomes fully material-dependent. The progressive transition from a universal to a non-universal behavior of $C^{RS}(\omega)$ correlates with the frequency of the Ioffe–Regel crossover, providing further evidence of the close link between the strong scattering regime of acoustic branches and the boson peak maximum.

Until recently, infrared spectroscopy has been barely used to study the BP modes due to the technical difficulty of reaching low frequencies. The recent development of THz light sources (far IR) in laboratories or of soft x-rays spectrometer at synchrotrons enabled us to overcome this issue. Time-domain or THz-spectroscopies (THz-TDS) are now mature for measuring the absorption coefficient $\alpha(\omega)$ in the far infrared, i.e., down to a few wavenumbers.[170] Within the linear response theory, $\alpha(\omega)$ relates to the density of vibrational states $g(\omega)$ by $\alpha(\omega) = C^{IR}(\omega)g(\omega)$, where $C^{IR}(\omega)$ is the infrared coupling coefficient. Since the BP is seen in a $g(\omega)/\omega^2$ plot, one simply has

$$\frac{\alpha(\omega)}{\omega^2} = C^{\mathrm{IR}}(\omega)\frac{g(\omega)}{\omega^2}. \tag{19}$$

In glasses, Taraskin *et al.*[171] proposed a model where $C^{\mathrm{IR}}(\omega) = A + B\omega^{\beta}$ for frequencies below the Ioffe–Regel crossover. A and B are material-dependent constants related to long-range uncorrelated charge fluctuations and correlated short-range (inter-atomic) charge fluctuations, respectively. $\beta = 2$ in the model of Ref. [171], but exponents as low as $\beta = 1$ have been measured in molecular glasses, e.g., glucose.[172] This departure has been interpreted in terms of acousto-optic mode couplings and bending of the acoustic branches which are not taken into account in Taraskin and Elliott model.

Hyper-Raman scattering (HRS) is a nonlinear spectroscopy where two incident photons scatter one photon after interaction with a vibration in the medium. It corresponds to the first nonlinear term in Eq. (16), and hence the signal originates from the fluctuations of the hyper-polarizability tensor β.[173] β' has different symmetry properties than α' yielding to different selection rules. For example, polar vibrations are always active in HRS whatever the crystalline or molecular symmetry is. Conversely, there exist modes that are active in HRS but inactive in both RS and IR, improperly named *silent modes*. The relation between the measured intensity I_{HRS} and the density of state $g(\omega)$ is similar to that of Eqs. (17) and (18) replacing $C_{\sigma}^{\mathrm{RS}}(\omega)$ by $C_{\sigma}^{\mathrm{HR}}(\omega)$, the latter mirroring the properties of the hyper-polarizability tensor.

Figure 22 compiles the boson peaks obtained in vitreous silica and vitreous borate glasses using neutron scattering, Raman scattering, hyper-Raman scattering, and THz-TDS spectroscopy. The different spectral shapes arise from the different frequency dependence of the coupling coefficients combined with the sensitivity of each technique to specific vibrations. In v-B_2O_3, see Figure 22(a), the Raman and hyper-Raman BP are rather similar and coincide with the quasi-incoherent INS contribution as shown in Figure 17(c). The combined analysis revealed that the latter could correspond to local vibrations involving out-of-plane librations of rigid BO_3 triangles

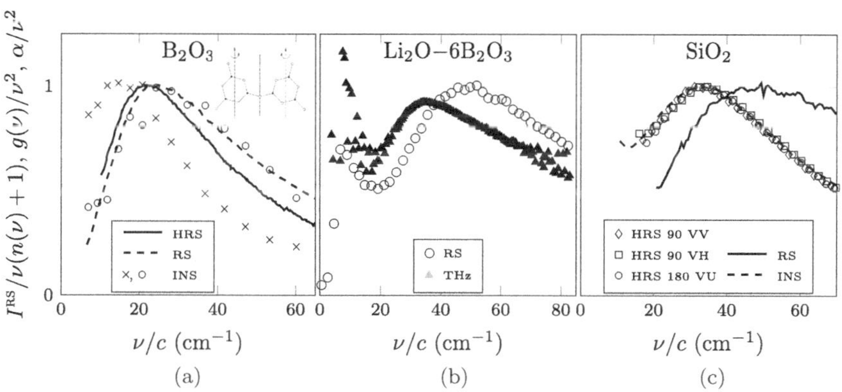

Fig. 22. (a) Boson peak of v-B_2O_3: HRS (solid line), RS (dashed line), out-of-phase INS component (circles), and in-phase INS component (crosses).[161] A schematic representation of E''-symmetry librations of boroxol rings is also shown. (b) Boson peak in a lithium borate glass with 14 mol% Li_2O content in RS and THz-TDS spectroscopy.[174] (c) Boson peak of v-SiO_2: HRS (symbols), RS (solid line), and INS (dashed line).[175]

and B_3O_6 rings whose schematic representation is also shown in Figure 22(a). Within the D_{3h} point group of these molecular units, this corresponds to E''-symmetry modes which are indeed active in both RS and HRS.[161,167] THz-TDS in pure v-B_2O_3 is very challenging, but data obtained in a lightly doped lithium borate glass $Li_2O-6B_2O_3$[174] highlight a BP at significantly lower frequency than the RS one (see Figure 22(b)) and suggest that the IR response rather associates with the in-phase INS contribution. In silica, the HRS BP has been associated with local or quasi-local libration motions involving rigid SiO_4 units[162,175] introduced by Buchenau *et al.* (see Figure 17(c)). Within the T_d point group symmetry of SiO_4 tetrahedra, this vibration is inactive in both IR (thus non-polar) and RS, and only active in HRS (*silent mode* of F_1-symmetry). This is likely the reason why RS and HRS responses are so different. The former has been ascribed to leakage of these modes due to the structural disorder and dipolar-type contributions.[175] Moreover, since the F_1-symmetry modes are non-polar, they do not carry an electric charge and, therefore, do not participate in the local contribution of charge fluctuations (B term). This is consistent with the observed constant behavior of $C^{\rm IR}$ in vitreous silica.[171] Finally, the fact that the HRS-BP perfectly matches the INS one supports INS assumption that these modes play an important role in the excess of modes at low frequency in vitreous silica. Besides, libration motion of SiO_4 tetrahedra is the soft mode of the structural α-β instability at 846 K in quartz.[160] Its frequency extrapolates to $\sim$36 cm^{-1} at $T_g \sim 1600$ K,[176] i.e., to the frequency of the maximum of $g(\omega)/\omega^2$ measured by INS. Librations of rigid units are also low-lying vibrations in β-cristobalite but are located at the zone boundary of a transverse acoustic branch. In glass, these modes are characterized by weak bondings (those creating the disorder). It is, therefore, natural that they vibrate at low frequency. In a connected network, they also hybridize with transverse and longitudinal acoustic phonons of similar frequency. This source of scattering combines with the incoherent scattering of acoustic modes to built up the reservoir of boson peak modes. The loss of the Brillouin zone in the glass combined with the localization of plane wave makes identification of these vibrations as acoustic- or optic-like useless. On heating, the inter-unit bonds gradually break and these libration motions (vibrations) transform into rotation motions (relaxations) in the liquid state.

4.3. *Nuclear inelastic scattering*

Phonon spectra can be obtained with a Mössbauer radioisotope source by analyzing the frequency dependence of its nuclear resonance fluorescence with recoil. The principles were formulated soon after the discovery of the Mössbauer effect,[177] but nuclear inelastic scattering (NIS) setups have emerged only in the 90's with the development of synchrotron radiation offering highly monochromatic x-rays beams, i.e., compatible with the phonon energy spectra. Albeit the technique is limited to elements possessing a Mössbauer isotope, this can turn into an important advantage, for example, when the dynamics of the nuclei must be studied separately.

In molecular glasses, the probes were neutral ferrocene molecules with the central ^{57}Fe nucleus as a Mössbauer isotope.[178] In that case,

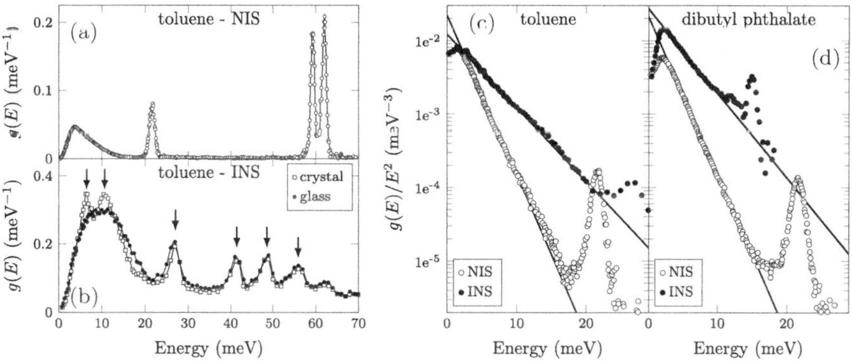

Fig. 23. Density of states of toluene glass measured by (a) nuclear inelastic scattering at 22 K from ferrocene molecules and (b) inelastic neutron scattering. The arrows indicate local optic modes not seen in the NIS vDOS. Reduced NIS and INS vDOS in (c) toluene and (d) dibutyl phthalate in semi-log plots.[178]

neither rotations of the probe nor intramolecular modes of the glassy matrix are seen. At boson peak frequencies, the probe selects only few modes among the total vDOS, mostly displacements of the rigid probe driven by the translational collective motions of the glass network. The coherent length of such vibrations was estimated to be about 20 Å. The net difference between the NIS vDOS and the INS vDOS of toluene is shown in Figures 23(a) and 23(b). The latter is much more complex due to the participation of rotations and librations of the methyl groups, as well as other local modes at high frequency. The comparison of the NIS and INS boson peaks of toluene and dibutyl phthalate is also very informative (see Figures 23(c) and 23(d)). The higher number of vibrational states in the total vDOS (INS) as compared to the NIS vDOS (by a factor of 2) in dibutyl phthalate at and below the BP maximum was explained by modes with coherent lengths shorter than 20 Å probed by the technique, i.e., quasi-local or local vibrations. NIS was also performed in a $Na_2FeSi_3O_8$ glass with 95% enrichment of ^{57}Fe isotope.[151,179] The structural analysis has shown that iron is mostly in tetrahedral coordination and, therefore, plays a similar role as silicon. A comparative analysis of the NIS vDOS of the glass and of the crystalline polymorph of $NaFeSi_2O_6$ revealed that the number of vibrational states of the boson peak is similar to that of the Van Hove singularity of the TA branch in the crystal. It also suggested that the latter are redistributed at lower frequency in the glass and construct the boson peak. It is worth reminding, however, that in these pioneered NIS campaigns, the Mössbauer isotope ^{57}Fe, positioned at the center of mass of the ferrocene molecule and of the FeO_4 tetrahedra, is insensitive to libration motions which are supposed to play an important role at the boson peak frequencies in silicates.

4.4. *Inelastic X-ray scattering*

The densities of vibrational states of pristine and densified silica glasses have been collected by inelastic X-ray scattering with setups equipped either with a crystal-type or by a Mössbauer-type analyzer.[18] The data have been compared with those obtained in various crystalline SiO_2 polymorphs. Calculating the absolute value of the DOSs, the authors have shown that the number of modes in excess relative to the Debye level is similar in the glass and in the crystalline polymorph with a comparable density. In addition, the shape of $g(\omega)/\omega^2$ in the glass (BP) appears as a smeared out version of $g(\omega)/\omega^2$ in the corresponding crystal which shows the Van Hove singularity. Despite these striking resemblances, it would be misleading to conclude that the BP simply identifies the Van Hove singularity of the acoustic branches in the corresponding crystal. The nature of the vibrations underlying the low-frequency vDOS in glasses and crystals is indeed very different. In the latter, the vibrations are plane waves (acoustic and optic) propagating in an ordered material, while in the former, the acoustic modes are scattered by the disorder. These differences produce the shadowed region in $C_P(T)/T^3$ (cf. Figure 1) and may appear as a *detail* when comparing the total vDOS of crystals and glasses. It is, however, these specific properties which lead to the scattering regime of phonons at the origin, e.g., of the plateau in the thermal conductivity, and to the boson peak observed in inelastic light spectroscopies.

References

1. N. Bilir and W. A. Phillips. (1975). Phonons in SiO_2: The low-temperature heat capacity of cristobalite. *Phil. Mag.* **32**(1), 113–122.
2. M. T. Dove, M. J. Harris, A. C. Hannon, J. M. Parker, I. P. Swainson, and M. Gambhir. (1997). Floppy modes in crystalline and amorphous silicates. *Phys. Rev. Lett.* **78**(6), 1070–1073.
3. V. N. Sigaev, E. N. Smelyanskaya, V. G. Plotnichenko, V. V. Koltashev, A. A. Volkov, and P. Pernice. (1999). Low-frequency band at $50\,cm^{-1}$ in the raman spectrum of cristobalite: Identification of similar structural motifs in glasses and crystals of similar composition. *J. Non-Cryst. Solids.* **248**(2), 141–146.
4. R. Wietzel. (1921). The stability conditions of the glass and crystal phase of silicon dioxide. *Z. Anorg. Allg. Chem.* **116**, 71–95.
5. F. Simon. (1922). Analysis of specific heat capacity in low temperatures. *Annalen der Physik.* **68**(11), 241–280.
6. R. B. Sosman. (1927). *The Properties of Silica: An Introduction to the Properties of Substances in the Solid Non-conducting State.* Vol. 37, *American Chemical Society. Monograph series,* p. 362. New York, NY: The Chemical Catalog Co.
7. P. Flubacher, A. J. Leadbetter, J. A. Morrison, and B. P. Stoicheff. (1959). The low-temperature heat capacity and the raman and brillouin spectra of vitreous silica. *J. Phys. Chem. Solids.* **12**(1), 53–65.
8. R. C. Zeller and R. O. Pohl. (1971). Thermal conductivity and specific heat of noncrystalline solids. *Phys. Rev. B.* **4**(6), 2029–2041.
9. U. Buchenau, M. Prager, N. Nücker, A. J. Dianoux, N. Ahmad, and W. A. Phillips. (1986). Low-frequency modes in vitreous silica. *Phys. Rev. B.* **34**(8), 5665–5673.
10. M. A. Ramos and U. Buchenau. (1997). Low-temperature thermal conductivity of glasses within the soft-potential model. *Phys. Rev. B.* **55**(9), 5749–5754.
11. J. C. Lasjaunias, A. Ravex, M. Vandorpe, and S. Hunklinger. (1975). The density of low energy states in vitreous silica: Specific heat and thermal conductivity down to 25 mk. *Solid State Commun.* **17**(9), 1045–1049.
12. D. G. Cahill and R. O. Pohl. (1987). Thermal conductivity of amorphous solids above the plateau. *Phys. Rev. B.* **35**(8), 4067–4073.
13. A. Eucken. (1911). über die temperaturabhängigkeit der wärmeleitähigkeit fester nichtmetalle. *Annalen der Physik.* **339**(2), 185–221.

14. G. W. C. Kaye, W. F. Higgins, and J. E. Petavel. (1926). The thermal conductivity of vitreous silica, with a note on crystalline quartz. *Proc. R. Soc. Lond. Series A.* **113**(764), 335–351.
15. F. Birch and H. Clark. (1940). The thermal conductivity of rocks and its dependence upon temperature and composition. *Am. J. Sci.* **238**(8), 529–558.
16. M. Kunugi, N. Soga, H. Sawa, and A. Konishi. (1972). Thermal conductivity of cristobalite. *J. Am. Ceram. Soc.* **55**(11), 580–580.
17. D. J. Safarik, R. B. Schwarz, and M. F. Hundley. (2006). Similarities in the C_p/T^3 peaks in amorphous and crystalline metals. *Phys. Rev. Lett.* **96**(19), 195902.
18. A. I. Chumakov, G. Monaco, A. Fontana, A. Bosak, R. P. Hermann, D. Bessas, B. Wehinger, W. A. Crichton, M. Krisch, R. Rüffer, G. Baldi, G. Carini Jr., G. Carini, G. D'Angelo, E. Gilioli, G. Tripodo, M. Zanatta, B. Winkler, V. Milman, K. Refson, M. T. Dove, N. Dubrovinskaia, L. Dubrovinsky, R. Keding, and Y. Z. Yue. (2014). Role of disorder in the thermodynamics and atomic dynamics of glasses. *Phys. Rev. Lett.* **112**(2), 025502.
19. M. Rothenfusser, W. Dietsche, and H. Kinder. (1983). Linear dispersion of transverse high-frequency phonons in vitreous silica. *Phys. Rev. B.* **27**(8), 5196–5198.
20. W. A. Phillips. (1972). Tunneling states in amorphous solids. *J. Low Temp. Phys.* **7**(3), 351–360.
21. P. W. Anderson, H. B. I, and V. C. M. (1972). Anomalous low-temperature thermal properties of glasses and spin glasses. *Philos. Mag.* **25**(1), 1–9.
22. J. E. Graebner, B. Golding, and L. C. Allen. (1986). Phonon localization in glasses. *Phys. Rev. B.* **34**(8), 5696–5701.
23. W. Schirmacher. (2006). Thermal conductivity of glassy materials and the Boson peak. *Europhys. Lett.* **73**(6), 892–898.
24. Y. M. Beltukov, C. Fusco, D. A. Parshin, and A. Tanguy. (2016). Boson peak and ioffe-regel criterion in amorphous siliconlike materials: The effect of bond directionality. *Phys. Rev. E.* **93**(2), 023006.
25. A. J. Martin and W. Brenig. (1974). Model for Brillouin-scattering in amorphous solids. *Phys. Status Solidi B.* **64**(1), 163–172.
26. H. J. Maris. (1971). Interaction of sound waves with thermal phonons in dielectric crystals. In W. Mason and R. Thurston (eds.), *Principles and Methods*, vol. 8, *Physical Acoustics*, Chapter 6, pp. 279–345. New York, NY: Academic Press.
27. A. N. Akhiezer. (1939). On the absorption of sound in solids. *J. Phys. Acad. Sci. USSR.* **1**, 277–287.

28. H. E. Bömmel and K. Dransfeld. (1960). Excitation and attenuation of hypersonic waves in quartz. *Phys. Rev.* **117**(5), 1245–1252.
29. R. Vacher, E. Courtens, and M. Foret. (2005). Anharmonic versus relaxational sound damping in glasses. ii. vitreous silica. *Phys. Rev. B.* **72**(21), 214205.
30. J. S. Blinick and H. J. Maris. (1970). Velocities of first and zero sound in quartz. *Phys. Rev. B.* **2**(6), 2139–2146.
31. J. T. Krause and C. R. Kurkjian. (1968). Vibrational anomalies in inorganic glass formers. *J. Am. Ceram. Soc.* **51**(4), 226–227.
32. T. N. Claytor and R. J. Sladek. (1978). Ultrasonic velocities in amorphous As_2S_3 and As_2Se_3 between 1.5 K and 296 K. *Phys. Rev. B.* **18**(10), 5842–5850.
33. R. Vacher, J. Pelous, F. Plicque, and A. Zarembowitch. Ultrasonic and Brillouin-scattering study of the elastic properties of vitreous silica between 10 K and 300 K. *J. Non-Cryst. Solids.* **45**(3), 397–410. (1981).
34. J. Jäckle. (1972). Ultrasonic attenuation in glasses at low-temperatures. *Z. Phys.* **257**(3), 212–223.
35. J. Jäckle, L. Piché, W. Arnold, and S. Hunklinger. (1976). Elastic effects of structural relaxation in glasses at low temperatures. *J. Non-Cryst. Solids.* **20**(3), 365–391.
36. K. S. Gilroy and W. A. Phillips. (1981). An asymmetric double-well potential model for structural relaxation processes in amorphous materials. *Philos. Mag. B.* **43**(5), 735–746.
37. D. Tielbürger, R. Merz, R. Ehrenfels, and S. Hunklinger. (1992). Thermally activated relaxation processes in vitreous silica — an investigation by Brillouin-scattering at high-pressures. *Phys. Rev. B.* **45**(6), 2750–2760.
38. S. Hunklinger and W. Arnold. (1976). Ultrasonic properties of glasses at low temperatures. In W. Mason and R. Thurston (eds.), *Principles and Methods*, Vol. 12, *Physical Acoustics*, Chapter 3, pp. 155–215. New York, NY: Academic Press.
39. O. L. Anderson and H. E. Bömmel. (1955). Ultrasonic absorption in fused silica at low temperatures and high frequencies. *J. Am. Ceram. Soc.* **38**(4), 125–131.
40. R. Keil, G. Kasper, and S. Hunklinger. (1993). Distribution of barrier heights in a-sio_2 and a-se. *J. Non-Cryst. Solids.* **164–166**, 1183–1186.
41. T. Damart and D. Rodney. (2018). Atomistic study of two-level systems in amorphous silica. *Phys. Rev. B.* **97**(1), 014201.

42. G. Guimbretière, B. Rufflé, and R. Vacher. (2012). Acoustic damping and dispersion in vitreous germanium oxide. *Phys. Rev. B.* **86**(9), 094304.
43. G. Carini, G. Carini, G. Tripodo, G. Di Marco, and E. Gilioli. (2012). Elastic and anelastic properties of densified vitreous B_2O_3: Relaxations and anharmonicity. *Phys. Rev. B.* **85**(9), 094201.
44. G. Carini, G. Carini, G. D'Angelo, D. Fioretto, and G. Tripodo. (2014). Identification of relaxing structural defects in densified B_2O_3 glasses. *Phys. Rev. B.* **90**(14), 140204.
45. G. Carini, G. Tripodo, and L. Borjesson. (2009). Hypersonic attenuation in cesium borate glasses: Relaxation and anharmonicity. *Mater. Sci. Eng. A.* **521–522**, 247–250.
46. G. Carini, Jr., G. Tripodo, and L. Borjesson. (2008). Thermally activated relaxations and vibrational anharmonicity in alkali-borate glasses: Brillouin scattering study. *Phys. Rev. B.* **78**(2), 024104.
47. S. M. Lindsay, M. W. Anderson, and J. R. Sandercock. (1981). Construction and alignment of a high performance multipass vernier tandem Fabry-Perot interferometer. *Rev. Sci. Instrum.* **52**(10), 1478–1486.
48. R. Vacher and J. Pelous. (1976). Behavior of thermal phonons in amorphous media from 4 to 300 K. *Phys. Rev. B.* **14**(2), 823–828.
49. B. Rufflé, S. Ayrinhac, E. Courtens, R. Vacher, M. Foret, A. Wischnewski, and U. Buchenau. (2010). Scaling the temperature-dependent boson peak of vitreous silica with the high-frequency bulk modulus derived from Brillouin scattering data. *Phys. Rev. Lett.* **104**(6), 067402.
50. G. Carini, G. Carini, G. D'Angelo, G. Tripodo, A. Bartolotta, and G. Salvato. (2005). Ultrasonic relaxations, anharmonicity, and fragility in lithium borate glasses. *Phys. Rev. B.* **72**(1), 014201
51. J. Wachtman, W. Tefft, D. Lam, and C. Apstein. (1961). Exponential temperature dependence of Youngs modulus for several oxides. *Phys. Rev.* **122**(6), 1754–1759.
52. S. Ayrinhac, B. Rufflé, M. Foret, H. Tran, S. Clément, R. Vialla, R. Vacher, J. C. Chervin, P. Munsch, and A. Polian. (2011). Dynamical origin of anomalous temperature hardening of elastic modulus in vitreous silica. *Phys. Rev. B.* **84**, 024201.
53. G. Carini, G. Carini, G. Tripodo, A. Bartolotta, and G. D. Marco. (2006). Effect of cation sizes on tunnelling states, relaxations and anharmonicity of alkali borate glasses. *J. Phys.-Condens. Matter.* **18**(12), 3251–3262.
54. G. Federle and S. Hunklinger. (1982). Ultrasonic studies of some polymers at low temperatures. *J. Phys. Colloques.* **43**(C9), 505–508.

55. S. Federle and G. Hunklinger. (1982). Ultrasonic absorption in polymethylmethacrylate at low temperatures. In G. Hartwig and D. Evans (eds.), *Nonmetallic Materials and Composites at Low Temperatures*, pp. 49–57. Boston, MA: Springer US.
56. M. Schmidt, R. Vacher, J. Pelous, and S. Hunklinger. (1982). Brillouin scattering from amorphous polymers at low temperatures. *J. Phys. Colloques.* **43**(C9), 501–504.
57. E. Rat, M. Foret, G. Massiera, R. Vialla, M. Arai, R. Vacher, and E. Courtens. (2005). Anharmonic versus relaxational sound damping in glasses. i. brillouin scattering from densified silica. *Phys. Rev. B.* **72**, 214204.
58. R. Vacher and J. Pelous. (1985). Rayleigh brillouin scattering and molecular disorder in glassy crystals and molecular glasses. *J. Chim. Phys.* **82**, 311–316.
59. B. Rufflé. (2007). unpublished.
60. G. Caliskan, A. Kisliuk, V. N. Novikov, and A. P. Sokolov. (2001). Relaxation spectra in poly(methylmethacrylate): Comparison of acoustic attenuation and light scattering data. *J. Chem. Phys.* **114**(22), 10189–10195.
61. G. Weiss, A. Daum, M. Sohn, and J. Arndt. (1996). Vibrating reed experiments on compacted vitreous silica. *Physica B.* **219–220**, 290–292.
62. C. Masciovecchio, A. Gessini, S. Di Fonzo, L. Comez, S. C. Santucci, and D. Fioretto. (2004). Inelastic ultraviolet scattering from high frequency acoustic modes in glasses. *Phys. Rev. Lett.* **92**(24), 247401
63. P. Benassi, S. Caponi, R. Eramo, A. Fontana, A. Giugni, M. Nardone, M. Sampoli, and G. Viliani. (2005). Sound attenuation in a unexplored frequency region: Brillouin ultraviolet light scattering measurements in v-SiO_2. *Phys. Rev. B.* **71**(17), 172201.
64. R. Vacher, S. Ayrinhac, M. Foret, B. Rufflé, and E. Courtens. (2006). Finite size effects in brillouin scattering from silica glass. *Phys. Rev. B.* **74**(1), 012203.
65. W. T. Grubbs and R. A. MacPhail. (1994). Dynamics in supercooled glycerol by high resolution stimulated brillouin gain spectroscopy. *J. Chem. Phys.* **100**(4), 2561–2570.
66. C. Masciovecchio, G. Baldi, S. Caponi, L. Comez, S. Di Fonzo, D. Fioretto, A. Fontana, A. Gessini, S. C. Santucci, F. Sette, G. Viliani, P. Vilmercati, and G. Ruocco. (2006). Evidence for a crossover in the frequency dependence of the acoustic attenuation in vitreous silica. *Phys. Rev. Lett.* **97**(3), 035501.

67. B. Rufflé, E. Courtens, and M. Foret. (2011). Inelastic ultraviolet Brillouin scattering from superpolished vitreous silica. *Phys. Rev. B.* **84**(13), 132201.
68. J. Sandercock. (1972). Brillouin-scattering measurements on silicon and germanium. *Phys. Rev. Lett.* **28**(4), 237–240.
69. F. Inam, J. P. Lewis, and D. A. Drabold. (2010). Hidden structure in amorphous solids. *Phys. Status Solidi A-Appl. Mat.* **207**(3), 599–604.
70. C. Thomsen, J. Strait, Z. Vardeny, H. Maris, J. Tauc, and J. Hauser. (1984). Coherent phonon generation and detection by picosecond light-pulses. *Phys. Rev. Lett.* **53**(10), 989–992.
71. T. Zhu, H. Maris, and J. Tauc. (1991). Attenuation of longitudinal-acoustic phonons in amorphous SiO_2 at frequencies up to 440 GHz. *Phys. Rev. B.* **44**(9), 4281–4289.
72. C. Thomsen, H. Grahn, and J. Tauc. (1986). Surface generation and detection of phonons by picosecond light-pulses. *Phys. Rev. B.* **34**(6), 4129–4138.
73. P. Benassi, M. Krisch, C. Masciovecchio, V. Mazzacurati, G. Monaco, G. Ruocco, F. Sette, and R. Verbeni. (1996). Evidence of high frequency propagating modes in vitreous silica. *Phys. Rev. Lett.* **77**(18), 3835–3838.
74. A. Devos, M. Foret, S. Ayrinhac, P. Emery, and B. Rufflé. (2008). Hypersound damping in vitreous silica measured by picosecond acoustics. *Phys. Rev. B.* **77**(10), 100201.
75. S. Ayrinhac, M. Foret, A. Devos, B. Rufflé, E. Courtens, and R. Vacher. (2011). Subterahertz hypersound attenuation in silica glass studied via picosecond acoustics. *Phys. Rev. B.* **83**(1), 014204.
76. C. Klieber, E. Peronne, K. Katayama, J. Choi, M. Yamaguchi, T. Pezeril, and K. A. Nelson. (2011). Narrow-band acoustic attenuation measurements in vitreous silica at frequencies between 20 and 400 ghz. *Appl. Phys. Lett.* **98**(21), 211908.
77. Y.-C. Wen, S.-H. Guol, H.-P. Chen, J.-K. Sheu, and C.-K. Sun. (2011). Femtosecond ultrasonic spectroscopy using a piezoelectric nanolayer: Hypersound attenuation in vitreous silica films. *Appl. Phys. Lett.* **99**(5), 051913.
78. H. Lin, R. Stoner, H. Maris, and J. Tauc. (1991). Phonon attenuation and velcoity-measurements in transparent materials by picosecond acosutic interferometry. *J. Appl. Phys.* **69**(7), 3816–3822.
79. C. Morath and H. Maris. (1996). Phonon attenuation in amorphous solids studied by picosecond ultrasonics. *Phys. Rev. B.* **54**(1), 203–213.

80. A. Mermet, A. Cunsolo, E. Duval, M. Krisch, C. Masciovecchio, S. Perghem, G. Ruocco, F. Sette, R. Verbeni, and G. Viliani. (1998). Pressure-induced in-glass structural transformation in the amorphous polymer poly(methylmethacrylate). *Phys. Rev. Lett.* **80**, 4205–4208.
81. A. Huynh, E. Péronne, C. Gingreau, X. Lafosse, A. Lemaître, B. Perrin, R. Vacher, B. Rufflé, and M. Foret. (2017). Temperature dependence of hypersound attenuation in silica films via picosecond acoustics. *Phys. Rev. B.* **96**, 174206.
82. C. Masciovecchio, G. Ruocco, F. Sette, M. Krisch, R. Verbeni, U. Bergmann, and M. Soltwisch. (1996). Observation of large momentum phononlike modes in glasses. *Phys. Rev. Lett.* **76**(18), 3356–3359.
83. M. Foret, B. Hehlen, G. Taillades, E. Courtens, R. Vacher, H. Casalta, and B. Dorner. (1998). Neutron Brillouin and umklapp scattering from glassy selenium. *Phys. Rev. Lett.* **81**(10), 2100–2103.
84. M. Foret, E. Courtens, R. Vacher, and J. Suck. (1996). Scattering investigation of acoustic localization in fused silica. *Phys. Rev. Lett.* **77**(18), 3831–3834.
85. G. Polatsek and O. Entin-Wohlman. (1988). Effective-medium approximation for a percolation network: The structure factor and the ioffe-regel criterion. *Phys. Rev. B.* **37**(13), 7726–7730.
86. U. Buchenau. (2014). Evaluation of x-ray brillouin scattering data. *Phys. Rev. E.* **90**(6), 062319.
87. E. Courtens, R. Vacher, J. Pelous, and T. Woignier. (1988). Observation of fractons in silica aerogels. *Europhys. Lett.* **6**(3), 245–250.
88. C. Masciovecchio, G. Ruocco, F. Sette, P. Benassi, A. Cunsolo, M. Krisch, V. Mazzacurati, A. Mermet, G. Monaco, and R. Verbeni. (1997). High-frequency propagating modes in vitreous silica at 295 K. *Phys. Rev. B.* **55**(13), 8049–8051.
89. C. Masciovecchio, G. Monaco, G. Ruocco, F. Sette, A. Cunsolo, M. Krisch, A. Mermet, M. Soltwisch, and R. Verbeni. (1998). High frequency dynamics of glass forming liquids at the glass transition. *Phys. Rev. Lett.* **80**(3), 544–547.
90. F. Sette, M. Krisch, C. Masciovecchio, G. Ruocco, and G. Monaco. (1998). Dynamics of glasses and glass-forming liquids studied by inelastic x-ray scattering. *Science.* **280**(5369), 1550–1555.
91. G. Monaco, C. Masciovecchio, G. Ruocco, and F. Sette. (1998). Determination of the infinite frequency sound velocity in the glass former o-terphenyl. *Phys. Rev. Lett.* **80**(10), 2161–2164.
92. D. Fioretto, U. Buchenau, L. Comez, A. Sokolov, C. Masciovecchio, A. Mermet, G. Ruocco, F. Sette, L. Willner, B. Frick, D. Richter, and L. Verdini. (1999). High-frequency dynamics of glass-forming polybutadiene. *Phys. Rev. E.* **59**(4), 4470–4475.

93. O. Pilla, A. Cunsolo, A. Fontana, C. Masciovecchio, G. Monaco, M. Montagna, G. Ruocco, T. Scopigno, and F. Sette. (2000). Nature of the short wavelength excitations in vitreous silica: An x-ray Brillouin scattering study. *Phys. Rev. Lett.* **85**(10), 2136–2139.
94. A. Matic, L. Borjesson, G. Ruocco, C. Masciovecchio, A. Mermet, F. Sette, and R. Verbeni. (2001). Contrasting behaviour of acoustic modes in network and non-network glasses. *Europhys. Lett.* **54**(1), 77–83.
95. J. Mattsson, A. Matic, G. Monaco, D. Engberg, and L. Borjesson. (2003). High-frequency collective excitations in a molecular glass-former. *J. Phys.-Condens. Matter.* **15**(11, SI), S1259–S1267
96. T. Scopigno, R. Di Leonardo, G. Ruocco, A. Baron, S. Tsutsui, F. Bossard, and S. Yannopoulos. (2004). High frequency dynamics in a monatomic glass. *Phys. Rev. Lett.* **92**(2), 025503.
97. A. Matic, C. Masciovecchio, D. Engberg, G. Monaco, L. Borjesson, S. Santucci, and R. Verbeni. (2004). Crystal-like nature of acoustic excitations in glassy ethanol. *Phys. Rev. Lett.* **93**(14), 145502.
98. L. E. Bove, E. Fabiani, A. Fontana, F. Paoletti, C. Petrillo, O. Pilla, and I. C. V. Bento. (2005). Brillouin neutron scattering of v-GeO_2. *Europhys. Lett.* **71**(4), 563–569.
99. E. Rat, M. Foret, E. Courtens, R. Vacher, and M. Arai. (1999). Observation of the crossover to strong scattering of acoustic phonons in densified silica. *Phys. Rev. Lett.* **83**(7), 1355–1358.
100. A. Matic, D. Engberg, C. Masciovecchio, and L. Borjesson. (2001). Sound wave scattering in network glasses. *Phys. Rev. Lett.* **86**(17), 3803–3806.
101. M. Foret, R. Vacher, E. Courtens, and G. Monaco. (2002). Merging of the acoustic branch with the boson peak in densified silica glass. *Phys. Rev. B.* **66**(2), 024204.
102. S. N. Taraskin and S. R. Elliott. (1999). Low-frequency vibrational excitations in vitreous silica: The ioffe-regel limit. *J. Phys.-Condens. Matter.* **11**(10A), A219–A227.
103. S. N. Taraskin and S. R. Elliott. (2000). Ioffe-regel crossover for plane-wave vibrational excitations in vitreous silica. *Phys. Rev. B.* **61**(18), 12031–12037.
104. B. Rufflé, M. Foret, E. Courtens, R. Vacher, and G. Monaco. (2003). Observation of the onset of strong scattering on high frequency acoustic phonons in densified silica glass. *Phys. Rev. Lett.* **90**(9), 095502.
105. B. Rufflé, G. Guimbretière, E. Courtens, R. Vacher, and G. Monaco. (2006). Glass-specific behavior in the damping of acousticlike vibrations. *Phys. Rev. Lett.* **96**(4), 045502.

106. R. Vacher, B. Rufflé, B. Hehlen, G. Guimbretière, G. Simon, and E. Courtens. (2008). The vibrational excitations of glasses in the boson-peak region: Application to borates. *Phys. Chem. Glasses-Eur. J. Glass Sci. Technol. Part B.* **49**(1), 19–25.
107. E. Courtens, B. Ruffl, and R. Vacher. (2006). Low-q brillouin scattering from glasses. *J. Neut. Res.* **14**(4), 361–366.
108. R. Vacher, E. Courtens, and M. Foret. (1999). Are high frequency acoustic modes in glasses dominated by strong scattering or by lifetime broadening? *Philos. Mag. B.* **79**(11–12), 1763–1774.
109. B. Rufflé, G. Guimbretière, E. Courtens, R. Vacher, and G. Monaco. (2007). Comment on "glass-specific behavior in the damping of acousticlike vibrations" - reply. *Phys. Rev. Lett.* **98**(7), 079602
110. W. Schirmacher, G. Ruocco, and T. Scopigno. (2007). Acoustic attenuation in glasses and its relation with the boson peak. *Phys. Rev. Lett.* **98**(2), 025501.
111. H. R. Schober. (2004). Vibrations and relaxations in a soft sphere glass: Boson peak and structure factors. *J. Phys. Condens. Matter.* **16**(27), S2659–S2670.
112. T. Scopigno, J. Suck, R. Angelini, F. Albergamo, and G. Ruocco. (2006). High-frequency dynamics in metallic glasses. *Phys. Rev. Lett.* **96**(13), 135501.
113. E. Courtens, M. Foret, B. Rufflé, and R. Vacher. (2007). Comment on "high-frequency dynamics in metallic glasses". *Phys. Rev. Lett.* **98**(7), 079603.
114. P. Bruna, G. Baldi, E. Pineda, J. Serrano, J. B. Suck, D. Crespo, and G. Monaco. (2011). Communication: Are metallic glasses different from other glasses? A closer look at their high frequency dynamics. *J. Chem. Phys.* **135**(10), 101101.
115. V. Karpov, M. Klinger, and F. Ignatiev. (1983). Theory of low-temperature anomalies in thermal-properties of amorphic structures. *Zh. Eksp. Teor. Fiz.* **84**(2), 760–775.
116. V. Karpov and D. Parshin. (1985). Heat-conductivity of glasses at temperatures below the Debye temperature. *Zh. Eksp. Teor. Fiz.* **88**(6), 2212–2227.
117. D. A. Parshin, H. R. Schober, and V. L. Gurevich. (2007). Vibrational instability, two-level systems, and the boson peak in glasses. *Phys. Rev. B.* **76**(6), 064206.
118. U. Buchenau, Y. Galperin, V. Gurevich, D. Parshin, M. Ramos, and H. Schober. (1992). Interaction of soft modes and sound-waves in glasses. *Phys. Rev. B.* **46**(5), 2798–2808.
119. D. Parshin and C. Laermans. (2001). Interaction of quasilocal harmonic modes and boson peak in glasses. *Phys. Rev. B.* **63**(13), 132203.

120. W. Schirmacher and M. Wagener. (1993). Vibrational anomalies and phonon localization in glasses. *Solid State Commun.* **86**(9), 597–603.
121. W. Schirmacher, G. Diezemann, and C. Ganter. (1998). Harmonic vibrational excitations in disordered solids and the boson peak. *Phys. Rev. Lett.* **81**, 136–139.
122. H. R. Schober. (2011). Quasi-localized vibrations and phonon damping in glasses. *J. Non-Cryst. Solids.* **357**(2), 501–505.
123. G. Ruocco, A. Matic, T. Scopigno, and S. N. Yannopoulos. (2007). Comment on "glass-specific behavior in the damping of acousticlike vibrations". *Phys. Rev. Lett.* **98**, 079601.
124. J. P. Wittmer, A. Tanguy, J.-L. Barrat, and L. Lewis. (2002). Vibrations of amorphous, nanometric structures: When does continuum theory apply? *Europhysics Lett.* **57**(3), 423–429.
125. A. Tanguy, J. P. Wittmer, F. Leonforte, and J.-L. Barrat. (2002). Continuum limit of amorphous elastic bodies: A finite-size study of low-frequency harmonic vibrations. *Phys. Rev. B.* **66**(17), 174205.
126. K. Yoshimoto, T. S. Jain, K. V. Workum, P. F. Nealey, and J. J. de Pablo. (2004). Mechanical heterogeneities in model polymer glasses at small length scales. *Phys. Rev. Lett.* **93**(17), 175501.
127. F. Leonforte, R. Boissière, A. Tanguy, J. P. Wittmer, and J.-L. Barrat. (2005). Continuum limit of amorphous elastic bodies. iii. three-dimensional systems. *Phys. Rev. B.* **72**(22), 224206.
128. H. Mizuno, S. Mossa, and J.-L. Barrat. (2014). Acoustic excitations and elastic heterogeneities in disordered solids. *Proc. Natl. Acad. Sci. USA.* **11**(33), 11949–11954.
129. B. B. Laird and H. R. Schober. (1991). Localized low-frequency vibrational modes in a simple model glass. *Phys. Rev. Lett.* **66**(5), 636–639.
130. J. Hafner and M. Krajci. (1994). Propagating and localized vibrational modes in Ni-Zr glasses. *J. Phys.-Condens. Matter.* **6**(25), 4631–4654.
131. S. N. Taraskin and S. R. Elliott. (1999). Anharmonicity and localization of atomic vibrations in vitreous silica. *Phys. Rev. B.* **59**, 8572–8585.
132. E. Lerner, G. Düring, and E. Bouchbinder. (2016). Statistics and properties of low-frequency vibrational modes in structural glasses. *Phys. Rev. Lett.* **117**(3), 035501.
133. H. Mizuno, H. Shiba, and A. Ikeda. (2017). Continuum limit of the vibrational properties of amorphous solids. *Proc. Natl. Acad. Sci. USA.* **114**(46), E9767–E9774.
134. H. Mizuno and A. Ikeda. (2018). Phonon transport and vibrational excitations in amorphous solids. *Phys. Rev. E.* **98**, 062612.

135. M. Shimada, H. Mizuno, and A. Ikeda. (2018). Anomalous vibrational properties in the continuum limit of glasses. *Phys. Rev. E.* **97**, 022609.
136. M. Shimada, H. Mizuno, M. Wyart, and A. Ikeda. (2018). Spatial structure of quasilocalized vibrations in nearly jammed amorphous solids. *Phys. Rev. E.* **98**, 060901.
137. G. Kapteijns, E. Bouchbinder, and E. Lerner. (2018). Universal nonphononic density of states in 2d, 3d, and 4d glasses. *Phys. Rev. Lett.* **121**(5), 055501.
138. S. Bonfanti, R. Guerra, C. Mondal, I. Procaccia, and S. Zapperi. (2020). Universal low-frequency vibrational modes in silica glasses. *Phys. Rev. Lett.* **125**(8), 085501.
139. D. Richard, K. González-López, G. Kapteijns, R. Pater, T. Vaknin, E. Bouchbinder, and E. Lerner. (2020). Universality of the nonphononic vibrational spectrum across different classes of computer glasses. *Phys. Rev. Lett.* **125**(8), 085502.
140. B. Rufflé, D. A. Parshin, E. Courtens, and R. Vacher. (2008). Boson peak and its relation to acoustic attenuation in glasses. *Phys. Rev. Lett.* **100**(1), 015501.
141. W. Schirmacher, B. Schmid, C. Tomaras, G. Viliani, G. Baldi, G. Ruocco, and T. Scopigno. (2008). Vibrational excitations in systems with correlated disorder. *Phys. Stat. Sol. (c).* **5**(3), 862–866.
142. S. Köhler, G. Ruocco, and W. Schirmacher. (2013). Coherent potential approximation for diffusion and wave propagation in topologically disordered systems. *Phys. Rev. B.* **88**(6), 064203.
143. G. Monaco and V. M. Giordano. (2009). Breakdown of the debye approximation for the acoustic modes with nanometric wavelengths in glasses. *Proc. Natl. Acad. Sci. USA.* **106**(10), 3659–3663.
144. W. Schirmacher. (2013). The boson peak. *Phys. Stat. Sol. (b).* **250**(5), 937–943.
145. G. Baldi, V. M. Giordano, G. Monaco, and B. Ruta. (2010). Sound attenuation at terahertz frequencies and the boson peak of vitreous silica. *Phys. Rev. Lett.* **104**(19), 195501.
146. B. Ruta, G. Baldi, V. M. Giordano, L. Orsingher, S. Rols, F. Scarponi, and G. Monaco. (2010). Communication: High-frequency acoustic excitations and boson peak in glasses: A study of their temperature dependence. *J. Chem. Phys.* **133**(4), 041101.
147. B. Ruta, G. Baldi, F. Scarponi, D. Fioretto, V. M. Giordano, and G. Monaco. Acoustic excitations in glassy sorbitol and their relation with the fragility and the boson peak. *J. Chem. Phys.* **137**(21), 214502 (2012).

148. G. Baldi, V. M. Giordano, B. Ruta, R. Dal Maschio, A. Fontana, and G. Monaco. (2014). Anharmonic damping of terahertz acoustic waves in a network glass and its effect on the density of vibrational states. *Phys. Rev. Lett.* **112**.
149. G. Baldi, V. M. Giordano, and G. Monaco. (2011). Elastic anomalies at terahertz frequencies and excess density of vibrational states in silica glass. *Phys. Rev. B.* **83**(17), 174203.
150. A. Monaco, A. I. Chumakov, Y. Z. Yue, G. Monaco, L. Comez, D. Fioretto, W. A. Crichton, and R. Ruffer. (2006). Density of vibrational states of a hyperquenched glass. *Phys. Rev. Lett.* **96**(20), 205502.
151. A. Monaco, A. I. Chumakov, G. Monaco, W. A. Crichton, A. Meyer, L. Comez, D. Fioretto, J. Korecki, and R. Ruffer. (2006). Effect of densification on the density of vibrational states of glasses. *Phys. Rev. Lett.* **97**(13), 135501.
152. G. Baldi, A. Fontana, G. Monaco, L. Orsingher, S. Rols, F. Rossi, and B. Ruta. (2009). Connection between boson peak and elastic properties in silicate glasses. *Phys. Rev. Lett.* **102**(19), 195502.
153. K. Niss, B. Begen, B. Frick, J. Ollivier, A. Beraud, A. Sokolov, V. N. Novikov, and C. Alba-Simionesco. (2007). Influence of pressure on the boson peak: Stronger than elastic medium transformation. *Phys. Rev. Lett.* **99**(5), 055502.
154. L. Hong, B. Begen, A. Kisliuk, C. Alba-Simionesco, V. N. Novikov, and A. P. Sokolov. (2008). Pressure and density dependence of the boson peak in polymers. *Phys. Rev. B.* **78**(13), 134201.
155. L. Orsingher, A. Fontana, E. Gilioli, G. Carini, G. Carini, G. Tripodo, T. Unruh, and U. Buchenau. (2010). Vibrational dynamics of permanently densified GeO_2 glasses: Densification-induced changes in the boson peak. *J. Chem. Phys.* **132**(12), 124508.
156. E. DeGiuli, A. Laversanne-Finot, G. Dring, E. Lerner, and M. Wyart. (2014). Effects of coordination and pressure on sound attenuation, boson peak and elasticity in amorphous solids. *Soft Matter.* **10**, 5628–5644.
157. E. Fabiani, A. Fontana, and U. Buchenau. (2008). Neutron scattering study of the vibrations in vitreous silica and germania. *J. Chem. Phys.* **128**(24), 244507.
158. D. Engberg, A. Wischnewski, U. Buchenau, L. Börjesson, A. J. Dianoux, A. P. Sokolov, and L. M. Torell. (1999). Origin of the boson peak in a network glass B_2O_3. *Phys. Rev. B.* **59**(6), 4053–4057.
159. U. Buchenau, A. Wischnewski, D. Richter, and B. Frick. (1996). Is the fast process at the glass transition mainly due to long wavelength excitations? *Phys. Rev. Lett.* **77**(19), 4035–4038.

160. Y. Tezuka, S. Shin, and M. Ishigame. (1991). Observation of the silent soft phonon in β-quartz by means of hyper-raman scattering. *Phys. Rev. Lett.* **66**(18), 2356–2359.
161. G. Simon, B. Hehlen, E. Courtens, E. Longueteau, and R. Vacher. (2006). Hyper-Raman scattering from vitreous boron oxide: Coherent enhancement of the boson peak. *Phys. Rev. Lett.* **96**(10), 105502.
162. B. Hehlen and B. Ruffl. (2020). Atomic vibrations in glasses. In P. Richet (ed.), *Encyclopedia of Glass Science, Technology, History, and Culture.* in press, John Wiley & Sons, Inc.
163. R. Shuker and R. Gammon. (1970). Raman-scattering selection-rule breaking and density of states in amorphous materials. *Phys. Rev. Lett.* **25**(4), 222–225.
164. A. P. Sokolov, A. Kisliuk, M. Soltwisch, and D. Quitmann. (1992). Medium-range order in glasses: Comparison of raman and diffraction measurements. *Phys. Rev. Lett.* **69**(10), 1540–1543.
165. E. Duval, A. Boukenter, and T. Achibat. (1990). Vibrational dynamics and the structure of glasses. *J. Phys.-Condens. Matter.* **2**(51), 10227–10234.
166. N. V. Surovtsev and A. P. Sokolov. (2002). Frequency behavior of raman coupling coefficient in glasses. *Phys. Rev. B.* **66**(5), 054205.
167. G. Simon, B. Hehlen, R. Vacher, and E. Courtens. (2007). Hyper-raman scattering analysis of the vibrations in vitreous boron oxide. *Phys. Rev. B.* **76**(5), 054210.
168. I. Prots, V. Malinovsky, and N. Surovtsev. (2008). Investigation of the fast relaxation in glass-forming selenium by low-frequency raman spectroscopy. *Glass Phys. and Chem.* **34**(1), 30–36.
169. A. Fontana, F. Rossi, and E. Fabiani. (2006). The raman coupling function in v-GeO_2 and v-SiO_2: A new light and neutron scattering study. *J. Non-Cryst. Solids.* **352**(42–49), 4601–4605.
170. J. Sibik, S. R. Elliott, and J. A. Zeitler. (2014). Thermal decoupling of molecular-relaxation processes from the vibrational density of states at teraherz frequencies in supercooled hydrogen-bonded liquids. *J. Phys. Chem. Lett.* **5**(11), 1968–1972.
171. S. N. Taraskin, S. I. Simdyankin, S. R. Elliott, J. R. Neilson, and T. Lo. (2006). Universal features of terahertz absorption in disordered materials. *Phys. Rev. Lett.* **97**(5), 055504.
172. M. Kabeya, T. Mori, Y. Fujii, A. Koreeda, B. Wan Lee, J.-H. Ko, and S. Kojima. (2016). Boson peak dynamics of glassy glucose studied by integrated terahertz-band spectroscopy. *Phys. Rev. B.* **94**(22), 224204.

173. V. Denisov, B. Mavrin, and V. Podobedov. (1987). Hyper-Raman scattering by vibrational excitations in crystals, glasses and liquids. *Phys. Rep.* **151**(1), 1–92.
174. Y. Iijima, T. Mori, S. Kojima, Y. Fujii, A. Koreeda, S. Kitani, H. Kawaji, and J. Ko. (2018). Terahertz time-domain spectroscopy and low-frequency raman scattering of boson peak dynamics of lithium borate glasses. In *43rd International Conference on Infrared, Millimeter, and Terahertz Waves*, pp. 957–958, Institute of Electrical and Electronics Engineers.
175. B. Hehlen and G. Simon. (2012). The vibrations of vitreous silica observed in hyper-raman scattering. *J. Raman Spectrosc.* **43**(12), 1941–1950.
176. G. Dolino, B. Berge, M. Vallade, and F. Moussa. (1992). Origin of the incommensurate phase of quartz .1. inelastic neutron-scattering study of the high-temperature beta-phase of quartz. *J. Phys. I.* **2**(7), 1461–1480.
177. W. M. Visscher. (1960). Study of lattice vibrations by resonance absorption of nuclear gamma rays. *Ann. Phys.* **9**(2), 194–210.
178. A. I. Chumakov, I. Sergueev, U. van Bürck, W. Schirmacher, T. Asthalter, R. Rüffer, O. Leupold, and W. Petry. (2004). Collective nature of the boson peak and universal transboson dynamics of glasses. *Phys. Rev. Lett.* **92**(24), 245508.
179. A. I. Chumakov, G. Monaco, A. Monaco, W. A. Crichton, A. Bosak, R. Rüffer, A. Meyer, F. Kargl, L. Comez, D. Fioretto, H. Giefers, S. Roitsch, G. Wortmann, M. H. Manghnani, A. Hushur, Q. Williams, J. Balogh, K. Parliński, P. Jochym, and P. Piekarz. (2011). Equivalence of the boson peak in glasses to the transverse acoustic van hove singularity in crystals. *Phys. Rev. Lett.* **106**(22), 225501.

https://doi.org/10.1142/9781800612587_0008

Chapter 8

The Soft-Potential Model and Its Extensions

U. Buchenau

Forschungszentrum Jülich GmbH, Jülich Centre for Neutron Science (JCNS-1) and Institute for Complex Systems (ICS-1), 52425 Jülich, Germany
buchenau-juelich@t-online.de

The soft-potential model focuses on the crossover from tunneling states to soft vibrations in glasses, considering both kinds of modes as members of a continuous distribution. Since a tunneling state has a negative potential curvature at the top of its energy barrier, one needs to postulate a stabilizing fourth-order potential term to bring both kinds of modes into a common picture. There is numerical evidence for the existence of a stabilizing fourth-order term and of an unstable inner core in soft, quasi-localized vibrations. The soft-potential parameters can be calculated for numerical soft modes from their fourth-order potential term and their long-range elastic distortion field. The soft-potential model describes the crossover region successfully, reproducing the Standard Tunneling Model on the tunneling side, and the plateau in the thermal conductivity as well as the rise of the soft mode vibrational density of states with the fourth power of the frequency on the other. More distant excitations like the boson peak on the vibrational side and higher-barrier thermally activated relaxations beyond the tunneling regime require extensions of the model.

1. Introduction

As explained in the preceding chapters, the low-temperature properties of glasses and disordered crystals — heat capacity, thermal conductivity, and sound absorption — do not follow the predictions of the Debye model, according to which the only relevant low-temperature excitations should be longitudinal and transverse sound waves. The deviations below 1 K seem to be due to tunneling states, and the deviations above 1 K due to the additional vibrations of the boson peak.

Though we do not yet have a quantitative picture of the glass transition, it is clear that both groups of excitations form in the cooling process close to the glass temperature. The soft-potential model[1–13] is based on the assumption that the distinction between a tunneling state and a soft quasi-localized vibration is irrelevant at their formation temperature.

But, this assumption is not possible without a stabilization of all soft modes, tunneling states, and vibrations, by a common positive anharmonic fourth-order term in the mode potential, with only the lower potential terms varying from mode to mode. The existence of a stabilizing fourth-order term is to be expected in a solid consisting of atoms with anharmonic interatomic potentials and has in fact been found in numerical simulations of frozen simple liquids[14,15] (see also Chapter 10).

Another common feature of tunneling states and soft vibrations is their strong resonant interaction with the sound waves, leading to a much lower thermal conductivity in insulating glasses at low temperature than the one of the corresponding crystals. Obviously, the displacement of the mode is accompanied by a strong elastic distortion of the surroundings. This kind of behavior was first observed, studied, and analyzed for defects in crystals, in particular in the dynamics of an interstitial atom in fcc close packing.[16] Since the elastic distortion of the surroundings provides a positive contribution to the force constant of the mode, there must be a negative contribution from the central core of the soft mode to get the total force constant down to zero.

Unstable inner cores of soft modes have been first seen numerically[17] in a glass of particles interacting with a soft sphere potential. In this glass, unstable cores of about fifty atoms were found, much larger than the gliding triangle transition between two close packing configurations of six atoms, which has been discussed as a possible unstable soft mode core in simple liquids.[18]

The numerical evidence indicates unstable cores in simple liquids which are less compact than the gliding triangle. There seems to be a string-like parallel motion of atoms in these cores, first discovered in the soft vibrations and low-barrier relaxations of a soft sphere glass[19] and later in the atomic motion at the glass transition.[20]

The unstable core concept for the quasi-localized soft vibrational modes has recently been corroborated numerically again.[21] Moreover, as will be seen in the discussion of the boson peak in Section 4.1, the understanding of the surprisingly strong temperature dependence of the boson peak in some glasses (within the glass phase) does require an unstable core even at the boson peak itself.[22]

An important result of the soft-potential model, the vibrational density of states of the soft modes proportional to ω^4 (ω frequency),[23] has been corroborated by the Q^4-dependence of the damping (Q wavevector) in X-ray Brillouin scattering described in Chapter 7 and the recent numerical work described in Chapter 10.

2. The Soft-Potential Model

2.1. *The anharmonic quartic potential*

The crossover from tunneling states to vibrations is a quantum mechanical problem. Let us first consider the level splitting in a purely quartic potential (see Figure 1). If all modes have the same quartic potential term in their normal coordinate, the crossover between tunneling states and soft vibrations is determined by the zero-point energy W of the purely quartic potential.

One can convince oneself of this fact by a simple quantum-mechanical argument: A small negative quadratic term broadens the potential and creates a small potential barrier at $x = 0$, causing the distance between the zero-point level and the first excited level in Figure 1 to decrease. Further increase of the barrier causes the distance of the two levels to decrease further, until they settle to the right and to the left of the barrier as a tunneling state.

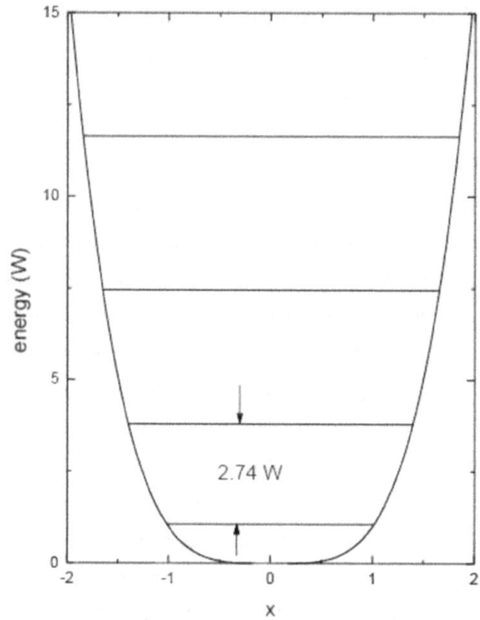

Fig. 1. The first four energy levels of the purely quartic potential $V(x) = Wx^4$ of the soft-potential model.

Conversely, if one adds a positive quadratic term, the potential gets steeper, more harmonic, and the level distance increases. The same happens if one adds a positive or negative linear term to the potential, as illustrated in the insert on the upper left of Figure 2.

The zero-point energy is not only determined by the potential energy but also by the kinetic energy of the atoms which move in the mode. For a quantitative access, one defines the normal coordinate A via the normalized eigenvector $\mathbf{e}_i$ with $i = 1, \ldots, N$ for the N atoms of the glass

$$\sum_{i=1}^{N} \mathbf{e}_i^2 = 1 \tag{1}$$

and the displacement $\mathbf{u}_i$ of atom i in the mode

$$\mathbf{u}_i = \frac{\mathbf{e}_i}{M_i^{1/2}} A \tag{2}$$

where M_i is the mass of atom i.

The definition ensures that the kinetic energy of the mode is given by

$$E_{\mathrm{kin}} = \frac{1}{2}\dot{A}^2 \tag{3}$$

The zero-point energy W in the purely quartic potential is obtained by equating the kinetic confinement energy and the potential energy at the normal coordinate displacement A_0

$$\frac{\hbar^2}{2A_0^2} = \chi_4 A_0^4 \equiv W \tag{4}$$

where χ_4 is the prefactor of the quartic potential for the normal coordinate, leading to

$$W = \frac{\hbar^{4/3}\chi_4^{1/3}}{2^{2/3}} \qquad A_0 = \frac{\hbar^{1/3}}{2^{1/6}\chi_4^{1/6}} \tag{5}$$

Defining the coordinate $x = A/A_0$, one gets the soft potential for the general case

$$V(x) = W\left(D_1 x + D_2 x^2 + x^4\right) \tag{6}$$

with D_1 and D_2 as the coefficients of the linear and the quadratic term, respectively.

It should be noted that the setting of the third-order term to zero corresponds to a choice, namely, to start from the origin of the quartic potential. An equivalent choice is to start from the potential minima.[4,5,11–13] The equivalence of the two choices has been demonstrated numerically.[24,25]

In the choice of a vanishing third-order term,[6,7,9,10] the soft-potential model assumes a constant distribution of soft modes in the $D_1 - D_2$ plane (see Figure 2). In this plane, one gets tunneling states only in the very small rectangle to the left of the origin of Figure 2, because a very small D_1 suffices to push the tunneling state to such a big level separation that is does not tunnel anymore.

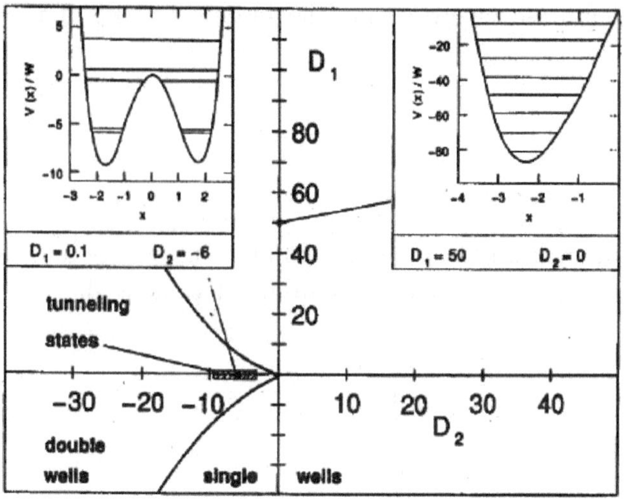

Fig. 2. The location of tunneling states and soft vibrational states in the $D_1 - D_2$ plane.

2.2. *Assumptions*

The soft-potential model has the following assumptions:

(1) All soft modes can be characterized by a single W.
(2) D_1 and D_2 are randomly distributed around the origin with a constant density $P(D_1, D_2) = P_s$.
(3) The soft modes have a bilinear interaction with the uniaxial strain field ε_l of the longitudinal sound waves and the shear strain ε_t of the transverse sound waves

$$\delta V_l = \Lambda_l \varepsilon_l x \quad \delta V_t = \Lambda_t \varepsilon_t x \tag{7}$$

with the coupling constants Λ_l and Λ_t for longitudinal and transverse sound waves, respectively.

With these assumptions, the soft-potential model has four parameters: the crossover energy W between tunneling states and vibrations, the density P_s, and the two coupling constants Λ_l and Λ_t which characterize the interaction between soft modes and sound waves. These last two are not really independent parameters, because they often show the relation $\Lambda_l^2/\Lambda_t^2 = 3$ expected for a single local elastic dipole proportional to the displacement of the soft mode.

Before working out the quantitative predictions of the soft-potential model, let us first summarize what is known about the nature of the soft modes from numerical work[14,15,19,21,24–36] and from theoretical considerations,[11–13,18] in Section 2.3.

2.3. *The nature of the soft modes*

While it is easy to consider a localized mode in the case of a tunneling state, a soft localized vibrational state is bound to lose its localization by its resonant interaction with the sound waves.[11–13] In fact, the numerical studies do either consider vibrational eigenmodes below the lowest sound wave frequency given by the cell dimensions[15,19,21,24–36] or a decomposition[14] of local mode and sound wave parts of the vibration. That is the reason why one calls these modes quasi-localized vibrations.

On the other hand, there is general agreement that the soft modes are due to local structures with a soft mode, "soft spots", even in the alternative explanation of the boson peak of Chapter 9 in terms of the elastic heterogeneity of the glass. The existence of tunneling states shows that at least some of these soft modes must have a structurally unstable core, unstable enough to overcome the strong elastic restoring forces of the surroundings.

An increasing part of our knowledge about the soft modes is due to numerical work,[14,15,17,19–21,24–36] the unstable core,[17,21] a new ingenious characterization[24,25] of the modes in terms of eigenvectors of the third- and fourth-order term in the mode potential, the density $p(V_b) \sim V_b^{1/4}$ of soft modes vibrating in double-well potentials[30] with a barrier height V_b (this feature will be shown again to be a soft-potential prediction in Section 3.4), and the strong decrease[29] in the number density of the modes in glasses cooled with the swap mechanism,[37] showing that the number density of soft modes increases with increasing temperature in the undercooled liquid.

Very recently, it has been established[35] that the long-range elastic distortion around a soft mode is describable in terms of the elastic Greens function of an isotropic elastic medium, assuming vibrating elastic dipoles in the center of the mode. For this binary glass, the distortion inside was essentially a planar shear.

The work[35] has been extended[36] to investigate whether the outer elastic deformation energy holds the unstable core on its saddle point in energy as hypothesized earlier.[18] It turned out that the long-range elastic deformation only explains one third of the positive restoring forces holding the core. For its full stabilization, one needs to invoke additional medium-range interactions between neighboring soft spots.

The long-range elastic deformation determines the bilinear coupling of the mode to the sound waves. Knowing the deformation, and using Eq. (5) to calculate W from χ_4 for many modes, one has the soft-potential parameters W, Λ_l, and Λ_t for this binary glass.[36] Scaling the results with the shear modulus, the atomic mass, and the atomic volume, one gets quantitative agreement with measured low-temperature sound absorption and X-ray Brillouin data in metallic glasses within the soft-potential model.

Equation (5) predicts W proportional to $\chi_4^{1/3}$. This means that W has a much narrower distribution than χ_4. This is indeed necessary for the soft-potential fit of the heat capacity, as will be seen in Section 3.2.

3. Achievements of the Soft-Potential Model

3.1. *Tunneling states below 1 K*

The first and most important achievement is that the soft-potential model reproduces the standard tunneling model[38] described in Chapters 1–5, at least to a degree where experiments below 1 K cannot distinguish between the two models. The standard tunneling model assumes a constant distribution

$$P(\Delta, \lambda) = P_0 \tag{8}$$

where Δ is the energy difference between the two minima of the double-well potential, and the tunnel splitting is

$$\Delta_0 = \hbar\omega_0 e^{-\lambda} \tag{9}$$

with $\hbar\omega_0$ of the order of the zero-point energy in one of the wells and a total level splitting

$$E = \sqrt{\Delta^2 + \Delta_0^2} \tag{10}$$

In the soft-potential model, a tunneling state has a negative D_2 value. If D_1 is zero (symmetric case), one gets a double-well potential with minima at

$$x = \pm\sqrt{|D_2|/2} \tag{11}$$

and a barrier height of

$$V_b = WD_2^2/4 \tag{12}$$

For a small D_1, the energy difference Δ between the two minima is

$$\Delta = WD_1\sqrt{2|D_2|} \tag{13}$$

An example for such a potential and its numerically calculated tunneling levels is shown in the insert at the upper left of Figure 2.

The first formulations of the soft-potential model used the half-classical Wentzel–Kramers–Brillouin approximation to calculate the tunnel splitting Δ_0 for a given negative D_2. But, this is not good enough for a comparison with experiment, which requires accuracy on the 10% level.

To get the required accuracy, one can calculate the tunnel splitting and the asymmetry numerically and then search for appropriate analytical approximations. This was done[7] and led to the equations

$$\Delta_0 = W|D_2|^{3/2}\exp\left(1 - \frac{\sqrt{2}}{3}|D_2|^{3/2}\right) \tag{14}$$

and

$$\Delta = WD_1\sqrt{2(|D_2| - 1)} \tag{15}$$

Note that Eq. (15) differs from Eq. (13). The first refers to potential minima, the second to tunneling levels.

With these equations, the Jacobian of the transformation to λ and Δ of the tunneling model leads to

$$P_0 = \frac{2P_s}{W(|D_2| - 1)} f(D_2) \tag{16}$$

with

$$f(D_2) = \sqrt{|D_2| - 1} \Big/ \left(\sqrt{D_2} - \frac{3}{\sqrt{2} D_2} \right) \tag{17}$$

In the relevant range, $f(D_2)$ is close to unity. This implies that P_0 is not a constant as in the tunneling model, predicting a slightly larger decrease of the heat capacity with decreasing temperature. However, within the accuracy of experimental data, one cannot find a preference for the one or the other model.

The coupling between tunneling states and sound waves in the tunneling model[38] is defined over the asymmetry

$$\gamma_j = \frac{1}{2} \frac{\partial \Delta}{\partial \varepsilon_j} \tag{18}$$

where $j = l, t$ for longitudinal and transverse waves, respectively. Applying the same recipe to the tunneling states of the soft-potential model, one gets

$$\gamma_j = \Lambda_j \sqrt{(|D_2| - 1)/2} \tag{19}$$

With these definitions, one gets the relation between the coupling parameters of the two models

$$P_0 \gamma_j^2 = \frac{P_s \Lambda_j^2}{W} f(D_2) \tag{20}$$

For the tunneling states, the lifetime calculation of the tunneling model transforms into

$$\tau^{-1} = A_t \frac{|D_2| - 1}{2} \Delta_0^2 E \coth\left(\frac{E}{2k_B T} \right) \tag{21}$$

with

$$A_t = \frac{1}{2\pi \rho \hbar^4} \left(\frac{\Lambda_l^2}{\mathrm{v}_l^5} + \frac{2\Lambda_t^2}{\mathrm{v}_t^5} \right) \tag{22}$$

For the sound waves, the influence of the tunneling states on the acoustic properties is determined[38] by the parameter combination

$$C_j^{\mathrm{tunn}} = \frac{P_0 \gamma_j^2}{\rho \mathrm{v}_j^2} \tag{23}$$

where v_j is the respective sound velocity and ρ is the density. The corresponding definition in the soft-potential model

$$C_l = \frac{P_s \Lambda_l^2}{W \rho \mathrm{v}_l^2} \quad C_t = \frac{P_s \Lambda_t^2}{W \rho \mathrm{v}_t^2} \tag{24}$$

Integrating over all tunneling states, one finds at not too low temperature the tunneling plateau in the sound absorption, which around 1 K has the height

$$Q^{-1} = \frac{\pi C_j^{\mathrm{tunn}}}{2} = \frac{\pi C_j}{2} f(D_2) \tag{25}$$

Since $f(D_2)$ is close to unity in the relevant range, one gets again the same answer of a frequency-independent tunneling plateau height from both models within experimental error (the relevant D_2 depends on the tunneling states which fulfill $\omega\tau = 1$).

With the relations (8) to (25), one can transform tunneling model predictions into soft-potential predictions and vice versa.

3.2. *Vibrational soft modes above 1 K*

While the tunneling model is clearly very successful in explaining experimental results below 1 K, it fails to describe the strong rise of the vibrational density of states, starting at zero frequency and culminating at the boson peak. To explain this rise in terms of a vibrational density of states increasing with the fourth power of the frequency (in agreement with the new numerical results[15,21,24,25,27–30]) is another important achievement of the soft-potential model.

In 1971, the famous paper by Zeller and Pohl[39] (see Chapter 1) demonstrated to the world that glasses have much more low-frequency excitations than just the sound waves, as shown in Figure 3 in a plot of the heat capacity $C_p(T)$ divided by T^3. One finds not only

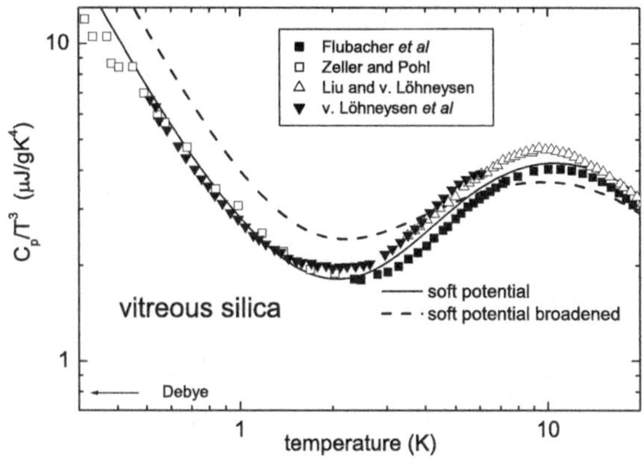

Fig. 3. Heat capacity measurements in vitreous silica,[39–42] together with a soft-potential fit (the continuous line). The dashed line shows what happens if one allows for a logarithmic broadening of the crossover energy W of a decade.

the linear or quasilinear term in $C_p(T)$, which the tunneling model ascribes to tunneling states, but also a T^5 term, which leads up to a maximum in C_p/T^3, the boson peak.

A rise of $C_p(T)$ with T^5 corresponds to a density of states $g(\omega)$ which rises with ω^4. This rise is exactly the one predicted by the basic soft-potential model for the soft vibrations[6,7]

$$g_s(\omega) = \frac{1}{8}\frac{P_s}{W}\left(\frac{\hbar\omega}{W}\right)^4 \tag{26}$$

The identification of the T^5-term with vibrations is corroborated by numerous Raman and neutron scattering investigations, which find essentially the temperature dependence expected for harmonic vibrations (this is the origin of the name "boson peak").[10]

Naturally, the soft-potential model does only explain the ω^4-rise and not the boson peak. In order to explain the boson peak, one needs an extension of the soft-potential model (see Section 4.1).

The soft-potential fit of the data in Figure 3 uses the theoretical boson peak expression for the extended soft-potential model[13] and has three parameters: P_s, the density of states of the soft-potential model, the crossover energy W (both from Table 1), and the boson peak frequency ω_b of 3.9 meV.

Figure 3 also illustrates what happens to the soft-potential fit if one assumes a broad distribution of W-values, here for a logarithmic full width at half maximum of a decade, like the width of the numerical χ_4-data.[15] The broadening tends to increase the minimum in C_p/T^3 between tunneling states and soft vibrations away from the measured one, a discrepancy which is impossible to repair by a readjustment of the parameters.

This is a quadratic effect, so one can exclude a half width of a decade, but not one third of this value, as suggested by the relation $W \sim \chi_4^{1/3}$ in Eq. (5).

Important information from the neutron data is that these additional soft vibrations do not have a sound wave eigenvector. This is demonstrated[43] in Figure 4(b). If they were long wavelength

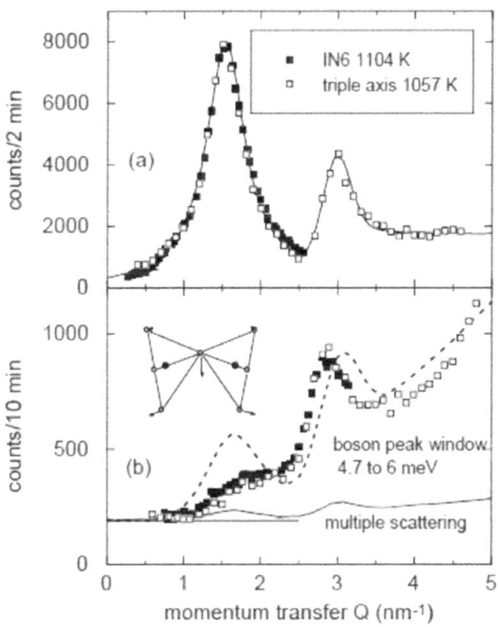

Fig. 4. Vitreous silica neutron[43] (a) elastic structure factor, (b) dynamic structure factor of soft vibrations in silica, demonstrating that they do not have the dynamic structure factor of sound waves (dashed line), but rather the one of coupled SiO_4-tetrahedra librations in the insert. The continuous line is the expected sound wave scattering.

sound waves, the wave-vector dependence of the signal would be Q^2 times the elastic structure factor. Instead, they show the wave-vector dependence of coupled SiO_4-tetrahedra librations, the mode which becomes soft in the structural transformation from α-quartz to β-quartz. This picture was later corroborated by Hyper-Raman-data[44] in vitreous silica.

The dynamic structure factor of the boson peak vibrations has also been measured in boron trioxide[45] and in polybutadiene.[46] In these two cases, one finds a mixture of the sound wave structure factor with a simple Q^2-dependence. Below the boson peak, the sound wave structure factor fraction exceeds the Debye density of states, but extrapolates to the Debye expectation at zero frequency. This shows again that the excess modes are accompanied by an elastic distortion of the surrounding medium.

An interesting case is amorphous germanium, which does not have a boson peak, but rather a low-frequency tail[47] of the lowest van-Hove singularity of the crystal at 2 THz. In this tail, neutron data[48] find a mixture of sound waves and of the bond bending motion of the lowest van-Hove singularity.

3.3. *Plateau in the thermal conductivity*

The tunneling model explains many puzzling experimental data below 1 K, but does not explain one of the most striking findings, namely, the plateau in the thermal conductivity at about 10 K (see Chapter 3). At the plateau, the thermal conductivity of the glass can be many orders of magnitude smaller than the one of the corresponding crystal,[49] as shown in Figure 5.

The soft-potential model is able to explain the plateau in the thermal conductivity in terms of the resonant interaction between the sound waves and the soft vibrations. Above the crossover frequency $W/\hbar$, the model predicts a rise of the density of soft modes with ω^4 (see Section 3.2), which leads to a strong resonant scattering of the sound waves.

In an insulating glass, where there is no electronic thermal conductivity, the heat is carried by the sound waves propagating

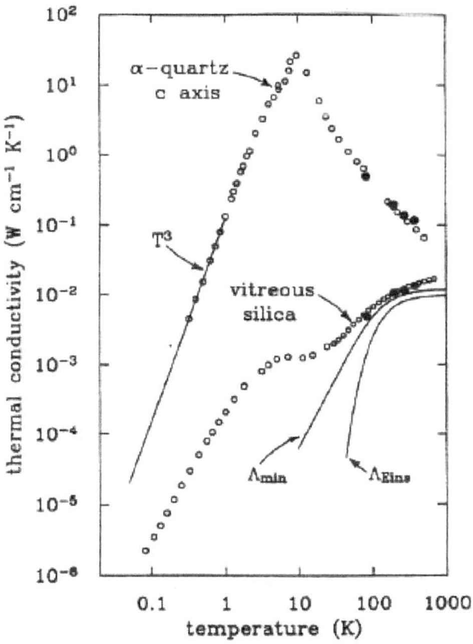

Fig. 5. Thermal conductivity of vitreous silica compared with the one of crystalline quartz.[49]

through the sample, and the strong reduction of the conductivity at the plateau must be ascribed to the scattering of the sound waves, leading to a short mean free path l_j for $j = l$ (longitudinal) or $j = t$ (transverse).

Below the plateau, the frequency of the relevant phonons is already high, so the tunneling model[38] and, correspondingly, the soft-potential model predict resonant scattering from the tunneling states

$$l^{-1}_{\text{res, tunn}} = \frac{\pi\omega C_j}{\text{v}_j} \tanh\left(\frac{\hbar\omega}{2k_B T}\right) \tag{27}$$

which dominates the thermal conductivity below 1 K.

Above 1 K, one has two effects, namely, the scattering from the soft vibrations[9] at higher frequencies

$$l^{-1}_{\text{res, vib}} = \frac{\pi\omega C_j}{\text{v}_j}\frac{1}{8}\left(\frac{\hbar\omega}{W}\right)^3 \tag{28}$$

and the scattering from classical thermal activation[9] in the double-well part of the potentials

$$l_{\text{rel, class}}^{-1} = \frac{\pi\omega C_j}{\mathrm{v}_j}\left(\frac{k_B T}{W}\right)^{3/4} \tag{29}$$

at lower frequencies.

To get all three effects, one has to add the three inverse mean free paths together and to calculate the resulting thermal conductivity. In Ref. [9], this was done for ten glasses. Figure 6 shows the result for three of them, demonstrating that the soft-potential model is indeed able to explain the plateau in the thermal conductivity of glasses.

The fit of the thermal conductivity yields two parameters, the crossover energy W and an average value C for the two C_j, which are usually close to each other. One can check the value of W by comparing it to the value fitted to the heat capacity of the glass. In all ten glasses,[9] the values agreed within experimental error.

Table 1 lists the sound wave and soft-potential parameters obtained by combining heat capacity and thermal conductivity measurements.[9, 50]

At this point, it is necessary to point out the limitations of the soft-potential model. It is able to describe the onset of the plateau, because it predicts correctly the initial ω^4-rise of the additional

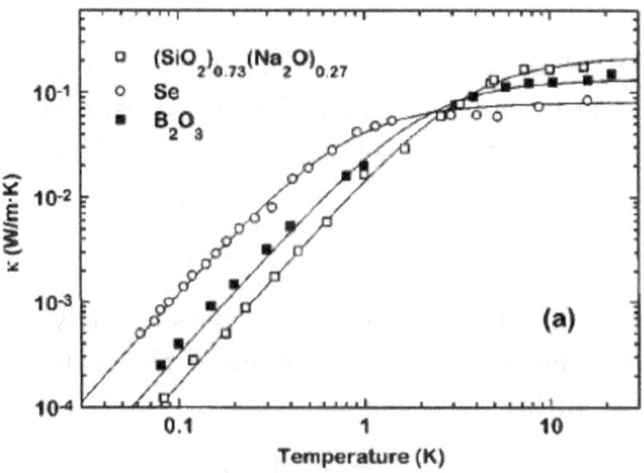

Fig. 6. Soft-potential fit of thermal conductivity plateau data in three glasses.[9]

Table 1. Sound wave and soft-potential parameters of five glasses, determined from the crossover from tunneling states to vibrations in the heat capacity and from the onset of the plateau in the thermal conductivity[0] (glycerol from Ref. [50]).

Glass	ρ (kg/m^3)	v_l (m/s)	v_t (m/s)	W/k_B (K)	P_s/ρ (10^{19}/kg)	C (10^{-4})
SiO_2	2200	5800	3800	3.8	6.3	2.6
GeO_2	3600	3680	2410	3.8	3.9	2.2
B_2O_3	1810	3390	1870	2.6	2.3	3.1
Polystyrene	1050	2800	1570	1.8	11.0	7.1
Glycerol	1380	3600	1920	4.3	1.6	1.9

density of states. But, without extensions, it is not able to predict how things proceed at higher frequencies, why one observes a boson peak, or why the thermal conductivity again shows a rise after the plateau (see Figure 5). We will come back in Section 4 to some of these problems in the discussion of the limitations and extensions of the model.

3.4. *Sound absorption at low frequency*

The plateau in the thermal conductivity discussed in Section 3.3 is an effect which integrates over many frequencies as well as over longitudinal and transverse sound waves.

But, it is naturally possible to get much more detailed information by measuring the sound absorption at a given frequency and for a given kind of sound wave. This allows for the additional checks of the soft-potential model to be discussed in this and the next subsection (for a review of experiments connecting thermal conductivity and sound absorption in glasses and disordered crystals see Ref. [51]).

For sound waves with low frequencies, below 100 MHz, the resonant scattering contributions from the soft vibrations and the tunneling states can both be neglected. The resonant vibration scattering is too small; the resonant tunneling scattering requires very low temperatures for its measurement (otherwise, one has equal population of the two levels and the scattering disappears).

But, what is very clearly seen is the relaxational scattering from the tunneling states, namely, from those tunneling states with a relaxation rate τ^{-1} close to the frequency ω of the sound waves. This relaxation rate can be calculated from Eq. (21) and is due to the interaction of the tunneling state with all sound waves. The relaxational scattering from the tunneling leads to the tunneling plateau of Eq. (25).

The tunneling plateau has a low-temperature end and a high-temperature end. The low-temperature end is reached when the cotangens hyperbolicus in Eq. (21) begins to approach the value 1 for the relevant tunneling states at the given frequency; then, the plateau drops and begins to approach a T^3 decrease.[52] The high-temperature end is reached when the classical Kramers-Arrhenius relaxation process[53] via thermal activation of states near the saddle point of the barrier begins to overcome the tunneling relaxation rate; then, the absorption begins to rise[9] proportional to $T^{3/4}$. The explanation of this rise is again an important achievement of the soft-potential model (though it should be mentioned that there is an alternative explanation[54] in terms of tunneling states with larger splitting, which according to the soft-potential model do not exist).

As pointed out in Section 2.4, the $T^{3/4}$-rise is consistent with the new numerical finding[30] of a double-well potential density $p(V_b) \sim V_b^{1/4}$. To see this, note that the minimum distance of Eq. (11) is proportional to $V_b^{1/4}$ as well according to Eq. (12). The sound absorption increases with the product of $p(V_b)$ with the square of the coupling constant, so it increases proportional to $V_b^{3/4}$, from which the $T^{3/4}$ dependence follows.

The soft-potential model predicts a classical relaxation damping of the sound waves[9] at higher temperatures, given by

$$Q^{-1} = \tan \delta = \left(\frac{k_B T}{W}\right)^{3/4} \ln^{-1/4}\left(\frac{1}{\omega\tau_0}\right) \qquad (30)$$

For a frequency of some kHz and a τ_0 of 10^{-13} s, the logarithmic factor in that expression is about 0.47, leading to[9]

$$Q^{-1} = 0.86 Q_{\text{plateau}} \left(\frac{k_B T}{W}\right)^{3/4} \qquad (31)$$

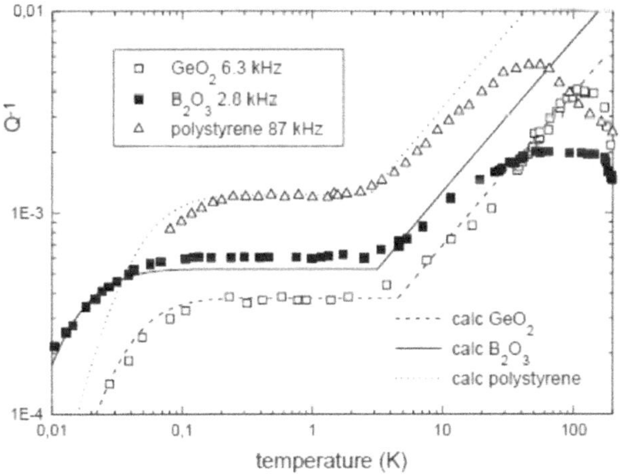

Fig. 7. The tunneling plateau and the crossover to the scattering from thermally activated relaxation processes[55,56] in the three glasses GeO_2, B_2O_3, and polystyrene, together with the soft-potential calculation. Note that the soft-potential parameters C_t and W are not adapted to the data, but taken from those adapted to the heat capacity and the thermal conductivity in Table 1.

This classical relaxation reaches the plateau level at a crossover temperature of about 1.2 W/k_B. As shown in Ref. [9], one gets a good description of measured data by calculating only the tunneling contribution below this crossover temperature and only the classical relaxation above this temperature.

Figure 7 shows a comparison of calculated and measured data for three glasses. The soft-potential parameters for this calculation were taken from Table 1, in order to see whether the parameters determined from the heat capacity and the thermal conductivity can also describe the sound absorption.

Figure 7 shows that this is indeed the case. The rise to the plateau at very low temperature is determined by the value of A_t in Eq. (22), which in its turn is determined by Λ_l and Λ_t. To calculate them from W, P_s, and C in Table 1, one assumes $C_l = C_t = C$ and uses Eq. (24).

While the rise to the plateau is determined exclusively by the bilinear coupling, the height of the plateau is determined by C.

Finally, the onset of the relaxation rise is again given by W. As pointed out in Ref. [9], the agreement of the two W values, one determined from the crossover from tunneling to vibration in thermal properties and the other from the crossover from tunneling to relaxation in acoustic properties, is another important achievement of the soft-potential model.

Within the relaxation regime, the measured sound wave damping does not follow the $T^{3/4}$ rise up to very high temperatures, but drops below, leading to a peak around 50 to 100 K. This shows that the soft-potential assumption of a constant $P(D_1, D_2)$ begins to fail toward higher barriers. In fact, in B_2O_3 and polystyrene, one already notes a slope T^α with an α slightly smaller than $3/4$.

This tendency is even stronger in polymethylmethacrylate,[57] where the relaxation rise disappears completely, and in selenium[58] with an α of about 0.2. Another example[52,59] is the metallic glass PdSiCu in Figure 8, with $\alpha = 0.05$.

Though the slope of the relaxation rise is small, the crossover itself is clearly identifiable and allows determining the value of W/k_B at 3.2 K This value can be compared to the numerical result scaled[36] to PdSiCu of $W/k_B = 3.85$ K, in good agreement to the value found here.

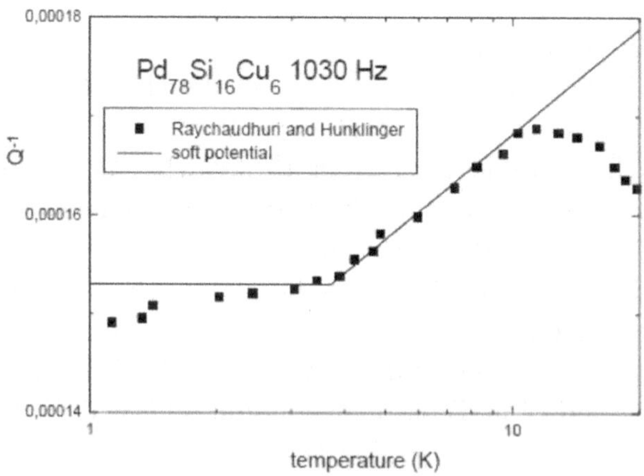

Fig. 8. The tunneling plateau and the crossover to the scattering from thermally activated relaxation processes in the sound wave absorption of PdSiCu. The data[52] show a crossover corresponding to $W/k_B = 3.2$ K.

3.5. *Sound absorption at high frequency*

Looking at Eq. (29) for the sound wave absorption by classical relaxation and Eq. (28) for the sound wave absorption by resonant scattering from vibrational soft modes, one realizes that one can measure them separately. If one wants to see only thermally activated relaxations in double-well potentials, one chooses a low frequency, because then the influence of the vibrations is negligible. The opposite is true for high frequencies, though there one has to vary the temperature as well, to check whether the absorption is temperature independent as one expects for the scattering from harmonic vibrations.

On the high-frequency side, X-ray Brillouin scattering offers the possibility to measure the sound absorption at THz frequencies. But, it took some time before the technique developed a good enough resolution to see the soft-potential Q^4-term in the sound absorption (Q wavevector of the longitudinal Brillouin sound wave).

The first glass in which the Q^4-dependence could be demonstrated was densified vitreous silica,[59] the second one lithium borate (see Figure 9),[60] both substances with a high boson peak frequency. Unfortunately, there is no soft-potential fit to the low-temperature properties in both cases, so one cannot check the soft-potential model quantitatively.

This is different for the third example of a Q^4-term in X-ray Brillouin scattering, glycerol,[61] where the soft-potential parameters were determined from the heat capacity and the thermal conductivity[50] and are listed in Table 1.

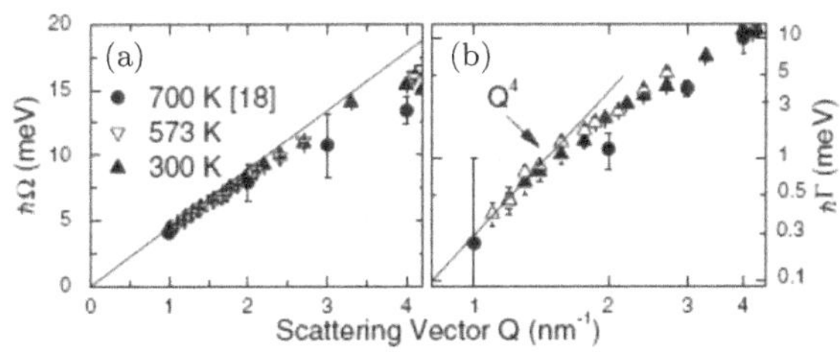

Fig. 9. Evidence for a Q^4-dependence in X-ray Brillouin scattering of a lithium borate glass.[60]

For the X-ray Brillouin data, the soft-potential model[9] predicts a damping Γ proportional to ω^4, which results from the rise of the vibrational density of states of the soft modes

$$\Gamma(\omega) = v_l l^{-1}_{res,vib} = \frac{\pi\omega C_l}{8}\left(\frac{\hbar\omega}{W}\right)^3 = \frac{\omega^4}{\pi\omega_{IR}^3} \tag{32}$$

where $\Gamma(\omega)$ is the half width at half maximum of the Brillouin line and ω_{IR} is the frequency of the Ioffe–Regel limit, where the mean free path equals half the wavelength of the longitudinal sound wave. In Ref. [60], it was argued that ω_{IR} is close to the boson peak frequency ω_b in many glasses (see also Chapter 7).

Assuming again $C_l = C$, one calculates $\hbar\omega_{IR} = 6.0$ meV for glycerol. The X-ray Brillouin measurements[61] show a clearly defined Q^4-dependence of Γ, which extrapolates to an Ioffe–Regel limit of 6.3 meV, in close agreement with the soft-potential value, and not too far from the boson peak[62] at 4.1 meV.

4. Limitations and Extensions

4.1. *Boson peak*

As has been demonstrated in the preceding sections, the soft-potential postulate of a constant density $P(D_1, D_2)$around the purely quartic potential leads to a surprisingly satisfactory description of the low-temperature glass anomalies.

But, the postulate has its limits. For example, the density of soft vibrations is not proportional to the fourth power of the frequency without end. Instead, the vibrational density of states $g(\omega)$ shows the boson peak, a peak in $g(\omega)/\omega^2$, which appears as a peak in the heat capacity $C_p(T)/T^3$ (see Figure 3) as well as in inelastic neutron, X-ray, or Raman scattering.

A first attempt to extend the soft-potential model to an understanding of the boson peak was the assumption of a thermal limitation[6,7] of the D_1-distribution at the glass temperature. The scheme yields the right order of magnitude for the boson peak frequency, but neglects the effect of the interaction at the boson peak frequency.

A theoretically better founded approach to understand both the occurrence of the soft modes around the quartic potential and the boson peak in a unified theoretical picture has been developed by Gurevich, Parshin, and Schober.[11–13] They consider the vibrational instability of the spectrum of weakly interacting quasi-local harmonic modes, limited by the anharmonicity. The theory supplies both the soft-potential modes around the restoring force zero and the boson peak. This theory has been used to describe the data in Figure 3.

The existence of an unstable core in the boson peak modes can be demonstrated convincingly from the temperature dependence of the boson peak in some glasses. The best example is vitreous silica, where the core motion is a coupled rotation of corner-connected SiO_4-tetrahedra, and the positive contribution from the outside is due to the elastic constants. Figure 10 shows the temperature dependence of the Boson peak frequency ω_b in silica,[43,63] together with a fit which assumes that only the elastic constants depend on temperature, while the core retains a temperature-independent negative restoring force constant. The temperature dependence of the elastic constants is taken from Brillouin measurements.[64,65]

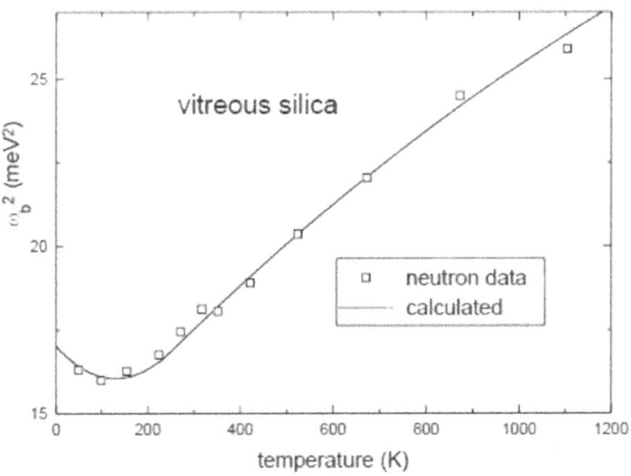

Fig. 10. Temperature dependence of the boson peak in vitreous silica,[43,63] explained in terms of an unstable core of coupled SiO_4-tetrahedra surrounded by a stabilizing elastic medium.

For the theoretical fit, one assumes the mode potential

$$V(A) = -v_{\text{int}}\frac{A^2}{2} + v_{\text{ext}}\frac{A^2}{2} \tag{33}$$

in the normal coordinate A of the mode, with a negative internal and a positive external restoring force constant. The external force constant v_{ext} is supposed to be proportional to the shear modulus G. This leads to a harmonic vibration with the boson peak frequency

$$\omega_b^2 = (v_{\text{ext}} - v_{\text{int}}) \tag{34}$$

so that one gets the relation

$$\frac{\omega_b^2(T)}{\omega_b^2(T_0)} = \frac{v_{\text{ext}}(T) - v_4}{v_{\text{ext}}(T_0) - v_4} = \frac{G(T) - G_4}{G(T_0) - G_4} \tag{35}$$

where $G(T)$ is the temperature-dependent shear modulus, and $G_4 = v_{\text{int}}G/v_{\text{ext}}$. G_4 is a property of a local saddle point, independent of temperature because the structure of the glass is independent of temperature. The fit of the vitreous silica data in Figure 10 requires $G_4 = 25$ GPa, while the shear modulus G itself varies between 30.7 and 35 GPa.

This demonstrates clearly that the boson peak modes in silica have an unstable core, with a negative restoring force constant, which is 70–80% of the positive restoring forces from the elastic surroundings.

In vitreous silica, the boson peak frequency grows by 30% between 100 K and the glass temperature. In the same range, the density decreases by 0.25%. That makes it hard to believe that a lower density is the only reason for a lower boson peak frequency, as postulated on the basis of a comparison of quartz, cristobalite, silica, and densified silica[66] (see also Chapter 6). In these four substances, it happens to be approximately fulfilled, but the strong temperature dependence of the boson peak in the glass phase shows that it is certainly not the whole story. Also, one should be aware that quartz and cristobalite have a structural phase transformation[67,68] involving a soft mode which is again a coupled rotation of corner-connected SiO_4-tetrahedra, explaining why the lowest van-Hove singularity of these two crystals is exceptionally low.

The unstable core argument is also needed to explain why the boson peak frequency in the glass phase of two polycarbonates[69] decreases by a factor of two from 50 to 300 K, while the shear modulus decreases only by about thirty percent.[70,71] Obviously, the temperature dependence of the negative unstable core force constant is considerably weaker than the one of the elastic constants outside.

The concept of an unstable core of the boson peak modes, moving in phase with the movement of the surrounding stabilizing elastic medium, also explains the dynamic structure factor findings in vitreous silica and GeO_2,[72] in B_2O_3,[45] and in polybutadiene,[46] where one has more long-wavelength sound wave motion between zero frequency and the boson peak than the one expected from the Debye density of states. This additional sound wave signal stems from the elastic deformation outside the core of the quasi-localized modes.

It is clear that the unstable core concept fits better into the theoretical vibrational instability picture of Gurevich, Parshin, and Schober[11–13] than in the otherwise rather similar theoretical picture of Schirmacher[73] of elastic inhomogeneities, which is described in detail in Chapter 9.

In this context, it should be noted that in fragile glass formers the boson peak softens dramatically above the glass temperature,[74,75] a softening which is accompanied by a strong rise of the mean square displacement and an equally strong decrease of the shear modulus,[76] consistent with the creation of a larger number of unstable cores with increasing temperature.[29]

4.2. *Thermally activated relaxation*

The validity of the simplest form of the soft-potential model with a constant density $P(D_1, D_2)$ is not only limited at larger positive force constants D_2 but at larger negative values as well. This was already demonstrated in the discussion of the crossover from tunneling states to the classical Kramers-Arrhenius relaxation in Section 3.4, where in all investigated glasses the initial $T^{3/4}$ rise of the sound absorption dwindles to a peak and a following decrease.

There is a second empirical model for this temperature region which is the natural continuation of the soft-potential model, the Gilroy-Phillips model.[77,78]

The Gilroy-Phillips model is the proper choice if one wants to study what happens at higher barriers and at the glass transition. It describes the sound wave absorption in terms of a barrier density function $f(V)$, which can be adapted to the data of a given glass.[58] It is based on the same concept as the tunneling model and allows a continuous and flexible continuation of soft-potential fits toward higher barriers.

Figure 11 shows the application of the Gilroy-Phillips model to the polymer polymethylmethacrylate (PMMA) (data taken

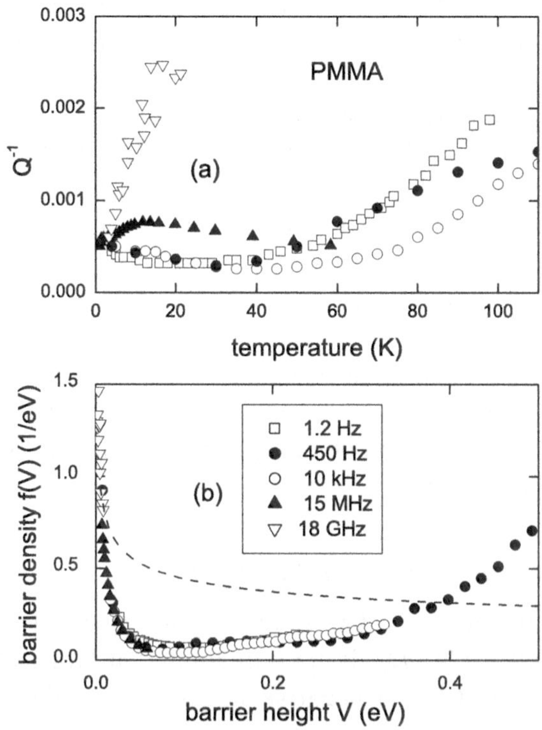

Fig. 11. (Top) Thermally activated low-temperature relaxation in polymethylmethacrylate for different frequencies.[79] (Bottom) The same data evaluated in the Gilroy-Phillips model. The dashed line shows the expectation of the soft-potential model,[78] which is only fulfilled for the lowest barriers.

from Ref. [79]). The lower part of the figure compares the evaluated barrier density $f(V)$ to the soft-potential expectation[78]

$$f(V) = \frac{2C}{W^{3/4}V^{1/4}} \tag{36}$$

The comparison shows that the low-temperature peak in the sound absorption of many glasses (in PMMA only weakly visible at 15 MHz) is not a real peak in the barrier density of the energy landscape, but rather the limit of the validity of the soft-potential model.

There are many successful applications of the Gilroy-Phillips model to experimental glass data, not only mechanical ones but also Raman[80] and dielectric data.[81]

This discussion of the appropriate Gilroy-Phillips extension toward classical relaxation in higher barriers terminates the present review of the soft-potential model, a model which is able to describe quantitatively a large number of different low-temperature experimental data with only the three parameters P_s, C, and W.

References

1. V. G. Karpov, M. I. Klinger, and F. N. Ignatiev. (1982). Atomic tunneling states and low-temperature anomalies of thermal properties in amorphous materials. *Solid State Commun.* **44**, 333–337.
2. V. G. Karpov and D. A. Parshin. (1985). The thermal conductivity of glasses at temperatures below the Debye temperature. *Sov. Phys. JETP.* **61**, 1308–1317.
3. M. A. Il'in, V. G. Karpov, and D. A. Parshin. (1987). Parameters of soft potentials in glasses. *Sov. Phys. JETP.* **65**, 165–174.
4. U. Buchenau, Yu. M. Galperin, V. L. Gurevich, and H. R. Schober. (1991). Anharmonic potentials and vibrational localization in glasses. *Phys. Rev. B.* **43**, 5039–5045.
5. U. Buchenau, Yu. M. Galperin, V. L. Gurevich, D. A. Parshin, M. A. Ramos, and H. R. Schober. (1992). Interaction of soft modes and sound waves in glasses. *Phys. Rev. B.* **46**, 2798–2808.
6. L. Gil, M. A. Ramos, A. Bringer, and U. Buchenau. (1993). Low temperature specific heat and thermal conductivity of glasses. *Phys. Rev. Lett.* **70**, 182–185.
7. M. A. Ramos, L. Gil, A. Bringer, and U. Buchenau. (1993). The density of tunneling and vibrational states of glasses within the soft-potential model. *Phys. Stat. Sol.*(a). **135**, 477.

8. D. A. Parshin. (1994). Model of soft atomic potentials and universality of low-temperature properties of glasses. *Phys. Solid State* **36**, 991–1024.
9. M. A. Ramos and U. Buchenau. (1997). Low-temperature thermal conductivity of glasses within the soft-potential model. *Phys. Rev. B.* **55**, 5749–5754.
10. M. A. Ramos and U. Buchenau. (1998). Beyond the standard tunneling model: The soft-potential model. In P. Esquinazi (ed.) *Tunneling Systems in Amorphous and Crystalline Solids*, pp. 528–569. Springer, New York (1998).
11. V. L. Gurevich, D. A. Parshin, and H. R. Schober. (2003). Anharmonicity, vibrational instability, and the Boson peak in glasses. *Phys. Rev. B.* **67**, 094203.
12. V. L. Gurevich, D. A. Parshin, and H. R. Schober. (2005). Pressure dependence of the Boson peak in glasses. *Phys. Rev. B.* **71**, 014209.
13. D. A. Parshin, H. R. Schober, and V. L. Gurevich. (2007). Vibrational Instability, two-level systems, and the Boson peak in glasses. *Phys. Rev. B.* **76**, 064206.
14. H. R. Schober and C. Oligschleger. (1996). Low frequency vibrations in a model glass. *Phys. Rev. B.* **53**, 11469–11480.
15. E. Lerner, G. Düring, and E. Bouchbinder. (2016). Statistics and properties of low-frequency vibrational modes in structural glasses. *Phys. Rev. Lett.* **117**, 035501.
16. P. H. Dederichs, C. Lehmann, H. R. Schober, A. Scholz, and R. Zeller. (1978). Lattice theory of point defects. *J. Nucl. Mat.* **69/70**, 176–199.
17. V. A. Luchnikov, N. N. Medvedev, Yu. I. Naberukhin, and H. R. Schober. (2000). Voronoi-Delaunay analysis of normal modes in a simple model glass. *Phys. Rev. B.* **62**, 3181–3189.
18. U. Buchenau and H. R. Schober. (2008). An atomic mechanism for the boson peak in metallic glasses. *Phil. Mag.* **88**, 3885–3990.
19. H. R. Schober, C. Oligschleger, and B. B. Laird. (1993). Low-frequency vibrations and relaxations in glasses. *J. Non-Cryst. Solids* **156–158**, 965–968.
20. C. Donati, J. F. Douglas, W. Kob, S. J. Plimpton, P. H. Poole, and S. C. Glotzer. (1998). Stringlike cooperative motion in a supercooled liquid. *Phys. Rev. Lett.* **80**, 2338–2341.
21. M. Shimada, M. Mizuno, M. Wyart, and A. Ikeda. (2018). Spatial structure of quasilocalized vibrations in nearly jammed solids. *Phys. Rev. E.* **98**, 060901.
22. U. Buchenau, Anharmonic soft modes in glasses, arXiv:1907.04848.

23. V. Gurarie and J. T. Chalker. (2003). Bosonic excitations in random media. *Phys. Rev. B.* **68**, 134207.
24. L. Gartner and E. Lerner. (2016). Nonlinear plastic modes in disordered solids. *Phys. Rev. E.* **93**, 011001(R).
25. L. Gartner and E. Lerner. (2016). Nonlinear modes disentangle glassy and Goldstone modes in structural glasses. *Sci. Post Phys.* **1**, 016.
26. S. Wijtmans and L. M. Manning. (2017). Disentangling defects and sound modes in disordered solids. *Soft Matter* **13**, 5649–5645.
27. G. Kapteijns, E. Bouchbinder, and E. Lerner. (2018). Universal nonphononic density of states in 2D, 3D and 4D glasses. *Phys. Rev. Lett.* **121**, 055501.
28. H. Mizuno and A. Ikeda. (2018). Phonon transport and vibrational excitations in amorphous solids. *Phys. Rev. E.* **98**, 062612.
29. L. Wang, A. Ninarello, P. Guan, L. Berthier, G. Szamel, and E. Flenner. (2019). Low-frequency vibrational modes of stable glasses. *Nat. Commun.* **10**, 26.
30. G. Kapteijns, D. Richard, and E. Lerner. (2020). Nonlinear quasilocalized excitations in glasses I. True representatives of soft spots. *Phys. Rev. E.* **101**, 032130.
31. H. Mizuno, H. Tong, A. Ikeda, and S. Mossa. (2020). Intermittent rearrangements accompanying thermal fluctuations distinguish glasses from crystals. *J. Chem. Phys.* **153**, 154501.
32. H. Mizuno, M. Shimada, and A. Ikeda. (2020). Anharmonic properties of vibrational excitations in amorphous solids. *Phys. Rev. Res.* **2**, 013215.
33. C. Rainone, E. Bouchbinder, and E. Lerner. (2020). Statistical mechanics of local force dipole responses in computer glasses. *J. Chem. Phys.* **152**, 194503.
34. D. Richard, K. Gonzalez-Lopez, G. Kapteijns, R. Pater, T. Vaknin, E. Bouchbinder, and E. Lerner. (2020). Universality of the nonphononic vibrational spectrum across different classes of computer glasses. *Phys. Rev. Lett.* **125**, 085502.
35. A. Moriel, Y. Lubomirsky, E. Lerner, and E. Bouchbinder. (2020). Extracting the properties of quasilocalized modes in computer glasses: Long-range continuum fields, contour integrals and boundary effects. *Phys. Rev. E.* **102**, 033008.
36. U. Buchenau, Strain field of soft modes in glasses. arXiv: 2010.10870.
37. A. Ninarello, L. Berthier, and D. Coslovich. (2017). Models and algorithms for the next generation of glass transition studies. *Phys. Rev. X.* **7**, 021039.

38. W. A Phillips. (1987). Two-level states in glasses. *Rep. Prog. Phys.* **50**, 1657–1708.
39. R. C. Zeller and R. O. Pohl. (1971). Thermal conductivity and specific heat of non-crystalline solids. *Phys. Rev. B.* **4**, 2029–2041.
40. P. Flubacher, A. J. Leadbetter, J. A. Morrison, and B. P. Stoicheff. (1959). The low-temperature heat capacity and the Raman and Brillouin spectra of vitreous silica. *J. Phys. Chem. Solids* **12**, 53–65.
41. H. v. Löhneysen, H. Rüsing, and W. Sander. (1985). Structural relaxation effects on the low-temperature properties of vitreous silica. *Z. Phys. B — Condens. Matter* **60**, 323–330.
42. X. Liu and H. v. Löhneysen. (1996). Specific-heat anomaly of amorphous solids at intermediate temperatures (1 to 30 K). *Europhys. Lett.* **33**, 617–622.
43. A. Wischnewski, U. Buchenau, A. J. Dianoux, W. A. Kamitakahara, and J. L. Zarestky. (1998). Sound wave scattering in silica. *Phys. Rev. B.* **57**, 2663–2666.
44. B. Hehlen, E. Courtens, R. Vacher, A. Yamanaka, M. Takaoka, and K. Inoue. (2000). Hyper-Raman scattering observation of the boson peak in vitreous silica. *Phys. Rev. Lett.* **84**, 5355–5358.
45. D. Engberg, A. Wischnewski, U. Buchenau, L. Börjesson, A. J. Dianoux, A. P. Sokolov, and L. M. Torell. (1999). Origin of the boson peak in a network glass B_2O_3. *Phys. Rev. B.* **59**, 4053–4057.
46. U. Buchenau, A. Wischnewski, D. Richter, and B. Frick. (1996). Is the fast process at the glass transition mainly due to long wavelength excitations? *Phys. Rev. Lett.* **77**, 4035–4038.
47. C. N. King, W. A. Phillips, and J. P. de Neufville. (1974). Low-temperature heat capacity of amorphous germanium. *Phys. Rev. Lett.* **32**, 538–541.
48. U. Buchenau, M. Prager, W. A. Kamitakahara, H. R. Shanks, and N. Nücker. (1988). Neutron scattering from low-frequency excitations in amorphous germanium. *Europhys. Lett.* **6**, 695–700.
49. D. G. Cahill. (1989). Lattice vibrations in glasses, Ph. D. thesis, Cornell Univ., USA.
50. C. Talon, Q. W. Zhou, M. A. Ramos, R. Villar, and S. Vieira. (2001). Low temperature specific heat and thermal conductivity of glycerol. *Phys. Rev. B.* **65**, 012203.
51. R. O. Pohl, X. Liu, and E. J. Thompson. (2002). Low-temperature thermal conductivity and acoustic attenuation in amorphous solids. *Rev. Mod. Phys.* **74**, 991–1013.
52. A. K. Raychaudhuri and S. Hunklinger. (1987). Low frequency elastic properties of glasses at low temperatures — Implications on the tunneling model. *Z. Phys. B — Condens. Matter* **57**, 113–125.

53. H. A. Kramers. (1940). Brownian motion in a field of force and the diffusion model of chemical reactions. *Physica* **7**, 284–304.
54. P. Neu and A. Würger. (1994). Mode coupling theory for the dynamics of tunneling systems in dielectrics. *Z. Phys. B — Condens. Matter* **95**, 385–395.
55. S. Rau, Ch. Enss, S. Hunklinger, P. Neu, and A. Würger. (1995). Acoustic properties of oxide glasses at low temperatures. *Phys. Rev. B.* **52**, 7179–7194.
56. K. A. Topp and D. A. Cahill. (1996). Elastic properties of several amorphous solids and disordered crystals below 100 K. *Z. Phys. B.* **101**, 235–245.
57. A. Nittke, M. Scherl, P. Esquinazi, W. Lorenz, J. Li, and F. Pobell. (1995). Low temperature heat release, sound velocity and attenuation, specific heat and thermal conductivity in polymers. *J. Low Temp. Phys.* **98**, 517–547.
58. U. Buchenau, G. D. D'Angelo, G. Carini, X. Liu, and M. A. Ramos, Sound absorption in glasses, arXiv:2012.10139.
59. B. Rufflé, M. Fouret, E. Courtens, R. Vacher, and G. Monaco. (2003). Observation of the onset of strong scattering on high frequency acoustic phonons in densified silica glass. *Phys. Rev. Lett.* **90**, 095502.
60. B. Rufflé, G. Guimbretière, E. Courtens, R. Vacher, and G. Monaco. (2006). Glass-specific behavior in the damping of acoustic-like vibrations. *Phys. Rev. Lett.* **96**, 045502.
61. G. Monaco and V. M. Giordano. (2009). Breakdown of the Debye approximation for the acoustic modes with nanometric wavelengths in glasses. *PNAS* **106**, 3659–3663.
62. J. Wuttke, W. Petry, G. Coddens, and F. Fujara. (1995). Fast dynamics of glass-forming glycerol, *Phys. Rev. E.* **52**, 4026–4034.
63. A. Wischnewski. (1998). Kohärent inelastische Neutronenstreuung an Gläsern, Ph. D. thesis, University of Düsseldorf, Germany.
64. R. Vacher, J. Pelous, F. Plicque, and A. Zarembowitch. (1981). Ultrasonic and Brillouin scattering study of the elastic properties of vitreous silica between 10 and 300 K. *J. Non-Cryst. Solids* **45**, 397–410.
65. J. A. Bucaro and H. D. Dardy. (1974). High-temperature Brillouin scattering in fused quartz. *J. Appl. Phys.* **45**, 5324–5329.
66. A. I. Chumakov, G. Monaco, A. Fontana, A. Bosak, R. P. Hermann, D. Bessas, B. Wehinger, W. A. Crichton, M. Krisch, R. Rüffer, G. Baldi, G. Carini Jr, G. Carini, G. D'Angelo, E. Gilioli, G. Tripodo, M. Zanatta, B. Winkler, V. Milman, K. Refson, M. Dove, N. Dubrovinskaia, L. Dubrovinsky,R. Keding, and Y. Z. Yue. (2014). The role of disorder in the dynamics and thermodynamics of glasses. *Phys. Rev. Lett.* **112**, 025502.

67. H. Grimm and B. Dorner. (1975). On the mechanism of the α-β phase transformation of quartz. *J. Phys. Chem. Solids* **36**, 407–413.
68. D. M. Hatch and S. Ghose. (1991). The α-β phase transformation in cristobalite, SiO_2. *Phys. Chem. Minerals* **17**, 554–562.
69. U. Buchenau, C. Schönfeld, D. Richter, T. Kanaya, K. Kaji, and R. Wehrmann. (1994). Neutron scattering study of the vibration-relaxation crossover in amorphous polycarbonates. *Phys. Rev. Lett.* **73**, 2344–2347.
70. G. Scieloszyk, M. T. Cruz, and G. L. Salinger. (1973). Thermal properties of dielectric solids below 4 K. I. Polycarbonate. *Cryogenics* **13**, 718–721.
71. G. D. Patterson. (1976). Brillouin scattering from amorphous bisphenol-A polycarbonate. *J. Polym. Sci.: Polym. Phys. Ed.* **14**, 741–748.
72. E. Fabiani, A. Fontana, and U. Buchenau. (2008). Neutron scattering study of the vibrations in vitreous silica and germania. *J. Chem. Phys.* **128**, 254507.
73. W. Schirmacher. (2006). Thermal conductivity of glassy materials and the "Boson peak". *Europhys. Lett.* **73**, 892–898.
74. W. A. Phillips, U. Buchenau, N. Nücker, A. J . Dianoux, and W. Petry. (1989). Dynamics of glassy and liquid selenium. *Phys. Rev. Lett* **63**, 2381–2384.
75. B. Frick and D. Richter. (1995). The microscopic basis of the glass transition in polymers from neutron scattering studies. *Science* **267**, 1939–1945.
76. U. Buchenau, R. Zorn, and M. A. Ramos. (2014). Probing cooperative liquid dynamics with the mean square displacement. *Phys. Rev. E.* **90**, 043312.
77. K. S. Gilroy and W. A. Phillips. (1981). An asymmetric double-well potential model for structural relaxation processes in amorphous materials. *Phil. Mag. B.* **43**, 735–746.
78. U. Buchenau. (2001). Mechanical relaxation in glasses and at the glass transition. *Phys. Rev. B.* **63**, 104203.
79. U. Buchenau, A. Wischnewski, R. Zorn, and N. Hadjichristides. (2002). Relaxations in the glass phase of silica and of polymethylmethacrylate. *Phil. Mag. B.* **82**, 209–222.
80. J. Wiedersich, S. V. Adichtchev, and E. Rössler. (2000). Spectral shape of relaxations in silica glass. *Phys. Rev. Lett.* **84**, 2718–2721.
81. C. Gainaru, R. Böhmer, R. Kahlau, and E. Rössler. (2010). Energy landscape in molecular glasses probed by high-resolution dielectric experiments. *Phys. Rev B.* **82**, 104205.

https://doi.org/10.1142/9781800612587_0009

Chapter 9

Heterogeneous Elasticity: The Tale of the Boson Peak

Walter Schirmacher[*,†] and Giancarlo Ruocco[†,‡]

[*] *Institut für Physik, Universität Mainz, Staudinger Weg 7, Mainz D-55099, Germany*
walter.schirmacher@uni-mainz.de
[†] *Fondazione Istituto Italiano di Tecnologia (IIT), Center for Life Nano Science, Viale Regina Elena 291, Roma I00161, Italy*
[‡] *Department of Physics, University of Rome 'La Sapienza', Piazzale Aldo Moro, 5, Rome I00185, Italy*

The vibrational anomalies of glasses, in particular the boson peak, are addressed from the standpoint of heterogeneous elasticity, namely the spatial fluctuations of elastic constants caused by the structural disorder of the amorphous materials.

In the first part of this review article, a mathematical analogy between diffusive motion in a disordered environment and a scalar simplification of vibrational motion under the same condition is employed. We demonstrate that the disorder-induced long-time tails of diffusion correspond to the Rayleigh scattering law in the vibrational system and that the crossover from normal to anomalous diffusion corresponds to the boson peak. The anomalous motion arises as soon as the disorder-induced self-energy exceeds the frequency-independent diffusivity/elasticity. For this model, a variational scheme is employed for deriving two mean-field theories of disorder: the self-consistent Born approximation (SCBA) and coherent potential approximation (CPA). The former applies if the fluctuations are weak and Gaussian, and the latter applies for stronger and non-Gaussian fluctuations.

In the second part, the vectorial theory of heterogeneous elasticity is presented and solved in SCBA and CPA, introduced for the scalar model. Both approaches predict and explain the boson peak and the associated anomalies, namely a dip in the acoustic phase velocity and a characteristic strong increase of the acoustic attenuation below the boson peak. Explicit expressions for the density of states and the inelastic Raman, neutron, and X-ray scattering laws are given. Recent conflicting ways of explaining the boson peak anomalies are discussed.

1. Introduction

The vibrational properties of disordered solids are quite different from those of crystals.[1,2] While the harmonic vibrational spectra of crystals are given by the dispersion relations obtained from the crystalline symmetry groups, those of amorphous solids, in particular glasses, exhibit anomalous, continuous spectra which have been a matter of debate for 60 years.[2–7]

The first evidence of something unusual in the vibrational spectra of glasses came from Raman scattering.[8–13] In the THz or 100 wavenumber regime, one observed a broad maximum, where usually either no intensity or very sharp peaks due to low-lying optical modes were observed. Because this maximum obeyed the frequency/temperature dependence of the Bose function $n(\omega) + 1 = [1 - exp\{-\hbar\omega/k_BT\}]^{-1}$ ($\omega/2\pi$ is the frequency and k_BT is the Boltzmann constant times the temperature), one called this maximum "boson peak".[14] Here we note that because the Raman scattering intensity is proportional to $n(\omega) + 1$ times the Raman spectral function $\chi''_R(\omega)$,[13,15,16] the latter must be temperature independent if the entire temperature dependence comes from the Bose function. This points to a harmonic origin of the boson peak and questions all interpretations in terms of anharmonic interactions.

A maximum in the frequency range $\sim$5 meV $\sim$1 THz was observed in inelastic coherent neutron scattering data of several glasses.[17] This maximum appeared to be related to an excess of the vibrational density of states (DOS) $g(\omega)$ with respect to Debye's $g(\omega) \propto \omega^2$ law which appears as a peak in the reduced DOS $g(\omega)/\omega^2$. This was confirmed experimentally by means of inelastic incoherent neutron[18] and nuclear scattering,[19] which both measures directly the DOS. Furthermore, the boson peak turned out to be related to a maximum in the "reduced" specific heat $C(T)/T^3$ in the 10K regime[18,20] which is also called "boson peak".

Right at the boson peak temperature, the thermal conductivity of the same glasses exhibits a characteristic shoulder or dip in its temperature dependence[21] which turned out to be an "upside-down boson peak".[22]

These are not the only low-temperature thermal anomalies of glasses. Below the boson peak, in the ~1 K regime, the specific heat does not show Debye's T^3 law, but varies approximately linearly with temperature. The thermal conductivity varies quadratically in this regime. These findings have been attributed to the existence of bistable structural arrangements, which allow for tunneling between the two positions, giving rise to a tunnel splitting (two-level systems (TLS), tunneling systems).[23–27] If the energy separations of the TLS are assumed to have a broad distribution, the 1 K anomalies can be explained. Independent evidence for the existence of the TLS comes from ultrasonic and nuclear magnetic resonance data.[28,29]

Inelastic X-ray scattering contributed more anomalous features in the boson peak regime.[30] In particular, the group velocity of longitudinal sound (dispersion of the Brillouin line position) was found to exhibit a minimum at the wavenumber corresponding to the boson peak, and the sound attenuation (width of the Brillouin line) was found to increase strongly with frequency in the boson peak regime (Rayleigh scattering).[31–36] These anomalies were confirmed in computer simulations.[37,38] We call all these features the boson peak related anomalies.

The theoretical interpretation of the boson peak has a very diverse history and gave rise to several controversies.[3] In one line of argumentation for explaining the boson peak related anomalies, the authors tried to find a classical analogy of the tunneling model, the soft-potential model.[39–41] The bistable configurations, which can be described by an anharmonic double-well potential with a rather shallow barrier between the wells, were assumed to have a distribution of their characteristic parameters, in particular the second-order coefficient. It was found that, in general, the statistics of the vibrational excitations of such defect potentials would give rise to a constant density of eigenvalues $g(\omega^2)$, which leads to a DOS $g(\omega) \propto \omega$. On the other hand, so the argument, at lower frequencies (below a crossover frequency ω_c), the balance between unstable configurations with negative second-order coefficients and the anharmonic fourth-order ones gives rise to a defect DOS $\propto \omega^4$. Therefore, the reduced DOS $g(\omega)/\omega$ increases $\propto \omega^2$ below ω_c and decreases $\propto \omega^{-1}$ above ω_c. In between, i.e., near ω_c is then the boson peak.[39,42]

Arguments that the boson peak, by its temperature characteristics, is not an anharmonic phenomenon were met by the authors of the soft-potential model by noting that the anharmonic interaction only acts in producing the soft configurations in the quenching process.[43] In the quenched state, they are supposed to act like local harmonic oscillators, similar to heavy-mass atoms coupled to the acoustic harmonic degrees of freedom.[44–47]

Another attempt to explain the boson peak anomaly was the phonon–fracton model.[48–52] It was postulated that disordered solids exhibit a certain degree of fractal structure. A fractal is a self-similar structure,[53,54] which has a non-integer dimensionality D_0, which is smaller than the embedding dimensionality d. Real fractals like sponges or trees have smallest and largest scales in which the self-similarity holds. The smallest scale is, e.g., the smallest pore diameter of a sponge, and the largest is the correlation length ξ. For scales larger than ξ, the object looks like an ordinary material in which the mass scales as L^d, where L is the size. For scales smaller than ξ, the mass scales as L^{D_0}. Alexander and Orbach[48] have shown that the vibrational degrees of freedom scale with a fractal dimensionality $d_s < D_0$ (spectral dimensionality) and that the DOS of such an object obeys a Debye law below $\omega_\xi = 2\pi v/\xi$, where v is the sound velocity. Above ω_ξ, the DOS behaves as $g(\omega) \propto \omega^{d_s-1}$. They found that in all fractal structures they investigated, $d_s \approx 4/3$. The specific model employed by the phonon–fracton supporters was a percolating lattice, i.e., a cubic lattice in which a certain percentage of bonds $1-p$ (carrying nearest-neighbor force constants) was missing. If the bond concentration p is larger than the critical concentration p_c, which determines the connectedness of the structure, a finite correlation length $\xi \propto (p - p_c)^\nu$ exists (ν is the order-parameter exponent[52]). Calculations using the coherent potential approximation (CPA)[55–57] showed that in between the Debye and the fracton regime, an enhancement over the Debye $g(\omega) \propto \omega^2$ law was present.[49,50] This was, for the time being, a satisfactory explanation of the boson peak. However, numerical simulations of the percolation phonon–fracton model[52] showed that the phonon–fracton crossover in the DOS of this model occurs very smooth without any excess over the Debye law. Obviously, the excess in the calculations[49,50] had been an artifact of the CPA. Another argument against the phonon–fracton model as candidate for explaining the boson peak anomalies is that apart from aerogels,[58,59] glasses do not show any self-similarity, which should show up (but does not) as an enhanced small-angle scattering in neutron or X-ray diffraction data.

In other articles reflecting on the vibrational anomalies of glasses, the boson peak and the anomalous shoulder in the temperature dependence of the thermal conductivity were also considered to be related to Anderson localization of sound.[51,60,61] It was observed that near the boson peak frequency, the mean free path of the acoustic excitations is of the order of the sound wavelength. According to a rule, coined by Ioffe and Regel in their survey of electronic conduction in semiconductors,[62] the notion of a mean free path, which implies a wave, which is occasionally scattered by an inhomogeneity, breaks down once the mean free path becomes equal to the wavelength (Ioffe–Regel limit). Mott[63] conjectured that Anderson localization occurs for electrons near the Ioffe–Regel energy. For phonon, it was assumed[61,64] that the crossover from extended to Anderson-localized states would take place near the Ioffe–Regel frequency and that the boson peak would mark the onset of localized states. In particular, it was thought that the presence of localized states would cause the dip in the thermal conductivity.[61] Later, more detailed theoretical investigations showed, however, that the Anderson transition in realistic solids does not occur near the Ioffe–Regel crossover but in a much higher frequency range near the Debye frequency.[65–71]

Later, many researchers were interested in whether the boson peak frequency coincides exactly with the Ioffe–Regel frequency given by the implicit relation $\omega_{IR} = \pi\Gamma(\omega_{IR})$, where $\Gamma(\omega)$ is the sound attenuation coefficient.[72,73] $\Gamma(\omega)$ and hence ω_{IR} is, of course, different for longitudinal and transverse sound waves. It appeared that in materials governed by hard sphere-like potentials, the boson peak coincides with the transverse Ioffe–Regel frequency,[38,74] whereas in network glasses, with the longitudinal one.[75–78]

In 1991, Schirmacher and Wagener[79] exploited the mathematical analogy between a single-particle random walk and harmonic phonons[80] using an off-lattice version of the CPA. They demonstrated that the crossover from a frequency-independent conductivity/diffusivity to a frequency-dependent one[81,82] corresponds to the onset of the frequency dependence of the complex sound velocity. The latter was found to lead to a boson peak in the vibrational DOS.

Because understanding this analogy is essential for grasping the essence of the BP-related anomalies, we devote a whole section of the present review to this analogy.

However, because the CPA in the case of the percolating lattice predicted a boson peak[49,50] which did not exist in the simulation of the same system,[52] a check was needed whether the resulting boson peak enhancement of the DOS was not an artifact of the CPA like in the phonon–fracton model. Therefore, in Ref. [68], the lattice version of the CPA[83–85] was compared to a numerical calculation for a cubic lattice with fluctuating nearest-neighbor force constants. Both simulation and CPA calculation showed a boson peak, and good agreement between CPA and the numerical spectrum was found. So it was demonstrated that effective-medium calculations are reliable for investigating the influence of disorder on the harmonic spectrum of a model solid. The breakdown of the CPA in the case of the phonon–fracton model was obviously due to the critical fluctuations in this model which are not generic for disordered solids.

The boson peak in the model calculations of Ref. [68] was identified to be caused by very small positive and negative force constants, and being a precursor of an instability, which happens for stronger disorder.[69,70]

With the help of the numerical calculation in Ref. [68], it could also be decided whether the vibrational states near and above the boson peak were localized or extended. This was achieved by means of the statistics of the distances between the eigenvalues. Near and above the boson peak, they showed the so-called Gaussian orthogonal ensemble (GOE) statistics of random matrix theory,[86,87] which proves that the corresponding states are delocalized. At high frequencies, near the upper band edge, a transition to Poissonian statistics was observed, which is evidence for a delocalization-localization transition in this regime, in agreement with earlier estimates.[65,66] On the other hand, the GOE statistics is a sign for the so-called level repulsion showing that each eigenvalue is non-degenerate due to the absence of symmetries in the disordered system. As the generic spectrum of random matrices is not a Debye spectrum (which is highly degenerate), the boson peak marks the transition from a Debye to a random matrix-type spectrum.[6]

The spectra calculated in Ref. [68] for a cubic lattice with very small disorder exhibited the usual Van Hove singularities which appear as a result of the leveling-off of the crystalline phonon dispersions $\omega(k)$ at the Brillouin zone boundary.[88] With increasing disorder, the data showed that the sharp Van Hove peak became rounder and is shifted downwards and gradually transformed to the low-frequency boson peak. This led Taraskin *et al.*[89] to the conclusion that the boson peak is just a crystal-like Van Hove peak modified by disorder. This appeared as a rather unexpected conclusion because a Van Hove singularity is a typical signature of a crystalline structure with long-range order and was not known to exist in glasses. However, until now, the boson peak explanation as a glassy version of a Van Hove singularity is still considered to be an alternative to the disorder explanation.[90–92] We further comment on this in Section 4.

Quite recently, in an experimental study of a macroscopic disordered model glass, it was shown that the disorder-induced maximum of the reduced DOS and that induced by the Van Hove mechanism are, in fact, two different phenomena.[93]

The random matrix aspect of the disorder-vibration problem was elaborated further in the literature,[94–104] in particular, by means of the euclidean random matrix theory.[96–103]

As in Ref. [68], a disorder-induced boson peak was found and shown to be the precursor of instability ("phonon-saddle transition"[100]).

A quite different and interesting approach to the boson peak anomaly was worked out[105] in the framework of the mode-coupling theory of the glass transition. This theory of glassy freezing in its original form[106, 107] describes the idealized glass transition as a dynamical transition towards a non-ergodic state leading to a frozen-in additional contribution to the static longitudinal susceptibility. This in turn leads to a characteristic hump in the density fluctuation spectrum of the idealized glass (proportional to the neutron scattering law $S(k, \omega)$) which was identified with the boson peak anomaly found in neutron scattering experiments. Interestingly, this theory already predicted the characteristic dip in the longitudinal group velocity, which was found later to be associated with the boson peak, as mentioned earlier.[31, 35, 37, 38]

We feel that an important step in the understanding of the boson peak was achieved by working out heterogeneous elasticity theory.[22] In this phenomenological theory, it is assumed that the shear modulus in ordinary elasticity theory[108] is assumed to exhibit spatial fluctuations. The resulting stochastic equations were solved by field theoretical techniques,[65, 109] resulting in a mean field theory called self-consistent Born approximation (SCBA). The SCBA is obtained from assuming a Gaussian distribution of the elasticity fluctuations and that the relative width of the Gaussian is a small parameter ("disorder parameter" γ). Again, if the disorder parameter becomes larger than a critical one, an instability occurs which is due to too many regions with negative elastic constants.

Shortly before the heterogeneous elasticity theory had been worked out, a series of papers appeared in which molecular dynamics simulations of glasses were investigated for their elastic and vibrational properties.[110–113] It turned out that, by applying external forces, indeed heterogeneous shear deformations are present. The authors showed that in regions with strong deformations, the shear response is highly non-affine, i.e., the displacements do not follow the direction of the applied stress. In the non-affine regions, the local shear moduli were found to be very small and even negative, a finding which was observed also in other simulations.[38, 114–116] This nicely confirmed the model assumption of heterogeneous elasticity theory. In Ref. [38], a direct comparison between the elasticity fluctuations in a simulated glass and the theory of Ref. [22] was made and good agreement was found. We shall comment on this article in more detail as follows.

Recently the non-affine reduction of the macroscopic shear modulus was successfully used as disorder descriptor of glasses.[117] The authors introduced quantitative measures of this inversion symmetry breaking and found a unique correlation with the height of the boson peak. This was also found in the quoted investigation of a macroscopic model of a disordered solid in which soft spots with strong inversion symmetry breaking were shown to contribute predominantly to the boson peak.[93]

An important aspect of the boson peak appeared when it was shown[118] with the help of heterogeneous elasticity theory that the excess DOS $\Delta g(\omega) = g(\omega) - g_D(\omega)$ with respect to the Debye DOS $g_D(\omega)$ is proportional to the sound attenuation in the boson peak frequency range. Heterogeneous elasticity provides an expression for the disorder-induced sound attenuation as the imaginary part of frequency-dependent elastic coefficients (see Section 1). These enter into the spectral functions, and the DOS enhancement is just produced by the sound attenuation, which increases rapidly in the boson peak frequency regime due to Rayleigh scattering.

The rest of this contribution is organized as follows: In Section 2, the mathematical analogy between diffusion and scalar elasticity is investigated in detail and two mean field or effective medium theories, the SCBA and the Coherent Potential Approximation (CPA), for this model are introduced and solved. A pedagogical derivation, which is more simple than the original field theoretical one, is presented. We show that the boson peak related anomalies are the analogy of the crossover from a frequency-independent diffusivity/conductivity to a frequency-dependent one.

In Section 3, the full vectorial heterogeneous elasticity theory is presented and solved in SCBA and CPA. The salient features of the boson peak related anomalies of glasses are discussed with the help of these mean field theories. In Section 4, we discuss recent conflicting theories of these anomalies.

2. "Scalar Elasticity" and Diffusion-vibration Analogy

2.1. *Diffusion-vibration analogy*

A simplified version of heterogeneous elasticity theory, which proved to be helpful in understanding the spectral properties of disordered solids,[46, 79, 100, 119–121] is represented by a scalar wave equation with a spatially fluctuating elastic constant $K(\mathbf{r}) \equiv \rho\widetilde{K}(\mathbf{r})$ (ρ is the mass density)

$$\frac{\partial^2}{\partial t^2}u(\mathbf{r},t) = \nabla\widetilde{K}(\mathbf{r})\nabla u(\mathbf{r},t). \tag{1}$$

The elastic coefficient $\widetilde{K}(\mathbf{r})$, which is the square of the local sound velocity $v(\mathbf{r})$, may be subdivided into an average elasticity $\widetilde{K}_0 = \langle\widetilde{K}\rangle$ and deviations $\Delta\widetilde{K}(\mathbf{r})$

$$\widetilde{K}(\mathbf{r}) = v^2(\mathbf{r}) = \widetilde{K}_0 + \Delta\widetilde{K}(\mathbf{r}). \tag{2}$$

If we replace the second time derivative in Eq. (1) by a first one, we arrive at a diffusion equation, which describes the random walk of a particle, which encounters a spatially varying diffusivity, e.g., due to a spatially varying activation energy

$$\begin{aligned}\frac{\partial}{\partial t}n(\mathbf{r},t) &= \nabla D(\mathbf{r})\nabla n(\mathbf{r},t)\\ &= \nabla\Big[D_0 + \Delta D(\mathbf{r})\Big]\nabla n(\mathbf{r},t),\end{aligned} \tag{3}$$

with D_0, again, denoting the average diffusivity and $\Delta D(\mathbf{r})$ the fluctuation. $n(\mathbf{r},t)$ is the probability density for finding the particle within a volume element around $\mathbf{r}$ at time t. Green's function of Eq. (1) in frequency space obeys the equation

$$\begin{aligned}\Big(-z^2 - \nabla\widetilde{K}(\mathbf{r})\nabla\Big)G(\mathbf{r},\mathbf{r}') &= \Big(-z^2 - \nabla[\widetilde{K}_0 + \Delta\widetilde{K}(\mathbf{r})]\nabla\Big)G(\mathbf{r},\mathbf{r}')\\ &\equiv A[z,\mathbf{r},\widetilde{K}(\mathbf{r})]G(\mathbf{r},\mathbf{r}') = \delta(\mathbf{r}-\mathbf{r}')\end{aligned} \tag{4}$$

with $z = \omega + i\epsilon, \epsilon \to +0$. The operator $A[z, \mathbf{r}, \widetilde{K}(\mathbf{r})]$ is the operator inverse of Green's function. On the other hand, Green's function corresponding to the heterogeneous diffusion equation (3) in frequency space is

$$s - \nabla D(\mathbf{r}) \nabla G(\mathbf{r}, \mathbf{r}') = \delta(\mathbf{r} - \mathbf{r}') \tag{5}$$

with $s = i\omega + \epsilon$.

So all calculations done for the scalar vibration problem (4) can be taken over for the diffusion problem (5) provided we identify $D \leftrightarrow \widetilde{K}$, $s \leftrightarrow -z^2$, or $i\omega \leftrightarrow -\omega^2$.

In the next three subsections, we shall demonstrate by means of different approximation schemes (Born approximation, self-consistent Born approximation, and coherent potential approximation) that the quenched glassy disorder induces a characteristic frequency dependence to the macroscopic diffusivity/elasticity. So, these approximation schemes act as a coarse-graining scheme which converts spatial fluctuations to frequency dependence.

In this context, we may distinguish between three characteristic scales: the microscopic scale, the mesoscopic scale, and the macroscopic one. The microscopic scale is the molecular one and may be described by microscopic quantum or classical equations of motions. The mesoscopic scale is a scale of 5 or 6 atomic or molecular diameter. This is the minimal scale at which one may define local diffusivities or elastic constants[38, 114, 115, 121, 122] which exhibit spatial fluctuations in structurally disordered materials. The macroscopic scale is the experimental one (mm or cm) in which the macroscopic diffusivities or elastic coefficients are frequency dependent.

2.2. *Low-frequency limit: Born approximation, Rayleigh scattering, and long-time tails*

The solution of Eqs. (4) or (5) without fluctuations is given in **k** space by (**k** is the wave vector corresponding to $\mathbf{r} - \mathbf{r}' \equiv \tilde{\mathbf{r}}$)

$$G_0(k, z) = \frac{1}{-z^2 + \widetilde{K}_0 k^2}. \tag{6}$$

The disorder-averaged full Green's function should also only depend on $|\tilde{\mathbf{r}}|$ only and, therefore, may be represented as[120]

$$\langle G(z)\rangle_{\mathbf{k}} = \mathsf{G}(\mathbf{k},z) = \frac{1}{-z^2 + k^2(\widetilde{K}_0 - \Sigma(z))} \equiv \frac{1}{-z^2 + k^2 Q(z)}, \tag{7}$$

where $\Sigma(z) = \Sigma'(\omega) + i\Sigma''(\omega)$ is the self-energy function which describes the influence of the fluctuations $\Delta\widetilde{K}(\mathbf{r})$ or $\Delta\widetilde{D}(\mathbf{r})$. Here, we have defined a frequency-dependent elasticity $Q(z) = \widetilde{K}_0 - \Sigma(z)$ corresponding to a frequency-dependent diffusivity $D(s) = D_0 - \Sigma(s)$. The former may be identified with the square of a frequency-dependent sound velocity $v(z)$, i.e., $Q(z) = v(z)^2$.

To the lowest order in the fluctuations, one obtains by straightforward perturbation theory[120] the Born approximation

$$\Sigma(z) = \gamma \frac{1}{V} \sum_{\mathbf{k}} k^2 G_0(\mathbf{k}, z), \tag{8}$$

where $\sum_{\mathbf{k}} \equiv \frac{V}{(2\pi)^3} \int d^3\mathbf{k}$, V is the sample volume, and

$$\gamma = \langle \Delta\widetilde{K}^2 \rangle V_c\,, \tag{9}$$

i.e., the variance of $\widetilde{K}(\mathbf{r})$ times a coarse-graining volume V_c, which serves to calculate the local elastic coefficient.[38, 120, 122]

If one imposes an upper cutoff $k_{\max}$ in the wavenumber integration, the integral in Eq. (8) can be done exactly. For small frequencies, we obtain

$$\Delta\Sigma(s) = \Sigma(z) - \Sigma(0) \propto s^{3/2}. \tag{10}$$

For the vibrational problem, $s^{3/2} \to i\omega^3$.

We now show that this leads to Rayleigh's ω^4 scattering law[123, 124]:

We may be interested in the wave intensity given by the modulus of Eq. (7)

$$|G(\tilde{\mathbf{r}}, z)|^2 = \left(\frac{1}{4\pi\widetilde{K}_0}\right)^2 e^{-\tilde{r}/\ell(\omega)} \tag{11}$$

with the mean free path

$$\frac{1}{\ell(\omega)} = 2\mathrm{Im}\left\{\frac{\omega}{v(\omega)}\right\}$$
$$\approx \frac{\Sigma''(\omega)\omega}{v_0^3} = \frac{\gamma}{12\pi}\left(\frac{\omega}{v_0}\right)^4, \tag{12}$$

where $v_0 = \sqrt{\widetilde{K}_0}$.

Equation (12) constitutes the *Rayleigh scattering law.*[123,124] It holds for harmonic excitations in the presence of quenched disorder[103,120] provided the disorder fluctuations do not exhibit long-range order.[120,125,126]

Rayleigh scattering in glasses is usually obscured by anharmonic sound attenuation which prevails in the sub-THz frequency range. It has been observed experimentally in the THz regime in some glasses[31,35] as well as in computer simulations.[37,38]

In the diffusion problem, $D(s)$ is the frequency-dependent diffusivity which can be shown[127] to be the Laplace transform of the velocity autocorrelation function $Z(t)$ of the moving particle $D(s) = \int_0^\infty dte^{-st}Z(t)$.

We apply the Tauberian theorem[128]

$$\lim_{s\to 0} D(s) \propto s^{-\rho} \quad \Leftrightarrow \quad \lim_{t\to\infty} D(t) \propto t^{\rho-1} \tag{13}$$

from which we conclude

$$\lim_{t\to\infty} Z(t) \propto t^{-5/2}, \tag{14}$$

a behavior well known for particles performing a random walk in a quenched-disordered environment.[129,130]

On the other hand, by the Nernst–Einstein relation

$$\sigma(s) = \sigma'(\omega) + i\sigma''(\omega) = \frac{ne^2}{k_B T}, \tag{15}$$

the frequency-dependent diffusivity is related to the dynamic conductivity, the real part of which $\sigma'(\omega)$ is the alternate current (AC) conductivity. Therefore, one expects a non-analytic low-frequency dependence of the AC conductivity increment $\sigma'(\omega) - \sigma(0) \propto \omega^{3/2}$.

Indeed, such a behavior has been observed in amorphous semiconductors.[131] We come back to this in the subsection on strong disorder.

2.3. *Weak disorder and the SCBA*

A well-known characteristic of the AC conductivity in disordered materials is that beyond a characteristic frequency ω^*, it starts to increase with frequency, in many cases with a characteristic power law $\sigma'(\omega) \propto \omega^\alpha$, where α is smaller than 1 and takes values around 0.8.[81,82,132]

For the random walker in the disordered environment, this means that the mean square distance walked does not increase linearly with time but sublinearly with exponent $1-\alpha$. Such a behavior has been termed as anomalous diffusion. So, the crossover at ω^* corresponds to a transition from anomalous diffusion for times $t < 1/\omega^*$ to normal diffusion for $t > 1/\omega^*$.

As pointed out in Refs. [79, 121], the crossover at ω^* — if transformed from the diffusion to the scalar-vibrational system — corresponds to the boson peak. In other words: it corresponds to the begin of the frequency dependence of $\widetilde{K}(\omega) = v(\omega^2)$. By the Kramers–Kronig correspondence, this implies the onset of an imaginary part of $\widetilde{K}(\omega)$ which becomes of the order of its real part.

A minimal theory for the boson peak, in fact, can be obtained by the *self-consistent* version of the Born approximation. It is obtained by replacing the bare Green's function in Eq. (8) by the full Green's function:

$$\Sigma(z) = \gamma \frac{1}{V} \sum_{|\mathbf{k}| \leq k_{\max}} k^2 \mathsf{G}(\mathbf{k}, z) \tag{16}$$

with $\mathsf{G}(k, z)$ given by Eq. (7). For the ultraviolet cutoff $k_{\max}$, one should take the inverse of the length scale which is the diameter of the coarse-graining volume V_c used to define the local elastic constant/local diffusivity.[120] On the other hand, this length scale should be of the order of the correlation length ξ of the elasticity/diffusivity fluctuations.[125] In this work, we treat these two length scales as being the same. We, therefore, call the cutoff $k_{\max} = k_\xi$.

While the "derivation" of Eq. (16) is, of course, just an *ad hoc* replacement, a proper derivation is achieved by field theoretical techniques[46] in analogy to the derivation of the nonlinear sigma model for electrons,[109,133–135] classical sound waves,[65,125] and electromagnetic waves.[136,137]

A royal road for this derivation is to minimize the following (dimensionless, frequency-dependent) mean field free energy or effective action with respect to $\Sigma(z)$:

$$S_{\text{eff}}[\Sigma(z)] = S_{\text{med}}[\Sigma(z)] + S_{\text{SCBA}}[\Sigma(z)] \tag{17}$$

with

$$\begin{aligned} S_{\text{med}}[\Sigma(z)] &= S_{\text{med}}[Q(z)] \\ &= \operatorname{Tr}\ln\Big(A\big[z,\mathbf{r},\underbrace{\widetilde{K}_0-\Sigma(z)}_{Q(z)}\big]\Big) = \sum_{\mathbf{k}}\ln\Big(A\big[z,\mathbf{k},Q(z)\big]\Big) \end{aligned} \tag{18}$$

and

$$S_{\text{SCBA}}[\Sigma(z)] = \frac{1}{2}\frac{V}{\gamma}\Sigma(z)^2. \tag{19}$$

The first term, $S_{\text{med}}[\Sigma(z)]$, is the generalized free energy of the effective medium[138] which is a medium without disorder, in which the fluctuating force constants in the operator $A[z,\mathbf{r},\widetilde{K}(\mathbf{r})]$ are replaced by the homogeneous (but frequency-dependent) force constant $Q(z) = K_0 - \Sigma(z)$. The trace can, therefore, be calculated in $\mathbf{k}$ space. The Fourier transform of $A[z,\mathbf{r},Q(z)]$ is given by

$$A[z,\mathbf{k},Q(z)] = -z^2 + k^2 Q(z). \tag{20}$$

This is no more the operator inverse but just the ordinary inverse of the mean field Green's function

$$\begin{aligned} G(\mathbf{k},z) &= \frac{1}{A[z,\mathbf{k},Q(z)]} \\ &= \frac{1}{-z^2 + k^2 Q(z)}. \end{aligned} \tag{21}$$

The second term, $S_{\text{SCBA}}[\Sigma(z)]$, arises[22,65] from a Gaussian configuration average of the full Green's function with distribution density

$$P[\Delta\widetilde{K}(\mathbf{r})] = P_0 \exp\left\{-\frac{1}{2\gamma}\int d^3\mathbf{r}[\Delta\widetilde{K}(\mathbf{r})]^2\right\}. \tag{22}$$

It is easily verified that the SCBA (16) is obtained by minimizing S_{eff} of Eq. (18). This corresponds to the saddle-point approximation of the effective field theory derived for the appropriate stochastic Helmholtz equation.[22,65,109,119] The saddle-point approximation relies on the large prefactor of $S_{SCBA} \propto 1/\gamma$, i.e., the SCBA has its validity range for

(i) small disorder $\langle K^2\rangle/K_0^2 \ll 1$;
(ii) Gaussian disorder, Eq. (22).

As in our calculation of the Rayleigh scattering and the non-analyticity of $D(s)$, the SCBA transforms the microscopic wave equation with fluctuating elastic coefficient

$$\left[-z^2 - \nabla\widetilde{K}(\mathbf{r})\nabla\right]u(\mathbf{r},z) \tag{23}$$

into a macroscopic mean field wave equation

$$\left[-z^2 - \widetilde{K}(z)\nabla^2\right]u(\mathbf{r},z). \tag{24}$$

The fluctuations of $\widetilde{K}(\mathbf{r})$ (represented by the variance) determine the frequency dependence of $\widetilde{K}(z)$.

It is useful to formulate the SCBA, Eq. (16), in dimensionless units (indicated by a hat). We measure velocities in units of $\sqrt{\widetilde{K}_0}$, lengths in units of k_ξ, and angular frequencies in units of $k_\xi\sqrt{\widetilde{K}_0}$. (In the diffusion problem, diffusivities are measured in units of D_0 and angular frequencies in units of $D_0 k_\xi^2$.) In these units, the SCBA, Eq. (16), takes the form

$$\widehat{\Sigma}(\hat{z}) = \frac{\Sigma(z)}{\widetilde{K}_0} = 3\widehat{\gamma}\int_0^1 d\hat{k}\hat{k}^2 \frac{\hat{k}^2}{-\hat{z}^2 + \hat{k}^2[1-\widehat{\Sigma}(\hat{z})]} \tag{25}$$

with the dimensionless "disorder parameter"

$$\widehat{\gamma} = \gamma\frac{\nu}{\widetilde{K}_0^2} = V_c\nu\frac{\langle(\Delta\widetilde{K})^2\rangle}{\widetilde{K}_0^2}. \tag{26}$$

For $\hat{z} = 0$, Eq. (25) takes the form

$$\widehat{\Sigma}(0) = \frac{\widehat{\gamma}}{1 - \widehat{\Sigma}(0)}. \tag{27}$$

This quadratic equation has the solution

$$\widehat{\Sigma}(0) = \frac{1}{2}\Big[-1 + \sqrt{1 - 4\widehat{\gamma}}\Big]. \tag{28}$$

We observe that for $\widehat{\gamma} > \widehat{\gamma}_c = 1/4$, no real solution is obtained. This indicates an instability, i.e., for $\widehat{\gamma} > \widehat{\gamma}_c$, the SCBA predicts eigenvalues $\omega < 0$. This can be rationalized by the fact that the SCBA is obtained from assuming a Gaussian distribution (22). If the width of this distribution exceeds a critical value, there exist too many local regions with negative elastic coefficients which then leads to the instability. Within this model, a strong boson peak is obtained if the disorder parameter $\widehat{\gamma}$ approaches the critical value $\widehat{\gamma}_c$.

If we assume that the Debye cutoff $k_D = \sqrt[3]{6\pi^2 N/V}$ coincides with k_ξ (N is the number of atoms or molecular units and V is the sample volume), the density of states is given by

$$g(\hat{\omega}) = \frac{2\hat{\omega}}{\pi}3\int_0^1 d\hat{k}\hat{k}^2 \mathrm{Im}\left\{\frac{1}{-\hat{z}^2 + k^2\big[1 - \widehat{\Sigma}(\hat{z})\big]}\right\}. \tag{29}$$

In Figure 1, we have plotted the frequency-dependent diffusivity (panel a) together with the DOS, divided by the Debye DOS $g_D\hat{\omega} = 3\hat{\omega}^2/\hat{\omega}_D^3$ (reduced DOS), with $\hat{\omega}_D = \sqrt{1 - \widehat{\Sigma}_0}$, for different values of $\widehat{\gamma}$ near $\widehat{\gamma}_c$. We see that the boson peaks indeed coincide with the onset of the frequency dependence of $D(\omega)$.

2.4. *Strong disorder and the CPA*

It is clear from Figure 1 that the frequency dependence of the diffusivity (and hence of the AC conductivity), predicted by the SCBA, is rather weak as compared to the strong frequency dependence of the conductivity in ionic conducting glasses or amorphous semiconductors.[81,82] This is so because the SCBA is restricted to very weak disorder. In materials with activated ionic or electronic hopping

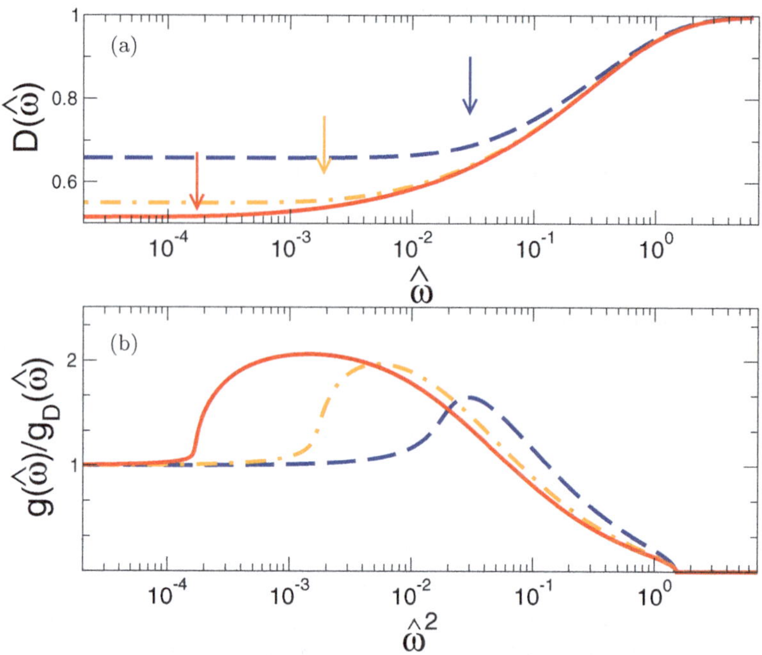

Fig. 1. Panel a: Frequency-dependent diffusivity calculated in SCBA, Eq. (25), with disorder parameters $(\hat{\gamma} - \hat{\gamma}_c)/\hat{\gamma}_c = 10^{-1}$ (blue dashes), 10^{-2} (orange dash-dots), and 10^{-3} (red line). Panel b: Reduced density of states $g(\omega)/g_D(\omega)$ for the same parameters as in panel a and setting $k^{\xi} = k_D$.

conduction, on the other hand, the local diffusivities fluctuate very strongly because they depend exponentially on local activation energies and local tunneling distances.[139,140]

A rather successful and reliable mean field theory for strong disorder is the CPA. It was widely used for electronic structure calculations of disordered crystals[141–143] and later to diffusion and vibrational properties of disordered crystals.[68,83–85,89] In this lattice version of the CPA, one considers a certain lattice site of the ordered effective medium in which the potentials,[141–143] the force constants, or the diffusivities[68,83–85,89] are homogeneous ("coherent") but frequency dependent. At this special site, the effective medium is replaced by the real medium causing a "perturbation" of the effective medium. Enforcing now the averaged T matrix of this perturbation

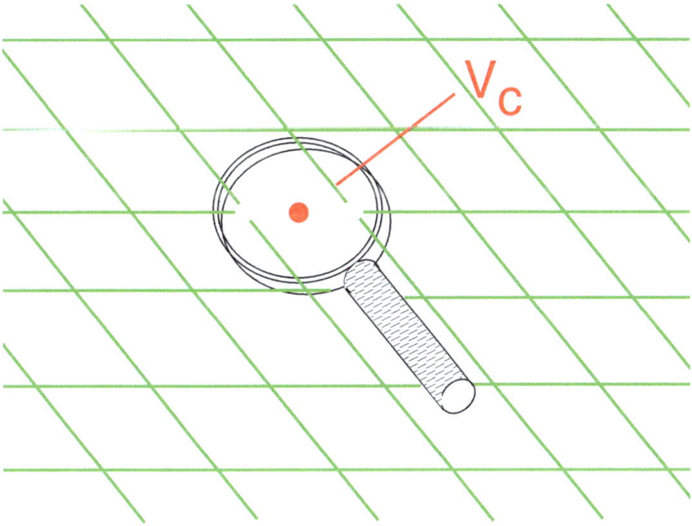

Fig. 2. Visualization of the continuum version of the CPA: Inside a volume V_c of the effective medium with homogeneous elasticity $Q(z)$, the homogeneous one is replaced by the fluctuating one $\widetilde{K}(\mathbf{r})$ which gives rise to a perturbation $\widetilde{K}(\mathbf{r}) - Q(z)$. The CPA postulate is to minimize the influence of this perturbation, i.e., forcing the averaged T matrix of the perturbation to be equal to zero.

to vanish gives the self-consistent CPA equation for the coherent potential.

A version suitable for non-crystalline materials has been worked out by S. Köhler and the present authors using field theoretical techniques.[121] The resulting CPA equation may be pedagogically visualized in the following way, see Figure 2: in a certain region of the effective medium, the frequency-dependent elasticity $Q(z)$ is replaced by the fluctuating one $\widetilde{K}(\mathbf{r}_i)$, where $\mathbf{r}_i$ is the mid-point of the region. The CPA postulate takes the form

$$\left\langle \frac{\widetilde{K}(\mathbf{r}_i) - Q(z)}{1 + \Lambda(z)[\widetilde{K}(\mathbf{r}_i) - Q(z)]} \right\rangle = 0. \tag{30}$$

The average is over the distribution of the elasticity values $\widetilde{K}(\mathbf{r}_i) \equiv \widetilde{K}_i$ with distribution density $P(\widetilde{K}_i)$. The latter may be taylored to the statistics of the disordered material at hand.

$\Lambda(z)$ generalizes the effective-medium propagator in the lattice CPA. Both $\Lambda(z)$ and $Q(z)$ may be obtained by minimizing the following mean field action:

$$S_{\text{eff}}[Q(z), \Lambda(z)] = S_{\text{med}}[Q(z)] + S_{\text{CPA}}[Q(z), \Lambda(z)], \tag{31}$$

where $S_{\text{med}}[Q(z)]$ is given by Eq. (18) and the CPA action by

$$S_{CPA}[Q(z), \Lambda(z)] = \frac{V}{V_c}\left\langle \ln\left(1 + \Lambda(z)[\widetilde{K}(\mathbf{r}_i) - Q(z)]\right)\right\rangle. \tag{32}$$

The large parameter of the saddle-point approximation, which validates the mean field approximation, is now *not* the inverse disorder parameter, as in the case of the SCBA, but the parameter V/V_c. This is the reason why the CPA is not restricted to small disorder. Varying the action with respect to $\Lambda(z)$, we obtain the CPA equation (30) which can be put into the equivalent forms

$$\left\langle \frac{1}{1 + \Lambda(z)[\widetilde{K}_i - Q(z)]}\right\rangle = 1 \tag{33}$$

$$Q(z) = \left\langle \frac{\widetilde{K}_i}{1 + \Lambda(z)[\widetilde{K}_i - Q(z)]}\right\rangle. \tag{34}$$

Varying the action with respect to $Q(z)$ gives a relation for $\Lambda(z)$:

$$\begin{aligned}\sum_{\mathbf{k}} \frac{k^2}{-z^2 + k^2 Q(z)} &= \frac{V}{V_c}\Lambda(z)\left\langle \frac{1}{1 + \Lambda(z)[\widetilde{K}_i - Q(z)]}\right\rangle \\ &= \frac{V}{V_c}\Lambda(z),\end{aligned} \tag{35}$$

where the second line follows from Eq. (33).

It is advantageous to normalize the **k** integration to 1

$$\Lambda(z) = \frac{V_c}{V}\sum_{\mathbf{k}} \frac{k^2}{-z^2 + k^2 Q(z)} = p\chi^{\xi}(z) \tag{36}$$

with

$$\chi^{\xi}(z) = \frac{3}{k_\xi^3}\int_0^{k_\xi} dk k^2 \frac{k^2}{-z^2 + k^2 Q(z)} \tag{37}$$

and

$$p = \frac{V_c k_\xi^3}{6\pi^2}. \tag{38}$$

p should be smaller than 1, and it has been argued in Ref. [121] that one may identify p with the continuum percolation concentration p_c. One has $\chi^\xi(z=0) = 1/Q(0)$.

The CPA equation (34) now takes the form

$$Q(z) = \left\langle \frac{\widetilde{K}_i}{1 + p\chi^\xi(z)[\widetilde{K}_i - Q(z)]} \right\rangle. \tag{39}$$

As in the case of the SCBA, the CPA turns the wave equation with fluctuating elasticity into a mean field wave equation with frequency-dependent elastic coefficient $Q(z)$ or a diffusion equation with frequency-dependent diffusivity $D(s)$.

Let us now identify $\widetilde{K}_i$ with a spatially fluctuating diffusivity D_i and assume that the fluctuations are caused by a fluctuating activation energy

$$D_i = D_0 e^{-E_i/k_B T}. \tag{40}$$

If we now impose a constant distribution density of activation energies

$$P(E_i)\frac{1}{E_c}\theta(E_c - E_i), \tag{41}$$

i.e., a flat distribution with cutoff E_c. For the diffusivities D_i (or elasticities $\widetilde{K}_i$), this transforms to a truncated inverse-power distribution[121]

$$P(\widetilde{K}_i) = \frac{1}{\mu_1/\mu_2}\frac{1}{\widetilde{K}_i} \qquad \mu_1 < \widetilde{K}_i \le \mu_2 \tag{42}$$

with $\mu_1 = \mu_2 e^{-E_c/k_B}$, and we have

$$\frac{1}{\langle\widetilde{K}\rangle^2}\langle\widetilde{K}^2\rangle - \langle\widetilde{K}\rangle^2) = \frac{1}{2}E_c/k_B T \equiv \widetilde{\gamma}. \tag{43}$$

Low temperatures in the diffusion problem obviously mean strong disorder.

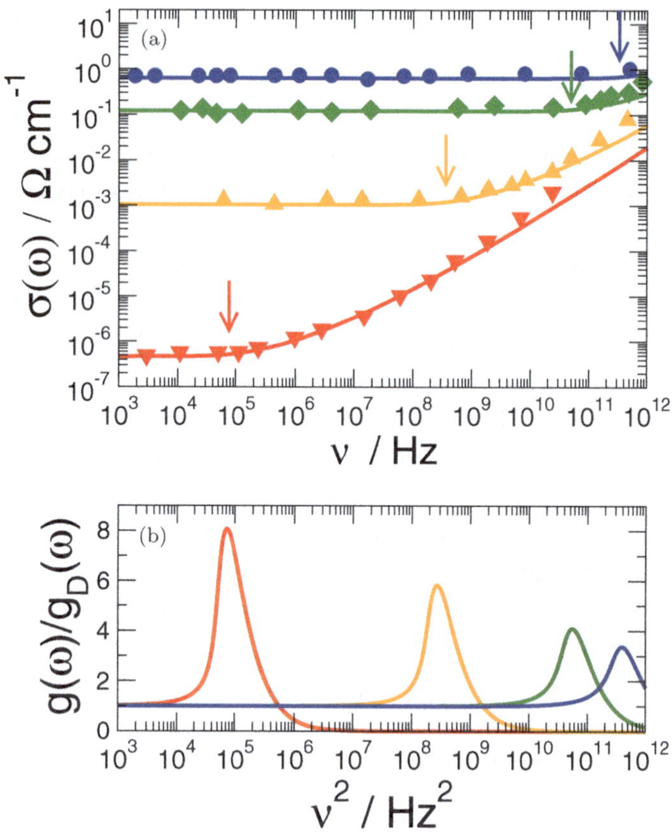

Fig. 3. Panel a: Full lines: Frequency-dependent diffusivity calculated in CPA, Eq. (74), with a flat distribution of activation energies, Eq. (41). Symbols: AC conductivity data of the ionic-conducting glass sodium trisilicate, compiled by Wong and Angell,[144] with disorder parameters $\tilde{\gamma} = 7.92$ ($\leftrightarrow T = 1673\,\mathrm{K}$), 10.41 ($\leftrightarrow T = 1273\,\mathrm{K}$), 17.53 ($\leftrightarrow T = 756\,\mathrm{K}$), and 29.06 ($\leftrightarrow T = 456\,\mathrm{K}$) (from top to bottom). Panel b: Reduced density of states $g(\hat{\omega})/g_D(\hat{\omega})$ for the equivalent distribution density (42) for the same parameters as in panel a.

In panel a of Figure 3, we have plotted AC conductivity data for the glassy ionic conductor sodium trisilicate.[144] Together with these data, we plot the CPA result with a distribution of activation energies as given in Eq. (41). In panel b of this figure, we have plotted the reduced DOS of the equivalent vibrational problem with inverse-power distribution (42). The DOS for the scalar phonon problem is given by

$$g(\omega) = \frac{2\omega}{\pi}\mathrm{Im}\left\{\frac{1}{N}\sum_{|\mathbf{k}|\leq k_D}\frac{1}{-z^2+k^2Q(z)}\right\}. \tag{44}$$

Apart from the perfect agreement between the conductivity data and the CPA curves, we note that the (rather strong) boson peak precisely marks the crossover from frequency-independent to frequency-dependent conductivity of the diffusion problem. In the vibrational problem, this crossover means a transition from a regime with Debye waves and Rayleigh scattering to a non-Debye regime in which the vibrational excitations are not waves. These excitations have been called "diffusions"[145] because their intensity obeys a diffusion equation like light in turbid media.[146] On the other hand, the vibrational excitations in this regime show the statistical properties of random matrices.[68,94,104] Therefore, the frequency regime above the boson peak may also be called "random matrix regime".[6,7]

Returning to the diffusion problem, we pointed out that in the frequency range below the crossover (the frequency-independent regime), the Rayleigh scattering corresponds to a contribution to the frequency dependent conductivity with $\Delta\sigma'(\omega) \propto \omega^{3/2}$. This may be experimentally verified by considering the dielectric loss function

$$\epsilon''(\omega) \propto \frac{1}{\omega}\big[\sigma'(\omega) - \sigma(0)\big]. \tag{45}$$

In the frequency-independent regime, this function *increases* with frequency $\epsilon'' \propto \omega^{1/2}$, whereas in the frequency-dependent regime, it *decreases* due to the sublinear behavior of $\sigma'(\omega)$. The maximum of the loss function corresponds to the maximum of the reduced DOS in the vibrational problem (boson peak).

It has been noted in the literature[132] that the AC conductivity data taken at different temperatures show universal behavior if the data are divided by the DC conductivity $\sigma(0,T)$ and the frequency by the crossover frequency $\omega_0(T) \propto \sigma(0,T)$. For the loss function, this means that $\epsilon''(\omega/\omega_0)$ should also be the same for different temperatures.

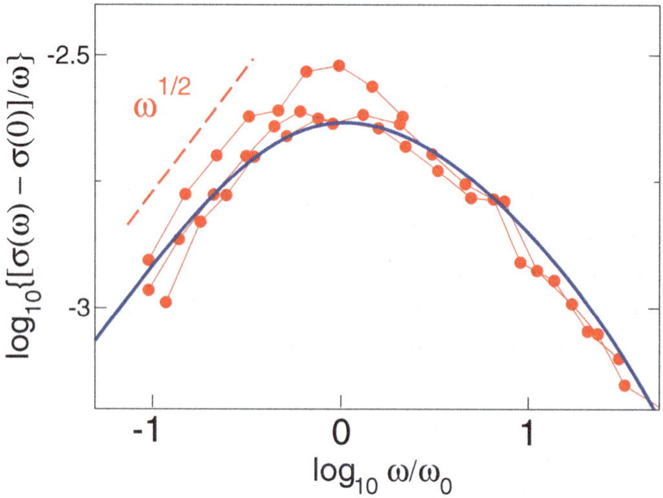

Fig. 4. Connected symbols: AC loss function $[\sigma(\omega) - \sigma(0)]/\omega$ measured in sputtered amorphous silicon[131] for three different temperatures ($T = 51\,\mathrm{K}$, $77\,\mathrm{K}$, and $154\,\mathrm{K}$). The scaling frequency $\omega_0(T)$ is proportional to $\sigma(0, T)$. Full thick line: CPA calculation for a constant distribution of activation energies.

In Figure 4, we show the loss function of sputtered amorphous silicon[131] against ω/ω_0 for three temperatures, together with the result of the CPA, which predict the scaling and the crossover from $\epsilon'' \propto \omega^{1/2}$ to decrease with frequency. We discuss the corresponding scaling of the vibrational DOS in Section 2.4.

3. Heterogeneous Elasticity Theory

3.1. *Model*

We now formulate the full heterogeneous elasticity theory for vector displacements $\mathbf{u}(\mathbf{r}, t)$. The equation of motion for an elastic medium with spatially fluctuating shear modulus $G(\mathbf{r})$[a]

$$\left[\frac{\partial^2}{\partial t^2} - \nabla \cdot \widetilde{M}(\mathbf{r})\nabla \cdot + \nabla \times \widetilde{G}(\mathbf{r})\nabla\times\right] \mathbf{u}(\mathbf{r}, \omega) = 0 \qquad (46)$$

[a]As locally the translational and rotational invariance is broken, one should, in principle, work with the full fourth-rank Hooke tensor $C_{ijkl}(\mathbf{r})$ instead of Eq. (46). The latter is an approximation to keep the model tractable.

with the reduced shear modulus $\widetilde{G}(\mathbf{r}) = G(\mathbf{r})/\rho$ and the reduced longitudinal modulus

$$\widetilde{M}(\mathbf{r}) = M(\mathbf{r})/\rho = \widetilde{K} + \frac{4}{3}\widetilde{G}(\mathbf{r}). \tag{47}$$

In this formulation, the dilatational (bulk) modulus $K = \widetilde{K}/\rho$ is assumed not to exhibit spatial fluctuations, i.e., the fluctuations of the shear modulus $G(\mathbf{r}) = \widetilde{G}(\mathbf{r})/\rho$ are assumed to affect the traceless stress and strain tensors only.[7,121]

Equation (46) may be decoupled by introducing longitudinal and transverse displacements $\mathbf{u}_L(\mathbf{r},t)$ and $\mathbf{u}_T(\mathbf{r},t)$ with

$$\nabla \times \mathbf{u}_L(\mathbf{r},t) = 0 \qquad \text{and} \qquad \nabla \cdot \mathbf{u}_T(\mathbf{r},t) = 0. \tag{48}$$

In the frequency domain, we then have

$$\begin{aligned} 0 &= \Big(-z^2 - \nabla \cdot \widetilde{M}(\mathbf{r}) \nabla \cdot \Big) \mathbf{u}_L(\mathbf{r},z) \\ &\equiv A_L[z,\mathbf{r},\widetilde{G}(\mathbf{r})]\mathbf{u}_L(\mathbf{r},z), \end{aligned} \tag{49}$$

$$\begin{aligned} 0 &= \Big(-z^2 + \nabla \times \widetilde{G}(\mathbf{r}) \nabla \times \Big) \mathbf{u}_T(\mathbf{r},z) \\ &\equiv A_T[z,\mathbf{r},\widetilde{G}(\mathbf{r})]\mathbf{u}_T(\mathbf{r},z). \end{aligned} \tag{50}$$

3.2. *SCBA*

As in the case of the scalar model, the SCBA and the CPA (see the following subsection) serve to calculate the frequency dependence of the reduced frequency-dependent shear modulus

$$\begin{aligned} Q(z) &= G(z)/\rho = Q'(\omega) - iQ''(\omega) \\ &= v_T(z)^2 = \widetilde{G}_0 - \Sigma(z) \end{aligned} \tag{51}$$

and longitudinal modulus

$$\begin{aligned} \widetilde{M}(z) &= M(z)/\rho = \widetilde{K} + \frac{4}{3}Q(z) = \widetilde{M}'(\omega) - i\widetilde{M}''(\omega) \\ &= v_L(z)^2 = \widetilde{K} + \frac{4}{3}[\widetilde{G}_0 - \Sigma(z)] \end{aligned} \tag{52}$$

which enter into macroscopic mean field equations of motion

$$0 = -z^2 - \widetilde{M}(z)\nabla^2\mathbf{u}_L(\mathbf{r},z) \equiv A_L[z,\mathbf{r},Q(z)]\mathbf{u}_L(\mathbf{r},z), \tag{53}$$

$$0 = -z^2 - Q(z)\nabla^2\mathbf{u}_T(\mathbf{r},z) \equiv A_T[z,\mathbf{r},Q(z)]\mathbf{u}_T(\mathbf{r},z). \tag{54}$$

The effective action for deriving the SCBA is

$$S_{\text{eff}}[\Sigma(z)] = S_{\text{med}}[\Sigma(z)] + S_{SCBA}[\Sigma(z)]. \tag{55}$$

Here, $S_{SCBA}[\Sigma(z)]$ is given by Eq. (19) with

$$\gamma = V_c\langle[\widetilde{G}(\mathbf{r}) - \widetilde{G}_0]^2\rangle. \tag{56}$$

The difference from the scalar model is that we now deal with three-dimensional vectors. Therefore, the trace includes a sum over the three cartesian degrees of freedom. This means that one has to sum the longitudinal contribution once and that of the transverse twice. Explicitly, $S_{\text{med}}[\Sigma(z)]$ takes the form

$$\begin{aligned} S_{\text{med}}[\Sigma(z)] &= \operatorname{Tr}\ln\left(A\big[z,Q(z)\big]\right) \\ &= \sum_{\mathbf{k}} \ln\left(A_L\big[z,\mathbf{k},\widetilde{G}_0 - \Sigma(z)\big]\right) \\ &\quad + 2\sum_{\mathbf{k}} \ln\left(A_T\big[z,\mathbf{k},\widetilde{G}_0 - \Sigma(z)\big]\right). \end{aligned} \tag{57}$$

If we now vary S_{eff} of Eq. (55) with respect to $\Sigma(z)$, i.e., $\frac{\partial S_{\text{eff}}}{\partial \Sigma(z)} = 0$, we get

$$\Sigma(z) = \gamma \frac{k_\xi^3}{6\pi^2}\chi^\xi(z) \tag{58}$$

with

$$\chi^\xi(z) = \frac{4}{3}\chi_L^\xi(z) + 2\chi_T^\xi(z). \tag{59}$$

This weighted susceptibility is given in terms of the local longitudinal and transverse susceptibilities

$$\chi_{L,T}^\xi(z) = \frac{3}{k_\xi^3}\int_0^{k_\xi} dk k^2 \chi_{L,T}(k,z) \tag{60}$$

with the k dependent susceptibilities

$$\chi_{L,T}(k,z) = k^2 \mathsf{G}_{L,T}(k,z) = \frac{k^2}{-z^2 + k^2 v_{L,T}(z)^2}. \tag{61}$$

$\mathsf{G}_{L,T}(k,z)$ are the longitudinal and transverse Green's functions[b] Eq. (58) together with Eqs. (51), (52), (59), (60), and (61) establish the self-consistent vector SCBA equations for calculating the frequency-dependent reduced shear modulus $Q(z)$, and from this, the relevant measurable quantities:

- Vibrational density of states

$$g(\omega) = \frac{2\omega}{3\pi} \frac{3}{k_D^3} \int_0^{k_D} dk k^2 \Big(\mathsf{G}''_L(k,\omega) + 2\mathsf{G}''_T(k,\omega) \Big) \tag{62}$$

- Specific heat

$$C(T) \propto \int_0^\infty d\omega g(\omega)(\omega/T)^2 \frac{e^{\hbar\omega/k_B T}}{[e^{\hbar\omega/k_B T} - 1]^2} \tag{63}$$

- Longitudinal and transverse sound attenuation $\Gamma_{L,T}$

$$\begin{aligned} \Gamma_L(\omega) &= \omega M''(\omega)/M'(\omega) \\ \Gamma_T(\omega) &= \omega G''(\omega)/G'(\omega) \\ \Leftrightarrow \quad v_{L,T}(z)^2 &= \mathrm{Re}\{v_{L,T}^2\}\left[1 - i\Gamma_{L,T}/\omega\right] \end{aligned} \tag{64}$$

- Longitudinal and transverse mean free paths $\ell_{L,T}$

$$\frac{1}{\ell_{L,T}(\omega)} = \frac{\Gamma_{L,T}(\omega)}{2v_{L,T}(0)} \tag{65}$$

- Thermal conductivity[22]

$$\kappa(T) \propto \int_0^\infty d\omega \ell_T(\omega)(\omega^4/T^2) \frac{e^{\hbar\omega/k_B T}}{[e^{\hbar\omega/k_B T} - 1]^2} \tag{66}$$

[b]We use sans serif for Green's functions $\mathsf{G}_{L,T}(k,z)$ in order to distinguish them from the shear modulus G.

- Coherent inelastic neutron and X-ray scattering intensity

$$S(k,\omega) \propto \underbrace{\frac{1}{1-e^{\hbar\omega/k_BT}}}_{n(\omega)+1} \operatorname{Im}\{\chi_L(k,\omega)\} \tag{67}$$

- Depolaritzed Raman scattering intensity[7,147]

$$I_{VH}(\omega) \propto [n(\omega)+1]\operatorname{Im}\{\chi^{k_P}(\omega)\} \tag{68}$$

It should be noted that the same susceptibility combination, namely $\chi^{\xi}(\omega)$, which enters into the SCBA equation (58), appears also in the Raman intensity, Eq. (68), albeit with a cutoff k_P given by the fluctuations of the Pockels constants and not the elastic constants.[7,147]

It should further be noted that the Raman intensity is *not* given by the Shuker–Gammon formula $I_{VH}(\omega) \propto [n(\omega)+1]g(\omega)/\omega$.[13] If one divides the expression of Eq. (68) for $I_{VH}(\omega)$ by Shuker and Gammon's expression, one obtains the frequency-dependent coupling constant $C(\omega)$ which had been inserted as additional factor into the Shuker–Gammon formula in order to reconcile neutron and Raman scattering results for the vibrational DOS.[14,16,147–150]

3.3. *General features of the disorder-induced vibrational anomalies: Comparison of the SCBA version of heterogeneous elasticity theory with a simulation*

As stated in the introduction, the disorder-induced vibrational anomalies of glasses and other disordered solids feature three phenomena which are related to each other:

(i) The crossover from Debye to non-Debye behavior of the vibrational DOS leading to a maximum in the reduced DOS $g(\omega)/\omega^2$, the boson peak;
(ii) a pronounced dip in the real part of the elastic coefficients (and their square root, the frequency-dependent velocities) near the boson peak frequency;

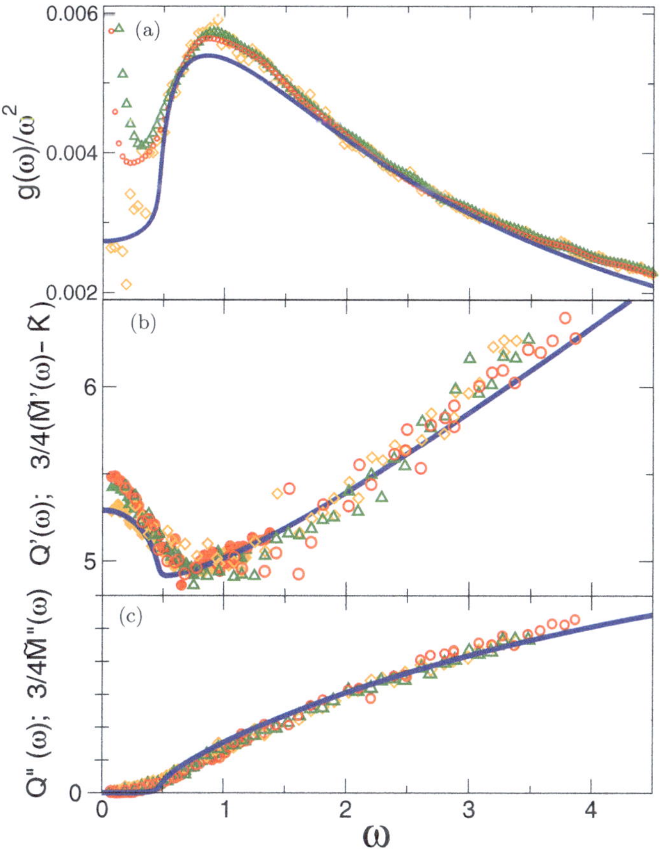

Fig. 5. Comparison of results of a soft-sphere molecular dynamics simulation (symbols) with the prediction of heterogeneous elasticity theory in SCBA. We show the real parts (upper panel) and imaginary parts (lower panel) of the frequency-dependent shear modulus $G(\Omega_T)$ and the quantity $3/4(M(\Omega_L) - \tilde{K})$ for three temperatures ($5 \cdot 10^{-3}$, $5 \cdot 10^{-4}$, and $5 \cdot 10^{-5}$, in Lennard-Jones units) with $K = 30.4$. The SCBA parameters are $\gamma - \gamma_c = 0.08$ and $K/G_0 = 3.166$ from Ref. [38].

(iii) a strong increase of the sound attenuation below the boson peak frequency (Rayleigh scattering) which enters into the imaginary parts of the elastic coefficients via Eq. (64).

These three anomalies are displayed in Figure 5 in which the results of a molecular dynamic simulation are compared with the prediction of heterogeneous elasticity theory solved in SCBA.[38] In this simulation, a binary soft-sphere potential (i.e., a Kob-Andersen-type[151] binary Lennard-Jones potential without the attractive part) was taken for ten million particles. Such a potential mimics a metallic glass. The longitudinal and transverse frequency-dependent moduli were obtained from determining the longitudinal and transverse current correlation functions

$$C_{L,T}(k,\omega) = \frac{K_B T\omega}{\pi m}\mathsf{G}''(k,\omega), \tag{69}$$

where m is the particles' mass. The simulation was run for three very different temperatures deep in the glassy state (see the figure caption). The very large particle number made it possible to avoid finite-size effect in the boson peak frequency range. The data for the complex longitudinal modulus were converted to $Q(z) = G(z)/\rho$ via the inverse of Eq. (52)

$$Q(z) = \frac{3}{4}[\widetilde{M}(z) - \widetilde{K}] \tag{70}$$

in order to compare the data with the measured $Q(z)$. It is seen in the figure that — as the longitudinal and transverse data on top of each other — the bulk modulus $\widetilde{K}$ indeed does not appreciably depend on frequency and correspondingly has a negligible imaginary part.

It is the strength of the present theory that it easily allows to explain how the three anomalies are related to each other. It was pointed out by Schirmacher *et al.* 2007[118] that one can deduce from Eq. (62) and using the 3rd line of (64) a relation between the DOS and $\Gamma(\omega)$:

$$g(\omega) - g_D(\omega) \propto \Gamma(\omega) \propto \omega\Sigma''(\omega). \tag{71}$$

This means that the disorder-induced frequency dependence of the elastic coefficient controlled by $\Sigma(\omega)$ is responsible for the boson peak. This can already understood using lowest-order perturbation theory which leads to Rayleigh scattering.

On the other hand, the real parts and the imaginary parts of the frequency-dependent moduli are related to those of the response functions $\chi(z) = \chi'(\omega) - i\chi''(\omega)$. These have, due to the causality requirement (the answer must occur at a time later than the question), a one-to-one correspondence by the Kramers–Kronig relation[152]

$$\chi'(\omega) = \frac{1}{\pi} P \int_0^\infty d\bar{\omega}^2 \frac{\chi''(\bar{\omega})}{\bar{\omega}^2 - \omega^3}. \tag{72}$$

This is most easily visualized by looking at the simulated data of the real and imaginary parts of $Q(z) = \widetilde{G}_0 - \Sigma(z)$ displayed in panels b and c. Usually, the real part of an analytic function, such as $\chi^\xi(z)$, has a maximum near the bottom of the spectrum $[\chi^\xi]''(\omega)$ and a minimum near the top. Now, because not $Q(z)$ but $\Sigma(z)$ is proportional to $\chi^\xi(z)$ via the SCBA relation (58), $Q'(\omega)$ displays a *minimum* near the begin of the "disorder spectrum", or "random matrix spectrum" $[\chi^\xi]''(\omega)$, obviously marked by the boson peak displayed in panel a.

In the boson peak regime and above, the data for very different temperatures collapse which proves that the anomalies must be of harmonic origin. This is not so in the very low frequency regime where anharmonic effects become visible.[153, 154]

3.4. *CPA*

The vector CPA may be derived in the same way as the vector SCBA from the effective action

$$S_{\text{eff}}[Q(z)] = S_{\text{med}}[Q(z)] + S_{CPA}[Q(z)], \tag{73}$$

where $S_{\text{med}}[Q(z)]$ is given by Eq. (57) of the vector SCBA and $S_{CPA}[Q(z)]$ by Eq. (32) of the scalar CPA. The vector CPA equation takes the same form as the scalar one:

$$Q(z) = \left\langle \frac{\widetilde{G}_i}{1 + p\chi^\xi(z)[\widetilde{G}_i - Q(z)]} \right\rangle \tag{74}$$

but now with fluctuating shear moduli $\widetilde{G}_i$ and the vector version of $\chi^\xi(z)$, Eq. (59).

As in the scalar model, the CPA has several advantages compared to the SCBA:

* one can treat arbitrary distributions $P(G)$;
* one is not restricted to weak disorder;
* one need not (but can) take negative values of G into account.

For the calculations presented in Figures 6–8, we used the CPA equation (74) together with (62) with a truncated Gaussian distribution of shear moduli of the form

$$P(G) = P_0\,\theta(G - G_{\min})\,e^{-(G-G_0)^2/2\gamma}, \tag{75}$$

where $\theta(x)$ is the Heaviside step function and $G_{\min}$ is the lower cutoff. In these calculations, we used the *renormalized* value of G (i.e., the self-consistently calculated one) for evaluating the Debye frequency ω_D and Debye DOS $g_D(\omega)$ in terms of the longitudinal and transverse

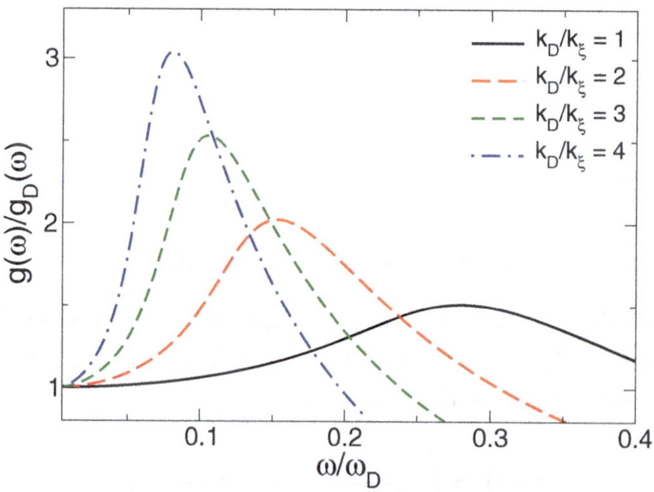

Fig. 6. Reduced density of states $g(\omega)/g_D(\omega)$ vs. the rescaled frequency $(\omega/\omega_D)(k_D/k_xi) = \omega/v_D k_\xi$ for different values of the ratio k_ξ/k_D. The other parameters are $\gamma/G_0^2 = 1$, $G_{\min} = 0$.

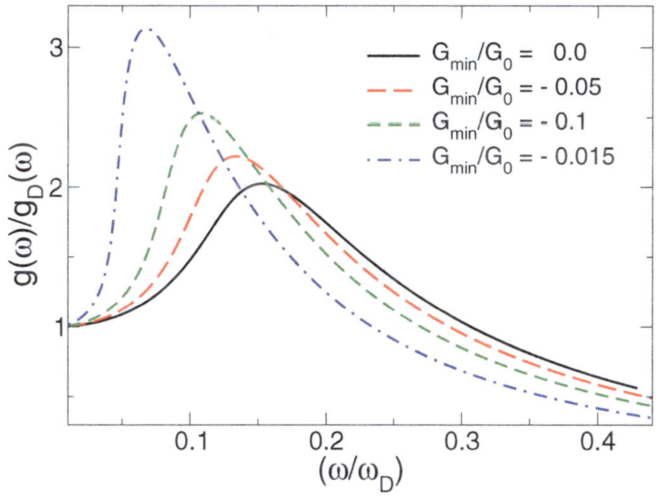

Fig. 7. Reduced density of states $g(\omega)/g_D(\omega)$ vs. the rescaled frequency (ω/ω_D) for different values of the lower cutoff $G_{\min}$ of the Gaussian distribution $P(G)$. The other parameters are $\gamma = /G_0^2 = 1$, $k_D/k_\xi = 2$.

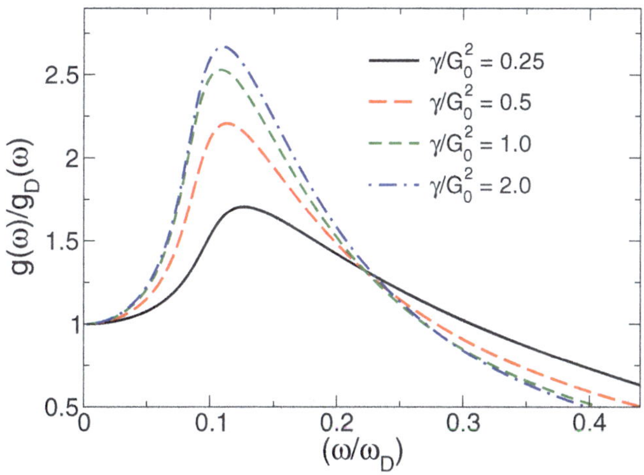

Fig. 8. Reduced density of states $g(\omega)/g_D(\omega)$ vs. the rescaled frequency (ω/ω_D) for different values of the width parameter γ. The other parameters are $G_{\min}/G_0 = -0.1$ and $k_D/k_\xi = 2$.

sound velocities $v_L^2 = K + \frac{4}{3}Q(0)$, $v_T^2 = Q(0)$

$$\omega_D = k_D \left[\frac{1}{3} \left(\frac{1}{v_L^3} + \frac{2}{v_T^3} \right) \right]^{-1/3}, \tag{76}$$

$$g_D(\omega) = 3\omega^2/\omega_D^3. \tag{77}$$

For the bulk modulus of the calculations, we used the value $K = 3.3\,G_0$, and for the cutoff parameter, the value $\widetilde{\nu} = 1$, which implies $k_D/k_\xi = \sqrt[3]{3}\xi/a$, where $a = \sqrt[3]{V/N}$ is the mean intermolecular distance. The distribution of shear moduli (75) involves three parameters G_0, γ, and $G_{\rm min}$. Because G_0 is used to fix the elastic-constant scale, there remain three adjustable parameters to fix the *state of elastic disorder* of the material, namely $k_D/k_\xi \sim \xi/a$, γ, and $G_{\rm min}$. The latter (which we used with negative values or equals zero) specify the amount of regions with negative shear modulus (soft regions) in the material. As can be seen from Figures 6–8, increasing ξ and $|G_{\rm min}|$ enhances the BP *and* shifts its position to lower frequencies, whereas increasing γ just leads to an enhancement while keeping the BP position constant. It has been pointed out in the literature[155–158] that the position of the boson peak in relation to the Debye frequency correlates with the inverse correlation length of density and elasticity fluctuations.

Let us discuss our findings further in terms of measured vibrational spectra of materials in which an external parameter (temperature, pressure, or the amount of polymerization) is changed. If the Debye frequency (depending on the moduli K and G) is changed, this leads to a modification of the spectrum which has been called *elastic-medium transformation.* This transformation is taken care of if the DOS is represented in a normalized way, as is the case in Figures 6–8. A number of boson peak data, if normalized in this way, lead to a universal curve, i.e., all data points fall onto the same curve if replotted, taking the elastic transformation into account.[32,34,59,90,92,121,159] Other investigation reveal a *deviation* from this scaling.[160–168] In terms of our model calculations, this means if the state of disorder is not changed, but just the value of the mean elastic constants or the density, which go into the Debye frequency, this corresponds to elastic transformation scaling. In the other cases obviously, the state of disorder is changed by changing the external conditions.

A very interesting case in which the elastic transformation scaling does not hold has been reported recently: the case of prehistoric amber. Reference [168] measured the temperature dependence of the specific heat of the hyperaged and rejuvenated material. The height of the boson peak, taken from a $C(T)/T^3$ curve is by 22% lower in the hyperaged material, compared with the rejuvenated one. An elastic transformation using the change in the Debye frequency determined by the authors would only lead to a difference by 7.4%.

4. Discussion and Conclusions

Reading the text of this article, one could be convinced that the boson peak in glasses and the associate vibrational anomalies can be satisfactory explained by the presence of the structural disorder, leading to spatial fluctuations of elastic coefficient, in particular the shear modulus. However, in the community working experimentally and theoretically on the vibrational properties of disordered and complex condensed matter, there is no agreement about this. Whereas many authors agree that the boson peak anomalies in glasses are due to the structural disorder, others maintain that this is not so and that the anomalies may be explained by conventional crystalline solid-state theory. Crystalline phonon theory is based on the phonon dispersions,[169] reflecting the crystal symmetry group and anharmonic interactions, leading to renormalization and viscous damping.[170–172]

As mentioned in the introduction, Chumakov *et al.*[90,92] argue that the boson peak is "identical"[90] to a washed-out Van Hove singularity, i.e., just the result of the bending over of the lowest (transverse) phonon dispersion near the first Brillouin zone. In fact, there is experimental evidence for crystal-like features in the spectrum of glasses and even liquids.[173,174] These are caused by the short-range order, which is documented in the static structure factor $S(k)$ of glasses. The peak position k_0 of $S(k)$ corresponds to the diameter of the first Brillouin zone of a crystal. Correspondingly, $k_0/2$ corresponds to the radius of the Brillouin zone where the phonon dispersions become horizontal and cause Van Hove singularities in the crystalline DOS.

In their inelastic X-ray study of polycrystalline α-Quartz and glassy SiO_2, densified to match the crystalline density, accompanied with a numerical lattice-dynamics calculation of the crystalline phonon dispersions, Baldi *et al.*[174] find a crossover of the sound attenuation of the glass from a Rayleigh $\Gamma \propto \omega^4$ behavior to a quadratic one at $\hbar\omega_c \sim 9$ meV, where the boson peak of densified silica has been reported.[92] At the same frequency, the Van Hove singularity of the polycrystal is located. On the other hand, in the polycrystal, instead of Rayleigh scattering, there is a linear behavior. Above ω_c, the two attenuation coefficients match.

The interpretation of the authors is that below ω_c, the glassy attenuation is due to elastic heterogeneities, and above ω_c, the spectrum is essentially the same of that of the polycrystal. These findings are certainly at variance with the claim that the boson peak-type scenario leading to the $\Gamma \propto \omega^4 \to \omega^2$ crossover would have something to do with the bending over of the transverse dispersion of the crystal. Obviously, in this material, the boson peak frequency and the Van Hove singularity are the same. This might also be the case for the other examples found by the authors of Refs. [90, 92].

As indicated in the introduction, in the investigation of a two-dimensional macroscopic model glass, the co-existence of crystal-like dispersions with the boson peak was observed,[93] but the frequencies of the boson peak and that of the transverse Van Hove singularity are quite different. The boson peak was shown to show all salient features which identify its origin arising from the disorder. The conclusion is that the boson peak, namely the disorder-induced peak in the reduced DOS, is not "identical" with a washed-out Van Hove singularity. In some glasses, their frequencies are just very near to each other.

In conclusion, the boson peak related vibrational anomalies in glasses can consistently be explained by the presence of spatial fluctuations of elastic coefficients (elastic heterogeneity) with the help of the SCBA and CPA.

References

1. S. R. Elliott. (1984). *The Physics of Amorphous Materials*. New York: Longman.
2. K. Binder and W. Kob. (2011). *Glassy Materials and Disordered Solids: An Introduction*. London: World Scientific.
3. See Refs. [2, 4–7] for recent reviews.
4. T. Nakayama. (2002). *Rep. Prog. Phys.* **65**, 1195.
5. M. Klinger. (2010). *Phys. Rep.* **492**, 111.
6. W. Schirmacher. (2013). *Phys. Stat. Sol. (b)* **250**, 937.
7. W. Schirmacher, T. Scopigno, and G. Ruocco. (2014). *J. Noncryst. Sol.* **407**, 133.
8. P. Flubacher, A. J. Leadbetter, and J. A. Morris. (1959). *J. Phys. Chem. Solids* **12,** 53.
9. A. J. Leadbetter and J. A. Morrison. (1963). *Phys. Chem. Glasses* **4**, 188.
10. A. J. Leadbetter. (1968). *Phys. Chem. Glasses* **9**, 1.
11. A. J. Leadbetter. (1969). *J. Chem. Phys.* **51**, 779.
12. R. Shuker and R. W. Gammon. (1970). *Phys. Rev. Lett.* **25**, 222.
13. R. Shuker and R. W. Gammon. (1971). In M. Balkanski (ed.), *Proceedings of the Second International Conference on Light Scattering in Solids*, p. 334. Paris: Flammarion.
14. J. Jäckle. (1981). In W. A. Phillips (ed.), *Amorphous Solids: Low-Temmperature Properties*, p. 135. Heidelberg: Springer.
15. A. J. Martin and W. Brenig. (1974). *Phys. Status Solidi (b)* **64**, 163.
16. B. Schmid and W. Schirmacher. (2009). *Phys. Rev. Lett.* **103**, 169702.
17. U. Buchenau, N. Nücker, and A. J. Dianoux. (1984). *Phys. Rev. Lett.* **53**, 2316
18. J. Wuttke, W. Petry, G. Coddens, and F. Fujar. (1995). *Phys. Rev. E.* **52**, 4026.
19. A. I. Chumakov, *et al.* (2004). *Phys. Rev. Lett.* **92**, 245508.
20. R. C. Zeller and R. O. Pohl. (1971). *Phys. Rev. B.* **4**, 2029.
21. J. J. Freeman and A. C. Anderson. (1986). *Phys. Rev. B.* **34**, 2726.
22. W. Schirmacher. (2006). *Europhys. Lett.* **73**, 892.
23. W. A. Phillips. (1972). *J. Low Temp. Phys.* **7**, 351.
24. P. W. Anderson, B. I. Halperin, and C. M. Varma. (1972). *Philos. Mag.* **25**, 1.
25. W. A. Phillips. (1987). *Rep. Prog. Phys.* **50**, 1557.
26. C. C. Yu and A. J. Leggett. (1988). *Comm. Cond. Matter. Phys.* **14**, 231.
27. A. Würger. (1996). From coherent tunneling to relaxation. In *Springer Tracts in Modern Physics*, vol. 135. Heidelberg: Springer.

28. S. Hunklinger and A. K. Raychaudhuri. (1986). In D. F. Brewer, (ed.), *Progress in Low-Temperature Physics*, vol. 9, p. 265. Amsterdam: North-Holland.
29. S. Hunklinger. (1997). Tunneling states. In M. F. Thorpe and M. I. Mitkova (eds.), *Amorphous Insulators and Semiconductors.* Dordrecht: Kluwer Acad. Publ.
30. F. Sette, M. H. Krisch, C. Masciovecchio, G. Ruocco, and G. Monaco. (1998). *Science* **280**, 1550.
31. G. Monaco and V. M. Giordano. (2009). *PNAS* **106**, 3659.
32. G. Baldi, A. Fontana, G. Monaco, L. Orsingher, S. Rols, F. Rossi, and B. Ruta. (2009). *Phys. Rev. Lett.* **102**, 195502.
33. G. Baldi, V. M. Giordano, G. Monaco, and B. Ruta. (2010). *Phys. Rev. Lett.* **104**, 195501.
34. B. Ruta, G. Baldi, V. M. Giordano, L. Orsingher, S. Rols, F. Scarponi, and G. Monaco. (2010). *J. Chem. Phys.* **113**, 041101.
35. G. Baldi, V. M. Giordano, and G. Monaco. (2011). *Phys. Rev. B.* **83**, 174203.
36. B. Ruta, G. Baldi, F. Scarponi, D. Fioretto, M. Giordano, and G. Monaco. (2012). *J. Chem. Phys.* **137**, 214502.
37. G. Monaco and S. Mossa. (2009). *PNAS* **106**, 16907.
38. A. Marruzzo, W. Schirmacher, A. Fratalocchi, and G. Ruocco. (2013). *Sci. Rep.* **3**, 1407.
39. V. G. Karpov, M. I. Klinger, and F. N. Ignatiev. (1983). *Sov. Phys. JETP* **84**, 760.
40. U. Buchenau, Yu. M. Galperin, V. L. Gurevich, and H. R. Schober. (1991). *Phys. Rev. B.* **43**, 5039.
41. See Refs. [5, 42] for the extended literature on soft potentials and quasi-local oscillators.
42. H. R. Schober, U. Buchenau, and V. L. Gurevich. (2014). *Phys. Rev. B.* **89**, 014204.
43. V. L. Gurevich, D. A. Parshin, and H. R. Schober. (2003). *Phys. Rev. B.* **67**, 094203.
44. E. N. Economou. (1971). *Green's Function in Quantum Physics.* Heidelberg: Springer-Verlag.
45. A. L. Burin, L. A. Maksimov, and I. Ya. Polishchuk. (1995). *Physica B.* **210**, 15.
46. E. Maurer and W. Schirmacher. (2004). *J. Low Temp. Phys.* **137**, 453.
47. W. Schirmacher. (2011). *J. Non-Cryst. Sol.* **357**, 518.
48. S. Alexander and R. Orbach. (1982). *J. Phys. (Paris) Lett.* **43**, L625.
49. B. Derrida, R. Orbach, and K.-W. Yu. (1984). *Phys. Rev. B.* **29**, 6645.

50. O. Entin-Wohlman, S. Alexander, R. Orbach, and K.-W. Yu. (1984). *Phys. Rev. B.* **29**, 4588.
51. S. Alexander. (1986). *Physica* **140A**, 397.
52. T. Nakayama, K. Yakubo, and R. L. Orbach. (1994). *Rev. Mod. Phys.* **66**, 381.
53. B. B. Mandelbrot. (1989). *The Fractal Geometry of Nature.* New York: Freeman.
54. W. Schirmacher. (2015). Theory of liquids and other disordered media. In *Lecture Notes in Physics*, Vol. 887. Heidelberg: Springer.
55. I. Webman. (1981). *Phys. Rev. Lett.* **47**, 1496.
56. S. Summerfield. (1981). *Sol. State Comm.* **39**, 401.
57. T. Odagaki and M. Lax. (1981). *Phys. Rev. B.* **24**, 5284.
58. E. Courtens, J. Pelous, J. Phalippou, R. Vacher, and T. Woignier. (1987). *Phys. Rev. Lett.* **58**, 128.
59. S. Caponi *et al.* (2009). *Phys. Rev. Lett.* **102**, 027402.
60. E. Akkermans and R. Maynard. (1985). *Phys. Rev. B.* **32**, 7850.
61. J. E. Graebner, B. Golding, and L. C. Allen. (1986). *Phys. Rev. B.* **34**, 5696.
62. A. F. Ioffe and A. R. Regel. (1960). *Prog. Semicond.* **4**, 237.
63. N. F. Mott. (1990). *Metal-insulator Transitions.* 2nd Ed. London: Taylor & Francis.
64. E. Akkermans and R. Maynard. (1985). *Phys. Rev. B.* **32**, 7850.
65. S. John, H. Sompolinky, and M. J. Stephen. (1983). *Phys. Rev. B.* **28**, 5592.
66. P. B. Allen and J. L. Feldman. (1989). *Phys. Rev. Lett.* **62**, 645.
67. W. Schirmacher and M. Wagener. (1993). *Sol. State Comm.* **86**, 597.
68. W. Schirmacher, G. Diezemann, and C. Ganter. (1998). *Phys. Rev. Lett.* **81**, 136.
69. W. Schirmacher, S. Pinski, and R. A. Rmer. (2012). *Europhys. Lett.* **97**, 16007.
70. W. Schirmacher, S. Pinski, T. Whall, and R. A. Rmer. (2012). *J. Condens. Matt.* **24**, 405401.
71. C. Tomaras and W. Schirmacher. (2013). *J. Phys. Condens. Matter* **25**, 495402.
72. S. N. Taraskin and S. R. Elliott. (1999). *J. Phys. Condens. Matter* **11**, A219.
73. B. Rufflé, G. Guimbretire, E. Courtens, R. Vacher, and G. Monaco. (2006). *Phys. Rev. Lett.* **96**, 045502.
74. H. Shintani and H. Tanaka. (2008). *Nature Mater.* **7**, 870.
75. M. Foret, E. Courtens, R. Vacher, and J.-B. Suck. (1996). *Phys. Rev. Lett.* **77**, 3831.

76. M. Foret, E. Courtens, R. Vacher, and J.-B. Suck. (1997). *Phys. Rev. Lett.* **78**, 4669.
77. E. Rat, M. Foret, E. Courtens, R. Vacher, and M. Arai. (1999). *Phys. Rev. Lett.* **83**, 1355.
78. S. N. Taraskin and S. R. Elliott. (2000). *Phys. Rev. B.* **61**, 12017.
79. W. Schirmacher and M. Wagener. (1992). *Philos. Mag. B.* **65**, 861.
80. S. Alexander, J. Bernasconi, W. R. Schneider, and R. Orbach. (1981). *Rev. Mod. Phys.* **53**, 175.
81. A. K. Jonscher. (1977). *Nature* **267**, 673.
82. A. R. Long. (1982). *Adv. Phys.* **31**, 553.
83. I. Webman. (1981). *Phys. Rev. Lett.* **47**, 1496.
84. T. Odagaki and M. Lax. (1981). *Phys. Rev. B.* **24**, 5284.
85. S. Summerfield. (1981). *Sol. State Comm.* **39**, 401.
86. M. L. Mehta. *Random Matrices.*
87. F. M. Izrailev. (1990). *Phys. Rep.* **196**, 300.
88. N. W. Ashcroft and N. D. Mermin. (1976). *Solid State Physics.* Philadelphia: Saunders College.
89. S. N. Taraskin, S. Elliott, Y. H. Loh, and G. Nataranjan. (2001). *Phys. Ref. Lett.* **86**, 1255.
90. A. I. Chumakov *et al.* (2011). *Phys. Rev. Lett.* **106**, 8519.
91. R. Zorn. (2011). *Physics* **4**, 44.
92. A. I. Chumakov *et al.* (2014). *Phys. Rev. Lett.* **112**.
93. Y. Wang, L. Hong, Y. Wang, W. Schirmacher, and J. Zhang. (2018). *Phys. Rev. B.* **98**, 174207.
94. S. K. Sarkar, G. S. Matharoo, and A. Pandey. (2004). *Phys. Rev. Lett.* **92**, 215502.
95. R. Kühn and U. Horstmann. (1997). *Phys. Rev. Lett.* **78**, 4067.
96. M. Mézard, G. Parisi, and A. Zee. (1999). Spectra of euclidean random matrices. *Nucl. Phys. B.* **559**, 689.
97. V. Martín-Mayor, G. Parisi, and P. Verroccio. (2000). *Phys. Rev. E.* **62**, 2373.
98. T. S. Grigera, V. Martín-Mayor, G. Parisi, and P. Verrocchio. (2001). *Phys. Rev. Lett.* **87**, 085502.
99. V. Martín-Mayor, M. Mézard, G. Parisi, and P. Verrocchio. (2001). *J. Chem. Phys.* **114**, 8068.
100. T. S. Grigera, V. Martín-Mayor, G. Parisi, and P. Verrocchio. (2003). *Nature* **422**, 289.
101. S. Ciliberti, T. S. Grigera, V. Martín-Mayor, G. Parisi, and P. Verrocchio. (2003). *J. Chem. Phys.* **119**, 8577.
102. C. Ganter and W. Schirmacher. (2011). *Philos. Mag.* **91**, 1894.

103. T. S. Grigera, V. Martín-Mayor, G. Parisi, P. Urbani, and P. Verrocchio. (2011). *J. Statist. Mech.* **2011**, P02015.
104. Y. M. Beltukov, V. I. Kozub, and D. A. Parshin. (2013). *Phys. Rev. B.* **87**, 134203.
105. W. Götze and M. R. Mayr. (1999). *Phys. Rev. E.* **61**, 578.
106. U. Bengtzelius, W. Götze, and A. Sjölander. (1984). Dynamics of supercooled liquids and the glass transition. *J. Phys. C.* **17**, 5915.
107. W. Götze. (2009). *Complex Dynamics of Glass-Forming Liquids.* Oxford: Oxford University Press.
108. L. D. Landau and E. M. Lifshitz. (1959). *Theory of Elasticity.* Oxford: Pergamon Press.
109. A. J. McKane and M. Stone. (1981). Localization as an alternative to Goldstone's theorem. *Ann. Phys. (N. Y.)* **131**, 36.
110. J. P. Wittmer, A. Tanguy, J.-L. Barrat, and L. Lewis. (2002). *Europhys. Lett.* **57**, 423.
111. F. Léonforte, R. Boissière, A. Tanguy, J.-P. Wittmer, and J.-L. Barrat. (2005). *Phys. Rev. B.* **72**, 224206.
112. F. Léonforte, A. Tanguy, J.-P. Wittmer, and J.-L. Barrat. (2006). *Phys. Rev. Lett.* **97**, 055501.
113. M. Tsamadosand, A. Tanguy, C. Goldenberg, and J.-L. Barrat. (2009). *Phys. Rev. B.* **80**, 026112.
114. S. G. Mayr. (2009). *Phys. Rev. B.* **79**, 060201.
115. H. Mizuno, H. Mossa, and J.-L. Barrat. (2013). *Europhys. Lett.* **104**, 56001.
116. A. Lemaitre and C. Maloney. (2006). *J. Stat. Phys.* **123**, 415.
117. Z. Pam, O. Benzine, Sh. Sawamura, R. Limbach, A. Koike, Th. Bennet, G. Wilde, W. Schirmacher, L. Wondraczek. (2021). *Phys. Rev.* **104**, 134106.
118. W. Schirmacher, G. Ruocco, and T. Scopigno. (2007). *Phys. Rev. Lett.* **98**, 025501.
119. W. Schirmacher, E. Maurer, and M. Pöhlmann. (2004). *Phys. Stat. Sol. (c)* **1**, 17.
120. C. Ganter and W. Schirmacher. (2010). *Phys. Rev. B.* **82**, 094205.
121. S. Köhler, R. Ruocco, and W. Schirmacher. (2013). *Phys. Rev. B.* **88**, 064203.
122. J. F. Lutsko. (1988). *J. Appl. Phys.* **65**, 2991.
123. J. W. Strutt Lord Rayleigh. (1871). *Philos. Mag.* **41**, 241.
124. J. W. Strutt Lord Rayleigh. (1899). *Philos. Mag.* **47**, 375.
125. S. John and M. J. Stephen. (1983). *Phys. Rev. B.* **28**, 6358.
126. S. Gelin, H. Tanaka, and A. Lemaître. (2016). *Nature Mat.* **15**, 1177.
127. J. P. Hansen and I. R. McDonald. (2006). *Theory of Simple Liquids.* Amsterdam: Elsevier Science.

128. W. Feller. (1966). *An Introduction to Probability Theory and its Applications*. Vol. II. New York: Wiley.
129. M. H. Ernst, J. Machta, J. R. Dorfman, and H. (1984). *J. Stat. Phys.* **34**, 413.
130. J. Machta, M. H. Ernst, and H. van Beijeren. (1984). *J. Stat. Phys.* **35**, 413.
131. A. R. Long, J. McMillan, N. Balkan, and S. Summerfield. (1988). *Philos. Mag. B.* **58**, 153.
132. J. C. Dyre and T. B. Schrøder. (2000). *Rev. Mod. Phys.* **72**, 873.
133. F. Wegner. (1979). The mobility edge problem: Continuous symmetry and a conjecture. *Z. Phys. B.* **35**, 207.
134. K. B. Efetov, I. Larkin A, and D. E. Khmelnitsk. (1980). *Soviet Phys. JETP* **52**, 568.
135. L. Schäfer and F. Wegner. (1980). Disordered system with n orbitals per site: Lagrange formulation, hyperbolic symmetry, and Goldstone modes. *Zeitschrift für Physik B Condens. Matter* **38**(2), 113–126.
136. J. Sajeev. (1987). Strong localization of photons in certain disordered dielectric superlattices. *Phys. Rev. Lett.* **58**, 2486–2489.
137. W. Schirmacher, B. Abaie, A. Mafi, G. Ruocco, and M. Leonetti. (2018). *Phys. Rev. Lett.* **120**, 067401.
138. D. Vollhardt. (2010). *AIP Conf. Proc.* **1297**, 339.
139. H. Böttger and V. V. Bryksin. (1985). Hopping conduction in solids, VCH-Verlagsgesellschaft, Weinheim.
140. A. I. Efros and B. I. Shklovskiĭ. (1984). *Electronic Properties of Doped Semiconductors*. Heidelberg: Springer-Verlag.
141. P. Soven. (1967). *Phys. Rev.* **156**, 809.
142. D. W. Taylor. (1967). *Phys. Rev.* **156**, 1017.
143. H. Ebert, D. Ködderitzsch, and J. Minár. (2011). *Rep. Prog. Phys.* **74**, 096501.
144. J. Wong and C. A. Angell. (1976). *Glass: Structure by Spectroscopy*. New York: M. Dekker.
145. P. B. Allen, J. L. Feldman, and J. Fabian. (1999). *Philos. Mag. B.* **79**, 1715.
146. A. Ishimaru. (1978). *Wave Propagation and Scattering in Random Media. V.* New York: Academic Press.
147. B. Schmid and W. Schirmacher. (2008). *Phys. Rev. Lett.* **100**, 137402.
148. V. K. Malinovsky and A. P. Sokolov. (1986). *Sol. St. Comm.* **57**, 757.
149. G. Viliani *et al.* (1995). *Phys. Rev. B.* **52**, 3346.
150. V. N. Novikov and E. Duval. (2009). *Phys. Rev. Lett.* **103**, 169701.
151. W. Kob and H. C. Andersen. (1994). *Phys. Rev. Lett.* **73**, 1376.
152. J. D. Jackson. (1962). *Classical Electrodynamics*. New York: Wiley.

153. A. Marruzzo, S. Köhler, A. Fratalocchi, G. Ruocco, and W. Schirmacher. (2013). *Eur. Phys. J. Special Topics* **216**, 83.
154. C. Ferrante, E. Pontecorvo, G. Cerullo, A. Chiasera, G. Ruocco, W. Schirmacher, and T. Scopigno. (2013). *Nature Comm.* **4**, 1793.
155. E. Duval, A. Boukenter, and T. Achibat. (1990). *J. Phys. Condens. Matter* **2**, 10227.
156. S. R. Elliott. (1992). *Europhys. Lett.* **19**, 201.
157. A. P. Sokolov, *et al.* (1992). *Phys. Rev. Lett.* **69**, 1540.
158. L. Hong, V. N. Novikov, and A. P. Sokolov. (2013). *J. Noncryst. Sol.* **117**, 357.
159. A. Monaco, *et al.* (2006). *Phys. Rev. Lett.* **97**, 135501.
160. M. Zanatta, *et al.* (2010). *Phys. Rev. B.* **81**, 212201.
161. M. Zanatta, *et al.* (2011). *J. Chem. Phys.* **135**, 174506.
162. L. Hong, *et al.* (2008). *Phys. Rev. B.* **78**, 134201.
163. S. Caponi, *et al.* (2007). *Phys. Rev. B.* **76**, 092201.
164. B. Rufflé, S. Ayrinhac, E. Courtens, R. Vacher, M. Foretand, A. Wischnewski, and U. Buchenau. (2010). *Phys. Rev. Lett.* **104**, 067402.
165. K. Niss, *et al.* (2007). *Phys. Rev. Lett.* **99**, 055502.
166. L. Orsingher, *et al.* (2010). *J. Chem. Phys.* **132**, 124508.
167. S. Corezzi, *et al.* (2013). *J. Phys. Chem. B.* **117**, 14477.
168. T. Pérez-Castañeda, *et al.* (2014). *Phys. Rev. Lett.* **112**, 165901.
169. A. A. Maradudin, E. W. Montroll, G. H. Weiss, and I. P. Ipatova. (1971). *Theory of Lattice Dynamics in the Harmonic Approximation, Solid-state Physics, Suppl.3.* New York: Academic Press.
170. W. Götze and K. H. Michel. (1974). In G. K. Horton and A. A. Maradudin (eds.), *Dynamical Properties of Solids*, Vol. I., Chapter 9. Amsterdam: North Holland.
171. H. Horner. (1974). In G. K. Horton and A. A. Maradudin (eds.), *Dynamical Properties of Solids*, Vol. I., Chapter 9. Amsterdam: North Holland.
172. J. Scheipers and W. Schirmacher. (1996). *Z. Phys. B.* **103**, 547.
173. V. M. Giordano and G. Monaco. (2010). *PNAS* **107**, 21985.
174. G. Baldi, M. Zanatta, E. Gilioli, V. Milman, K. Refson, B. Wehinger, B. Winkler, A. Fontana, and G. Monaco. (2013). *Phys. Rev. Lett.* **110**, 185503.

https://doi.org/10.1142/9781800612587_0010

Chapter 10

Computational Simulations of the Vibrational Properties of Glasses

Hideyuki Mizuno* and Atsushi Ikeda

Graduate School of Arts and Sciences,
The University of Tokyo, Tokyo 153-8902, Japan
**hideyuki.mizuno@phys.c.u-tokyo.ac.jp*

Glasses show vibrational properties that are markedly different to those of crystals which are known as phonons. For example, excess low-frequency modes (the so-called boson peak), vibrational localization, and strong scattering of phonons have been the most discussed topics, and a theoretical understanding of these phenomena is challenging. To address this problem, computational simulations are a powerful tool, which have been employed by many previous works. In this chapter, we describe simulation methods for studying the vibrational properties of glasses (and any solid-state materials). We first present a method for studying vibrational eigenmodes. Since vibrational motions of particles are excited along eigenmodes, the eigenmodes are fundamental to descriptions of vibrational properties. The eigenmodes in glasses are non-phonon modes in general, and some of them are even localized in space. We next present a method of analysing phonon transport, which is also crucial for understanding vibrational properties. Since phonons are not eigenmodes in glasses, they are decomposed into several different, non-phonon eigenmodes. As a result, phonons in glasses are strongly scattered. In addition, we describe how to analyse the elastic response. The elastic response of glasses is also anomalous with respect to that of crystals. Finally, we briefly introduce recent advances that have been achieved by means of large-scale computational simulations.

1. Introduction

Currently, there is no doubt that computational simulations play an important role in the development of fundamental science as well as engineering applications. Molecular Dynamics (MD) simulations and Monte Carlo (MC) simulations have been established to simulate the behaviours of materials in dense gas, liquid, and solid states at the microscopic, molecular level.[1,2] The importance of these molecular simulations comes from the fact that they provide exact, quasi-experimental data on well-defined simulation models of materials. The usefulness of simulations is also based on the fact that they can access data that cannot be obtained through experiments. From the theoretical point of view, exact data on prototypical models are valuable for understanding the fundamental mechanisms of phenomena as well as for testing the validity of proposed theories. In the past, molecular simulations have been employed to solve many important problems. For example, in statistical physics, phase transitions, such as the gas-liquid transition and the paramagnetic-ferromagnetic transition, are among the topics that are most widely studied by means of computer simulations.[3] Additionally, to address the problem of glass transition, many previous works have relied on simulations of, e.g., the drastically slowed dynamics near the glass transition.[4,5]

Computational simulations are also a powerful tool for studying the vibrational properties of glasses. For crystals, thanks to their periodicity and symmetry, analytical formulations can be obtained for the vibrational motions of the molecules, which give the concept of phonons (lattice waves).[a,6,7] Particularly, the Debye theory has been established to explain the behaviour of the vibrational density of states (vDOS) in a crystal. In contrast, for glasses, due to the lack of periodicity and symmetry, analytical calculations are much difficult to perform. Although some mean-field theories, such as replica theory[8] and effective medium theory,[9–12] have been proposed, it is

[a]Phonons are quantized lattice waves, but here we use the term "phonons" for general lattice waves. We can also consider phonons in glasses, which are sinusoidal waves propagating in disordered structures. Such phonons in glasses will be discussed in Section 4.

crucial to test their validity and extend them to the 3 dimensional case. Also, in many past works, e.g., Refs. [13–25], experiments have been performed to study the thermal and vibrational properties of glasses. Although the vDOS and dynamic structure factor can be measured in experiments by monitoring the responses of various probes, such as light, X-rays, and neutrons, it is generally difficult to directly observe the vibrational motions of molecules. Considering this situation of separate purely theoretical and purely experimental works, computational simulations take advantage of the ability to directly observe and understand the vibrational motions of particles in well-defined simulation models. There are many relevant problems regarding the vibrational properties of glasses, e.g., excess low-frequency modes (the so-called boson peak), vibrational localization, and the strong scattering of phonons. A theoretical understanding of these phenomena remains to be developed, and doing so will be challenging. To address these problems, computer simulations have been employed in many previous works, e.g., Refs. [26–62].

In this chapter, we review simulation methods for analysing the vibrational properties of glasses; these methods can also be applied for any solid-state materials. In Section 2, we describe simulation models of glasses. We first present a brief description of MD simulations, and we next explain how to obtain glass systems through simulations. Section 3 introduces a method of analysing vibrational eigenmodes. Since the vibrational motions of particles are excited along eigenmodes, the eigenmodes are fundamental to descriptions of vibrational properties. Section 4 provides a method of analysing phonon transport, which is also a fundamental property. In addition, we also present a method of analysing elastic responses and the corresponding elastic moduli in Section 5. Finally, Section 6 briefly introduces recent advances that have been achieved by means of large-scale computational simulations.

2. Simulation Models

Several excellent books (e.g., Refs. [1,2]) have already been published that describe molecular simulations, i.e., MD and MC simulations. Here, we provide only a brief description of MD simulations and explain how we can obtain glass systems through MD simulations (we can also use MC simulations to obtain glasses in the same way). In any solid-state material, including glasses and crystals, the constituent particles vibrate around the so-called inherent structure.[b] We also explain how to obtain this inherent structure through computational simulations.

2.1. *Molecular Dynamics (MD) simulations*

In an MD simulation, we define a system composed of an enormous number of particles and numerically solve the equations of motion to evolve the dynamics of those constituent particles. Here, let us consider a 3-dimensional system composed of N particles in a cubic box of length L and volume $V \equiv L^3$. The mass of particle i $(i = 1, 2, \ldots, N)$ is denoted by m_i, and the position of particle i is $\boldsymbol{r}_i = [r_{ix}, r_{iy}, r_{iz}]^T$ (a 3-dimensional vector), where T denotes transposition. To represent the configuration of all the constituent particles, we introduce a $3N$-dimensional vector $\boldsymbol{r} \equiv \left[\boldsymbol{r}_1^T, \boldsymbol{r}_2^T, \ldots, \boldsymbol{r}_N^T\right]^T$. In this chapter, we treat vectorial quantities, such as $\boldsymbol{r}_i$ and $\boldsymbol{r}$, as *vertical* vectors. We suppose that the total potential energy of the system, Φ, is a function of the positions of the particles: $\Phi(\boldsymbol{r}) \equiv \Phi(\boldsymbol{r}_1, \boldsymbol{r}_2, \ldots, \boldsymbol{r}_N)$. For the case of a pair-wise potential, where particles i and j interact through the potential ϕ_{ij}, $\Phi(\boldsymbol{r}) = \sum_{i<j} \phi_{ij}$. The equation of motion is then

$$m_i \frac{d^2 \boldsymbol{r}_i}{dt^2} = -\frac{\partial \Phi}{\partial \boldsymbol{r}_i} \, (i = 1, 2, \ldots, N), \tag{1}$$

or, in terms of the $3N$-dimensional vector $\boldsymbol{r}$,

$$\mathcal{M} \frac{d^2 \boldsymbol{r}}{dt^2} = -\frac{\partial \Phi}{\partial \boldsymbol{r}}, \tag{2}$$

where t is time and we introduce the mass matrix $\mathcal{M}$, which is a $3N \times 3N$ diagonal matrix:

$$\mathcal{M} \equiv \text{diag}(m_1, m_1, m_1, m_2, m_2, m_2, \ldots, m_N, m_N, m_N). \tag{3}$$

---

[b]The inherent structure of a crystal is a periodic lattice structure.

We start with some initial configuration of particles, $\boldsymbol{r}(t-0)$, and initial velocity, $d\boldsymbol{r}/dt|_{t=0}$, and solve Eq. (1) or Eq. (2). Here, we need to implement appropriate boundary conditions in the space. In this chapter, we consider periodic boundary conditions in all directions. Thermodynamic quantities such as temperature T and pressure p can be calculated from the trajectory $\boldsymbol{r}(t)$: for example, the temperature is calculated from the total kinetic energy of the particles. We may use a heat bath and/or a pressure bath to control the temperature and/or the pressure, respectively. As the simulation runs, the system approaches a steady, equilibrium state.

Several potentials $\Phi(\boldsymbol{r})$ have been proposed for modelling different types of glass systems. Here, we briefly introduce the potentials for modelling atomic glasses (packed glasses) and covalently bonded glasses (network glasses).

(I) Atomic glasses (packed glasses):
To model an atomic glass composed of rare gas atoms such as argons, we can employ the Lennard-Jones (LJ) potential:[63]

$$\Phi_{\mathrm{LJ}}(\boldsymbol{r}) = \sum_{i<j} 4\epsilon_{ij}\left[\left(\frac{\sigma_{ij}}{r_{ij}}\right)^{12} - \left(\frac{\sigma_{ij}}{r_{ij}}\right)^{6}\right], \tag{4}$$

where $r_{ij} \equiv |\boldsymbol{r}_i - \boldsymbol{r}_j|$ is the distance between particles i and j and σ_{ij} and ϵ_{ij} represent the length and energy scales, respectively. We also often employ the soft-core (SC) potential, which is the repulsive part of the LJ potential:

$$\Phi_{\mathrm{SC}}(\boldsymbol{r}) = \sum_{i<j} \epsilon_{ij}\left(\frac{\sigma_{ij}}{r_{ij}}\right)^{12}. \tag{5}$$

In addition, as the simplest model of a glass, a finite-range harmonic potential has been employed:

$$\Phi_{\mathrm{HA}}(\boldsymbol{r}) = \sum_{i<j} \frac{\epsilon_{ij}}{2}\left(1 - \frac{r_{ij}}{\sigma_{ij}}\right)^{2} H(\sigma_{ij} - r_{ij}), \tag{6}$$

where $H(x)$ is the Heaviside step function: $H(x) = 1$ for $x \geq 0$ and $H(x) = 0$ for $x < 0$. This harmonic potential was originally proposed for modelling granular materials, emulsions, foams, etc.[64]

(II) Covalently bonded glasses (network glasses):
To model silica (SiO_2) glass, which is one example of a covalently bonded glass, the Beest-Kramer-Santen potential[65] has been proposed:

$$\Phi_{\mathrm{BKS}}(\boldsymbol{r}) = \sum_{i<j} \left[\frac{q_i q_j e^2}{r_{ij}} + A_{ij} \exp\left(-B_{ij} r_{ij}\right) - \frac{C_{ij}}{r_{ij}^6} \right], \quad (7)$$

where particles i and j are Si or O atoms, $q_i e$ is the electric charge (e is elementary charge), and A_{ij}, B_{ij}, and C_{ij} are constants. The first term in Eq. (7) represents the Coulomb interaction. In addition, the Stillinger-Weber potential,[66] $\Phi_{\mathrm{SW}}(\boldsymbol{r})$, has been proposed for modelling amorphous silicon. $\Phi_{\mathrm{SW}}(\boldsymbol{r})$ includes three-body interactions to represent the bending rigidity. For the explicit equation for $\Phi_{\mathrm{SW}}(\boldsymbol{r})$, please see Ref. [66].

In addition, for metallic glasses, Daw and Baskes have proposed the embedded-atom method (EAM) based on density functional theory.[67] Additionally, for polymer glasses, we can perform coarse-grained simulations by using the Kremer-Grest model,[68] which treats polymer chains as linear series of monomer beads (particles). In the following section, we take a system with the SC potential (Eq. (5)) as an example. We present numerical results not only for the glass but also for the crystal, which are obtained from Refs. [40–42]. By comparing the glass and crystal, we discuss the characteristic features of glasses. The values of quantities are presented in units of the mass (m), length (σ), and energy (ϵ) scales that are typical of the constituent particles.

2.2. *Glass and inherent structure*

By means of an MD simulation, we obtain the glass as follows.[4,5] We first equilibrate the system in the liquid phase above the melting temperature, T_m. We next rapidly quench the system below T_m. If we

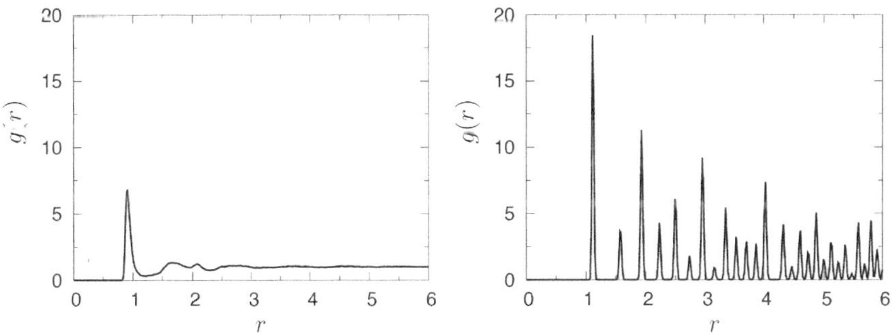

(a) Glass. (b) Crystal.

Fig. 1. Radial distribution frunction $g(r)$ of a soft-core(SC) system at a low temperature of $T = 10^{-2}$, well below T_g and T_m. (a) Glass. (b) Face-centred cubic (FCC) crystal.

avoid crystallization,[c] we can maintain the liquid phase even below T_m; such a phase is called a supercooled liquid phase. When we further quench the system below the glass transition temperature, T_g, the system is finally frozen in a disordered configuration, without crystallization. To study the structural properties of this system, we can measure the radial distribution function $g(r)$ or the static structure factor $S(q)$.[63] Figure 1 presents the $g(r)$ functions for two SC systems, (a) glass and (b) crystal, for comparison. We can see long-range order in the crystal, while the glass shows only short-range order. Although glasses are solids like crystals (i.e., they exhibit rigidity), they also possess disordered structures as liquids do.[4,5]

In the glass phase below T_g, the particles vibrate around the inherent structure, which is denoted by $\boldsymbol{R} \equiv \left[\boldsymbol{R}_1^T, \boldsymbol{R}_2^T, \ldots, \boldsymbol{R}_N^T\right]^T$.[d] In the inherent structure, the system is in a state of mechanical equilibrium, and the potential energy $\Phi(\boldsymbol{r})$ takes its minimum value

[c]To avoid crystallization, we may employ a poly-disperse system composed of a number of different kinds of particles with different mass, length, and energy scales.

[d]Glasses may show ageing phenomena at a time scale that is much longer than that of vibrations. Ageing induces rearrangements of some particles and alters the inherent structure. In this chapter, we will not consider this effect.

in the 3N-dimensional space:

$$\left.\frac{\partial \Phi}{\partial \boldsymbol{r}}\right|_{\boldsymbol{r}=\boldsymbol{R}} = 0. \tag{8}$$

Thus, we can numerically obtain $\boldsymbol{R}$ by minimizing $\Phi(\boldsymbol{r})$. For this purpose, several numerical techniques have been established, e.g., the steepest descent method,[69] the conjugate gradient method,[69] and the fire algorithm.[70] We note that minimizing the potential corresponds to quenching the system to zero temperature, $T = 0$, and the inherent structure is the configuration of the particles at $T = 0$.

3. Vibrational Eigenmodes

In a glass (or any solid-state material), the constituent particles vibrate around the inherent structure. These vibrational motions of particles are excited along vibrational eigenmodes. The eigenmodes are therefore fundamental to understanding the vibrational properties. In this section, we explain how to obtain the eigenmodes by means of computational simulations. We then describe the vDOS, which represents the statistics of the eigenmodes. We also explain order parameters used to characterize each vibrational eigenmode. In particular, we introduce the participation ratio, which measures the extent of localization, and the phonon order parameter, which measures the extent of phonon-like vibrations.

3.1. *General description*

Let us introduce the displacement vector (3N-dimensional vector) of the particles relative to the inherent structure $\boldsymbol{R}$ (multiplied by the mass factor $\sqrt{\mathcal{M}}$) as follows:

$$\boldsymbol{u} = \sqrt{\mathcal{M}}(\boldsymbol{r} - \boldsymbol{R}). \tag{9}$$

We then expand the potential $\Phi(\boldsymbol{r})$ around $\boldsymbol{R}$ in a power series of $\boldsymbol{u}$ as follows (T denotes transposition):

$$\Phi\left(\boldsymbol{r}\right) = \Phi\left(\boldsymbol{R}\right) + \frac{1}{2}\boldsymbol{u}^T \mathcal{D} \boldsymbol{u} + \mathcal{O}\left(|\boldsymbol{u}|^3\right), \tag{10}$$

where we note that the first derivative of $\Phi(\boldsymbol{r})$ is zero (mechanical equilibrium), as expressed in Eq. (8). $\mathcal{D}$ is the so-called dynamical matrix (a $3N \times 3N$ matrix) which is the second derivative of $\Phi(\boldsymbol{r})$:[6,7]

$$\mathcal{D} = \frac{1}{\sqrt{\mathcal{M}}} \left[\left. \frac{\partial^2 \Phi}{\partial \boldsymbol{r} \partial \boldsymbol{r}^T} \right|_{\boldsymbol{r}=\boldsymbol{R}} \right] \frac{1}{\sqrt{\mathcal{M}}}. \tag{11}$$

The matrix $\mathcal{D}$ contains 3×3 matrix elements, $(\mathcal{D})_{ij}$, corresponding to particles i and j:

$$(\mathcal{D})_{ij} = \frac{1}{\sqrt{m_i m_j}} \left[\left. \frac{\partial^2 \Phi}{\partial \boldsymbol{r}_i \partial \boldsymbol{r}_j^T} \right|_{\boldsymbol{r}=\boldsymbol{R}} \right]. \tag{12}$$

In the case of a pair-wise potential, $\phi_{ij}(r_{ij})$, which depends only on the distance r_{ij} between particles i and j, $(\mathcal{D})_{ij}$ is formulated as

$$(\mathcal{D})_{ij} = \begin{cases} \dfrac{1}{m_i} \displaystyle\sum_{k=1, k \neq i}^{N} \left[\dfrac{d^2 \phi_{ik}}{dr_{ik}^2} \dfrac{\boldsymbol{r}_{ik} \boldsymbol{r}_{ik}^T}{r_{ik}^2} + \dfrac{1}{r_{ik}} \dfrac{d\phi_{ik}}{dr_{ik}} \left(I_3 - \dfrac{\boldsymbol{r}_{ik} \boldsymbol{r}_{ik}^T}{r_{ik}^2} \right) \right] & (i = j), \\ -\dfrac{1}{\sqrt{m_i m_j}} \left[\dfrac{d^2 \phi_{ij}}{dr_{ij}^2} \dfrac{\boldsymbol{r}_{ij} \boldsymbol{r}_{ij}^T}{r_{ij}^2} + \dfrac{1}{r_{ij}} \dfrac{d\phi_{ij}}{dr_{ij}} \left(I_3 - \dfrac{\boldsymbol{r}_{ij} \boldsymbol{r}_{ij}^T}{r_{ij}^2} \right) \right] & (i \neq j), \end{cases} \tag{13}$$

where I_3 is the 3×3 unit matrix, $\boldsymbol{r}_{ij} \equiv \boldsymbol{r}_i - \boldsymbol{r}_j$, and $r_{ij} \equiv |\boldsymbol{r}_{ij}|$. Since $|\boldsymbol{u}| \ll 1$ at low temperatures, we neglect higher-order terms of $\mathcal{O}(|\boldsymbol{u}|^3)$ in $\Phi(\boldsymbol{r})$ as expressed in Eq. (10) and substitute this expression into Eq. (2) to obtain the linearized equation of motion (harmonic approximation):

$$\frac{d^2 \boldsymbol{u}(t)}{dt^2} = -\mathcal{D} \boldsymbol{u}(t). \tag{14}$$

To solve Eq. (14), we perform a Fourier transform and obtain

$$\omega^2 \tilde{\boldsymbol{u}}(\omega) = \mathcal{D} \tilde{\boldsymbol{u}}(\omega), \tag{15}$$

where ω is the frequency and $\tilde{\boldsymbol{u}}(\omega)$ is the Fourier transform of $\boldsymbol{u}(t)$:

$$\tilde{\boldsymbol{u}}(\omega) = \int \boldsymbol{u}(t) \exp(-\mathrm{i}\omega t)\, dt. \tag{16}$$

$u(t)$ is then obtained by taking the inverse Fourier transform of $\tilde{\boldsymbol{u}}(\omega)$:

$$\boldsymbol{u}(t) = \frac{1}{2\pi} \int \tilde{\boldsymbol{u}}(\omega) \exp\left(\mathrm{i}\omega t\right) d\omega. \tag{17}$$

We thus encounter the eigenvalue problem of the matrix $\mathcal{D}$ (Eq. (15)).

Since $\mathcal{D}$ is a symmetric matrix, it can be diagonalized by an orthonormal matrix X, and its eigenvalues λ^k ($k = 1, 2, \ldots, 3N$) are all real numbers:

$$X^T D X = \mathrm{diag}(\lambda^1, \lambda^2, \ldots, \lambda^{3N}), \tag{18}$$

$$X^T X = I_{3N}, \tag{19}$$

where I_{3N} is the $3N \times 3N$ unit matrix. We write the matrix in the form of $X \equiv [\boldsymbol{e}^1, \boldsymbol{e}^2, \ldots, \boldsymbol{e}^{3N}]$, thereby introducing $3N$-dimensional eigenvectors, $\boldsymbol{e}^k \equiv [\boldsymbol{e}_1^{kT}, \boldsymbol{e}_2^{kT}, \ldots, \boldsymbol{e}_N^{kT}]^T$ ($k = 1, 2, \ldots, 3N$). The orthonormality condition, Eq. (19), then gives

$$\boldsymbol{e}^k \cdot \boldsymbol{e}^l = \sum_{i=1}^{N} \boldsymbol{e}_i^k \cdot \boldsymbol{e}_i^l = \delta_{kl}, \tag{20}$$

where δ_{kl} is the Kronecker delta.[e] We thus obtain $3N$ sets of eigenvalues λ^k and eigenvectors $\boldsymbol{e}^k$ for the matrix $\mathcal{D}$:

$$\lambda^k \boldsymbol{e}^k = \mathcal{D}\boldsymbol{e}^k (k = 1, 2, \ldots, 3N), \tag{21}$$

and $\mathcal{D}$ can be described as

$$\mathcal{D} = \sum_{k=1}^{3N} \lambda^k \boldsymbol{e}^k \boldsymbol{e}^{kT}. \tag{22}$$

If the system is stable, the eigenvalues are all positive, $\lambda^k > 0$, and the eigenfrequencies, $\omega^k \equiv \sqrt{\lambda^k}$, are real numbers. A general solution

[e]The set of $3N$ eigenvectors $\{\boldsymbol{e}^1, \boldsymbol{e}^2, \ldots, \boldsymbol{e}^{3N}\}$ can be treated as an orthonormal basis in $3N$-dimensional space for $\boldsymbol{u}$.

for $\tilde{\boldsymbol{u}}(\omega)$ in Eq. (15) is thus obtained as a superposition of the $\boldsymbol{e}^k$:

$$\tilde{\boldsymbol{u}}(\omega) = \sum_{k=1}^{3N} A^k \delta(\omega - \omega^k) \boldsymbol{e}^k, \tag{23}$$

where $\delta(x)$ is the Dirac delta function, and the A^k are complex constants. Then, $\boldsymbol{u}(t)$ is obtained as

$$\boldsymbol{u}(t) = \frac{1}{2\pi} \sum_{k=1}^{3N} A^k \exp\left(\mathrm{i}\omega^k t\right) \boldsymbol{e}^k. \tag{24}$$

The values of the A^k are determined from the initial conditions $\boldsymbol{u} = \boldsymbol{u}_0$ and $d\boldsymbol{u}/dt = \dot{\boldsymbol{u}}_0$ at $t = 0$, as $\mathrm{Re}(A^k) = 2\pi(\boldsymbol{e}^k \cdot \boldsymbol{u}_0)$ and $\mathrm{Im}(A^k) = -2\pi(\boldsymbol{e}^k \cdot \dot{\boldsymbol{u}}_0)/\omega^k$ (where Re and Im denote the real and imaginary parts, respectively), and we finally obtain $\boldsymbol{u}(t)$ as follows:

$$\begin{aligned} \boldsymbol{u}(t) &= \sum_{k=1}^{3N} u^k(t) \boldsymbol{e}^k, \\ u^k(t) &= (\boldsymbol{e}^k \cdot \boldsymbol{u}_0) \cos(\omega^k t) + (\boldsymbol{e}^k \cdot \dot{\boldsymbol{u}}_0) \frac{\sin(\omega^k t)}{\omega^k}, \end{aligned} \tag{25}$$

which is a superposition of terms of the form $u^k(t)\boldsymbol{e}^k$, which represents the vibrations of particles along the eigenvector $\boldsymbol{e}^k$ (vibration of particle i along $\boldsymbol{e}_i^k$) with frequency ω^k and which we call the vibrational eigenmodes k.

The kinetic energy, K, and the potential energy, $\Delta\Phi \equiv \Phi - \Phi(\boldsymbol{R})$, are described as sums of the energies of the eigenmodes k:

$$\begin{aligned} K(t) &= \frac{1}{2} \frac{d\boldsymbol{u}(t)}{dt} \cdot \frac{d\boldsymbol{u}(t)}{dt} = \frac{1}{2} \sum_{k=1}^{3N} \left(\frac{du^k(t)}{dt} \right)^2, \\ \Delta\Phi(t) &= \frac{1}{2} \boldsymbol{u}(t)^T \mathcal{D} \boldsymbol{u}(t) = \frac{1}{2} \sum_{k=1}^{3N} \left(\omega^k u^k(t) \right)^2. \end{aligned} \tag{26}$$

The total energy, $K(t) + \Delta\Phi(t) = (1/2) \sum_{k=1}^{3N} [\omega^{k2} (\boldsymbol{e}^k \cdot \boldsymbol{u}_0)^2 + (\boldsymbol{e}^k \cdot \dot{\boldsymbol{u}}_0)^2]$, is constant, independent of time (energy conservation).

3.2. *Vibrational eigenmodes in elastic media (elastic waves)*

As the simplest case, let us consider the eigenmodes in an elastic medium, i.e., elastic waves. Elastic media are continuum systems (i.e., not particulate systems), and their eigenfrequencies and eigenvectors are determined by continuum mechanics.[6,7]

An elastic wave is specified by the wavevector $\boldsymbol{q}$ and the polarization α. In the 3-dimensional case, there are three polarizations: two transverse ($\alpha = T_1, T_2$) and one longitudinal ($\alpha = L$). An eigenvector of elastic waves is a continuous function of the position $\boldsymbol{r} \in V$ (where V is the volume of the system)[f]:

$$\boldsymbol{e}_{\text{el}}^{\boldsymbol{q},\alpha} \equiv \left[\boldsymbol{e}_{\text{el}}^{\boldsymbol{q},\alpha\,T}(\boldsymbol{r});\ \boldsymbol{r} \in V\right]^T \equiv \left[\boldsymbol{s}_{\alpha}^{T}(\hat{\boldsymbol{q}})\,\frac{\exp(\mathrm{i}\boldsymbol{q}\cdot\boldsymbol{r})}{\sqrt{V}};\ \boldsymbol{r} \in V\right]^T. \tag{27}$$

$\boldsymbol{s}_\alpha(\hat{\boldsymbol{q}})$ is the polarization vector, which depends on the direction of $\boldsymbol{q}$, i.e., $\hat{\boldsymbol{q}} \equiv \boldsymbol{q}/q$ (where $q \equiv |\boldsymbol{q}|$ is the wavenumber), and can be determined from the elastic equation of motion that corresponds to the eigen equation (21):

$$\omega^2 \boldsymbol{s}_\alpha(\hat{\boldsymbol{q}}) = \left(\frac{\mathcal{C}(\hat{\boldsymbol{q}})}{\rho}\right) q^2 \boldsymbol{s}_\alpha(\hat{\boldsymbol{q}}), \tag{28}$$

where ρ is the mass density and $\mathcal{C}(\hat{\boldsymbol{q}})$ (a 3×3 tensor) is the elastic modulus tensor in the $\hat{\boldsymbol{q}}$ direction.[g]

Since $(\mathcal{C}(\hat{\boldsymbol{q}})/\rho)q^2$ (a 3×3 matrix) is positive and symmetric, its eigenvalues ω^2 are real, positive numbers, and the eigenvectors $\boldsymbol{s}_\alpha(\hat{\boldsymbol{q}})$ are orthonormalized as $\boldsymbol{s}_\alpha(\hat{\boldsymbol{q}}) \cdot \boldsymbol{s}_{\alpha'}(\hat{\boldsymbol{q}}) = \delta_{\alpha\alpha'}$. The eigenvectors $\boldsymbol{e}_{\text{el}}^{\boldsymbol{q},\alpha}(\boldsymbol{r})$ are then orthonormalized as[h]

$$\boldsymbol{e}_{\text{el}}^{\boldsymbol{q},\alpha} \cdot \boldsymbol{e}_{\text{el}}^{\boldsymbol{q}',\alpha'*} = \int_V d^3\boldsymbol{r}\,\boldsymbol{e}_{\text{el}}^{\boldsymbol{q},\alpha}(\boldsymbol{r}) \cdot \boldsymbol{e}_{\text{el}}^{\boldsymbol{q}',\alpha'*}(\boldsymbol{r}) = \delta_{\boldsymbol{q}\boldsymbol{q}'}\delta_{\alpha\alpha'}, \tag{29}$$

---

[f] The continuous variable $\boldsymbol{r}$ for an elastic medium corresponds to the particle index "i" in a particulate system.

[g] Given the elastic modulus tensor $C_{\alpha\beta\gamma\delta}$ (a $3 \times 3 \times 3 \times 3$ tensor), where $\alpha, \beta, \gamma, \delta = x, y, z$, $\mathcal{C}(\hat{\boldsymbol{q}})$ is defined as $(\mathcal{C}(\hat{\boldsymbol{q}}))_{\alpha\delta} = \sum_{\beta,\gamma} C_{\alpha\beta\gamma\delta}\hat{q}_\beta\hat{q}_\gamma$, where $\hat{\boldsymbol{q}} \equiv (\hat{q}_x\,\hat{q}_y\,\hat{q}_z)$. We will describe the measurement of $C_{\alpha\beta\gamma\delta}$ in Section 5.

[h] For an elastic medium, we need to replace $\sum_{i=1}^{N}(\cdots)$ in Eq. (20) with $\int_V d^3\boldsymbol{r}(\cdots)$.

where $*$ denotes the complex conjugate. Equation (28) determines the eigenfrequency $\omega_\alpha(\boldsymbol{q})$ as a linear function of q:

$$\omega_\alpha(\boldsymbol{q}) = c_\alpha(\hat{\boldsymbol{q}})q, \tag{30}$$

where $c_\alpha(\hat{\boldsymbol{q}})$ is the sound speed, which is the square root of the eigenvalue of $\mathcal{C}(\hat{\boldsymbol{q}})/\rho$. We note that Eq. (28) is the so-called Christoffel equation, which can be solved analytically following the Every algorithm.[71]

In the case of an isotropic medium, the elastic modulus tensor contains two independent moduli: the shear modulus G and the bulk modulus K. $\boldsymbol{s}_{T_1}(\hat{\boldsymbol{q}})$ and $\boldsymbol{s}_{T_2}(\hat{\boldsymbol{q}})$ are perpendicular to $\boldsymbol{q}$, while $\boldsymbol{s}_L(\hat{\boldsymbol{q}})$ is parallel to $\boldsymbol{q}$: $\hat{\boldsymbol{q}} \cdot \boldsymbol{s}_{T_1}(\hat{\boldsymbol{q}}) = \hat{\boldsymbol{q}} \cdot \boldsymbol{s}_{T_2}(\hat{\boldsymbol{q}}) = 0$ and $\boldsymbol{s}_L(\hat{\boldsymbol{q}}) = \hat{\boldsymbol{q}}$. $c_\alpha(\hat{\boldsymbol{q}})$ does not depend on $\hat{\boldsymbol{q}}$, and it is calculated as

$$c_{T_1} = c_{T_2} = \sqrt{\frac{G}{\rho}}, \quad c_L = \sqrt{\frac{K + 4G/3}{\rho}}. \tag{31}$$

On the other hand, in the case of an anisotropic medium, both of $\boldsymbol{s}_\alpha(\hat{\boldsymbol{q}})$ and $c_\alpha(\hat{\boldsymbol{q}})$ depend on $\hat{\boldsymbol{q}}$ in a complicated manner. The simplest example is a medium with cubic symmetry, which has three independent moduli: the pure shear modulus G_p, the simple shear modulus G_s, and the bulk modulus K.[i] Figure 2 shows schematic illustrations of $\boldsymbol{s}_\alpha(\hat{\boldsymbol{q}})$ in three directions of $\hat{\boldsymbol{q}}$, namely, [100], [110], and [111], where we use Miller indices.[j] Additionally, $c_\alpha(\hat{\boldsymbol{q}})$ is calculated as

$$c_{T_1} = c_{T_2} = \sqrt{\frac{G_s}{\rho}}, \quad c_L = \sqrt{\frac{K + 4G_p/3}{\rho}} \quad \text{for } [100], \tag{32}$$

---

[i] G_p, G_s, and K correspond to the elastic moduli for pure shear, simple shear, and bulk deformations, respectively. These three deformations are illustrated in Fig. 11 of Section 5.

[j] $\hat{\boldsymbol{q}} = (1,0,0)$ and all equivalent vectors for [100], $\hat{\boldsymbol{q}} = (1/\sqrt{2}, 1/\sqrt{2}, 0)$ and all equivalent vectors for [110], and $\hat{\boldsymbol{q}} = (1/\sqrt{3}, 1/\sqrt{3}, 1/\sqrt{3})$ and all equivalent vectors for [111].

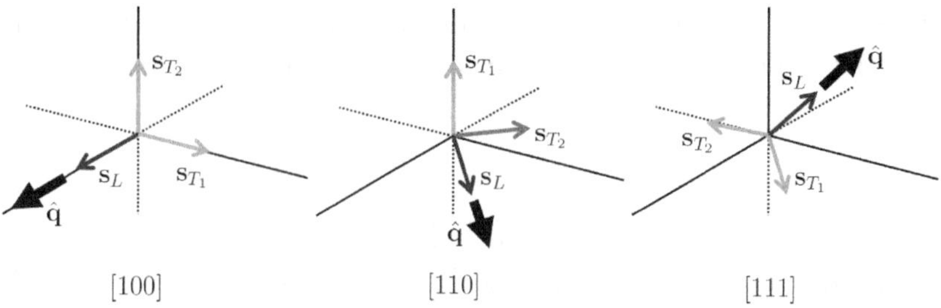

Fig. 2. Schematic illustration of the polarization vector $\boldsymbol{s}_\alpha(\widehat{\boldsymbol{q}})$ in a medium with cubic symmetry. We show $\boldsymbol{s}_{T1}(\widehat{\boldsymbol{q}})$, $\boldsymbol{s}_{T2}(\widehat{\boldsymbol{q}})$, and $\boldsymbol{s}_L(\widehat{\boldsymbol{q}})$ in three directions of $\widehat{\boldsymbol{q}}$: [100], [110], and [111] (Miller indices).

$$c_{T_1} = \sqrt{\frac{G_s}{\rho}}, \quad c_{T_2} = \sqrt{\frac{G_p}{\rho}}, \quad c_L = \sqrt{\frac{K + G_p/3 + G_s}{\rho}} \quad \text{for [110]}, \tag{33}$$

$$c_{T_1} = c_{T_2} = \sqrt{\frac{2G_p + G_s}{3\rho}}, \quad c_L = \sqrt{\frac{K + 4G_s/3}{\rho}} \quad \text{for [111]}. \tag{34}$$

A general solution for $c_\alpha(\hat{\boldsymbol{q}})$ as a function of $\hat{\boldsymbol{q}}$ is given in Ref. [72].

3.3. *Vibrational eigenmodes in crystals (phonons)*

We next consider the eigenmodes in crystals. The inherent structure $\boldsymbol{R}$ of a crystal is a periodic lattice structure (we show the radial distribution function $g(r)$ in Figure 1(b)). In this case, the eigenmodes have been established to be phonons (lattice waves).[6,7] Here, we consider single-component crystals, for which the eigenmodes are phonons of the acoustic type. For multi-component crystals, phonons of the optical type also appear, which are not considered below.

A phonon is specified by the wavevector $\boldsymbol{q}$ and the polarization α, as is an elastic wave. There are N wavevectors in the first Brillouin zone, $\boldsymbol{q} = \boldsymbol{q}_1, \boldsymbol{q}_2, \ldots, \boldsymbol{q}_N$, and three polarizations, $\alpha = T_1, T_2, L$; therefore, there are $3N$ phonons in total. The eigenvector $\boldsymbol{e}_{\text{ph}}^{\boldsymbol{q},\alpha}$ is described as

$$\boldsymbol{e}_{\text{ph}}^{\boldsymbol{q},\alpha} \equiv \left[\boldsymbol{s}_\alpha^T(\boldsymbol{q}) \frac{e^{\mathrm{i}\boldsymbol{q}\cdot\boldsymbol{R}_1}}{\sqrt{N}}, \boldsymbol{s}_\alpha^T(\boldsymbol{q}) \frac{e^{\mathrm{i}\boldsymbol{q}\cdot\boldsymbol{R}_2}}{\sqrt{N}}, \ldots, \boldsymbol{s}_\alpha^T(\boldsymbol{q}) \frac{e^{\mathrm{i}\boldsymbol{q}\cdot\boldsymbol{R}_N}}{\sqrt{N}} \right]^T, \tag{35}$$

where the factor $1/\sqrt{N}$ is necessary for the orthonormality condition, $\boldsymbol{e}_{\text{ph}}^{q,\alpha}\boldsymbol{e}_{\text{ph}}^{q',\alpha'*} = \delta_{qq'}\delta_{\alpha\alpha'}$. The eigen equation (21) then reduces to

$$\omega^2 \boldsymbol{s}_\alpha(\boldsymbol{q}) = D(\boldsymbol{q})\boldsymbol{s}_\alpha(\boldsymbol{q}), \tag{36}$$

where $D(\boldsymbol{q})$ (a 3×3 matrix) is the dynamical matrix of the unit cell of the lattice structure:

$$D(\boldsymbol{q}) = \frac{1}{N}\sum_{i,j=1}^{N} (\mathcal{D})_{ij} \exp(-\mathrm{i}\boldsymbol{q}\cdot(\boldsymbol{R}_i - \boldsymbol{R}_j)). \tag{37}$$

Since $D(\boldsymbol{q})$ is positive and symmetric, Eq. (36) gives the real number of eigenfrequencies $\omega_\alpha(\boldsymbol{q})$ and polarization vectors $\boldsymbol{s}_\alpha(\boldsymbol{q})$ with the orthonormality condition, $\boldsymbol{s}_\alpha(\boldsymbol{q}) \cdot \boldsymbol{s}_{\alpha'}(\boldsymbol{q}) = \delta_{\alpha\alpha'}$. The eigenvectors of the phonons in Eq. (35) are of the same form as those of the elastic waves in Eq. (27); however, we remark that they are discretized waves in a particulate system. As an example, we visualize a transverse phonon in Figures 3(b) and 3(d). We can see the vortex structure that is characteristic of transverse phonons.

An important feature is that the dynamical matrix, $D(\boldsymbol{q})$, converges to that of elastic waves, $(\mathcal{C}(\hat{\boldsymbol{q}})/\rho)q^2$, at low wavenumbers $q \ll 1$ with[k]

$$\frac{\mathcal{C}(\hat{\boldsymbol{q}})}{\rho} = -\frac{1}{2N}\sum_{i,j=1}^{N} (\mathcal{D})_{ij} \left[\hat{\boldsymbol{q}}\cdot(\boldsymbol{R}_i - \boldsymbol{R}_j)\right]^2, \tag{38}$$

and thus, the phonons smoothly converge to elastic waves.[1,6,7] In particular, the phonons in cubic crystals, such as FCC crystals, converge to the elastic waves in media with cubic symmetry. The eigenfrequency $\omega_\alpha(\boldsymbol{q})$ of the phonons converges to a linear function of q as in Eq. (30).

---

[k]We can show that $(\mathcal{C}(\hat{\boldsymbol{q}}))_{\alpha\delta} = \sum_{\beta,\gamma}\left(C^B_{\alpha\beta\gamma\delta} + C^C_{\alpha\beta\gamma\delta}\right)\hat{q}_\beta\hat{q}_\gamma$ with $\hat{\boldsymbol{q}} \equiv (\hat{q}_x, \hat{q}_y, \hat{q}_z)$, where $C^B_{\alpha\beta\gamma\delta}$ and $C^C_{\alpha\beta\gamma\delta}$ are respectively the Born term and the correction term of elastic modulus tensor, given in Eq. (85) of Section 5.

[l]Acoustic phonons (the type considered here) converge to elastic waves at $q \ll 1$; however, optical phonons (not considered here) do not. Optical phonons represent vibrations of particles within a unit cell (vibrations at microscopic scales).

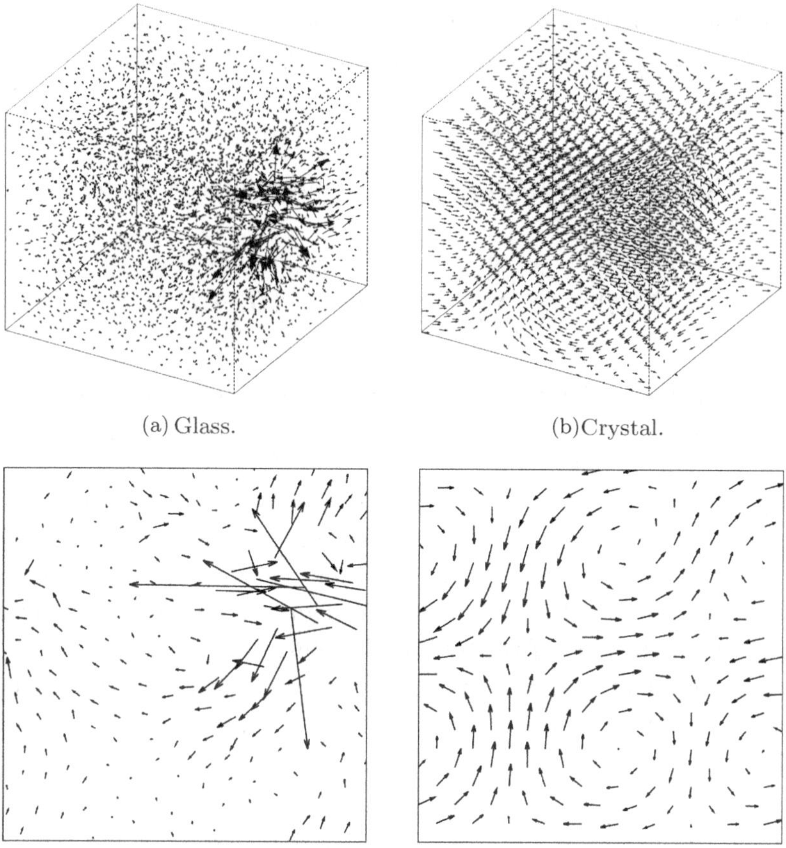

(a) Glass. (b)Crystal.

(c) Glass. (d)Crystal.

Fig. 3. Visualization of the eigenmodes in SC systems. (a),(c) Glass. (b),(d) FCC crystal. We show the eigenvector field, $\boldsymbol{e}^k \equiv \left[\boldsymbol{e}_1^{kT}, \boldsymbol{e}_2^{kT}, \ldots, \boldsymbol{e}_N^{kT}\right]^T$, as arrows. In (c),(d), the eigenvectors on some plane are presented in a two-dimensional plot.

3.4. *Vibrational eigenmodes in glasses*

Unlike for the phonons in crystals, we do not have any established formulation for the eigenmodes in glasses. In this case, computational simulations are a powerful tool. We can numerically diagonalize the dynamical matrix $\mathcal{D}$ (solve the eigen equation (21)) to obtain the eigenfrequencies, $\omega^k \equiv \sqrt{\lambda^k}$, and the eigenvectors, $\boldsymbol{e}^k \equiv [\boldsymbol{e}_1^{kT}, \boldsymbol{e}_2^{kT}, \ldots, \boldsymbol{e}_N^{kT}]^T$ $(k = 1, 2, \ldots, 3N)$. Note that when we diagonalize the $\mathcal{D}$ of a crystal, we obtain phonons, as presented in Figures 3(b) and 3(d). We also note that for a 3-dimensional system under periodic boundary conditions, there appear three zero-frequency modes that represent uniform translations of particles. These modes are known as Goldstone modes and are usually discarded.

Figures 3(a) and 3(c) present an example of eigenmodes in an SC glass. We can see localized vibration, in which some particles vibrate considerably while other particles vibrate much less. This localization is a characteristic feature of glasses (disordered systems).[26–28] Although phonon-like modes exist in glasses, the eigenmodes are non-phonon modes in general. We may expect that at low frequencies, glasses will behave as uniform elastic media and their eigenmodes will converge to elastic waves, as in the case of crystals and phonons. However, recent simulations[50–59] have demonstrated that this is not the case, as will be discussed in Section 6.

3.5. *Vibrational density of states*

Now that we have obtained information on the eigenfrequencies ω^k and eigenvectors $\boldsymbol{e}^k$, we can next analyse them to understand the vibrational properties of glasses (or any solids). One of the most important quantities is the vDOS $g(\omega)$. $g(\omega)$ measures how many eigenmodes exist around the frequency ω:

$$g(\omega) = \frac{1}{3N} \sum_{k=1}^{3N} \delta(\omega - \omega^k), \tag{39}$$

where $\delta(x)$ is the Dirac delta function. Figure 4(a) presents $g(\omega)$ for an SC glass. For comparison, we also plot $g(\omega)$ for an SC crystal in Figure 4(b). Before looking at the results in these figures, let us define the Debye vDOS $g_D(\omega)$,[6,7] which is the vDOS calculated by assuming glasses or crystals to be elastic media and is a useful reference for capturing the features of the vDOS.

The eigenfrequencies of elastic waves are given by $\omega_\alpha(\hat{\boldsymbol{q}}) = c_\alpha(\hat{\boldsymbol{q}})q$ as in Eq. (30). We can then calculate the vDOS by using Eq. (39) as follows (where we replace the summation $\sum_{k=1}^{3N}(\cdots)$ with the

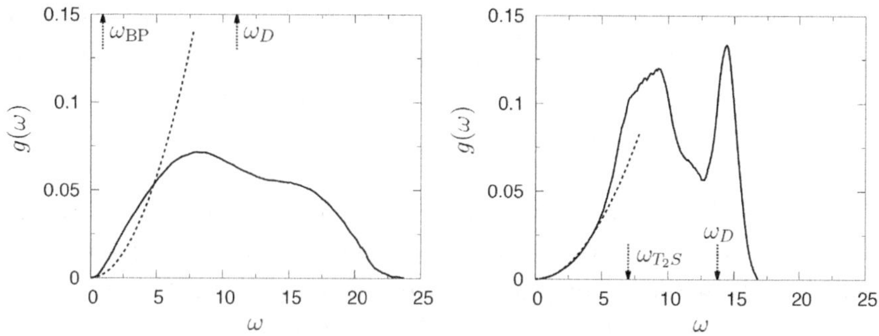

(a) Glass. (b) Crystal.

Fig. 4. The vDOS $g(\omega)$ of an SC system. (a) Glass. (b) FCC crystal. The dashed line shows the Debye vDOS, $g_D(\omega) = A_D\omega^2$. The Debye frequency ω_D, the boson peak (BP) frequency ω_{BP}, and the position of the lowest-frequency, van Hove singularity ω_{T_2S} are indicated by arrows.

integral $(L/2\pi)^3 \sum_{\alpha=T_1,T_2,L} \int d\boldsymbol{q}(\cdots)$):

$$g_D(\omega) = A_D\omega^2 = \frac{3}{\omega_D^3}\omega^2, \tag{40}$$

where $A_D \equiv 3/\omega_D^3$ is the Debye level and ω_D is the Debye frequency. $\omega_D \equiv \bar{c}k_D$, with $k_D \equiv (6\pi^2 N/V)^{1/3}$ being the Debye wavenumber and $\bar{c}$ being the average sound speed:

$$\bar{c} = \left(\frac{1}{3}\sum_{\alpha=T_1,T_2,L}\int \frac{d\Omega(\hat{\boldsymbol{q}})}{4\pi}\frac{1}{c_\alpha(\hat{\boldsymbol{q}})^3}\right)^{-1/3}, \tag{41}$$

where $d\Omega(\hat{\boldsymbol{q}})$ denotes the solid angle. In the case of an isotropic system such as a glass, ω_D is given in the following simple form:

$$\omega_D = \left[\frac{18\pi^2(N/V)}{2c_T^{-3} + c_L^{-3}}\right]^{1/3}. \tag{42}$$

For an anisotropic system, we need to solve Eq. (28) to obtain $c_\alpha(\hat{\boldsymbol{q}})$ as a function of $\hat{\boldsymbol{q}}$.[71] An analytical solution $c_\alpha(\hat{\boldsymbol{q}})$ for cubic crystals is given in Ref. [72]. Note that to calculate the Debye vDOS $g_D(\omega)$ (with Debye level A_D and Debye frequency ω_D), we need to measure the elastic moduli of the system, as described later in Section 5.

In Figure 4, the Debye vDOS $g_D(\omega)$ is plotted as a dashed line. Since the phonons in a crystal converge to elastic waves at low ω, the $g(\omega)$ function of the crystal converges to $g_D(\omega)$, as demonstrated in Figure 4(b). We can also clearly observe van Hove singularities.[6,7] In Figure 4(b), we use ω_{T_2S} to indicate the position of the lowest-frequency, van Hove singularity which corresponds to that of T_2 transverse phonons.

Compared to the crystal, the glass shows smoother variations and broader distributions of the eigenmodes. Remarkably, the low-frequency portion of $g(\omega)$ shows an enhancement in eigenmodes over the Debye vDOS. To demonstrate this point more clearly, Figure 5 presents the reduced vDOS, which is $g(\omega)$ divided by ω^2, i.e., $g(\omega)/\omega^2$. Note that $g_D(\omega)/\omega^2$ is equal to the Debye level A_D, i.e., a constant value, as shown by the dashed line in Figure 5. For the crystal, $g(\omega)/\omega^2$ coincides with A_D at low ω, whereas for the glass, $g(\omega)/\omega^2$ clearly shows an excess peak over A_D, which we call the boson peak (BP).[m,16,17,25] The frequency at which $g(\omega)/\omega^2$ reaches a maximum is called the BP frequency and is denoted by $\omega_{\rm BP}$ ($\omega_{\rm BP} \approx 0.9$ in Figure 5(a)). We remark that $g(\omega)/\omega^2$ also shows a peak for the crystal, as shown in Figure 5(b); however, this peak

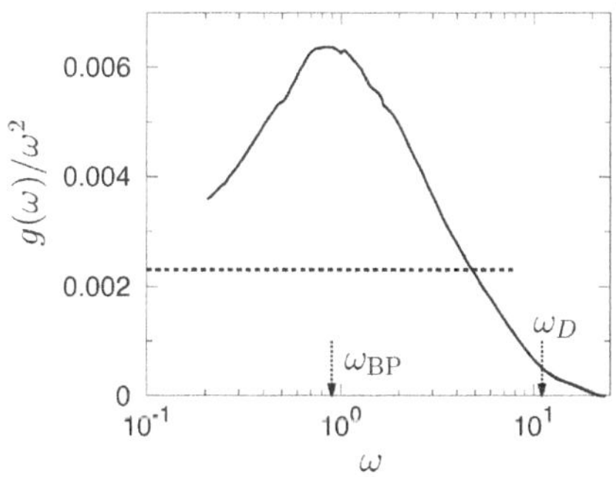

(a) Glass.

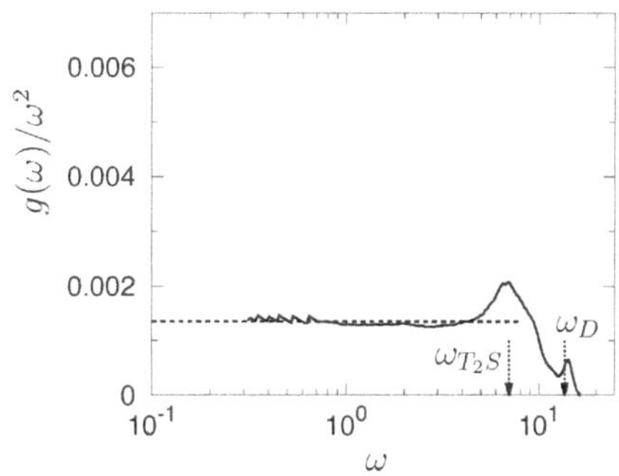

(b) Crystal.

Fig. 5. The reduced vDOS $g(\omega)/\omega^2$ of an SC system. (a) Glass. (b) FCC crystal. The dashed line shows the Debye level A_D. See also the caption of Figure 4.

[m]The excess eigenmodes result in enhancement of the specific heat of a glass. This excess specific heat is also called the boson peak.[13–15]

corresponds to a van Hove singularity and appears at a frequency much higher than ω_{BP} ($\omega_{T_2S} \approx 7 \gg \omega_{\mathrm{BP}} \approx 0.9$). The van Hove singularity picks up vibrations at a microscopic length scale on the order of the lattice constant,[6,7] whereas the boson peak exhibits vibrations at a much longer, mesoscopic length scale.[32,33]

As ω decreases below the BP frequency ω_{BP}, the $g(\omega)/\omega^2$ of the glass approaches the Debye level A_D, as shown in Figure 5(a). We might expect that $g(\omega)/\omega^2$ will converge to A_D at some frequency, where the eigenmodes will converge to elastic waves. However, this is not the case, as will be discussed in detail in Section 6.

3.6. *Characterization of vibrational eigenmodes*

We employ order parameters to characterize the vibrations of particles in each eigenmode k. Here, we introduce two order parameters. One is the participation ratio, which measures the extent of localization.[26–28] The other is the phonon order parameter, which measures the extent of phonon-like vibrations.[51,52]

3.6.1. *Participation ratio*

As already seen in Figure 3, vibrational localization occurs in glasses. The participation ratio quantitatively measures the extent of localization.[26–28] Given an eigenvector, $\mathbf{e}^k \equiv \left[\mathbf{e}_1^{kT}, \mathbf{e}_2^{kT}, \ldots, \mathbf{e}_N^{kT}\right]^T$, for eigenmode k, its participation ratio $\mathcal{P}^k$ is calculated as

$$\mathcal{P}^k \equiv \frac{1}{N}\left[\sum_{i=1}^{N} \mathbf{e}_i^k \cdot \mathbf{e}_i^k\right]^2 \left[\sum_{i=1}^{N} \left(\mathbf{e}_i^k \cdot \mathbf{e}_i^k\right)^2\right]^{-1} = \frac{1}{N}\left[\sum_{i=1}^{N} \left(\mathbf{e}_i^k \cdot \mathbf{e}_i^k\right)^2\right]^{-1}, \tag{43}$$

where the last equality comes from the orthonormality condition in Eq. (20). $\mathcal{P}^k$ measures the fraction of particles that participate in the vibrations. As extreme cases, $\mathcal{P}^k = \left(\sum_{i=1}^N m_i\right)^2 / \left(N \sum_{i=1}^N m_i^2\right)$ for an ideal mode in which all constituent particles vibrate equally,[n]

---

[n] For the case of identical mass $m_i \equiv m$, $\mathcal{P}^k = \left(\sum_{i=1}^N m\right)^2 / \left(N \sum_{i=1}^N m^2\right) = 1$.

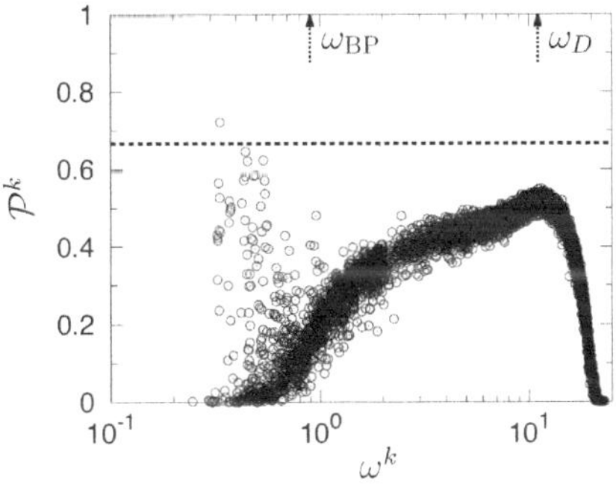

(a)Glass.

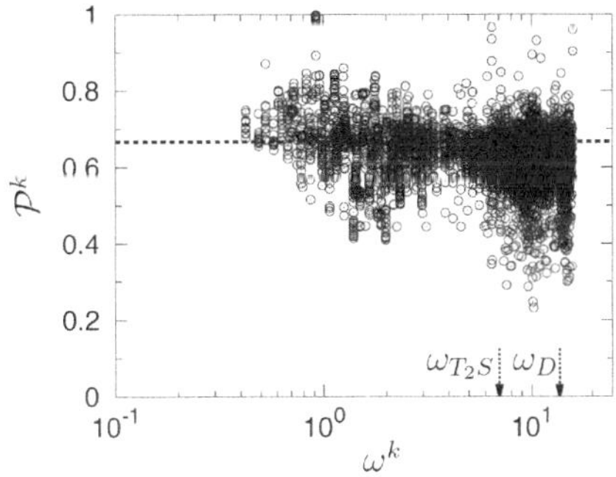

(b)Crystal.

Fig. 6. The participation ratio $\mathcal{P}^k$ of an SC system. (a) Glass. (b) FCC crystal. The dashed line corresponds to $\mathcal{P}^k = 2/3$, the value for elastic waves. See also the caption of Figure 4.

$\mathcal{P}^k = 1/N \ll 1$ for an ideal mode involving only one particle, and $\mathcal{P}^k = 2/3$ for elastic waves, $\boldsymbol{e}_{\rm el}^{q,\alpha}$, as given in Eq. (27).[o]

Figure 6 plots $\mathcal{P}^k$ as a function of the frequency ω^k for (a) an SC glass and (b) an SC crystal. We observe that the values of $\mathcal{P}^k$ in the crystal fluctuate around $\mathcal{P}^k = 2/3$, indicating that the vibrations are similar to elastic waves. Indeed, the eigenvectors of phonons, $\boldsymbol{e}_{\rm ph}^{q,\alpha}$, have the same form as those of elastic waves, $\boldsymbol{e}_{\rm el}^{q,\alpha}$ (compare Eqs. (35) and (27)). However, phonons are not continuum waves but discrete waves, and their discrete nature causes fluctuations in $\mathcal{P}^k$ around $\mathcal{P}^k = 2/3$.

On the other hand, the values of $\mathcal{P}^k$ in the glass are much different to those in the crystal. Remarkably, the glass shows localization at the low-ω and high-ω edges, which is a characteristic feature of disordered matter. For the low-frequency case, localization starts to occur near the BP frequency $\omega_{\rm BP}$. In addition to the participation ratio, we can also employ, for example, an effective mass[26] to measure the extent of localization.

[o]For elastic waves, we can calculate their participation ratio as $\mathcal{P}^k \equiv (1/V)$ $[\int_V d^3\boldsymbol{r}{\rm Re}(\boldsymbol{e}_{\rm el}^{q,\alpha}(\boldsymbol{r}))\cdot{\rm Re}(\boldsymbol{e}_{\rm el}^{q,\alpha}(\boldsymbol{r}))]^2[\int_V d^3\boldsymbol{r}\{{\rm Re}(\boldsymbol{e}_{\rm el}^{q,\alpha}(\boldsymbol{r}))\cdot{\rm Re}(\boldsymbol{e}_{\rm el}^{q,\alpha}(\boldsymbol{r}))\}^2]^{-1} = 2/3$.

3.6.2. *Phonon order parameter*

As the frequency decreases, the eigenmodes in glasses tend to show phonon-like vibrations. The phonon order parameter evaluates the extent to which an eigenmode $\boldsymbol{e}^k$ is similar to phonons $\boldsymbol{u}_{\text{ph}}^{q,\alpha}$:[51,52]

$$\boldsymbol{u}_{\text{ph}}^{q,\alpha} \equiv \left[\boldsymbol{s}_{\alpha}^{T}(\boldsymbol{q})\frac{e^{\text{i}\boldsymbol{q}\cdot\boldsymbol{R}_1}}{\sqrt{N}}, \boldsymbol{s}_{\alpha}^{T}(\boldsymbol{q})\frac{e^{\text{i}\boldsymbol{q}\cdot\boldsymbol{R}_2}}{\sqrt{N}}, \ldots, \boldsymbol{s}_{\alpha}^{T}(\boldsymbol{q})\frac{e^{\text{i}\boldsymbol{q}\cdot\boldsymbol{R}_N}}{\sqrt{N}}\right]^{T}. \tag{44}$$

Here we define $\boldsymbol{u}_{\text{ph}}^{q,\alpha}$ to extend the concept of phonons to any solid-states including amorphous states. The eigenvector $\boldsymbol{e}^k$ can always be expanded in terms of a series of $3N$ phonons $\boldsymbol{u}_{\text{ph}}^{q,\alpha}$, as $\boldsymbol{e}^k = \sum_{q,\alpha} A_{q,\alpha}^k \boldsymbol{u}_{\text{ph}}^{q,\alpha}$. In the case of a crystal where the inherent structure $\boldsymbol{R}$ is a lattice structure, $\boldsymbol{u}_{\text{ph}}^{q,\alpha}$ are exactly the same as the eigenvectors $\boldsymbol{e}_{\text{ph}}^{q,\alpha}$ in Eq. (35), which are orthonormalized as $\boldsymbol{u}_{\text{ph}}^{q,\alpha} \cdot \boldsymbol{u}_{\text{ph}}^{q',\alpha'*} = \delta_{qq'}\delta_{\alpha\alpha'}$. For the case of a glass where $\boldsymbol{R}$ is a disordered structure, the phonons $\boldsymbol{u}_{\text{ph}}^{q,\alpha}$ (which are not the eigenvectors in general) are not exactly identical to but can be approximately orthonormalized as $\boldsymbol{u}_{\text{ph}}^{q,\alpha} \cdot \boldsymbol{u}_{\text{ph}}^{q',\alpha'*} \approx \delta_{qq'}\delta_{\alpha\alpha'}$. Thus, we can calculate the projection of $\boldsymbol{e}^k$ onto one particular phonon, $\boldsymbol{u}_{\text{ph}}^{q,\alpha}$, as follows:

$$O_{q,\alpha}^k = \left|A_{q,\alpha}^k\right|^2 \approx \left|\boldsymbol{u}_{\text{ph}}^{q,\alpha} \cdot \boldsymbol{e}^k\right|^2. \tag{45}$$

Here, note that $\sum_{q,\alpha} O_{q,\alpha}^k \approx 1$ since $\boldsymbol{e}^k \cdot \boldsymbol{e}^k = 1$.

If eigenmode k shows the phonon vibration, then $\boldsymbol{e}^k$ can be described as the sum of a finite number of phonons with large overlap: $\boldsymbol{e}^k = \sum_{q,\alpha; O_{q,\alpha}^k \geq N_m/3N} A_{q,\alpha}^k \boldsymbol{u}_{\text{ph}}^{q,\alpha}$. Here, we define "large overlap" to mean that $O_{q,\alpha}^k \geq N_m/3N$, i.e., the phonons overlap by more than N_m modes. Considering this, we define the phonon order parameter O^k as

$$O^k = \sum_{q,\alpha;\, O_{q,\alpha}^k \geq N_m/3N} O_{q,\alpha}^k. \tag{46}$$

$O^k = 1$ for a phonon, whereas $O^k = 0$ for a mode that is considerably different to a phonon. To calculate the phonon order parameter in Eq. (46), we need to set an appropriate value of N_m; the obtained results and conclusions should not depend on the choice of N_m. $N_m = 100$

might be appropriate, but this depends on the considered system. Simulation data for the phonon order parameter will be presented in Section 6.

4. Phonon Transport

Phonon transport is a fundamental property of solid-state materials.[6,7] Phonons play a role in carrying heat energy. In crystals, phonons are the vibrational eigenmodes, and they propagate without any attenuation in the zero-temperature, harmonic limit. The scattering of phonons is induced through anharmonic effects at finite temperatures. This situation is drastically changed in glasses, where the eigenmodes are non-phonon modes, as seen in Section 3. In this case, a phonon will be decomposed into several different eigenmodes in some frequency range,[29] and as a result, it will be strongly scattered. The disordered structures therefore scatter phonons propagating in glasses.

A phonon is specified by the wavevector $\boldsymbol{q}$ (wavenumber $q \equiv |\boldsymbol{q}|$ and $\hat{\boldsymbol{q}} \equiv \boldsymbol{q}/q$) and the polarization α, where $\alpha = T_1, T_2$ (transverse) or L (longitudinal). Its transport is characterized by two quantities: the propagation frequency $\Omega(\boldsymbol{q})$ and the attenuation rate $\Gamma_\alpha(\boldsymbol{q})$. (For the frequency $\Omega(\boldsymbol{q})$, we do not explicitly write the polarization index α.) The propagation speed is given by $c_\alpha(\boldsymbol{q}) = \Omega(\boldsymbol{q})/q$.[p] Below, we will describe how to measure $\Omega(\boldsymbol{q})$ or $c_\alpha(\boldsymbol{q})$, and $\Gamma_\alpha(\boldsymbol{q})$ through simulations.

4.1. *Dynamic structure factor*

Let us introduce the dynamic structure factor to study phonon transport. The dynamic structure factor can be measured through scattering experiments involving light, X-rays, or neutrons. In the computational simulation approach, we first perform an MD simulation at a finite temperature $T > 0$ and obtain the corresponding

[p] $c_\alpha(\boldsymbol{q}) = \Omega(\boldsymbol{q})/q$ is the phase speed, while the group speed is given by $v_\alpha(\boldsymbol{q}) = d\Omega/dq|_{\hat{q}}$. In the low-wavenumber region, these two speeds coincide with each other.

trajectory data, $\boldsymbol{r}(t) \equiv \left[\boldsymbol{r}_1^T(t), \boldsymbol{r}_2^T(t), \ldots, \boldsymbol{r}_N^T(t)\right]^T$. From this trajectory, the dynamic structure factor, $S_\alpha(\boldsymbol{q}, \omega)$ (where $\alpha = T$ or L), is calculated as the Fourier transform of the current-current correlation function[37–39, 41]:

$$S_\alpha(\boldsymbol{q}, \omega) = \left(\frac{q}{\omega}\right)^2 \frac{1}{2\pi} \int \frac{1}{N} \langle \boldsymbol{j}_\alpha(\boldsymbol{q}, t) \cdot \boldsymbol{j}_\alpha^*(\boldsymbol{q}, 0)\rangle_0 \exp\left(\mathrm{i}\omega t\right) dt, \tag{47}$$

where $\langle \cdots \rangle_0$ denotes the ensemble average over configurations at $t = 0$. $\boldsymbol{j}_\alpha(\boldsymbol{q}, t)$ is the transverse ($\alpha = T$) or longitudinal ($\alpha = L$) current:

$$\begin{aligned} \boldsymbol{j}_T(\boldsymbol{q}, t) &= \sum_{i=1}^{N} \left[\boldsymbol{v}_i(t) - (\boldsymbol{v}_i(t) \cdot \hat{\boldsymbol{q}})\, \hat{\boldsymbol{q}}\right] \exp\left(\mathrm{i}\boldsymbol{q} \cdot \boldsymbol{r}_i(t)\right), \\ \boldsymbol{j}_L(\boldsymbol{q}, t) &= \sum_{i=1}^{N} (\boldsymbol{v}_i(t) \cdot \hat{\boldsymbol{q}})\, \hat{\boldsymbol{q}} \exp\left(\mathrm{i}\boldsymbol{q} \cdot \boldsymbol{r}_i(t)\right), \end{aligned} \tag{48}$$

where $\boldsymbol{v}_i \equiv d\boldsymbol{r}_i/dt$ is the velocity of particle i. For the longitudinal case, the conservation law relates the current $\boldsymbol{j}_L(\boldsymbol{q}, t)$ and the number density $\hat{\rho}(\boldsymbol{q}, t) \equiv \sum_{i=1}^{N} \exp\left(\mathrm{i}\boldsymbol{q} \cdot \boldsymbol{r}_i(t)\right)$ in the form:

$$\frac{\partial \hat{\rho}(\boldsymbol{q}, t)}{\partial t} = \mathrm{i}\boldsymbol{q} \cdot \boldsymbol{j}_L(\boldsymbol{q}, t), \tag{49}$$

and $S_L(\boldsymbol{q}, \omega)$ corresponds to the density-density correlation function.[q]

We can understand the dynamic structure factor $S_\alpha(\boldsymbol{q}, \omega)$ as follows. The phonon displacement vector, $\boldsymbol{u}_{\mathrm{ph}}^{\boldsymbol{q},\alpha}$, is given in Eq. (44). When we take the polarization vectors to be $\boldsymbol{s}_T(\boldsymbol{q}) = \hat{\boldsymbol{q}}^{\perp}$ (where $\perp$ denotes a perpendicular vector) and $\boldsymbol{s}_L(\boldsymbol{q}) = \hat{\boldsymbol{q}}$, the current-current correlation function $\langle \boldsymbol{j}_\alpha(\boldsymbol{q}, t) \cdot \boldsymbol{j}_\alpha^*(\boldsymbol{q}, 0)\rangle_0$ is formulated as

$$\begin{aligned} \frac{1}{N} \langle \boldsymbol{j}_\alpha(\boldsymbol{q}, t) \cdot \boldsymbol{j}_\alpha^*(\boldsymbol{q}, 0)\rangle_0 &= \left\langle \left(\boldsymbol{v}(t) \cdot \boldsymbol{u}_{\mathrm{ph}}^{\boldsymbol{q},\alpha}\right) \left(\boldsymbol{v}(0) \cdot \boldsymbol{u}_{\mathrm{ph}}^{\boldsymbol{q},\alpha}\right)^* \right\rangle_0 \\ &\quad \times \left[1 + \mathcal{O}\left(\sqrt{T}\right)\right], \end{aligned} \tag{50}$$

[q]The factor $(q/\omega)^2$ in $S_\alpha(\boldsymbol{q}, \omega)$ as expressed in Eq. (47) is necessary for $S_L(\boldsymbol{q}, \omega)$ to correspond to the density-density correlation function.

where $\boldsymbol{v}(t) \equiv \left[\boldsymbol{v}_1^T(t), \boldsymbol{v}_2^T(t), \ldots, \boldsymbol{v}_N^T(t)\right]^T$ is the velocity vector of the system. Therefore, $S_\alpha(\boldsymbol{q},\omega)$ at the low-T harmonic limit converges to the correlation function of the velocity projected onto $\boldsymbol{u}_{\mathrm{ph}}^{\boldsymbol{q},\alpha}$. Thermal energy activates many different phonons with different wavevectors and different polarizations. From among these activated phonons, $S_\alpha(\boldsymbol{q},\omega)$ picks up a specific phonon $\boldsymbol{u}_{\mathrm{ph}}^{\boldsymbol{q},\alpha}$ of wavevector $\boldsymbol{q}$ and polarization α. We remark that the actual polarization vectors of the phonons can deviate from $\boldsymbol{s}_T(\boldsymbol{q}) = \hat{\boldsymbol{q}}^\perp$ and $\boldsymbol{s}_L(\boldsymbol{q}) = \hat{\boldsymbol{q}}$. However, they still considerably overlap with $\boldsymbol{s}_T(\boldsymbol{q}) = \hat{\boldsymbol{q}}^\perp$ and $\boldsymbol{s}_L(\boldsymbol{q}) = \hat{\boldsymbol{q}}$, and we can properly analyse phonon transport by means of $S_\alpha(\boldsymbol{q},\omega)$.

Figure 7 shows $S_\alpha(\boldsymbol{q},\omega)$ in the [110] direction (Miller indices) for an SC system. The temperature is very low, $T = 10^{-2}$, well below

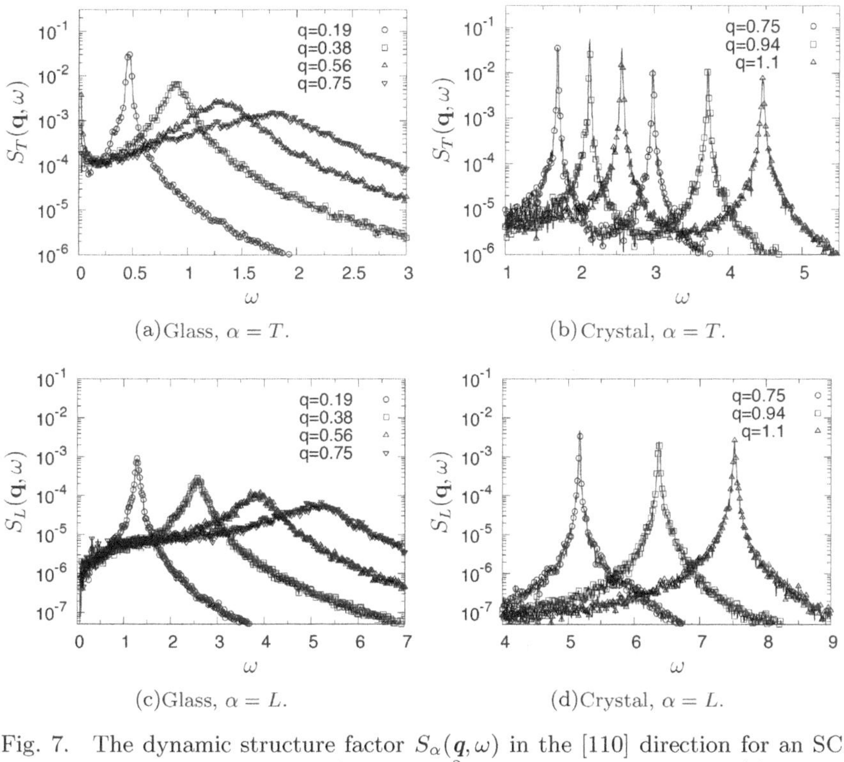

(a) Glass, $\alpha = T$. (b) Crystal, $\alpha = T$.

(c) Glass, $\alpha = L$. (d) Crystal, $\alpha = L$.

Fig. 7. The dynamic structure factor $S_\alpha(\boldsymbol{q},\omega)$ in the [110] direction for an SC system at a low temperature of $T = 10^{-2}$, well below T_g and T_m. (a) Glass, $\alpha = T$. (b) FCC crystal, $\alpha = T$. (c) Glass, $\alpha = L$. (d) FCC crystal, $\alpha = L$.

the glass transition temperature T_g and the melting temperature T_m. We plot $S_\alpha(\boldsymbol{q}, \omega)$ for a glass and for an FCC crystal. We can clearly observe the Brillouin peaks for both the glass and the crystal. For the case of the FCC crystal, $S_T(\boldsymbol{q}, \omega)$ for transverse phonons shows two peaks at a fixed q, as shown in Figure 7(b), which correspond to a T_2 phonon at the lower ω and a T_1 phonon at the higher ω (see Figure 2 for the polarization vectors $\boldsymbol{s}_{T_2}$ and $\boldsymbol{s}_{T_1}$). On the other hand, the longitudinal $S_L(\boldsymbol{q}, \omega)$ shows a single Brillouin peak, as shown in Figure 7(d).

Unlike the FCC crystal, since the glass is an isotropic system, it shows a single Brillouin peak in both the transverse and longitudinal dynamic structure factors, as shown in Figures 7(a) and 7(c). Notably, the Brillouin peaks are rather broad compared to those in the crystal. This finding indicates that phonon attenuation is considerably enhanced in glasses, as we will see below.

Based on the data for $S_\alpha(\boldsymbol{q}, \omega)$, the propagation frequency $\Omega(\boldsymbol{q})$ and the attenuation rate $\Gamma_\alpha(\boldsymbol{q})$ can be extracted by fitting the spectral region around the Brillouin peak to the damped harmonic oscillator model[37–39, 41]:

$$S_\alpha(\boldsymbol{q}, \omega) \propto \frac{\Gamma_\alpha(\boldsymbol{q})\Omega^2(\boldsymbol{q})}{\left(\omega^2 - \Omega^2(\boldsymbol{q})\right)^2 + \omega^2\Gamma_\alpha^2(\boldsymbol{q})}. \tag{51}$$

This fitting procedure in the frequency domain is equivalent to fitting the correlation function $\langle \boldsymbol{j}_\alpha(\boldsymbol{q}, t) \cdot \boldsymbol{j}_\alpha^*(\boldsymbol{q}, 0)\rangle_0$ to the following function in the time domain:

$$\frac{1}{N} \langle \boldsymbol{j}_\alpha(\boldsymbol{q}, t) \cdot \boldsymbol{j}_\alpha^*(\boldsymbol{q}, 0)\rangle_0 \propto \cos\left(\Omega(\boldsymbol{q})t\right) \exp\left(-\frac{\Gamma_\alpha(\boldsymbol{q})}{2}t\right). \tag{52}$$

In the following, we will show data for $\Omega(\boldsymbol{q})$, $c_\alpha(\boldsymbol{q}) \equiv \Omega(\boldsymbol{q})/q$, and $\Gamma_\alpha(\boldsymbol{q})$ that are extracted from the $S_\alpha(\boldsymbol{q}, \omega)$ presented in Figure 7.

4.2. *Propagation frequency and attenuation rate*

Figures 8 and 9 plot the sound speed $c_\alpha(\Omega)$ and the attenuation rate $\Gamma_\alpha(\Omega)$, respectively, as functions of the frequency Ω. Here, we treat c_α and Γ_α as functions of Ω by transforming $\boldsymbol{q}$ into Ω via the relation $\Omega = \Omega(\boldsymbol{q})$. For the FCC crystal, we focus on the lower-frequency

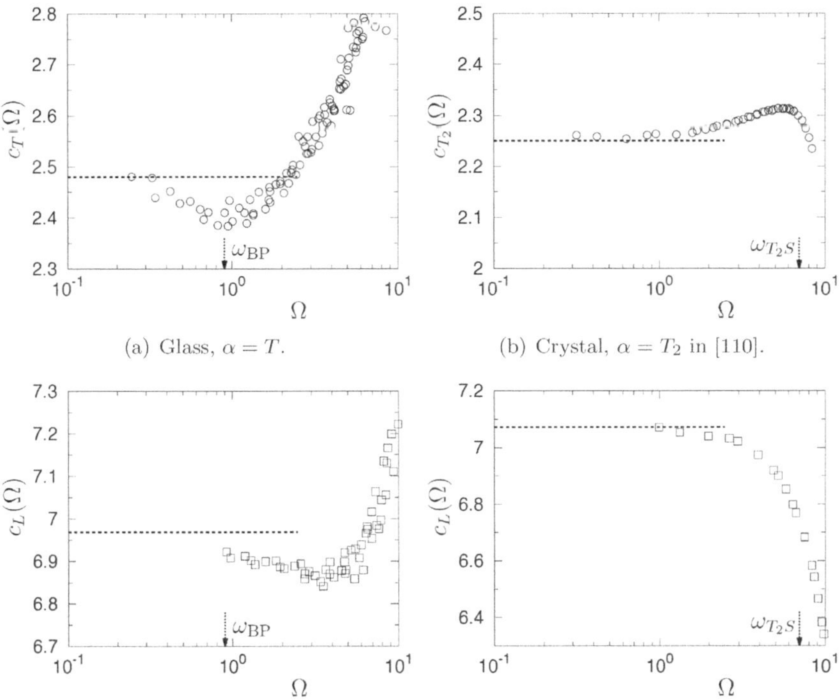

(a) Glass, $\alpha = T$.

(b) Crystal, $\alpha = T_2$ in [110].

(c) Glass, $\alpha = L$.

(d) Crystal, $\alpha = L$ in [110].

Fig. 8. The propagation speed $c_\alpha(\Omega)$ for an SC system. (a) Glass, $\alpha = T$. (b) FCC crystal, $\alpha = T_2$ in [110]. (c) Glass, $\alpha = L$. (d) FCC crystal, $\alpha = L$ in [110]. For the glass [(a),(c)] which is an isotropic system, we plot the data for the different directions ([100], [110], [111]) and for $\alpha = T_1$, T_2 all together, which coincide with each other. The dashed line indicates the macroscopic sound speed calculated from the elastic moduli, as $c_{T0} = \sqrt{G/\rho}$ and $c_{L0} = \sqrt{(K + 4G/3)/\rho}$ as in Eq. (31). For the crystal [(b),(d)], the macroscopic sound speed for [110] is plotted by the dashed line, as $c_{T_20} = \sqrt{G_p/\rho}$ and $c_{L0} = \sqrt{(K + G_p/3 + G_s)/\rho}$ as in Eq. (33). The BP frequency ω_{BP} and the position of the lowest-frequency, van Hove singularity ω_{T_2S} are indicated by arrows.

transverse ($\alpha = T_2$) and longitudinal ($\alpha = L$) phonons in the [110] direction. On the other hand, for the glass, which is an isotropic system, there are no distinctions among the different directions or between $\alpha = T_1$ and T_2, so we show the results for three different directions ([100], [110], [111]) and for $\alpha = T_1, T_2$ all together; these results indeed coincide with each other, as seen in Figures 8 and 9.

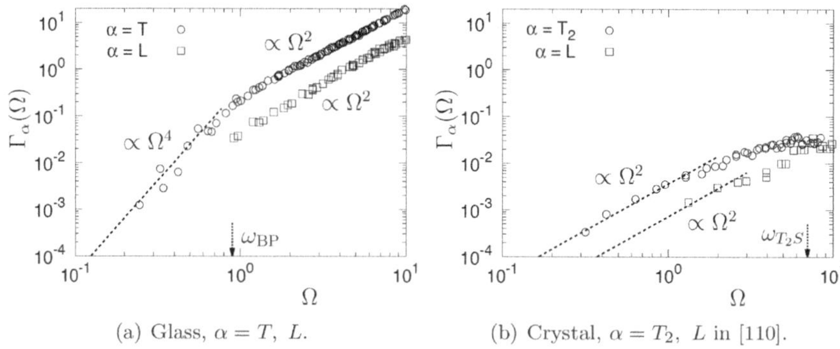

Fig. 9. The attenuation rate $\Gamma_\alpha(\Omega)$ for an SC system. (a) Glass, $\alpha = T,\ L$. (b) FCC crystal, $\alpha = T_2,\ L$ in [110]. The dashed line indicates the power-law scaling with the frequency Ω. See also the caption of Figure 8.

Let us first consider the data for the crystal. Since the phonons in crystals converge to elastic waves at low Ω, $c_\alpha(\Omega)$ converges to the macroscopic value $c_{\alpha 0}$, i.e., $c_{T_2 0} = \sqrt{G_p/\rho}$ or $c_{L0} = \sqrt{(K + G_p/3 + G_s)/\rho}$ (see Eq. (33)), as is indeed shown in Figures 8(b) and 8(d). In this frequency region, the dispersion curve is a straight line, $\Omega(\boldsymbol{q}) = c_{\alpha 0} q$, as in Eq. (30). It starts to deviate from linearity as Ω (or q) increases towards the edge of the first Brillouin zone, where the van Hove singularity occurs.[6,7] In addition, both the transverse and longitudinal $\Gamma_\alpha(\Omega)$ show an Ω^2 dependence, $\Gamma_\alpha \propto \Omega^2$, as seen in Figure 9(b). This damping originates from purely anharmonic couplings between phonons through the Umklapp process at finite $T > 0$. The Ω^2 dependence can be explained by thermo-elasticity theory or the Boltzmann equation analysis.[73,74] At $T = 0$, there are no anharmonic effects, and the attenuation becomes exactly zero.

The sound speed $c_\alpha(\Omega)$ of the glass also converges to its macroscopic value, $c_{T0} = \sqrt{G/\rho}$ or $c_{L0} = \sqrt{(K + 4G/3)/\rho}$ (see Eq. (31)), at low Ω, as shown in Figures 8(a) and 8(c).[r] As Ω increases, $c_\alpha(\Omega)$ decreases and reaches a minimum. This behaviour has been observed

---

[r]The longitudinal sound speed does not converge to c_{L0} in the frequency range presented in Figure 8(c). However, the recent work[57] has shown that it converges as Ω decreases.

in experiments[19,20] and simulations,[38,39,41] and it is called sound softening. The softening of transverse phonons occurs near the BP frequency ω_{BP}. Above the softening region, $c_\alpha(\Omega)$ increases with increasing Ω; this behaviour is called sound hardening.[44]

Notably, the attenuation rate $\Gamma_\alpha(\Omega)$ of the glass behaves much differently from that of the crystal, as shown in Figure 9(a). First, $\Gamma_\alpha(\Omega)$ takes much larger values than in the crystal. Second, $\Gamma_\alpha(\Omega)$ shows Rayleigh scattering, $\Gamma_\alpha \propto \Omega^4$, at low Ω below ω_{BP}, and it depends on Ω^2, i.e., $\Gamma_\alpha \propto \Omega^2$, at high Ω above ω_{BP}.[s] These anomalous attenuation behaviours have also been observed in experiments[18–20] and simulations.[38,39,41] There are two sources of scattering in glasses: one is anharmonic effects, as in crystals, and the other is the disordered structure. The latter effects are dominant over the former and induce much stronger sound damping compared to that in crystals; in particular, Rayleigh scattering at low Ω and an Ω^2 dependence at high Ω are induced. Although we cannot observe any evidence of them in Figure 9(a), the former, anharmonic effects should become visible as some dependence on Ω, possibly some power-law dependence, $\Gamma_\alpha \propto \Omega^a$, in the lower Ω regime. Indeed, both experimental works[21,22] and theoretical works[75–77] have studied these anharmonic effects, which add to the effects of the disordered structure. Notably, close to the glass transition, the fractal frequency dependence of the damping, $\Gamma_\alpha \propto \Omega^{3/2}$, has been reported by experiment[21] and predicted by theory.[77]

At low Ω below the boson peak, we observe a macroscopic sound speed and Rayleigh scattering in the attenuation behaviour. These phonon transport characteristics can be understood by considering the glass as an elastic medium with point defects. This analysis indicates that the disordered structure is not uniformly coarse-grained even at macroscopic length scales, but rather, it still plays a defect-type role. On the other hand, at shorter length scales, phonons are more strongly influenced by the disordered structure,

[s]The longitudinal phonons do not show Rayleigh scattering behaviour in the frequency range presented in Figure 9(b). However, the recent work[57] has shown that a Rayleigh scattering regime emerges as Ω decreases.

which induces sound softening and an Ω^2 dependence of damping. $\Gamma_\alpha \propto \Omega^2$ exhibits dynamics characteristic of viscous damping.[37] In this regime, the scattering is so strong that a phonon does not propagate as a plane wave but rather immediately attenuates to become diffusive. Refs. [78, 79] refer to this vibrational behaviour as diffusion (nonpropagating, delocalized vibration). We therefore expect the BP frequency ω_{BP} to represent the upper bound on the frequency at which phonons can propagate as plane waves, as will be discussed next.

4.3. *Ioffe-Regel (IR) frequency*

We introduce the Ioffe-Regel (IR) frequency, $\Omega_{\alpha\mathrm{IR}}$, which corresponds to an upper bound on the frequency at which a phonon can propagate as a plane wave.[37, 38, 41] For α phonons, $\Omega_{\alpha\mathrm{IR}}$ is defined as $\pi\Gamma_\alpha(\Omega = \Omega_{\alpha\mathrm{IR}})/\Omega_{\alpha\mathrm{IR}} = 1$. Above $\Omega_{\alpha\mathrm{IR}}$, the phonon decay time ($= \Gamma_\alpha^{-1}$) becomes shorter than half of the vibrational period ($= \pi/\Omega$), i.e., the phonon decays within half of the duration of one period. Figure 10 plots $\pi\Gamma_\alpha(\Omega)/\Omega$ as a function of Ω. The point of intersection with $\pi\Gamma_\alpha(\Omega)/\Omega = 1$ gives the value of $\Omega_{\alpha\mathrm{IR}}$. Note that the fitting functions in Eqs. (51) and (52) (the damped harmonic oscillator model) may

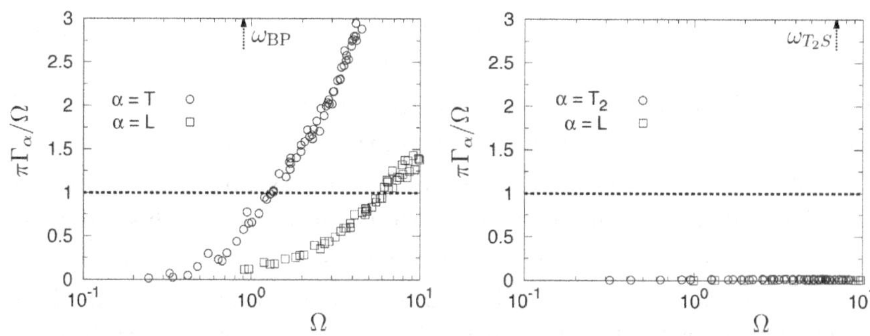

(a) Glass, $\alpha = T,\ L$.

(b) Crystal, $\alpha = T_2,\ L$ in [110].

Fig. 10. Ioffe-Regel (IR) frequency for an SC system. (a) Glass, $\alpha = T,\ L$. (b) FCC crystal, $\alpha = T_2,\ L$ in [110]. We plot $\pi\Gamma_\alpha/\Omega$ versus Ω. The data for Ω and Γ_α are taken from Figures 8 and 9. The point of intersection with $\pi\Gamma_\alpha/\Omega = 1$ as indicated by the dashed line gives the value of $\Omega_{\alpha\mathrm{IR}}$. See also the caption of Figure 8.

not be appropriate for measuring phonon transport properties above $\Omega_{\alpha\mathrm{IR}}$.[60]

For the crystal, $\pi\Gamma_\alpha(\Omega)/\Omega \ll 1$ in the whole Ω region, which means that phonons propagate as plane waves. In contrast, the glass shows a finite value of $\Omega_{\alpha\mathrm{IR}}$. In particular, the $\Omega_{T\mathrm{IR}}$ value for transverse phonons nearly coincides with the BP frequency, $\Omega_{T\mathrm{IR}} \approx \omega_{\mathrm{BP}}$. This coincidence has been observed in many glasses.[37,38,41,45] Moreover, this is consistent with the expectation that ω_{BP} should serve as the upper bound on the frequency at which phonons can propagate as plane waves.

However, the $\Omega_{L\mathrm{IR}}$ for longitudinal phonons in the glass is located at a much higher frequency, $\Omega_{L\mathrm{IR}} \gg \Omega_{T\mathrm{IR}} \approx \omega_{\mathrm{BP}}$,[37,38,41,45] meaning that longitudinal phonons can propagate even above ω_{BP}. This result originates from the fact that the attenuation of longitudinal phonons ($\Gamma_L(\Omega)$) is much lower than that of transverse phonons ($\Gamma_T(\Omega)$) at a fixed Ω, as shown in Figure 9(a), which can, in turn, be understood in terms of heterogeneous elasticity theory.[9,10] This theory assumes that shear modulus heterogeneity dominates compared with bulk modulus heterogeneity, which is true in the present SC system,[40–42] and predicts that the shear modulus heterogeneity induces anomalous behaviours for both transverse and longitudinal phonons. In this theoretical framework, we can understand that $\Gamma_T(\Omega)$ becomes larger than $\Gamma_L(\Omega)$ because transverse phonons are more sensitive to the shear modulus heterogeneity.

4.4. *Zero-temperature measurement*

As we have discussed so far, the disordered structure has a significant impact on the phonon transport in a glass. To study these structure-induced effects, it is most straightforward to measure phonon transport at zero temperature $T = 0$, without thermal fluctuations. We will present two methods of doing so below.

4.4.1. *Dynamic structure factor in the zero-temperature limit*

We first take the zero-temperature limit (as $T \to 0$) of the dynamic structure factor $S_\alpha(\boldsymbol{q}, \omega)$. As $T \to 0$ in the harmonic limit, the velocity vector $\boldsymbol{v}(t)$ converges as follows:

$$\boldsymbol{v}(t) = \frac{1}{\sqrt{\mathcal{M}}}\frac{d\boldsymbol{u}(t)}{dt} \longrightarrow \sum_{k=1}^{3N} \frac{du^k(t)}{dt}\left(\frac{\boldsymbol{e}^k}{\sqrt{\mathcal{M}}}\right), \tag{53}$$

where we apply Eq. (25) for $\boldsymbol{u}(t)$ and

$$\frac{du^k(t)}{dt} = -\left(\boldsymbol{e}^k \cdot \boldsymbol{u}_0\right)\omega^k \sin\left(\omega^k t\right) + \left(\boldsymbol{e}^k \cdot \dot{\boldsymbol{u}}_0\right)\cos\left(\omega^k t\right). \tag{54}$$

Equations (47) and (50) then lead to

$$S_\alpha(\boldsymbol{q},\omega) \longrightarrow \frac{T}{2}\left(\frac{q}{\omega}\right)^2 \sum_{k=1}^{3N}\left|\left(\frac{\boldsymbol{e}^k}{\sqrt{\mathcal{M}}}\right)\cdot \boldsymbol{u}_{\text{ph}}^{\boldsymbol{q},\alpha}\right|^2 [\delta(\omega-\omega^k)+\delta(\omega+\omega^k)], \tag{55}$$

where we use $\left\langle \left(\boldsymbol{e}^k \cdot \dot{\boldsymbol{u}}_0\right)^2 \right\rangle_0 = T$, which is the equipartition law for energy. By taking the polarization vector to be $\boldsymbol{s}_T(\boldsymbol{q}) = \hat{\boldsymbol{q}}^\perp$ or $\boldsymbol{s}_L(\boldsymbol{q}) = \hat{\boldsymbol{q}}$, we can also write Eq. (55) in a more familiar form:

$$S_\alpha(\boldsymbol{q},\omega) \longrightarrow \frac{T}{2N}\left(\frac{q}{\omega}\right)^2 \sum_{k=1}^{3N} F_\alpha^k(\boldsymbol{q})[\delta(\omega-\omega^k)+\delta(\omega+\omega^k)], \tag{56}$$

$$\begin{aligned} F_T^k(\boldsymbol{q}) &= \left|\sum_{i=1}^{N}\left(\frac{\boldsymbol{e}_i^k}{\sqrt{m_i}} \times \hat{\boldsymbol{q}}\right)\exp\left(\mathrm{i}\boldsymbol{q}\cdot\boldsymbol{R}_i\right)\right|^2, \\ F_L^k(\boldsymbol{q}) &= \left|\sum_{i=1}^{N}\left(\frac{\boldsymbol{e}_i^k}{\sqrt{m_i}} \cdot \hat{\boldsymbol{q}}\right)\exp\left(\mathrm{i}\boldsymbol{q}\cdot\boldsymbol{R}_i\right)\right|^2. \end{aligned} \tag{57}$$

For the case of single-component crystals with mass $m_i \equiv m$, $\boldsymbol{e}^k = \boldsymbol{u}_{\text{ph}}^{\boldsymbol{q},\alpha} = \boldsymbol{e}_{\text{ph}}^{\boldsymbol{q},\alpha}$, and we obtain

$$S_\alpha(\boldsymbol{q},\omega) \longrightarrow \frac{T}{2m}\left(\frac{q}{\omega}\right)^2 [\delta\left(\omega-\omega_\alpha(\boldsymbol{q})\right)+\delta\left(\omega+\omega_\alpha(\boldsymbol{q})\right)]. \tag{58}$$

We thus confirm that $S_\alpha(\boldsymbol{q},\omega)$ is the delta function as $T \to 0$; that the propagation frequency is the eigenfrequency, $\Omega(\boldsymbol{q}) = \omega_\alpha(\boldsymbol{q})$; and that the attenuation rate is exactly zero, $\Gamma_\alpha(\boldsymbol{q}) = 0$. For the case of a glass, it has been numerically verified that $S_\alpha(\boldsymbol{q},\omega)$ at low T coincides with the value at $T \to 0$.[30]

4.4.2. *Direct measurement at zero temperature*

The second method is to directly measure phonon transport at $T = 0$.[46,57] We start with the inherent structure $\boldsymbol{R} \equiv \left[\boldsymbol{R}_1^T, \boldsymbol{R}_2^T, \ldots, \boldsymbol{R}_N^T\right]^T$ and excite a phonon of wavevector $\boldsymbol{q}$ and polarization α by setting the initial conditions at $t = 0$ to $\boldsymbol{u} \equiv \boldsymbol{u}_0 = \boldsymbol{0}$ and $d\boldsymbol{u}/dt \equiv \dot{\boldsymbol{u}}_0 = \boldsymbol{u}_{\mathrm{ph}}^{\boldsymbol{q},\alpha}$ in Eq. (44). We next solve the linearized equation of motion, Eq. (14):

$$\frac{d^2\boldsymbol{u}(t)}{dt^2} = -\mathcal{D}\boldsymbol{u}(t) + \dot{\boldsymbol{u}}_0\delta(t). \tag{59}$$

From $\boldsymbol{u}(t)$, we calculate the (normalized) velocity-velocity time correlation function:

$$C(t) \equiv \left(\frac{\dot{\boldsymbol{u}}(t)}{\sqrt{\mathcal{M}}} \cdot \frac{\dot{\boldsymbol{u}}_0}{\sqrt{\mathcal{M}}}\right)\left(\frac{\dot{\boldsymbol{u}}_0}{\sqrt{\mathcal{M}}} \cdot \frac{\dot{\boldsymbol{u}}_0}{\sqrt{\mathcal{M}}}\right)^{-1}. \tag{60}$$

The function $C(t)$ represents the propagation and attenuation behaviours of the initially excited phonon $\dot{\boldsymbol{u}}_0$ with $\boldsymbol{q}$ and α. $\Omega(\boldsymbol{q})$ and $\Gamma_\alpha(\boldsymbol{q})$ are then extracted by fitting the simulated data for $C(t)$ to the damped harmonic oscillator model in Eq. (52):[46,57] $C(t) \equiv \cos\left(\Omega(\boldsymbol{q})t\right)\exp\left(-\Gamma_\alpha(\boldsymbol{q})t/2\right)$.

5. Elastic Deformation

In this section, we describe how to measure the elastic response, particularly the elastic moduli, of glasses and any solid-state materials. The elastic moduli are necessary for calculating the Debye vDOS and the macroscopic sound speed, which are useful references for understanding the vibrational properties of a material. There are two methods of measuring the elastic moduli. One is the so-called fluctuation formulation. In this method, we do not apply any external deformation but rather perform an equilibrium, molecular simulation and use the formulation developed based on linear response theory. The other method is to measure the elastic moduli directly. We apply an external strain to the system and measure the stress as a function of the strain, i.e., the stress-strain curve. The slope of this curve gives the value of the elastic modulus.

5.1. General description

We first provide a general explanation of the elastic response and modulus.

5.1.1. Elastic modulus tensor

The free energy is given by[63]

$$F = -T \ln Z = -T \ln \left\{ \mathrm{Tr} \exp \left[-\frac{1}{T} \left(\sum_{i=1}^{N} \frac{\boldsymbol{p}_i^2}{2m_i} + \Phi(\boldsymbol{r}) \right) \right] \right\}, \quad (61)$$

where we set the Boltzmann constant to $k_B = 1$ and $Z \equiv \mathrm{Tr} \exp(-H/T)$ is the partition function ($H \equiv \sum_{i=1}^{N} \boldsymbol{p}_i^2/(2m_i) + \Phi$ is the Hamiltonian, and Tr denotes the trace operator). Here, we consider the Helmholtz free energy and the canonical ensemble; however, the values of the elastic moduli are insensitive to the ensemble at low T.[t] The stress tensor $\sigma_{\alpha\beta}$ is defined as the first derivative of the free energy F with respect to the strain tensor $\epsilon_{\alpha\beta}$ as[80–84]

$$\sigma_{\alpha\beta} \equiv \frac{1}{V} \frac{\partial F}{\partial \epsilon_{\alpha\beta}} \bigg|_{\epsilon_{\alpha\beta} \to 0}, \quad (62)$$

where α, β (γ, δ) = x, y, z. There are two different definitions of $\epsilon_{\alpha\beta}$:[u] the linear strain (infinitesimal strain) tensor, $e_{\alpha\beta}$, and the Green-Lagrange strain (finite strain) tensor, $\eta_{\alpha\beta}$:

$$e_{\alpha\beta} \equiv \frac{1}{2} \left(\frac{\partial u_\alpha}{\partial r_\beta} + \frac{\partial u_\beta}{\partial r_\alpha} \right) \equiv \frac{1}{2} (u_{\alpha\beta} + u_{\beta\alpha}),$$

$$\eta_{\alpha\beta} \equiv \frac{1}{2} \left(\frac{\partial u_\alpha}{\partial r_\beta} + \frac{\partial u_\beta}{\partial r_\alpha} + \sum_{\gamma = x,y,z} \frac{\partial u_\gamma}{\partial r_\alpha} \frac{\partial u_\gamma}{\partial r_\beta} \right),$$

$$\equiv \frac{1}{2} \left(u_{\alpha\beta} + u_{\beta\alpha} + \sum_{\gamma = x,y,z} u_{\gamma\alpha} u_{\gamma\beta} \right), \quad (63)$$

---

[t]The isothermal moduli (in canonical ensemble) and the adiabatic moduli (in micro-canonical ensemble) coincide in the low-T harmonic limit.[6,7]
[u]The stress tensors defined by $\epsilon_{\alpha\beta} = e_{\alpha\beta}$ and $\eta_{\alpha\beta}$ in Eq. (62) coincide with each other.

where $[r_x, r_y, r_z]^T$ and $[u_x, u_y, u_z]^T$ represent the spatial coordinates and the displacement field, respectively, and $u_{\alpha\beta} \equiv \partial u_\alpha / \partial r_\beta$ is the displacement gradient tensor.

We then define the elastic modulus tensor in two different ways.[80–84] Firstly the modulus tensor $\tilde{C}_{\alpha\beta\gamma\delta}$ is defined as the second derivative of F with respect to the Green-Lagrange strain tensor $\eta_{\alpha\beta}$:

$$\tilde{C}_{\alpha\beta\gamma\delta} \equiv \frac{1}{V} \frac{\partial^2 F}{\partial \eta_{\alpha\beta} \partial \eta_{\gamma\delta}} \bigg|_{\eta_{\alpha\beta} \to 0,\, \eta_{\gamma\delta} \to 0}. \tag{64}$$

Secondly the modulus tensor $C_{\alpha\beta\gamma\delta}$ is defined as the first derivative of $\sigma_{\alpha\beta}$ with respect to the linear strain tensor $e_{\gamma\delta}$:

$$C_{\alpha\beta\gamma\delta} \equiv \frac{\partial \sigma_{\alpha\beta}}{\partial e_{\gamma\delta}} \bigg|_{e_{\gamma\delta} \to 0}. \tag{65}$$

These two modulus tensors do not coincide for the system under initial stress $\sigma^0_{\alpha\beta} \equiv \sigma_{\alpha\beta}(e_{\gamma\delta} = 0)$.[v] $\tilde{C}_{\alpha\beta\gamma\delta}$ and $C_{\alpha\beta\gamma\delta}$ are related as follows (please see Refs. [81, 84] for details):

$$\begin{aligned} C_{\alpha\beta\gamma\delta} &= \tilde{C}_{\alpha\beta\gamma\delta} + C^C_{\alpha\beta\gamma\delta}, \\ C^C_{\alpha\beta\gamma\delta} &= -\frac{1}{2}\left(2\sigma^0_{\alpha\beta}\delta_{\gamma\delta} - \sigma^0_{\alpha\gamma}\delta_{\beta\delta} - \sigma^0_{\alpha\delta}\delta_{\beta\gamma} - \sigma^0_{\beta\gamma}\delta_{\alpha\delta} - \sigma^0_{\beta\delta}\delta_{\alpha\gamma}\right). \end{aligned} \tag{66}$$

We refer to $C^C_{\alpha\beta\gamma\delta}$ as the correction term in the following. If the initial stress tensor $\sigma^0_{\alpha\beta}$ is zero, then $C^C_{\alpha\beta\gamma\delta}$ is zero, and $C_{\alpha\beta\gamma\delta}$ and $\tilde{C}_{\alpha\beta\gamma\delta}$ exactly coincide. Since $C_{\alpha\beta\gamma\delta}$ is necessary for calculations of the Debye vDOS and the macroscopic sound speed, we focus on $C_{\alpha\beta\gamma\delta}$ below.

Here we note that strictly speaking, for calculations of the Debye vDOS and the macroscopic sound speed, we should apply the modulus tensor $C'_{\alpha\beta\gamma\delta}$ which is defined as the first derivative of $\sigma_{\alpha\beta}$ with respect to the displacement gradient tensor $u_{\gamma\delta}$ (instead of $e_{\gamma\delta}$):

$$C'_{\alpha\beta\gamma\delta} \equiv \frac{\partial \sigma_{\alpha\beta}}{\partial u_{\gamma\delta}} \bigg|_{u_{\gamma\delta} \to 0}. \tag{67}$$

---

[v] $\tilde{C}_{\alpha\beta\gamma\delta}$ is defined in a Lagrangian framework, whereas $C_{\alpha\beta\gamma\delta}$ is based on an Eulerian framework.[81,84]

However, as long as we consider the situation under hydrostatic pressure where the initial stress is $\sigma^0_{\alpha\beta} = -p\delta_{\alpha\beta}$, $C_{\alpha\beta\gamma\delta}$ and $C'_{\alpha\beta\gamma\delta}$ are identical.[81,84]

5.1.2. *Bulk, pure shear, and simple shear elastic moduli*

We consider three types of deformations in particular, namely, a volume-changing bulk deformation and pure shear (plane or triaxial strains) and simple shear deformations, which are illustrated in Figure 11. Since there are two independent pure shear deformations and three independent simple shear deformations, we consider six independent deformations in total, including one bulk deformation.[85] The strain tensors for these six deformations are[w]

$$\epsilon_{xx} = \epsilon_{yy} = \epsilon_{zz} = \epsilon_b \quad \text{(bulk)}, \tag{68}$$

$$\begin{aligned} &\epsilon_{xx} = \epsilon_{p1}, \quad \epsilon_{yy} = -\epsilon_{p1} \quad \text{(pure shear 1)}, \\ &\epsilon_{xx} = \epsilon_{p2}, \quad \epsilon_{yy} = \epsilon_{p2}, \quad \epsilon_{zz} = -2\epsilon_{p2} \quad \text{(pure shear 2)}, \end{aligned} \tag{69}$$

$$\begin{aligned} &\epsilon_{xy} = \epsilon_{yx} = \epsilon_{s1} \quad \text{(simple shear 1)}, \\ &\epsilon_{xz} = \epsilon_{zx} = \epsilon_{s2} \quad \text{(simple shear 2)}, \\ &\epsilon_{yz} = \epsilon_{zy} = \epsilon_{s3} \quad \text{(simple shear 3)}, \end{aligned} \tag{70}$$

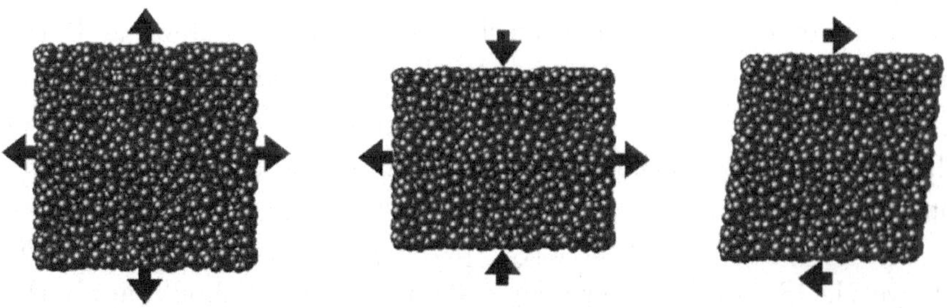

(a) Bulk (b) Pure shear (c) Simple shear

Fig. 11. Schematic illustrations of (a) bulk, (b) pure shear, and (c) simple shear deformations. The bulk (K), pure shear (G_p), and simple shear (G_s) moduli correspond to these deformations.

[w]We can also consider other pure shear deformations, which are described as superpositions of the present two independent pure shear deformations.

where no written components in $\epsilon_{\alpha\beta}$ are zero, and ϵ_b, ϵ_{p1}, ϵ_{p2}, ϵ_{s1}, ϵ_{s2}, ϵ_{s3} represent the applied strains corresponding to the deformations. $\delta V/V \equiv 3\epsilon_b$ represents the volume change, while $\gamma_{p1} \equiv 2\epsilon_{p1}$, $\gamma_{p2} \equiv 3\epsilon_{p2}$, $\gamma_{s1} = 2\epsilon_{s1}$, $\gamma_{s2} = 2\epsilon_{s2}$, $\gamma_{s3} \equiv 2\epsilon_{s3}$ present the shear strains for the corresponding shear deformations.

When we apply a deformation to the system, the corresponding stress varies with the applied strain. For the bulk deformation, the two pure shear deformations, and the three simple shear deformations, the applied stresses are the pressure, $p \equiv -(\sigma_{xx}+\sigma_{yy}+\sigma_{zz})/3$; two pure shear stresses, $\sigma_{p1} \equiv (\sigma_{xx} - \sigma_{yy})/2$ and $\sigma_{p2} \equiv (\sigma_{xx} + \sigma_{yy} - 2\sigma_{zz})/4$; and three simple shear stresses, $\sigma_{s1} \equiv \sigma_{xy}$, $\sigma_{s2} \equiv \sigma_{xz}$, and $\sigma_{s3} \equiv \sigma_{yz}$, respectively. Accordingly, the bulk modulus K, the two pure shear moduli G_{p1} and G_{p2}, and the three simple shear moduli G_{s1}, G_{s2}, and G_{s3} are defined as follows:

$$K \equiv -\left.\frac{\partial p}{\partial(\delta V/V)}\right|_{(\delta V/V)\to 0} = \left.\frac{\partial(\sigma_{xx}+\sigma_{yy}+\sigma_{zz})/3}{\partial(3\epsilon_b)}\right|_{\epsilon_b\to 0} \quad \text{(bulk)}, \tag{71}$$

$$G_{p1} \equiv \left.\frac{\partial\sigma_{p1}}{\partial\gamma_{p1}}\right|_{\gamma_{p1}\to 0} = \left.\frac{\partial(\sigma_{xx}-\sigma_{yy})/2}{\partial(2\epsilon_{p1})}\right|_{\epsilon_{p1}\to 0} \quad \text{(pure shear 1)},$$

$$G_{p2} \equiv \left.\frac{\partial\sigma_{p2}}{\partial\gamma_{p2}}\right|_{\gamma_{p2}\to 0} = \left.\frac{\partial(\sigma_{xx}+\sigma_{yy}-2\sigma_{zz})/4}{\partial(3\epsilon_{p2})}\right|_{\epsilon_{p2}\to 0} \quad \text{(pure shear 2)}, \tag{72}$$

$$G_{s1} \equiv \left.\frac{\partial\sigma_{s1}}{\partial\gamma_{s1}}\right|_{\gamma_{s1}\to 0} = \left.\frac{\partial\sigma_{xy}}{\partial(2\epsilon_{s1})}\right|_{\epsilon_{s1}\to 0} \quad \text{(simple shear 1)},$$

$$G_{s2} \equiv \left.\frac{\partial\sigma_{s2}}{\partial\gamma_{s2}}\right|_{\gamma_{s2}\to 0} = \left.\frac{\partial\sigma_{xz}}{\partial(2\epsilon_{s2})}\right|_{\epsilon_{s2}\to 0} \quad \text{(simple shear 2)}, \tag{73}$$

$$G_{s3} \equiv \left.\frac{\partial\sigma_{s3}}{\partial\gamma_{s3}}\right|_{\gamma_{s3}\to 0} = \left.\frac{\partial\sigma_{yz}}{\partial(2\epsilon_{s3})}\right|_{\epsilon_{s3}\to 0} \quad \text{(simple shear 3)}.$$

$K, G_{p1}, G_{p2}, G_{s1}, G_{s2}$, and G_{s3} can be formulated in terms of the elastic modulus tensor $C_{\alpha\beta\gamma\delta}$[85]:

$$K = \frac{\begin{array}{l}(C_{xxxx} + C_{yyyy} + C_{zzzz} + C_{xxyy} + C_{yyxx} + C_{xxzz} \\ \quad + C_{zzxx} + C_{yyzz} + C_{zzyy})\end{array}}{9}, \tag{74}$$

$$\begin{aligned} G_{p1} &= \frac{(C_{xxxx} + C_{yyyy} - C_{xxyy} - C_{yyxx})}{4}, \\ G_{p2} &= \frac{\begin{array}{c}(C_{xxxx} + C_{yyyy} + 4C_{zzzz} + C_{xxyy} + C_{yyxx} - 2C_{xxzz} \\ - 2C_{zzxx} - 2C_{yyzz} - 2C_{zzyy})\end{array}}{12}, \end{aligned} \tag{75}$$

$$G_{s1} = C_{xyxy}, \; G_{s2} = C_{xzxz}, \; G_{s3} = C_{yzyz}. \tag{76}$$

Note that $C_{xyxy} = C_{xyyx}$, $C_{xzxz} = C_{xzzx}$, $C_{yzyz} = C_{yzzy}$. For cubic crystals such as FCC crystals, the two pure shear moduli coincide, i.e., $G_p \equiv G_{p1} = G_{p2}$, and the three simple shear moduli also coincide, i.e., $G_s \equiv G_{s1} = G_{s2} = G_{s3}$; however, the pure shear and simple shear moduli generally take different values, i.e., $G_p \neq G_s$. On the other hand, for glasses, which are isotropic systems, all five shear moduli coincide: $G \equiv G_{p1} = G_{p2} = G_{s1} = G_{s2} = G_{s3}$.

5.1.3. *Affine and non-affine elastic moduli*

The elastic response can be decomposed into affine and non-affine components.[31,84,86,87] Affine deformation refers to particles that follow the applied affine strain field and are displaced affinely at all scales. More specifically, for a linear strain $\epsilon_{\alpha\beta} = e_{\alpha\beta}$, the affine deformation causes the displacement of the position of particle i, $\boldsymbol{r}_i = [r_{ix}, r_{iy}, r_{iz}]^T$, as follows:

$$r_{i\kappa} \rightarrow r_{i\kappa} + \frac{1}{2}\left(u_{\alpha\beta} r_{i\beta} \delta_{\alpha\kappa} + u_{\beta\alpha} r_{i\alpha} \delta_{\beta\kappa}\right), \tag{77}$$

where $\kappa = x, y, z$. Similarly, for the Green-Lagrange strain $\epsilon_{\alpha\beta} = \eta_{\alpha\beta}$, particles are displaced as follows:

$$r_{i\kappa} \rightarrow r_{i\kappa} + \frac{1}{2}\left(u_{\alpha\beta} r_{i\beta} \delta_{\alpha\kappa} + u_{\beta\alpha} r_{i\alpha} \delta_{\beta\kappa} + \sum_{\gamma=x,y,z} u_{\gamma\alpha} u_{\gamma\beta} r_{i\alpha} r_{i\beta} \delta_{\gamma\kappa}\right). \tag{78}$$

The affine strain tensors $\epsilon^A_{\alpha\beta} = e^A_{\alpha\beta}$ and $\eta^A_{\alpha\beta}$ are defined as displacing particles as expressed in Eqs. (77) and (78).

By applying the affine strain tensors to Eqs. (64) and (65), we can define the affine elastic modulus tensors as

$$\tilde{C}^A_{\alpha\beta\gamma\delta} \equiv \frac{1}{V} \left. \frac{\partial^2 F}{\partial \eta^A_{\alpha\beta} \partial \eta^A_{\gamma\delta}} \right|_{\eta^A_{\alpha\beta} \to 0,\, \eta^A_{\gamma\delta} \to 0}, \tag{79}$$

$$C^A_{\alpha\beta\gamma\delta} \equiv \left. \frac{\partial \sigma_{\alpha\beta}}{\partial e^A_{\gamma\delta}} \right|_{e^A_{\gamma\delta} \to 0}. \tag{80}$$

The elastic response of a crystal is characterized mostly by affine deformation and the corresponding affine modulus.[80,82] However, importantly, the elastic response of a glass does not follow solely the affine deformation. We additionally need to consider the non-affine deformation, which results in additional particle displacements at the microscopic scale and causes the displacements to deviate from the applied affine field.[31,84,86,87] Non-affine deformation contributes negatively to the overall modulus; thus, the corresponding modulus is defined as the non-affine modulus $-C^N_{\alpha\beta\gamma\delta}$. We therefore describe the elastic modulus tensor as

$$C_{\alpha\beta\gamma\delta} = C^A_{\alpha\beta\gamma\delta} - C^N_{\alpha\beta\gamma\delta}. \tag{81}$$

As we will see below, the non-affine component is important in glasses, being comparable in magnitude to the affine modulus.[31,84,86,87]

5.2. *Fluctuation formulation*

We now present how to measure the elastic moduli by means of computational simulations. We first introduce the fluctuation formulation. In the following, we consider a system with a pair-wise potential, where particles i and j interact through the potential $\phi_{ij}(r_{ij})$ (r_{ij} is the distance between them) and the total potential is $\Phi = \sum_{i<j} \phi_{ij}(r_{ij})$. We also employ the linear strain tensor $\epsilon_{\alpha\beta} = e_{\alpha\beta}$. Please see, e.g., Refs. [83, 84, 86], for detailed derivations of the following formulations.

5.2.1. *Finite temperature $T > 0$*

By using Eqs. (61) and (62), we can obtain the stress tensor in the form of

$$\sigma_{\alpha\beta} = \langle \hat{\sigma}_{\alpha\beta} \rangle,$$
$$\hat{\sigma}_{\alpha\beta} = \frac{1}{V}\left[-\sum_{i=1}^{N} m_i v_{i\alpha} v_{i\beta} + \sum_{i<j} \left(r_{ij} \frac{\partial \phi_{ij}}{\partial r_{ij}} \right) n_{ij\alpha} n_{ij\beta} \right], \tag{82}$$

where $\langle \cdots \rangle$ denotes the canonical ensemble average, and $\boldsymbol{n}_{ij} = [n_{ijx}, n_{ijy}, n_{ijz}]^T \equiv (\boldsymbol{r}_i - \boldsymbol{r}_j)/r_{ij}$ is the unit vector connecting particle j to particle i. The elastic modulus tensor $C_{\alpha\beta\gamma\delta}$ is formulated using Eqs. (64) and (66), which includes the affine modulus and the non-affine modulus as in Eq. (81). The affine modulus tensor $C^A_{\alpha\beta\gamma\delta}$ (Eq. (80)) is formulated as

$$C^A_{\alpha\beta\gamma\delta} = C^B_{\alpha\beta\gamma\delta} + C^K_{\alpha\beta\gamma\delta} + C^C_{\alpha\beta\gamma\delta},$$
$$C^B_{\alpha\beta\gamma\delta} = \frac{1}{V}\left\langle \sum_{i<j} \left(r_{ij}^2 \frac{\partial^2 \phi_{ij}}{\partial r_{ij}^2} - r_{ij} \frac{\partial \phi_{ij}}{\partial r_{ij}} \right) n_{ij\alpha} n_{ij\beta} n_{ij\gamma} n_{ij\delta} \right\rangle,$$
$$C^K_{\alpha\beta\gamma\delta} = 2\hat{\rho} T (\delta_{\alpha\gamma}\delta_{\beta\delta} + \delta_{\alpha\delta}\delta_{\beta\gamma}),$$
$$C^C_{\alpha\beta\gamma\delta} = -\frac{1}{2}\big(2\langle \hat{\sigma}_{\alpha\beta} \rangle \delta_{\gamma\delta} - \langle \hat{\sigma}_{\alpha\gamma} \rangle \delta_{\beta\delta} - \langle \hat{\sigma}_{\alpha\delta} \rangle \delta_{\beta\gamma} - \langle \hat{\sigma}_{\beta\gamma} \rangle \delta_{\alpha\delta} - \langle \hat{\sigma}_{\beta\delta} \rangle \delta_{\alpha\gamma}\big), \tag{83}$$

where $\hat{\rho} = N/V$ is the number density, $C^B_{\alpha\beta\gamma\delta}$ is the so-called Born term, and $C^K_{\alpha\beta\gamma\delta}$ is the kinetic contribution. Since these two terms are formulated by Eq. (79) based on the Green-Lagrange strain, we need the correction term $C^C_{\alpha\beta\gamma\delta}$ to obtain the modulus tensor based on the linear strain, as shown in Eq. (66).[81,84] In addition, the non-affine modulus is formulated as

$$C^N_{\alpha\beta\gamma\delta} = \frac{V}{T}\left[\langle \hat{\sigma}_{\alpha\beta} \hat{\sigma}_{\gamma\delta} \rangle - \langle \hat{\sigma}_{\alpha\beta} \rangle \langle \hat{\sigma}_{\gamma\delta} \rangle \right] = \frac{V}{T} \langle \delta\hat{\sigma}_{\alpha\beta} \delta\hat{\sigma}_{\gamma\delta} \rangle, \tag{84}$$

where $\delta\hat{\sigma}_{\alpha\beta} \equiv \hat{\sigma}_{\alpha\beta} - \langle \hat{\sigma}_{\alpha\beta} \rangle$ is the stress fluctuation. $C^N_{\alpha\beta\gamma\delta}$ is therefore formulated as the correlation function for stress fluctuations.

Table 1. The elastic moduli, K, G_p, and G_s, of SC systems: a glass and an FCC crystal. We present values obtained using two methods: the fluctuation formulation (FF) at a finite temperature of $T = 10^{-2}$ (almost at the zero-temperature limit) and direct measurement (DM) at zero temperature, $T = 0$. We present the average values of pure shear modulus, $G_p = (G_{p1}+G_{p2})/2$, and simple shear modulus, $G_s = (G_{s1} + G_{s2} + G_{s3})/3$. The affine and non-affine moduli are also presented.

	Method	K	K^A	K^N	G_p	G_p^A	G_p^N	G_s	G_s^A	G_s^N
Glass	FF	40.9	41.0	0.1	6.2	14.7	8.5	6.3	14.8	8.5
	DM	40.5	40.5	0.0	6.3	14.4	8.1	6.4	14.7	8.3
Crystal	FF	32.9	33.0	0.1	5.1	5.2	0.1	16.1	16.3	0.2
	DM	32.6	32.6	0.0	5.4	5.4	0.0	16.2	16.2	0.0

In our computer simulation, we first perform an equilibrium (NVT) MD or MC simulation at a finite $T > 0$ and generate configurations $\boldsymbol{r} \equiv \left[\boldsymbol{r}_1^T, \boldsymbol{r}_2^T, \ldots, \boldsymbol{r}_N^T\right]^T$. From these configurations, we evaluate $C_{\alpha\beta\gamma\delta}$ by using Eqs. (81–84). From $C_{\alpha\beta\gamma\delta}$, we then calculate the elastic moduli, K, G_p, and G_s, using Eqs. (74–76). Table 1 presents the values of K, G_p, and G_s for a glass and an FCC crystal with the SC potential. Here, we also present the values of the affine moduli K^A, G_p^A, and G_s^A and the non-affine moduli K^N, G_p^N, and G_s^N. The temperature is very low, $T = 10^{-2}$, almost at the zero-temperature limit. We confirm that the crystal shows negligible values of the non-affine components in both the bulk and shear moduli. The elastic response of the crystal is therefore characterized by affine deformation, which is due to the symmetry of the lattice structure.[80, 82]

In contrast to the crystal, the glass shows large values of the non-affine components in the shear moduli, comparable to those of the affine components. Therefore, non-affine deformation plays an important role in the elastic response of the glass, making it distinct from the elastic response of the crystal.[31, 84, 86, 87] Note that G_p and G_s coincide in the glass due to its isotropic structure. On the other hand, for bulk deformation, the glass also shows a very small non-affine component, which is due to the isotropic nature of bulk deformation. In particular, we can demonstrate that a mono-disperse system with an inverse-power-law potential, such as the present SC potential (Eq. (5)), shows a non-affine modulus of zero under bulk deformation at zero temperature.

5.2.2. *Zero-temperature limit* $T \to 0$

It is useful to take the limit as $T \to 0$ in the fluctuation formulation given in Eqs. (81–84).[83,84] As $T \to 0$, the configuration converges to the inherent structure, $\boldsymbol{r} \to \boldsymbol{R} = \left[\boldsymbol{R}_1^T, \boldsymbol{R}_2^T, \ldots, \boldsymbol{R}_N^T\right]^T$, and we obtain $C^B_{\alpha\beta\gamma\delta}$, $C^K_{\alpha\beta\gamma\delta}$, and $C^C_{\alpha\beta\gamma\delta}$ as follows:

$$
\begin{aligned}
C^B_{\alpha\beta\gamma\delta} &\longrightarrow \frac{1}{V}\sum_{i<j}\left(r_{ij}^2\frac{\partial^2\phi_{ij}}{\partial r_{ij}^2} - r_{ij}\frac{\partial\phi_{ij}}{\partial r_{ij}}\right) n_{ij\alpha}n_{ij\beta}n_{ij\gamma}n_{ij\delta}\Bigg|_{\boldsymbol{r}=\boldsymbol{R}}, \\
C^K_{\alpha\beta\gamma\delta} &\longrightarrow 0, \qquad\qquad (85) \\
C^C_{\alpha\beta\gamma\delta} &\longrightarrow -\frac{1}{2}\left(2\sigma_{\alpha\beta0}\delta_{\gamma\delta} - \sigma_{\alpha\gamma0}\delta_{\beta\delta} - \sigma_{\alpha\delta0}\delta_{\beta\gamma} - \sigma_{\beta\gamma0}\delta_{\alpha\delta} - \sigma_{\beta\delta0}\delta_{\alpha\gamma}\right),
\end{aligned}
$$

where $\sigma_{\alpha\beta0}$ is the stress tensor $\sigma_{\alpha\beta}$ at $T \to 0$,

$$
\sigma_{\alpha\beta} \longrightarrow \sigma_{\alpha\beta0} \equiv \frac{1}{V}\sum_{i<j}\left(r_{ij}\frac{\partial\phi_{ij}}{\partial r_{ij}}\right) n_{ij\alpha}n_{ij\beta}\Bigg|_{\boldsymbol{r}=\boldsymbol{R}}. \qquad (86)
$$

The non-affine term $C^N_{\alpha\beta\gamma\delta}$ is formulated in terms of the eigenfrequencies ω^k and the eigenvectors $\boldsymbol{e}^k$ as follows. The stress fluctuation is

$$
\delta\hat{\sigma}_{\alpha\beta} \longrightarrow \frac{\partial\sigma_{\alpha\beta0}}{\partial\boldsymbol{R}}\cdot\frac{\boldsymbol{u}(t)}{\sqrt{\mathcal{M}}} = \sum_{k=1}^{3N} u^k(t)\left(\frac{\partial\sigma_{\alpha\beta0}}{\partial\boldsymbol{R}}\cdot\frac{\boldsymbol{e}^k}{\sqrt{\mathcal{M}}}\right), \qquad (87)
$$

where we use $\boldsymbol{u}(t)$ in Eq. (25). We then obtain

$$
C^N_{\alpha\beta\gamma\delta} \longrightarrow V\sum_{k=1}^{3N}\frac{1}{\omega^{k2}}\left(\frac{\partial\sigma_{\alpha\beta0}}{\partial\boldsymbol{R}}\cdot\frac{\boldsymbol{e}^k}{\sqrt{\mathcal{M}}}\right)\left(\frac{\partial\sigma_{\gamma\delta0}}{\partial\boldsymbol{R}}\cdot\frac{\boldsymbol{e}^k}{\sqrt{\mathcal{M}}}\right), \qquad (88)
$$

where we use $\langle(\boldsymbol{e}^k\cdot\dot{\boldsymbol{u}}_0)^2\rangle = \omega^{k2}\langle(\boldsymbol{e}^k\cdot\boldsymbol{u}_0)^2\rangle = T$, which is the equipartition law for energy. It is worth noting that the affine component, $C^A_{\alpha\beta\gamma\delta} \equiv C^B_{\alpha\beta\gamma\delta} + C^K_{\alpha\beta\gamma\delta} + C^C_{\alpha\beta\gamma\delta}$, is determined by static structural properties, whereas the non-affine term, $C^N_{\alpha\beta\gamma\delta}$, reflects the vibrational properties. We also note that the non-affine term $C^N_{\alpha\beta\gamma\delta}$ in Eq. (88) can be written as

$$C^N_{\alpha\beta\gamma\delta} \longrightarrow V \frac{\partial \upsilon_{\alpha\beta 0}}{\partial \boldsymbol{R}^T} \left[\left. \frac{\partial^2 \Phi}{\partial \boldsymbol{r} \partial \boldsymbol{r}^T} \right|_{\boldsymbol{r}=\boldsymbol{R}} \right]^{-1} \frac{\partial \upsilon_{\alpha\beta 0}}{\partial \boldsymbol{R}}. \tag{89}$$

From Eqs. (85), (86), and (89), we can confirm that elastic modulus at zero temperature does not depend on mass of particles but rather it depends on only the potential.

5.3. *Direct measurement*

We can apply an external strain to the system to directly measure the elastic moduli. By increasing the strain, we obtain the stress-strain curve, and the slope of this curve yields the corresponding elastic modulus. Specifically, the bulk, pure shear 1, and simple shear 1 deformations, which are illustrated in Figure 11, yield the $p - \delta V/V$, $\sigma_{p1} - \gamma_{p1}$, and $\sigma_{s1} - \gamma_{s1}$ curves, respectively. Figure 12 (solid lines) plots these stress-strain curves, which we obtained by applying the linear strain $\epsilon_{\alpha\beta} = e_{\alpha\beta}$. The slopes of these curves give the bulk modulus K, the pure shear modulus G_{p1}, and the simple shear modulus G_{s1}, as shown in Eqs. (71–73). We note that other pure shear, G_{p2}, and simple shear, G_{s2} and G_{s3}, moduli can be also obtained in the same manner.

The obtained values of the elastic moduli are presented in Table 1. From Table 1, we can confirm that these values coincide well with those obtained from the fluctuation formulation. We note that the fluctuation formulation and direct measurement method were implemented at a finite temperature of $T = 10^{-2}$ and at zero temperature ($T = 0$), respectively. However, almost negligible differences are observed at these different Ts, since $T = 10^{-2}$ is almost at the zero-temperature limit and the elastic moduli are insensitive to T in the low-T regime.

To obtain the response under affine deformation, we can apply an affine strain, $\epsilon^A_{\alpha\beta} = e^A_{\alpha\beta}$, to displace the particles according to Eq. (77). After the application of an affine strain, the particles are generally not in mechanical equilibrium; however, we do not

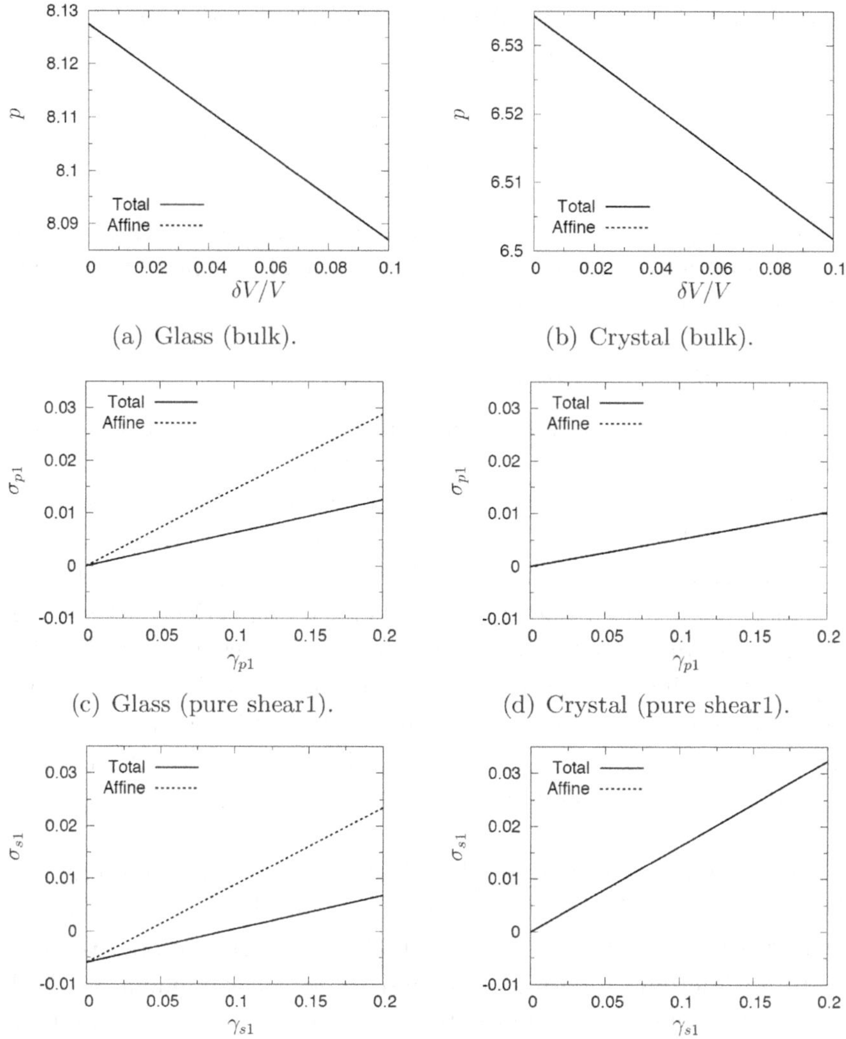

(a) Glass (bulk). (b) Crystal (bulk).

(c) Glass (pure shear1). (d) Crystal (pure shear1).

(e) Glass (simple shear1). (f) Crystal (simple shear1).

Fig. 12. The elastic response of an SC system. (a), (c), (e) Glass. (b), (d), (f) FCC crystal. We plot p versus $\delta V/V$ for bulk deformation in (a), (b), σ_{p1} versus γ_{p1} for pure shear 1 deformation in (c), (d), and σ_{s1} versus γ_{s1} for simple shear 1 deformation in (e), (f). The dashed line represents affine deformation, while the solid line represents the overall response, including non-affine deformation. The slopes of the stress-strain curves yield the values of the elastic moduli, K, G_{p1}, and G_{s1}, as in Eqs. (71–73). The obtained values are presented in Table 1.

permit the particles to relax to the equilibrium state. By doing so, we obtain the stress-strain curve under affine deformation, which is shown by the dashed lines in Figure 12. On the other hand, when we allow the particles to relax, we obtain the overall stress-strain curve, as shown by the solid lines in Figure 12. This relaxation process, i.e., non-affine deformation, generally reduces the stress and causes a reduction in the elastic modulus.

As shown in Figures 12(b), 12(d), and 12(f), for the crystal, the stress-strain curves of the overall response (solid line) and the affine response (dashed line) coincide. The elastic response of the crystal is therefore determined by the affine deformation without relaxation, as discussed above. In contrast to the crystal, the glass shows a large non-affine contribution in its shear deformations, as shown in Figures 12(c) and 12(e). For the bulk deformation of the glass, the non-affine relaxation is negligible, as shown in Figure 12(a); this is due to the isotropic nature of bulk deformation, as also discussed above.

6. Recent Advances in the Vibrational Properties of Glasses

At the end of this chapter, we will introduce recent advances in the understanding of the vibrational properties of glasses. Currently, the capability of computers is rapidly growing. This enables us to perform large-scale computational simulations and to understand the vibrations of the particles in glasses in greater detail. In particular, recent simulations[50–59] have revealed the vibrational properties in the low-frequency continuum limit. In this section, we present simulation data for vibrational eigenmodes and phonon transport, giving particular attention to the low-frequency regime. The data presented below were obtained in Refs. [51, 57] by simulating an atomic glass (packed glass) with a harmonic potential, Φ_{HA}, as expressed in Eq. (6).

As described in Section 3, the eigenmodes in a crystal, i.e., phonons, smoothly converge to elastic waves at low frequencies, and their vDOS converges to the Debye vDOS.[6,7] This means that crystals behave as elastic media at long length scales, where the microscopic lattice structures are uniformly coarse-grained. Similarly, we might expect that the disordered structures of glasses are also uniformly coarse-grained at long length scales and that they would therefore also behave as elastic media. The eigenmodes and vDOS of a glass would then be expected to converge to elastic waves and the Debye vDOS at low frequencies. However, in contrast to this expectation, we will show below that the disordered structures of glasses are not completely uniformly coarse-grained but rather act as defects even at macroscopic scales. Glasses therefore behave not as uniform elastic media but rather as elastic media with defects.

6.1. *Vibrational eigenmodes in glasses*

Figure 13 presents the results for the vibrational eigenmodes in a glass with a harmonic potential. The top panel of Figure 13(a) (circles) shows the reduced vDOS $g(\omega)/\omega^2$ together with the Debye level A_D. The reduced vDOS clearly exhibits a maximum, i.e., the boson peak (the arrow indicates the BP frequency, ω_{BP}). As the frequency decreases below ω_{BP}, $g(\omega)/\omega^2$ decreases toward but does not reach A_D in the present frequency region. We will carefully discuss this point below.

To enable characterization of the eigenmodes, Figure 13(a) also presents data on the phonon order parameter O^k defined in Eq. (46) (middle panel) and the participation ratio $\mathcal{P}^k$ defined in Eq. (43) (bottom panel). Let us first consider O^k, which measures the extent to which the eigenmode k exhibits phonon-like vibrations:[51,52] O^k takes values from 1 (phonon) to (non-phonon). At high ω, O^k is nearly zero, which confirms that these eigenmodes are considerably different to phonons. We may define the frequency, ω_*, at which O^k converges to zero.[x] It has been reported that at the high $\omega > \omega_*$, the eigenmodes are disordered and extended;[34–36] this report is

[x]The present glass system shows a characteristic plateau in the vDOS.[34–36] The value of ω_* is usually defined as the onset frequency of this plateau, which is plotted in Figure 13(a). However, we can confirm in Figure 13(a) that the frequency at which O^k converges to zero is consistent with the onset frequency ω_*.

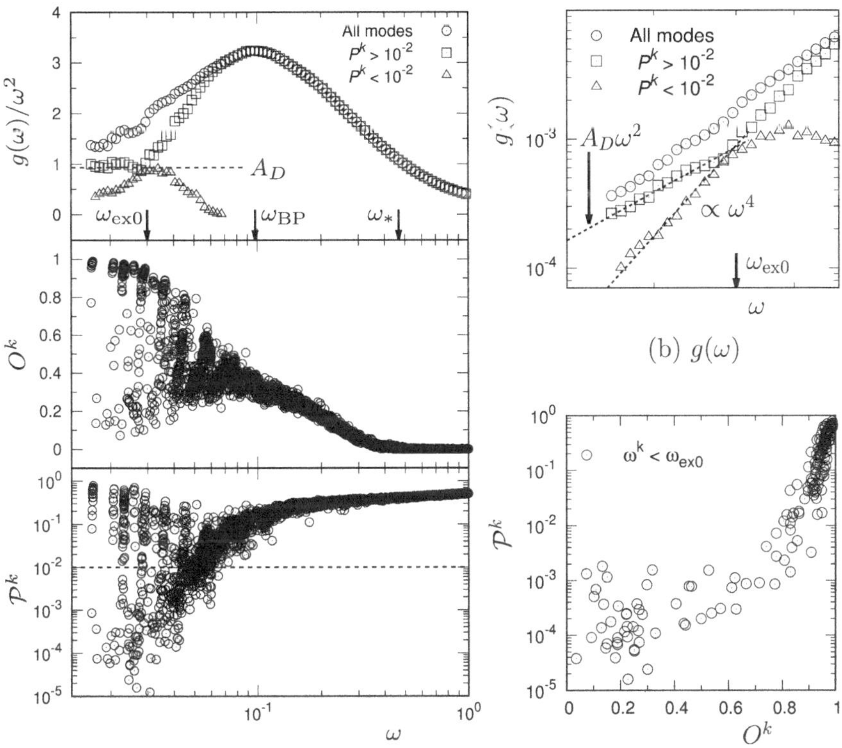

(a) $g(\omega)/\omega^2$, O^k, $\mathcal{P}^k$ (c) O^k versus $\mathcal{P}^k$

Fig. 13. Vibrational eigenmodes in a glass with a harmonic potential. (a) Plots of the reduced vDOS $g(\omega)/\omega^2$ and of the phonon order parameter O^k and participation ratio $\mathcal{P}^k$ for each eigenmode k as functions of the frequency ω. (b) The vDOS $g(\omega)$. (c) O^k versus $\mathcal{P}^k$ for the low-frequency eigenmodes with $\omega^k < \omega_{\text{ex0}}$. Regarding the vDOS, data are also presented for $g_{\text{ex}}(\omega)$, the vDOS of the extended modes ($\mathcal{P}^k > 10^{-2}$), and for $g_{\text{loc}}(\omega)$, the vDOS of the localized modes ($\mathcal{P}^k < 10^{-2}$). The horizontal lines at the top and bottom of (a) represent the Debye level A_D and $\mathcal{P}^k = 10^{-2}$, respectively. In the top panel of (a), we use arrows to indicate the characteristic frequencies, ω_{ex0}, ω_{BP}, and ω_*, as discussed in the main text.

consistent with the present result of $O^k \approx 0$. As ω decreases from ω_* to ω_{BP}, O^k smoothly increases to ≈ 0.3. This result indicates that the eigenmodes near ω_{BP} show phonon-like vibrations to some extent. Notably, as ω further decreases below ω_{BP}, the modes can be divided into two groups: O^k increases with decreasing ω in one group, whereas O^k decreases in the other group. In the former group, O^k converges to almost 1 at $\omega_{\text{ex}0}$ (we will provide the precise definition of $\omega_{\text{ex}0}$ later), indicating that these modes are phonon modes.[y]

We next consider $\mathcal{P}^k$, which evaluates the extent of spatial localization of eigenmode k.[26–28] $\mathcal{P}^k$ takes values from 1 (extended over all particles equally) to $1/N \ll 1$ (localized to one particle). $\mathcal{P}^k$ more clearly exhibits the division of the modes into two groups: one group approaches $\mathcal{P}^k = \mathcal{O}(1)$ with decreasing ω, and the other approaches $\mathcal{P}^k = \mathcal{O}(1/N)$. Figure 13(c) shows that the non-phonon modes (small O^k) are localized (small $\mathcal{P}^k$) at low $\omega < \omega_{\text{ex}0}$, whereas the phonon modes (large O^k) are extended (large $\mathcal{P}^k$). Therefore, the data for O^k and $\mathcal{P}^k$ unambiguously demonstrate that phonon modes and non-phonon localized modes coexist at the low $\omega < \omega_{\text{ex}0}$.

The distinction between phonon modes and localized modes enables us to separately consider the vDOSs for these two types of modes. We define $g_{\text{ex}}(\omega)$ as the vDOS for modes with $\mathcal{P}^k > \mathcal{P}_c$, and we define $g_{\text{loc}}(\omega)$ as the vDOS for modes with $\mathcal{P}^k < \mathcal{P}_c$. Here, we set a reasonable threshold value of $\mathcal{P}_c = 10^{-2}$, as in the data for $\mathcal{P}^k$ in Figure 13(a) (bottom panel).[z] We plot the reduced versions of $g_{\text{ex}}(\omega)$ and $g_{\text{loc}}(\omega)$ in Figure 13(a) (top panel). We also plot the vDOS itself in Figure 13(b). $g_{\text{ex}(\omega)}$ converges exactly to the Debye vDOS $A_D\omega^2$ at a finite value of ω, which we define as $\omega_{\text{ex}0}$. On the other hand, $g_{\text{loc}}(\omega)$ follows a different scaling law, $g_{\text{loc}}(\omega) \propto \omega^4$. Thus, we conclude that the phonon modes that follow the Debye law ($g_{\text{ex}}(\omega) = A_D\omega^2$) and the localized modes that follow the other, non-Debye law ($g_{\text{loc}}(\omega) \propto \omega^4$) coexist at $\omega < \omega_{\text{ex}0}$. Note that the ω^4 scaling is the same law proposed in the soft-potential model.[88–93] The total vDOS $g(\omega)$ can therefore be described as

$$g\left(\omega\right) = g_{\text{ex}}\left(\omega\right) + g_{\text{loc}}\left(\omega\right) = A_D\omega^2 + A_{\text{loc}}\omega^4. \tag{90}$$

[y] The values of O^k for these phonon modes are close to but not exactly 1, which indicates that they are weakly perturbed. An exact value of $O^k = 1$ may be realized only in the limit of $\omega \to 0$.

[z] However, the results are insensitive to the choice of $\mathcal{P}_c$ for $5 \times 10^{-3} < \mathcal{P}_c < 2 \times 10^{-2}$.[51]

Since $g_{\mathrm{loc}}(\omega)$ always takes finite values at $\omega < \omega_{\mathrm{ex0}}$, $g(\omega)$ coincides with the Debye vDOS only at zero frequency. This means that the disordered structures of glasses are not uniformly coarse-grained even at macroscopic scales but rather continue to produce localized vibrations.

Based on the vDOS in Eq. (90), the heat capacity $C(T)$ can be predicted within the harmonic approximation as follows[6,7]:

$$C(T) = 3 \int \hbar\omega \left(\frac{\partial f(\omega, T)}{\partial T} \right) g(\omega) d\omega, \tag{91}$$

where $f(\omega, T) = [\exp(\hbar\omega/k_B T) - 1]^{-1}$ is the Bose-Einstein distribution, k_B is the Boltzmann constant, and $\hbar = h/2\pi$ with h being the Plank constant. At the low temperature, $k_B T \lesssim \hbar\omega_{\mathrm{ex0}}$, $C(T)$ consists of two terms: the Debye term $C_{\mathrm{ex}}(T) \propto T^3$ and the non-Debye term $C_{\mathrm{loc}}(T) \propto T^5$. This result for $C(T)$ in the harmonic approximation cannot correctly capture experimental observations,[13–15] in which a linear T dependence appears at low temperatures. The present results therefore demonstrate that anharmonicities should play an important role in the low-T properties of glasses, e.g., a two-level system might give rise to the linear T dependence.[23,24,94] This is completely different to the case of crystals, which are well described within the harmonic approximation.[6,7]

6.2. *Phonon transport in glasses*

We next present the results for phonon transport in a glass with a harmonic potential. Figure 14 shows the sound speeds, $c_\alpha(\Omega)$, and the attenuation rates, $\Gamma_\alpha(\Omega)$, for transverse ($\alpha = T$, Figure 14(a)) and longitudinal ($\alpha = L$, Figure 14(b)) phonon transport. These results were obtained through direct measurement at zero temperature (see Section 4.4.2). In this figure, arrows indicate the characteristic frequencies, ω_*, ω_{BP}, and ω_{ex0}, all of which are obtained from the eigenmode results in Figure 13. In the boson peak regime, $\Omega \sim \omega_{\mathrm{BP}}$, c_α takes its minimum value, corresponding to sound softening. Γ_α shows an Ω^2 dependence, $\Gamma_\alpha \propto \Omega^2$. We also find that the IR limit for transverse phonons is $\Omega_{TIR} \approx \omega_{\mathrm{BP}}$. These results indicate that such a phonon does not propagate as a plane wave but rather exhibits

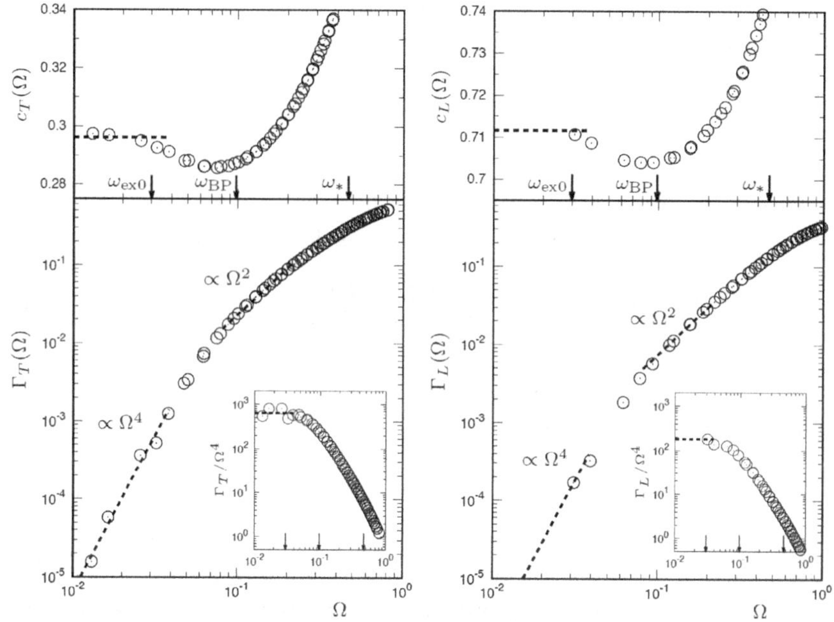

(a) Transverse phonon transport. (b) Longitudinal phonon transport.

Fig. 14. Phonon transport in a glass with a harmonic potential. (a) Transverse phonon transport. (b) Longitudinal phonon transport. We plot the sound speed $c_\alpha(\Omega)$ (top panel) and the attenuation rate $\Gamma_\alpha(\Omega)$ (bottom) as functions of the frequency Ω. The dashed line indicates the macroscopic sound speed (top panel) or the power-law scaling with Ω (bottom). The inset in the bottom panel plots Γ_α/Ω^4 versus Ω to clearly demonstrate the Rayleigh scattering law. We use arrows to indicate the characteristic frequencies, $\omega_{\rm ex0}$, $\omega_{\rm BP}$, and ω_*, that are obtained from the vibrational eigenmodes as shown in Figure 13.

dynamics characteristic of viscous damping.[37] As Ω decreases to $\Omega \lesssim \omega_{\rm ex0}$, we observe a clear crossover to Rayleigh scattering behaviour. c_α converges to its macroscopic value: $c_{T0} = \sqrt{G/\rho}$ or $c_{L0} = \sqrt{(K + 4G/3)/\rho}$. Additionally, Γ_α shows an Ω^4 dependence, $\Gamma_\alpha \propto \Omega^4$.

All of the above observations were previously discussed in Section 4. The crossover in the nature of phonon transport has been observed in both experiments[18–20] and simulations.[38,39,41,45,60] This crossover can be predicted by mean-field theories.[9–12] In particular, both transverse and longitudinal phonons show similar transport properties: we indeed observe the same crossover frequency ω_{ex0} for both of them. This result is consistent with the prediction of heterogeneous elasticity theory:[9,10] the shear modulus heterogeneity induces anomalous behaviours for both transverse and longitudinal phonon transport.

We can understand the crossover in phonon transport in terms of the underlying vibrational eigenmodes as follows. Because the eigenmodes at $\Omega \sim \omega_{\text{BP}}$ correspond to disordered and extended vibrations and an initially excited phonon is decomposed into these eigenmodes, it immediately attenuates to become diffusive. On the other hand, the eigenmodes at $\Omega \lesssim \omega_{\text{ex0}}$ consist of phonon modes and localized modes. Here, we measure the overlap O_{loc} between the initially excited phonon, $\dot{\boldsymbol{u}}_0 = \boldsymbol{u}_{\text{ph}}^{\boldsymbol{q},\alpha}$ (Eq. (44)), and the localized modes, $\boldsymbol{e}^k$, with participation ratio $\mathcal{P}^k < 10^{-2}$, by calculating

$$O_{\text{loc}} = \sum_{k;\,\mathcal{P}^k<10^{-2}} \left| \boldsymbol{u}_{\text{ph}}^{\boldsymbol{q},\alpha} \cdot \boldsymbol{e}^k \right|^2, \tag{92}$$

and we find that only a few % (at most) of the initially excited phonon is made up of localized modes. From this observation, we can understand that the initially excited phonon is decomposed mainly into phonon modes, so it attenuates only slowly. Therefore, the present results unambiguously link these two types of phonon transport to the eigenmodes in the corresponding frequency regimes. We are able to identify the crossover frequency as ω_{ex0}, at which the nature of the underlying eigenmodes changes.

As was already discussed in Section 4.3, we note that the IR limit for longitudinal phonons, $\Omega_{L\text{IR}}$, is much higher than that for transverse phonons: $\Omega_{L\text{IR}} \gg \Omega_{T\text{IR}} \approx \omega_{\text{BP}}$. Therefore, a longitudinal phonon can propagate even at $\Omega > \omega_{\text{BP}}$. This result indicates that although the disordered eigenmodes are dominant at $\Omega > \omega_{\text{BP}}$, longitudinal phonon modes also exist in this Ω regime, which support the propagation of longitudinal phonons.

In summary, we have shown that even in the low-frequency limit, the glass exhibits localized eigenmodes and Rayleigh scattering in phonon transport. The same conclusion has also been obtained for an LJ glass.[38,52] These results indicate that the disordered structures of glasses are not uniformly coarse-grained even at macroscopic scales but rather play the role of defects, influencing the vibrational properties of the material. Below the frequency ω_{ex0}, which corresponds to the continuum limit frequency, glasses behave as elastic media with defects.

As a final remark, the above results are $T = 0$ properties, in the harmonic approximation limit. Here we mention anharmonic properties of vibrational excitations. At finite temperatures $T > 0$, systems undergo anharmonic processes. In crystals phonons excited by thermal fluctuations couple through phonon–phonon interactions.[6,7] For example, in three-phonon processes, one phonon splits into two different ones, or conversely, two phonons combine into a single excitation. Similarly, in glasses, vibrational eigenmodes also show anharmonic processes due to mode–mode interactions.[95,96]

In contrast to crystals, however, it has been demonstrated that an additional anharmonic channel of different origin emerges in glasses, which induces unconventional intermittent rearrangements of particles.[97–100] The intermittent rearrangements can be considered as relics of the liquid state which survive the complete dynamic arrest taking place at the glass transition temperature.[101–103] It was also demonstrated that glasses exhibit anomalous temperature dependence of phonon attenuation at finite temperatures,[21,22,104–106] which can be attributed to this additional anharmonic channel.

References

1. M. P. Allen and D. J. Tildesley. (1986). *Computer Simulation of Liquids*. Oxford: Oxford University Press.
2. D. Frenkel and B. Smit. (2002). *Understanding Molecular Simulation*. 2nd ed. New York: Academic Press.
3. M. Newman and G. Barkema. (1999). *Monte Carlo Methods in Statistical Physics*. U.S.A.: Oxford University Press.
4. A. Cavagna. (2009). Supercooled liquids for pedestrians. *Phys. Rep.* **476**, 51.
5. L. Berthier and G. Biroli. (2011). Theoretical perspective on the glass transition and amorphous materials. *Rev. Mod. Phys.* **83**, 587.

6. N. W. Ashcroft and N. D. Mermin. (1976). *Solid State Physics*. New York: Harcourt College Publishers.
7. C. Kittel. (1996). *Introduction to Solid State Physics*. 7th ed. New York: John Wiley and Sons.
8. S. Franz, G. Parisi, P. Urbani, and F. Zamponi. (2015). Universal spectrum of normal modes in low-temperature glasses. *Proc. Nat. Acad. Sci.* **112**, 14539.
9. W. Schirmacher. (2006). Thermal conductivity of glassy materials and the "boson peak". *Europhys. Lett.* **73**, 892.
10. W. Schirmacher, G. Ruocco, and T. Scopigno. (2007). Acoustic attenuation in glasses and its relation with the boson peak. *Phys. Rev. Lett.* **98**, 025501.
11. M. Wyart. (2010). Scaling of phononic transport with connectivity in amorphous solids. *EPL (Europhysics Letters)* **89**, 64001.
12. E. DeGiuli, A. Laversanne-Finot, G. During, E. Lerner, and M. Wyart. (2014). Effects of coordination and pressure on sound attenuation, boson peak and elasticity in amorphous solids. *Soft Matter* **10**, 5628.
13. R. C. Zeller and R. O. Pohl. (1971). Thermal conductivity and specific heat of non-crystalline solids. *Phys. Rev. B* **4**, 2029.
14. W. A. Phillips. (1981). *Amorphous Solids: Low Temperature Properties*. 3rd edn. Berlin: Springer.
15. S. R. Elliott. (1990). *Physics of Amorphous Materials*. 2nd ed. New York: Longman.
16. U. Buchenau, N. Nucker, and A. J. Dianoux. (1984). Neutron scattering study of the low-frequency vibrations in vitreous silica. *Phys. Rev. Lett.* **53**, 2316.
17. O. Yamamuro, T. Matsuo, K. Takeda, T. Kanaya, T. Kawaguchi, and K. Kaji. (1996). Inelastic neutron scattering study of low energy excitations in glassy 1-butene. *J. Chem. Phys.* **105**, 732.
18. C. Masciovecchio, G. Baldi, S. Caponi, L. Comez, S. Di Fonzo, D. Fioretto, A. Fontana, A. Gessini, S. C. Santucci, F. Sette, G. Viliani, P. Vilmercati, and G. Ruocco. (2006). Evidence for a crossover in the frequency dependence of the acoustic attenuation in vitreous silica. *Phys. Rev. Lett.* **97**, 035501.
19. G. Monaco and V. M. Giordano (2009). Breakdown of the debye approximation for the acoustic modes with nanometric wavelengths in glasses. *Proc. Nat. Acad. Sci.* **106**, 3659.
20. G. Baldi, V. M. Giordano, G. Monaco, and B. Ruta. (2010). Sound attenuation at terahertz frequencies and the boson peak of vitreous silica. *Phys. Rev. Lett.* **104**, 195501.

21. C. Ferrante, E. Pontecorvo, G. Cerullo, A. Chiasera, G. Ruocco, W. Schirmacher, and T. Scopigno. (2013). Acoustic dynamics of network-forming glasses at mesoscopic wavelengths. *Nature Commun.* **4**, 1793.
22. G. Baldi, V. M. Giordano, B. Ruta, R. Dal Maschio, A. Fontana, and G. Monaco. (2014). Anharmonic damping of terahertz acoustic waves in a network glass and its effect on the density of vibrational states. *Phys. Rev. Lett.* **112**, 125502.
23. T. Perez-Castaneda, C. Rodrıguez-Tinoco, J. Rodrıguez-Viejo, and M. A. Ramos. (2014). Suppression of tunneling two-level systems in ultrastable glasses of indomethacin. *Proc. Nat. Acad. Sci.* **111**, 11275.
24. T. Perez-Castaneda, R. J. Jimenez-Rioboo, and M. A. Ramos. (2014). Two-level systems and boson peak remain stable in 110-million-year-old amber glass. *Phys. Rev. Lett.* **112**, 165901.
25. M. Kabeya, T. Mori, Y. Fujii, A. Koreeda, B. W. Lee, J.-H. Ko, and S. Kojima. (2016). Boson peak dynamics of glassy glucose studied by integrated terahertz-band spectroscopy. *Phys. Rev. B* **94**, 224204.
26. H. R. Schober and B. B. Laird. (1991). Localized low-frequency vibrational modes in glasses. *Phys. Rev. B* **44**, 6746.
27. V. Mazzacurati, G. Ruocco, and M. Sampoli. (1996). Low-frequency atomic motion in a model glass. *Europhys. Lett.* **34**, 681.
28. S. N. Taraskin and S. R. Elliott. (1999). Anharmonicity and localization of atomic vibrations in vitreous silica. *Phys. Rev. B* **59**, 8572.
29. S. N. Taraskin and S. R. Elliott. (2000). Propagation of plane-wave vibrational excitations in disordered systems. *Phys. Rev. B* **61**, 12017.
30. G. Ruocco, F. Sette, R. Di Leonardo, G. Monaco, M. Sampoli, T. Scopigno, and G. Viliani. (2000). Relaxation processes in harmonic glasses? *Phys. Rev. Lett.* **84**, 5788.
31. A. Tanguy, J. P. Wittmer, F. Leonforte, and J.-L. Barrat. (2002). Continuum limit of amorphous elastic bodies: A finite-size study of low-frequency harmonic vibrations. *Phys. Rev. B* **66**, 174205.
32. F. Leonforte, R. Boissiere, A. Tanguy, J. P. Wittmer, and J.-L. Barrat. (2005). Continuum limit of amorphous elastic bodies. iii. three-dimensional systems. *Phys. Rev. B* **72**, 224206.
33. F. Leonforte, A. Tanguy, J. P. Wittmer, and J.-L. Barrat. (2006). Inhomogeneous elastic response of silica glass. *Phys. Rev. Lett.* **97**, 055501.
34. M. Wyart, L. E. Silbert, S. R. Nagel, and T. A. Witten. (2005). Effects of compression on the vibrational modes of marginally jammed solids. *Phys. Rev. E* **72**, 051306.

35. L. E. Silbert, A. J. Liu, and S. R. Nagel. (2005). Vibrations and diverging length scales near the unjamming transition. *Phys. Rev. Lett.* **95**, 098301.
36. L. E. Silbert, A. J. Liu, and S. R. Nagel. (2009). Normal modes in model jammed systems in three dimensions. *Phys. Rev. E* **79**, 021308.
37. H. Shintani and H. Tanaka. (2008). Universal link between the boson peak and transverse phonons in glass. *Nature Mater.* **7**, 870.
38. G. Monaco and S. Mossa. (2009). Anomalous properties of the acoustic excitations in glasses on the mesoscopic length scale. *Proc. Nat. Acad. Sci.* **106**, 16907.
39. A. Marruzzo, W. Schirmacher, A. Fratalocchi, and G. Ruocco. (2013). Heterogeneous shear elasticity of glasses: The origin of the boson peak. *Sci. Rep.* **3**, 1407.
40. H. Mizuno, S. Mossa, and J.-L. Barrat. (2013). Elastic heterogeneity, vibrational states, and thermal conductivity across an amorphisation transition. *EPL (Europhysics Letters)* **104**, 56001.
41. H. Mizuno, S. Mossa, and J.-L. Barrat. (2014). Acoustic excitations and elastic heterogeneities in disordered solids. *Proc. Nat. Acad. Sci.* **111**, 11949.
42. H. Mizuno, S. Mossa, and J.-L. Barrat. (2016). Relation of vibrational excitations and thermal conductivity to elastic heterogeneities in disordered solids. *Phys. Rev. B* **94**, 144303.
43. A. Ikeda, L. Berthier, and G. Biroli. (2013). Dynamic criticality at the jamming transition. *J. Chem. Phys.* **138**, 12A507.
44. D. Crespo, P. Bruna, A. Valles, and E. Pineda. (2016). Phonon dispersion relation of metallic glasses. *Phys. Rev. B* **94**, 144205.
45. Y. M. Beltukov, C. Fusco, D. A. Parshin, and A. Tanguy. (2016). Boson peak and Ioffe-regel criterion in amorphous siliconlike materials: The effect of bond directionality. *Phys. Rev. E* **93**, 023006.
46. S. Gelin, H. Tanaka, and A. Lemaitre. (2016). Anomalous phonon scattering and elastic correlations in amorphous solids. *Nature Mater.* **15**, 1177.
47. R. Milkus and A. Zaccone. (2016). Local inversion-symmetry breaking controls the boson peak in glasses and crystals. *Phys. Rev. B* **93**, 094204.
48. J. Krausser, R. Milkus, and A. Zaccone. (2017). Non-affine lattice dynamics of defective fcc crystals. *Soft Matter* **13**, 6079.
49. K. Baumgarten, D. Vagberg, and B. P. Tighe. (2017). Nonlocal elasticity near jamming in frictionless soft spheres. *Phys. Rev. Lett.* **118**, 098001.

50. E. Lerner, G. During, and E. Bouchbinder. (2016). Statistics and properties of low-frequency vibrational modes in structural glasses. *Phys. Rev. Lett.* **117**, 035501.
51. H. Mizuno, H. Shiba, and A. Ikeda. (2017). Continuum limit of the vibrational properties of amorphous solids. *Proc. Nat. Acad. Sci.* **114**, E9767.
52. M. Shimada, H. Mizuno, and A. Ikeda. (2018). Anomalous vibrational properties in the continuum limit of glasses. *Phys. Rev. E* **97**, 022609.
53. E. Lerner and E. Bouchbinder. (2017) Effect of instantaneous and continuous quenches on the density of vibrational modes in model glasses. *Phys. Rev. E* **96**, 020104.
54. E. Lerner and E. Bouchbinder. (2018). Frustration-induced internal stresses are responsible for quasilocalized modes in structural glasses. *Phys. Rev. E* **97**, 032140.
55. L. Angelani, M. Paoluzzi, G. Parisi, and G. Ruocco. (2018). Probing the non-debye low-frequency excitations in glasses through random pinning. *Proc. Nat. Acad. Sci.* **115**, 8700.
56. E. Bouchbinder and E. Lerner. (2018). Universal disorder-induced broadening of phonon bands: From disordered lattices to glasses. *New J. Phys.* **20**, 073022.
57. H. Mizuno and A. Ikeda. (2018). Phonon transport and vibrational excitations in amorphous solids. *Phys. Rev. E* **98**, 062612 (2018).
58. M. Shimada, H. Mizuno, M. Wyart, and A. Ikeda. (2018). Spatial structure of quasilocalized vibrations in nearly jammed amorphous solids. *Phys. Rev. E* **98**, 060901.
59. L. Wang, A. Ninarello, P. Guan, L. Berthier, G. Szamel, and E. Flenner. (2019). Low-frequency vibrational modes of stable glasses. *Nature Commun.* **10**, 26.
60. Y. M. Beltukov, D. A. Parshin, V. M. Giordano, and A. Tanguy. (2018). Propagative and diffusive regimes of acoustic damping in bulk amorphous material. *Phys. Rev. E* **98**, 023005.
61. K. Saitoh, R. K. Shrivastava, and S. Luding. (2019). Rotational sound in disordered granular materials. *Phys. Rev. E* **99**, 012906.
62. H. Tong, H. Hu, P. Tan, N. Xu, and H. Tanaka. (2019). Revealing inherent structural characteristics of jammed particulate packings. *Phys. Rev. Lett.* **122**, 215502.
63. J. P. Hansen and I. R. McDonald. (2006). *Theory of Simple Liquids.* 3rd ed. London: Academic.
64. M. van Hecke. (2009). Jamming of soft particles: Geometry, mechanics, scaling and isostaticity. *J. Phys. Condens.Matter* **22**, 033101.

65. B. W. H. van Beest, G. J. Kramer, and R. A. van Santen. (1990). Force fields for silicas and aluminophosphates based on ab initio calculations. *Phys. Rev. Lett.* **64**, 1955.
66. F. H. Stillinger and T. A. Weber. (1985). Computer simulation of local order in condensed phases of silicon. *Phys. Rev. B* **31**, 5262.
67. M. S. Daw and M. I. Baskes. (1984). Embedded-atom method: Derivation and application to impurities, surfaces, and other defects in metals. *Phys. Rev. B* **29**, 6443.
68. K. Kremer and G. S. Grest. (1990). Dynamics of entangled linear polymer melts: A molecular-dynamics simulation. *J. Chem. Phys.* **92**, 5057.
69. W. Press, B. Flannery, S. Teukolsky, W. Vetterling, and B. Flannery. (2007). *Numerical Recipes: The Art of Scientific Computing.* 3rd ed. New York: Cambridge University Press.
70. E. Bitzek, P. Koskinen, F. Gahler, M. Moseler, and P. Gumbsch. (2006). Structural relaxation made simple. *Phys. Rev. Lett.* **97**, 170201.
71. A. G. Every. (1980). General closed-form expressions for acoustic waves in elastically anisotropic solids. *Phys. Rev. B* **22**, 1746.
72. C. Jasiukiewicz and V. Karpus. (2003). Debye temperature of cubic crystals. *Solid State Commun.* **128**, 167.
73. A. Akhieser. (1939). On the absorption of sound in solids. *J. Phys. — USSR* **1**, 277.
74. H. J. Maris. (1971). Interaction of sound waves with thermal phonons in dielectric crystals. *Phys. Acoust.* **8**, 279.
75. W. Schirmacher, C. Tomaras, B. Schmid, G. Baldi, G. Viliani, G. Ruocco, and T. Scopigno. (2010). Sound attenuation and anharmonic damping in solids with correlated disorder. *Condens. Matter Phys.* **13**, 23606.
76. C. Tomaras, B. Schmid, and W. Schirmacher. (2010). Anharmonic elasticity theory for sound attenuation in disordered solids with fluctuating elastic constants. *Phys. Rev. B* **81**, 104206.
77. A. Marruzzo, S. Kohler, A. Fratalocchi, G. Ruocco, and W. Schirmacher. (2013). Vibrational anomalies and marginal stability of glasses. *Eur. Phys. J. Spec. Top.* **216**, 83.
78. P. B. Allen and J. L. Feldman. (1993). Thermal conductivity of disordered harmonic solids. *Phys. Rev. B* **48**, 12581.
79. J. L. Feldman, M. D. Kluge, P. B. Allen, and F. Wooten. (1993). Thermal conductivity and localization in glasses: Numerical study of a model of amorphous silicon. *Phys. Rev. B* **48**, 12589.

80. M. Born and K. Huang. (1954). *Dynamical Theory of Crystal Lattices.* Oxford: Clarendon Press.
81. T. H. K. Barron and M. L. Klein. (1965). Second-order elastic constants of a solid under stress. *Proc. Phys. Soc. London* **85**, 523.
82. S. Alexander. (1998). Amorphous solids: Their structure, lattice dynamics and elasticity. *Phys. Rep.* **296**, 65.
83. J. F. Lutsko. (1989). Generalized expressions for the calculation of elastic constants by computer simulation. *J. Appl. Phys.* **65**, 2991.
84. A. Lemaitre and C. Maloney. (2006). Sum rules for the quasi-static and visco-elastic response of disordered solids at zero temperature. *J. Stat. Phys.* **123**, 415.
85. Y. C. Fung. (1977). *A First Course in Continuum Mechanics.* 2nd ed. New Jersey, U.S.A: Prentice-Hall.
86. J. P. Wittmer, H. Xu, P. Poli´nska, F. Weysser, and J. Baschnagel. (2013). Shear modulus of simulated glass-forming model systems: Effects of boundary condition, temperature, and sampling time. *J. Chem. Phys.* **138**, 12A533.
87. H. Mizuno, S. Mossa, and J.-L. Barrat. (2013). Measuring spatial distribution of the local elastic modulus in glasses. *Phys. Rev. E* **87**, 042306.
88. V. G. Karpov, M. I. Klinger, and F. N. Ignat'ev. (1983). Theory of the low temperature anomalies in the thermal properties of amorphous structures. *Sov. Phys. JETP.* **57**, 439.
89. U. Buchenau, Y. M. Galperin, V. L. Gurevich, and H. R. Schober. (1991). Anharmonic potentials and vibrational localization in glasses. *Phys. Rev. B* **43**, 5039.
90. U. Buchenau, Y. M. Galperin, V. L. Gurevich, D. A. Parshin, M. A. Ramos, and H. R. Schober. (1992). Interaction of soft modes and sound waves in glasses. *Phys. Rev. B* **46**, 2798.
91. V. L. Gurevich, D. A. Parshin, and H. R. Schober. (2003). Anharmonicity, vibrational instability, and the boson peak in glasses. *Phys. Rev. B* **67**, 094203.
92. V. L. Gurevich, D. A. Parshin, and H. R. Schober. (2005). Pressure dependence of the boson peak in glasses. *Phys. Rev. B* **71**, 014209.
93. D. A. Parshin, H. R. Schober, and V. L. Gurevich. (2007). Vibrational instability, two-level systems, and the boson peak in glasses. *Phys. Rev. B.* **76**, 064206.
94. P. W. Anderson, B. I. Halperin, and C. M. Varma. (1972). Anomalous low temperature thermal properties of glasses and spin glasses. *Philos. Mag.* **25**, 1.
95. J. Fabian and P. B. Allen. (1996). Anharmonic decay of vibrational states in amorphous silicon. *Phys. Rev. Lett.* **77**, 3839.

96. J. Fabian, J. L. Feldman, C. S. Hellberg, and S. M. Nakhmanson (2003). Numerical study of anharmonic vibrational decay in amorphous and paracrystalline silicon. *Phys. Rev. B* **67**, 224302.
97. A. Heuer. (2008). Exploring the potential energy landscape of glass-forming systems: From inherent structures via metabasins to macroscopic transport. *J. Phys. Condens. Matter* **20**, 373101.
98. N. Xu, V. Vitelli, A. J. Liu, and S. R. Nagel. (2010). Anharmonic and quasi-localized vibrations in jammed solids -Modes for mechanical failure. *EPL (Europhysics Letters)* **90**, 56001.
99. H. Mizuno, M. Shimada, and A. Ikeda. (2020). Anharmonic properties of vibrational excitations in amorphous solids. *Phys. Rev. Res.* **2**, 013215.
100. H. Mizuno, H. Tong, A. Ikeda, and S. Mossa. (2020). Intermittent rearrangements accompanying thermal fluctuations distinguish glasses from crystals. *J. Chem. Phys.* **153**, 154501.
101. L. Berthier, P. Charbonneau, Y. Jin, G. Parisi, B. Seoane, and F. Zamponi. (2016). Growing timescales and lengthscales characterizing vibrations of amorphous solids. *Proc. Natl. Acad. Sci. U. S. A.* **113**, 8397.
102. C. Scalliet, L. Berthier, and F. Zamponi. (2017). Absence of marginal stability in a structural glass. *Phys. Rev. Lett.* **119**, 205501.
103. C. Scalliet, L. Berthier, and F. Zamponi. (2019). Nature of excitations and defects in structural glasses. *Nat. Commun.* **10**, 5102.
104. H. Mizuno and S. Mossa. (2019). Impact of elastic heterogeneity on the propagation of vibrations at 1nite temperatures in glasses. *Condens. Matter Phys.* **22**, 43604.
105. H. Mizuno, G. Ruocco, and S. Mossa. (2020). Sound damping in glasses: Interplay between anharmonicities and elastic heterogeneities. *Phys. Rev. B* **101**, 174206.
106. L. Wang, G. Szamel, and E. Flenner. (2020). Sound attenuation in finite-temperature stable glasses. *Soft Matter* **16**, 7165.

https://doi.org/10.1142/9781800612587_0011

Chapter 11

Topological Phases of Amorphous Matter

Adolfo G. Grushin

Institut Néel, CNRS & Université Grenoble Alpes,
Grenoble 38042, France
adolfo.grushin@neel.cnrs.fr

Topological phases of matter are often understood and predicted with the help of crystal symmetries although they don't rely on them to exist. In this chapter, we review how topological phases have been recently shown to emerge in amorphous systems. We summarize the properties of topological states and discuss how disposing of translational invariance has motivated the surge of new tools to characterize topological states in amorphous systems, both theoretically and experimentally. The ubiquity of amorphous systems combined with the robustness of topology has the potential to bring new fundamental understanding in our classification of phases of matter and inspire new technological developments.

1. Introduction

Topological phases of matter are among the most robust phases of nature. Their many-body wave function cannot be described with purely localized orbitals, and the information that defines these phases is stored non-locally spread over the entire system. Therefore, their physical properties, such as their metallic boundary states or quantized responses to external fields, are protected from local perturbations, such as defects, impurities, or other material imperfections.

This chapter embraces two remarkable observations. First, it was recently predicted that around 30% of all non-magnetic crystals are in a topological phase,[1–3] and at least a similar proportion is expected for magnetic materials[4–6] (see Ref. [10] for a recent review). Second, the previous chapters in this book should have convinced the reader that nearly all materials can be prepared as amorphous solids if cooled fast enough and to low enough temperatures.[7] Here, we describe the blooming research on the interface between the above two observations. It demonstrates that topological properties can survive even under extreme structural disorder as in amorphous systems. We will describe the recent developments that have characterized theoretically and experimentally topological phases in amorphous solids. Along the way, we will mention other states that host topological phases without translational invariance, such as quasi-crystals, topological Anderson insulators, and fractals.

Amorphous and topological matter share a great technological potential.[8,9] Due to its versatility, the former is around us, in glass windows, plastics, fiber optics, and solar cells. Due to its robustness, the latter has been identified to hold the key to robust quantum computation, memory storage, chiral waveguides, spintronics, metrology, and thermo-electricity. Hence, it is likely that by combining these two forms of matter, we can develop a broader understanding and classification of the properties of solids, and a richer set of applications than those separately devised for the two fields.

Since this book focuses on amorphous matter and glasses and not topological states, we feel compelled to introduce the basics of topological matter to the reader. We will not give rigorous proofs but rather examples and plausibility arguments that illustrate the main properties associated with topological states. This introduction will contextualize the rupture that amorphous topological matter signifies for our current understanding of topological phases, especially due to the central role that symmetry plays in determining topological properties.

Before we commence, a word of warning. In this chapter, we will use the term topological phases to mean non-interacting topological phases. Interacting topological phases, such as fractional quantum Hall states, are usually referred to as *topologically ordered* states.[140] The extension of the developments presented in this chapter to interacting phases remains an open question to which we come back upon concluding.

2. A Primer on Topological Matter: The Role of Symmetry

The concept of symmetry is deeply rooted in our understanding of condensed matter, and topological insulators and metals are not an exception. Although topological phases can exist in the absence of any symmetry, symmetry enriches their classification, and allows to express the topological invariants that define these phases and their responses in mathematically simple ways. As we will see, the concept of translation symmetry, although strictly unnecessary, is central to our understanding.

The basic property of an insulator is the absence of electrically conductive states. The simplest of all insulators is the atomic insulator, a set of disconnected atoms with electrons forming closed shells. If these atoms form a crystal lattice, we may use Bloch's theorem to combine these shells into energy bands separated by finite gaps. Since electrons are bound to the atoms, they lack kinetic energy, and the bands $E_n(\mathbf{k})$ are exactly flat as a function of crystal momentum $\mathbf{k}$. Now, imagine slowly bringing these atoms together forming covalent bonds in the process. The electronic bands may now disperse with $\mathbf{k}$, and the gap sizes can change. If at all stages in the process we retained the existence of a band gap, we may say that the two insulators, the atomic limit and the covalently bonded solid, are smoothly connected to each other by varying a parameter, in this case, the strength of the covalent bonding. We call the insulators connected to the atomic insulator in this way topologically trivial insulators.

Topological insulators present fundamental obstructions to reach the atomic limit. The smooth process that connects them to the atomic insulator will always lead to a gap closing and, with it, fundamentally different responses. This defines topological insulators as insulators that cannot be smoothly transformed into an atomic limit. In many instances, this obstruction is due to an underlying symmetry; there is no path that simultaneously keeps the gap open and respects the symmetry.

2.1. *Basic properties of topological states: Atomic obstructions and quantized observables*

A one-dimensional chain with inversion symmetry ($\mathcal{I}$) is an illustrative example of how symmetry can distinguish two topologically distinct states. Consider placing one electron per unit cell in the chain of Figure 1 with lattice constant a. There are two possibilities consistent with $\mathcal{I}$ (recall that $\mathcal{I} : x \to -x$): centering the wave packet at an atomic site (Figure 1(a)) or centering it at the middle point between sites (Figure 1(b)). The first configuration is a good representation of the atomic limit, since the wave packet is centered on each atom, close to our intuitive picture of a trivial insulator. The second choice is also an insulator, but it is impossible to connect it to the first one respecting $\mathcal{I}$ and simultaneously preserving the gap. Indeed, if we rigidly shift the wave packet away from either of the two inversion-symmetric real-space positions in the unit cell (dashed lines), we would preserve the insulating character but break $\mathcal{I}$. If we

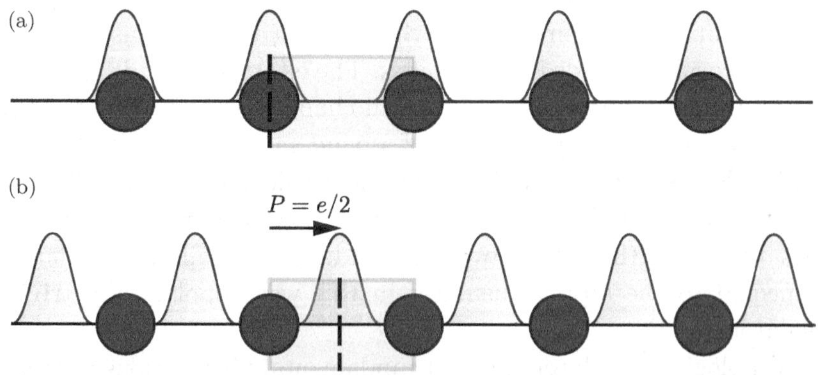

Fig. 1. **Example of one-dimensional topological phases.** (a) and (b) show two inversion symmetric positions (dashed lines) of a Wannier center (filled curve) in the unit cell (box). (a) is an atomic limit, while (b) is an obstructed topological phase because it cannot be deformed to the atomic limit in (a) without breaking inversion or closing the band gap. These two configurations differ by a shift of half a lattice site corresponding to a change in polarization of $P = e/2 \bmod e$.

instead preserve $\mathcal{I}$ by delocalizing the electron symmetrically from one of the two inversion symmetric points, we will pass through a delocalized state and close the band gap.

Based on inversion, these two insulators are in different classes; if we respect the inversion symmetry, they are connected by a gap closing. But how can we physically tell them apart? An important property of topological matter is that topologically distinct phases have different physical responses. Since we are discussing inversion symmetry, using the position operator seems like a natural choice. As it turns out, the position operator is central to describe most topological phases even without inversion. Before we get there, let us consider this operator in our 1D example.

The expectation value of the position operator over all occupied bands is the polarization, which can be defined, per unit volume, using Bloch's wave function $\psi_{\mathbf{k}}^n$ as

$$\mathbf{P} = \frac{e}{V}\langle\hat{\mathbf{r}}\rangle = -\frac{e}{2\pi}\sum_{n\in\text{occ.}}\int_{\text{BZ}} d^D k\, \langle\psi_{\mathbf{k}}^n|\, i\boldsymbol{\nabla}_{\mathbf{k}}\, |\psi_{\mathbf{k}}^n\rangle\,, \tag{1}$$

where V is the unit cell volume (equal to the lattice constant a for our 1D system), $\hat{\mathbf{r}}$ is the position operator, and e is the electron's electric charge. The integral is to be performed over the D dimensional Brillouin zone and summed over occupied bands. This expression should be intuitively reasonable. The polarization is the average of the wave packet's position, thus the first equality. For translationally invariant systems, we can define the crystal momentum $\mathbf{k}$. Quantum mechanics maps the position operator to a derivative, or a gradient in dimensions higher than one, with respect to the conjugate variable, the momentum $\mathbf{k}$, thus the second equality. Upon a momentum dependent phase shift of the wave function, $|\psi_{\mathbf{k}}^n\rangle \to e^{i\theta_{\mathbf{k},n}}\,|\psi_{\mathbf{k}}^n\rangle$, you can check, as noted first by Vanderbilt,[11] that the bulk polarization can only change by a unit of electric charge, and thus it is defined as only modulo e. The two insulators described above are, therefore, different because their bulk polarizations are $P = e/2$ and $P = 0$, mod e. This can be proven by direct calculation, but it should also be reasonable because the two insulators differ by a displacement of the electron by half a unit cell (see Figure 1).

Our 1D example already served to illustrate some of the main properties shared by all topological insulators. First, we saw that distinct topological phases can only be connected through a gap closing. Second, topological insulators define quantized observables, which depend only on fundamental constants like e, allowing us to classify them. In our 1D example, we have distinguished two insulators by their zero or non-zero polarizations, establishing a 0 or 1 type of classification, which is mathematically expressed as a $\mathbb{Z}_2$ classification. It is the first example of topological invariant we encounter because it stays fixed within a topological phase. Third, these observables, and more generally the topological invariants that define them, are determined by global properties of the wave functions. Indeed, note that Eq. (1) is an integral over the full Brillouin zone and involves all occupied states. Fourth, these invariants have simple expressions in the presence of symmetry. For example, Eq. (1) is calculated in momentum space if our system is translationally invariant. More generally, point group symmetries like mirror or inversion can be used to simplify expressions of topological invariants.

The two one-dimensional insulators in Figure 1 can be represented in the basis of localized wave functions (our wave packets were located either at the center of a bond or at an atomic site). The insulators that can be represented in the basis of localized wave functions, but which are not connected to the limit where the wave packet sits at an atomic site, are known as bulk obstructed insulators. By this nomenclature, Figure 1(a) is an atomic limit, while Figure 1(b) is a bulk obstructed topological insulator. A particularly famous system that realizes these phases is the Su–Schrieffer–Heeger chain,[12] a chain of carbon molecules with alternating hopping strength. The ratio between the two hoppings determines whether $P = 0$ or $P = e/2$.

2.2. *Strong topological insulators, gapless surface states, and symmetry classifications*

All one-dimensional insulators are either trivial atomic limits or topological in the sense that they are bulk obstructed insulators, which are not smoothly connected to atomic limits.[a] However, one may argue, trivial atomic limits and bulk obstructed insulators are not so different from each other since we can find a basis where the Fourier transforms of Bloch eigenstates, called Wannier functions, are exponentially localized while preserving all the symmetries. This suggests a stricter criterion to label an insulator as topological: a topological insulator is that which cannot be written in a basis of completely localized wave functions while preserving all symmetries. Thouless already noticed that it is impossible to "Wannierize" a two-dimensional insulator that has a non-zero quantum Hall effect written in terms of Landau levels.[14] In 2011, this intuition was extended to three-dimensional insulators,[15] and currently, this is a criterion used to classify non-interacting topological insulators.[1–3]

Topological insulators defined in this way are realized in various forms depending on the symmetry that protects them.[b] The basic classification of topology relies on 10 symmetry classes[16] that result from combining the three non-spatial symmetries: time-reversal, particle–hole, and chiral symmetry, the last being the product of the first two. For all these classes, it is possible to define a topological invariant analogous to our polarization example mentioned earlier: it is quantized in integers, determines an observable, and for translationally invariant systems, it can be expressed as a global property of Bloch eigenstates in momentum space. Those insulators that realize non-zero integers of these invariants are referred to as *strong* topological insulators.

All surfaces of strong topological insulators are conducting. So long as the symmetry that protects the topological state is respected and the bulk gap remains open, these conducting surface states cannot be turned insulating, for example, by adding impurities.

[a]We distinguish one-dimensional band insulators from one-dimensional superconductors since the Cooper pairs in topological one-dimensional superconductors are not exponentially localized.[13]

[b]Two-dimensional insulators with a Hall effect, known as Chern insulators, are special in the sense that they are protected by the absence of symmetry. They cannot be connected to the atomic limit without a gap closing transition.

This observation is known as the bulk-boundary correspondence and can be understood as follows. A boundary of a topological insulator is the collection of points in space between a topologically non-trivial interior (where the bulk topological invariant is non-zero by definition) and the vacuum, which for all intends and purposes is connected to an atomic insulator, which is trivial. If we were to interpolate between these two insulators with a single Hamiltonian, we can choose to vary a parameter that preserves the symmetry. Somewhere along the way, we have to close the gap because these two insulators belong to different topological classes. The place where this occurs is the boundary, and it is gapless. Since this argument only depends on symmetry and the presence of the bulk gaps, it is insensitive to local details. Thus, these gapless states cannot be gapped with local perturbations, such as the presence of impurities, so long as they respect the symmetry that protects the topological state.

Of course, in trivial insulators, metallic surface states can also exist, but there is nothing fundamental to them. We can always imagine that, by adding local perturbations at the surface, they can be gapped out and displaced in energy to overlap with bulk energy bands. In contrast, for topological phases, these metallic surface states remain protected against local perturbations, since it is a global, bulk property that guarantees their existence.

One important example of a three-dimensional strong topological insulator is the $\mathbb{Z}_2$ topological insulator protected by time-reversal symmetry. As before, the $\mathbb{Z}_2$ nomenclature refers to the fact that there are two types of insulators in this class, the trivial and the topological, with different electromagnetic responses. In three dimensions, the invariant, in this case $\theta = 0, \pi$, determines, respectively, a trivial or quantized magnetoelectric polarizability,[17,18] to which we will return later on in Section 3.3.2. Bi_2Se_3 crystals are the paradigmatic example of a $\mathbb{Z}_2$ topological insulator with $\theta = \pi$.

The gapless surface states of this $\mathbb{Z}_2$ time-reversal invariant topological insulator have remarkable properties. To understand them, consider a bulk insulator with time-reversal symmetry. If there is no spin-orbit coupling, time-reversal symmetry demands that the bands are even functions of momentum, $E_n(\mathbf{k}) = E_n(-\mathbf{k})$, and that they are doubly degenerate due to spin. This is a consequence of Kramers theorem which says that two eigenstates that are time-reversal partners of each other must have the same energy. With spin-orbit coupling, this degeneracy remains only at time-reversal invariant momenta, where $\mathbf{k} = -\mathbf{k}$ up to a lattice vector (e.g., $\mathbf{k} = (0, 0, \pi)$). Time-reversal symmetry also requires that states at momenta $\mathbf{k}$ and $-\mathbf{k}$ have opposite spin. This is true at the surface too, where the Hamiltonian must be gapless, as we argued above. Combining all of these arguments together, the simplest surface Hamiltonian that we can write has two bands and looks like this

$$H_{\text{surf}}(k) = v_{\text{F}}\mathbf{k}_{||} \cdot \sigma + \mu\sigma_0. \tag{2}$$

Here, μ is a chemical potential, $\mathbf{k}_{||} = (k_x, k_y)$ is the momentum parallel to the surface, and σ_0 is a 2×2 identity matrix. In the the simplest case, the Pauli matrices σ represent the spin degree of freedom, but in real materials, it may represent a degree of freedom that combines orbital and spin angular momenta. This hamiltonian, known as the Dirac Hamiltonian, is gapless because it describes quasi-particles with a relativistic-like conical dispersion relation $E_{\pm}(\mathbf{k}) = \pm v_{\text{F}}|\mathbf{k}| + \mu$ (see Figure 2(b)). Since time-reversal symmetry reverts the spin and momentum, a constant σ_z term would break this symmetry. The absence of this term keeps the dispersion gapless consistent with our previous bulk-based discussion. Additionally, time-reversal symmetry implies that states with opposite momenta have spins pointing in opposite directions. This requires that the spin rotates, pointing along $\mathbf{k}$, around the Fermi surface, a property known as spin-momentum locking. We will return to these surface properties in Section 3.4.

Another measurable consequence of these arguments is the number of Dirac cones existing at a given surface of a strong $\mathbb{Z}_2$ three-dimensional topological insulator. Note that there can be a Hamiltonian like $H_{\text{surf}}(\mathbf{k})$ around each time-reversal invariant momenta. However, time-reversal symmetry and a non-zero strong topological index require the number of Dirac cones to be odd.[19,20] This follows from noticing that if the invariant is non-trivial, the

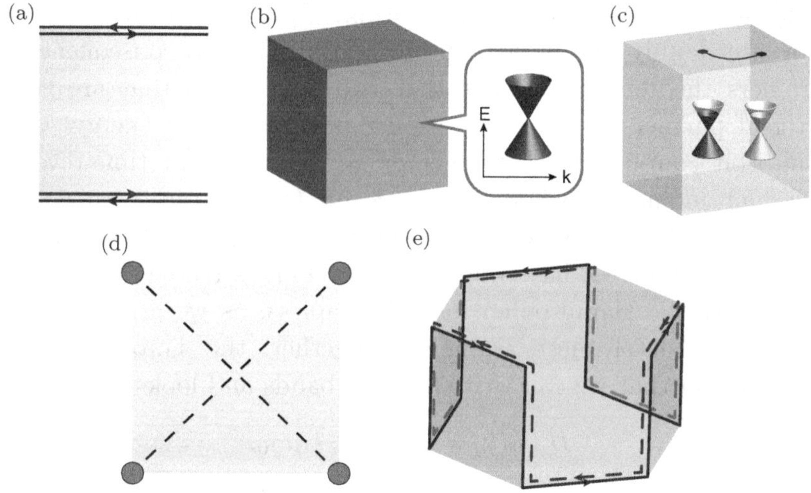

Fig. 2. **Schematics of different topological phases of matter** (a) $\mathbb{Z}_2$ topological insulator in two dimensions protected by time-reversal symmetry. It features an insulating bulk with two conducting helical edge states, observable for example in nano-SQUID measurements.[34] (b) $\mathbb{Z}_2$ topological insulator in three dimensions protected by time-reversal symmetry. It displays a surface Dirac cone observed in ARPES and spin polarized ARPES showing the spin asymmetry (see for example Ref. [35]). (c) In Weyl semimetals such as TaAs,[36] low-energy quasi-particles have a conical dispersion relation which comes in two flavors, or chiralities. At the surface, the projections of the Weyl nodes are connected by a surface state known as the Fermi arc. (d) An example of a two-dimensional higher-order topological insulator with corner modes (circles) protected by mirror symmetry (dashed lines). They have been observed for example in photonic systems.[37] (e) An example of a three-dimensional higher-order topological insulator with helical hinge states. This state is consistent with what is observed in Bismuth.[27]

only way to connect the topological insulator bulk and the vacuum, a trivial insulator, is by closing the gap at the surface. It can only close at an odd number of time-reversal invariant momenta because an even number implies an even number of copies of the surface Hamiltonian Eq. (2), which could be gapped out preserving time-reversal symmetry. Then, this tells us that, without any symmetry other than time-reversal symmetry, $\mathbb{Z}_2$ topological insulators come in two classes depending on whether the number of Dirac cones is even or odd ($\theta = 0, \pi$). An even number of Dirac cones is in this sense not protected, and disorder respecting time-reversal symmetry can gap them out.

It was soon noted that point group symmetries, such as rotations, mirrors, or inversion, can play a crucial role in defining and detecting topological phases.[21,22] First, these symmetries can greatly simplify the expressions for strong topological invariants. For example, in the absence of any crystal symmetry, the topological invariant θ is a complicated integral over the Brillouin zone that involves all filled states. In the presence of inversion symmetry, θ is given by the product of inversion eigenvalues of all filled bands at all time-reversal invariant momenta,[21] a much more practical expression. This is not an exception. For instance, the Hall conductivity of two-dimensional insulators with no time reversal symmetry, known as Chern insulators, can be calculated modulo an integer n without any integration if additional C_n rotational symmetries are present. More generally, the use of symmetry to simplify and classify topological phases has lead to the notion of symmetry indicators,[1–3,23–25] which are eigenvalues of point group operations that serve to define and classify different topological phases.

Crystal symmetries can protect topological states significantly enriching the classification of strong topological insulators. These topological insulators are known as crystalline topological insulators.[22] In these, only those surfaces that respect the crystal symmetry that protects the phase can hold gapless edge states. A paradigmatic example is SnTe.[26] It has a trivial strong topological index but has two Dirac cones at the surface protected by mirror symmetry along the $\{110\}$ mirror plane. Note that this is an even number of Dirac cones, which as mentioned above are only stable because of the extra mirror symmetry. The effect of disorder on topological crystalline insulators is, however, a subtle issue and an open research topic.

More generally, there can be boundary states protected in co-dimension higher than one. To understand what this means, note that so far we have explained how two (three)-dimensional topological insulators have gapless edges (surfaces). However, point group symmetries can protect gapless corner and hinge states in insulators that go under the name of higher-order topological insulators. For instance, a two-dimensional insulator with two diagonal mirror symmetries can host four corner modes (Figure 2(d)). In three dimensions, evidence of hinge states has been recently found in Bismuth[27] (see Figure 2(e)).

Although we have not mentioned them explicitly, most of our discussion applies to superconductors described by a Bogoliubov–de Gennes Hamiltonian. In this formulation, the Hamiltonian represents a gapped structure with particle–hole symmetry, and as such, it is subject to all the above classification schemes. The protected edge states of topological superconductors have a particular significance since particle–hole symmetry requires that there is a doubly degenerate state at zero energy, a combination between electrons and holes. In the particular case of a 1D superconducting chain, there are two doubly degenerate zero modes that are exponentially localized at opposite ends of the chain, known as Majorana zero modes. Two Majorana zero modes may be used to form a q-bit, and since they are topologically protected from decoherence into the bulk so long as the bulk gap remains open, they are a promising platform to realize robust (or topological) quantum computation.[28]

The classification of topological phases is still an open and thriving field of research. We note that not all of the properties we have discussed are realized in all topological insulators, yet the fact that distinct topological classes can only be connected by gap closings is universal. Defining all topological phases precisely and generically is currently an open problem. For example, recently, it was realized that some topological insulators, labeled as *fragile*, can be connected to a trivial state by adding more orbitals to the same lattice structure.[29–33] The presence of electron–electron interactions extends the topological classification further into states that are called topologically ordered.[140] By definition, these states host anyonic excitations which are quasi-particles with statistics that are neither bosonic nor fermionic. These states can be even further enriched by adding point group symmetries, which define symmetry enriched topological phases.

2.3. *Topological metals*

The concept of topological stability that we introduced in the previous section seemed to rely on symmetry but most importantly on the presence of a gap. Are all metals topologically trivial? The answer to this question is a very interesting no.

Although the first topological gapless state ever discovered was He-3,[38,39] the discovery of topological metals in the solid state[36] has fully uncovered their topological nature. The simplest class of topological metals are those in which two non-degenerate bands cross close to the Fermi level. These are called Weyl semimetals, and they are topological because if this crossing is isolated from other bands, it is topologically robust. To see this, note that we can write the crossing of two non-degenerate bands in three-dimensions to lowest order in momentum as

$$H_{\text{Weyl}}(\mathbf{k}) = v_{\text{F}} k_x \sigma_x + v_{\text{F}} k_y \sigma_y + v_{\text{F}} k_z \sigma_z + \mu \sigma_0, \tag{3}$$

which is called the Weyl Hamiltonian. It is gapless at $\mathbf{k} = 0$ because the dispersion relation is $E_{\mathbf{k}} = \pm v_{\text{F}} |\mathbf{k}| + \mu$. Equation (3) has the very important property that it exhausts all matrices that form the basis in this 2×2 sub-space of bands which are the Pauli matrices σ_i and the identity σ_0. Thus, any perturbation restricted to this sub-space will be, to the lowest order in momentum, a shift proportional to one of the matrices, say $\mathbf{b} \cdot \boldsymbol{\sigma} + b_0 \sigma_0 + \mathcal{O}(k)$. This will only shift the degeneracy point in momentum-energy space but will not alter the fact that the two bands touch at a point (now at $\mathbf{k} = -\mathbf{b}$). This point is called the Weyl point or Weyl node (see Figure 2(c)).

There is a deeper reason for the stability of a Weyl node. The quantity that can reveal this stability is built out of the Berry connection $\mathcal{A}^n_{\mathbf{k}} = \langle \psi^n_{\mathbf{k}} | \, i \boldsymbol{\nabla}_{\mathbf{k}} \, | \psi^n_{\mathbf{k}} \rangle$. As advertised around Eq. (1), this quantity is nothing but the position operator coming back again to describe topological states. As before, this quantity is not invariant by a change in the electron's phase $|\psi^n_{\mathbf{k}}\rangle \to e^{i\theta_{\mathbf{k},n}} |\psi^n_{\mathbf{k}}\rangle$. This tells us that the Berry connection is not an observable, much like the electromagnetic vector potential is not an observable; we can choose to shift it by the gradient of a scalar leaving Maxwell's equations invariant. Taking further the similarity with the vector potential, we can construct the rotational of $\mathcal{A}_{\mathbf{k}}^n$

$$\boldsymbol{\Omega}_{\mathbf{k}}^n = \boldsymbol{\nabla}_{\mathbf{k}} \times \mathcal{A}_{\mathbf{k}}^n = \boldsymbol{\nabla}_{\mathbf{k}} \times \langle \psi_{\mathbf{k}}^n | \, i\boldsymbol{\nabla}_{\mathbf{k}} \, | \psi_{\mathbf{k}}^n \rangle \,, \tag{4}$$

known as the Berry curvature. It is gauge invariant under changes in the phase of the wave function, much like a magnetic field is invariant under the choice of different gauges. For this reason, the Berry curvature is sometimes referred to as a magnetic field in momentum space.

By direct computation using Eq. (3), the Berry curvature around a Weyl node is a monopole in momentum space $\boldsymbol{\Omega}_{\mathbf{k}}^n = \frac{1}{2}\frac{\hat{k}}{|\mathbf{k}|^2}$. Attentive reader, you must frown here! In a lattice system, $|\psi_{\mathbf{k}}^n\rangle$ is periodic and so should $\boldsymbol{\Omega}_{\mathbf{k}}^n$. The only possible way out is if there is another anti-monopole $\boldsymbol{\Omega}_{\mathbf{k}}^n = -\frac{1}{2}\frac{\hat{k}}{|\mathbf{k}|^2}$ somewhere else in the Brillouin zone.

That Berry curvature monopoles exist in pairs is a general statement, and a consequence of the Nielsen–Ninomiya theorem: in the band structure of a local Hamiltonian, monopoles of Berry curvature come in pairs of opposite sign.[40] This sign is known as the chirality of the node, and so each monopole corresponds to a non-degenerate band crossing with a definite chirality. The chirality of a node is given by the sign of determinant of the velocity matrix $\mathrm{sgn}(\det(M_{ij}))$, where $H = M_{ij} k_i \sigma_j$.

Due to the restrictions in degeneracy imposed by time-reversal and inversion symmetries, the reader can be easily convinced that at least one of these two symmetries must be broken by the lattice structure to host pairs of Weyl nodes.[36] Without time-reversal symmetry, the minimum number of nodes is two, while with time-reversal symmetry, the minimum number of Weyl nodes is four and inversion is necessarily broken.

So long as the crossings between opposite chiralities are far in momentum space, they can be treated as independent and remain gapless. The degeneracy can only be lifted if two monopoles of opposite charge are brought together in momentum–energy space. This robustness is topological, since we cannot change the flux of the Berry curvature around a node by adding perturbations in the two-band sub-space. This is the reason why the flux over a closed momentum surface that encloses the Weyl point is a topological invariant called the Chern number. It equals $\pm C$, where $\pm$ is determined by the chirality and C is the monopole charge or Chern number. In our example Eq. (3), $C = 1$.

We arrive then to the definition of a Weyl semimetal, which is that material where the two nodes are close to the Fermi energy but separated in momentum space. But why are these crossings important? The existence of these node is fundamentally interesting because electrons close to it behave as pseudo-relativistic particles with no mass, like a photon, but with electrical charge. In this sense, the quasi-particles in Weyl semimetals behave similarly to those in graphene but moving in three dimensions and topologically protected. One of the main differences with graphene, that we shall not prove, is that the existence of Weyl nodes implies the existence of robust chiral surface states, known as Fermi arcs, that cannot be removed. In an heuristic way, these can be thought of as the Dirac string that connects both monopoles manifesting itself at the surface of the material (see Figure 2(c)).

When coupled to an electromagnetic field, the Weyl Hamiltonian Eq. (3) has a number of important physical consequences, some of great technological potential. A particularly surprising one is that under applied parallel electric and magnetic fields, the balance between the number of quasi-particles of different chiralities is destroyed. This phenomena is known as the chiral anomaly in high-energy physics and in condensed matter it significantly enhances the magnetoconductivity.[36] Weyl quasi-particles have significantly different responses to light including a universally quantized photocurrent.[41]

Weyl semimetals were discovered in the solid state in 2014 first in TaAs, and by now, many other materials have been identified or predicted to be in this phase.[36] Weyl semimetals exist both in magnetic and non-magnetic materials, and Fermi arcs, enhanced longitudinal magnetoresistence, and other signatures are by now routinely observed in experiment.

Finally, as with topological insulators, point group symmetries can enhance the protection and enrich the classification of topological semimetals. For example, in Dirac semimetals like Na_3Bi, two chiralities meet in momentum space without opening up a gap due to additional lattice symmetry that prevents it.[36] These symmetries can also protect nodal line crossings, for example, in non-symmorphic materials with negligible spin-orbit coupling. Recently, crossings of more than two bands, known as multi-fold fermions,[42–44] have been shown to be stabilized by screw rotations and other non-symmorphic symmetries. These have been found to exist in RhSi,[45,46] CoSi,[46] and AlPt,[47] among other materials, and share many of the exciting and useful electromagnetic responses of Weyl semimetals.[48]

2.4. *Topological phases in non-electronic systems*

Topological properties transcend electronic systems into the realm of the synthetic, the macroscopic, and the classical. As we have discussed, topological properties constrain the existence of degeneracies in band structures. Therefore, topological states can occur in any system where we can define a band structure: sound modes in acoustic 3D printed metamaterials,[49] waveguide modes in photonic metamaterials[50] (see Figure 2(c)), ultra-cold atomic lattices,[51] or the energy-phase relation in multi-terminal Josephson junctions.[52]

Topological states can be defined even for classical systems. Recently, the mechanical theory of stress and strain in mechanically constrained objects has also been connected to topological states.[53] For example, a mechanically constrained one-dimensional chain of rods can realize a mechanical analogue of the Su–Schrieffer–Heeger model[12] mentioned earlier. Perhaps even more striking is the connection of equatorial waves on Earth to topological boundary modes allowed by the Earth's rotation and the consequent breaking of time-reversal symmetry.[54]

Topological states seem ubiquitous. Our understanding of them is strongly centered on lattice periodicity. The remainder of this chapter dives into the fate of topology in the absence of translational invariance and the striking recent progress made to understand it.

3. Topological Phases Beyond Crystalline Solids

In our previous discussion, we purposely highlighted how periodicity and symmetry, in general, simplify the discussion, classification, and computation of topological properties. Expressions of topological invariants are succinctly written in momentum space. Crystalline symmetries further simplify these expressions and define new topological states. Is this all lost when the symmetry under lattice translations is broken? Although the reader might have guessed that the answer is no, one could still ask, is it worth the effort to find out? If topological matter already exists, why bother with translationally non-invariant systems? Here we give two reasons to keep on reading.

The first reason is conceptual. Topological states are robust, but how robust? We will see that on-site disorder on a crystalline lattice can induce topological phases but eventually will lead to a topologically trivial Anderson insulator. Can the absence of a crystalline lattice support topological states as well? If so, how can we find them systematically? And are they different from topological states of matter found in crystals? Even more fundamentally, how can we define and calculate invariants that signal the presence of non-crystalline topological states, in general, and with them their physical properties? More viscerally, what on earth does it mean to say we have a Dirac cone on the surface of a strong amorphous topological insulator if there is no momentum-space left to define it?

The second reason is technological. Topological insulators hold a great technological promise, rooted in their special properties, like their robust spin-momentum locking at the surface, useful for spintronic devices or topological superconductivity, which can open the doors to quantum computation. However, it is experimentally challenging to grow topological insulators where the chemical potential is in the gap, in order to clearly expose the topological properties. For example, heroic efforts were needed to turn the paradigmatic Bi_2Se_3 into a bulk insulator.[55] Moreover, it is desirable that topological properties are scalable. Among many things, this could mean that the growth process of topological solids should be as simple as possible while preserving the technologically useful properties. If all topological properties are observed only in carefully grown crystals, production costs might rise and technological impact would be less likely. Amorphous solids are easy to grow, which is one of the reasons behind their ubiquity in technology. Since translational symmetry is not a requirement for topological protection, combining amorphousness with useful topological properties seems a likely recipe for new states, new phenomena, and new applications.

The statement that amorphous matter can be topological may not be so surprising after all. We know that another key classification scheme, symmetry breaking, does not depend on crystallinity. For example, being a solid has nothing to do with being a crystal. A solid can be defined by its response properties to elastic shear, and since amorphous phases can present such properties, those that do can be called solids. Similarly, topological phases are characterized by their responses to external perturbations encoded mathematically in the form of topological invariants. Although both solidity and topology are conveniently expressed exploiting the benefits of periodicity, they do not rely on crystallinity to exist despite their shared absence in classical solid-state textbooks.[56]

3.1. *Half a step towards amorphous topological matter: Disorder, quasi-crystals, and Moiré lattices*

The question of the robustness of topological states under disorder is as old as the field of topological phases itself. It is now understood that topological properties do not require a crystalline lattice and can survive or even be enhanced by disorder.

A clear example is that of the topological Anderson insulator.[57,61,62] To understand it, take a step back and recall some of the main properties of the Anderson insulator state, which arises in a disordered solid. Anderson modelled a disordered solid by considering a randomly varying on-site potential drawn from a distribution of finite width w, representing the disorder strength. Upon increasing w, solids become insulators with exponentially localized wave functions. The susceptibility to localize depends, among other things, on the dimensionality of the systems in question: an infinitesimal amount of disorder localizes 1D states, while in 3D systems, a strong disorder strength is needed.

Are topological states different? Although strong disorder is expected to close the bulk gap and destroy the topological phase, topological properties can remain robust against a significant amount of disorder. For example, if disorder does not break the symmetry that protects the state, adding weak to intermediate disorder will not destroy the topological state. The most striking and technologically relevant example is once more the strong $\mathbb{Z}_2$ insulator protected by time-reversal invariance. If disorder is non-magnetic, it preserves time-reversal symmetry, and back-scattering is forbidden.

There is an even more surprising case in which disorder acts, for once, as an ally. The topological Anderson insulator is a state in which disorder, in moderate amounts, has turned a trivial insulator into a topological one. It happens when the trivial insulator is on the verge of turning into a topological insulator. On-site disorder might tip the balance by renormalizing the on-site energy such that the trivial insulator is pushed through a gapless state into a topological insulator (see Figure 3(a)). If disorder is strengthened even further, the topological Anderson insulator falls into a trivial Anderson insulator. This phenomenon has been observed in disordered atomic wires of trapped ultra-cold atoms.[63]

The search for topological states without crystalline order goes beyond the study of random disorder. Quasi-crystals are a recent addition to the collection of systems that can display topological phases.[59,64–71,138,139] They are ordered aperiodic structures that show sharp diffraction peaks which are incompatible with lattice translations.[72] Analogous to two-dimensional crystals, two-dimensional quasi-crystals are predicted to display a quantum Hall effect,[65] which can be engineered in photonic lattices by periodically modulating waveguides,[66] two-dimensional topological superconductivity,[67] two-dimensional quantum spin Hall phase,[68,69] and a topological Anderson insulator phase.[71] Notably, there are new topological phases predicted to occur only in quasi-crystals.[59,70] For example, eightfold rotations, possible only in quasi-crystalline lattices, can stabilize Majorana corner modes[59] in a higher-order

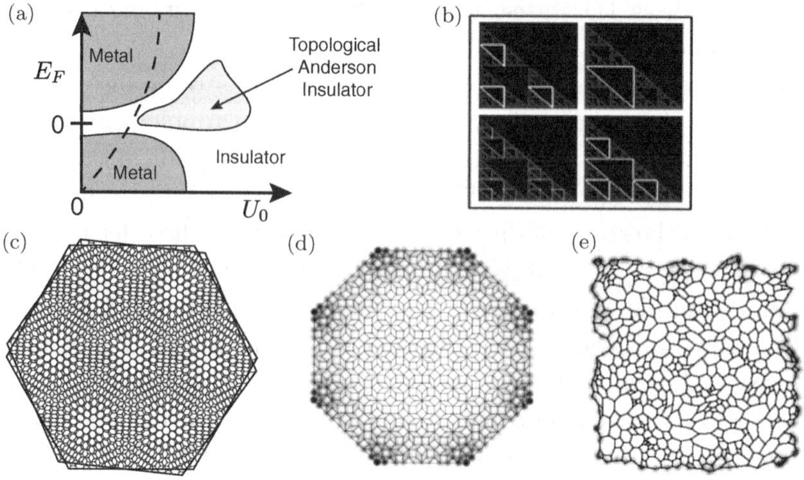

Fig. 3. **Examples of topological phases of matter in systems without translational symmetry** (a) Schematic phase diagram of a topological Anderson insulator. As disorder U_0 is increased, the average conductance plateaus at $G = 2e^2/h$, signaling a two-dimensional topological phase with two conducting edge channels known as topological Anderson insulator.[57] The dashed line signals where the effective topological gap inverts from trivial to topological. (b) Topological edge states in a fractal Sierpinski gasket under a magnetic field (adapted with permission from Ref. [58]). (c) Two layers of graphene twisted by a small angle θ can host a variety of topological phases including Chern insulators and higher-order topological phases with corner modes. (d) Higher-order quasi-crystalline topological insulator with corner states protected by C_8 symmetry (adapted with permission from Ref.[59]). (e) Amorphous topological Chern insulator phase showing the gapless edge state (adapted with permission from Ref. [60]).

topological quasi crystal (see Figure 3(d)). Additionally, some types of quasi-crystals can be constructed from projecting sites on a higher-dimensional crystal to a line or plane, sometimes referred to as the cut-and-project method.[65] This relationship to regular crystals maps certain quasi-crystalline topological phases into crystal ones, yet allowing in principle to reach higher-dimensional invariants beyond three dimensions. These ideas have been explicitly demonstrated in a one-dimensional Harper lattice realized in a photonic system.[64]

Topological states of matter are rapidly expanding to other systems without strict translational invariance. For example, it is currently investigated if topological phases can exist in fractal lattices

(Figure 3(b)) which have non-integer dimensionality. Numerical evidence supporting this possibility is currently being pursued in the particular case of quantum Hall-like states.[58,73–75] Although promising, the numerical simulations of fractals are necessarily of finite depth. A rigorous proof that fractals can host topological states is still absent.

Finally, we comment on another related and exciting search frontier: twisted bilayer graphene.[76,77] This system is composed of two perfectly hexagonal lattices twisted by a small angle $\theta \sim 1.1^o$ with respect to each other. This forms a Moiré superlattice that although not strictly periodic, it is close enough to treat the Moiré supercell as periodic (see Figure 3(c)). At low energies, the large real space unit cell folds the original Brillouin zone where the original bilayer graphene bands overlap. A minimal tight-binding model[78] predicts that the bandwidth of the relevant bands close to half-filling, graphene's natural state, is extremely flat, reaching close to exact flatness at certain angles known as magic angles (including $\theta \sim 1.1^o$). The band flatness enhances the effect of interactions compared to the typical kinetic energy scale set by the small bandwidth. Twisted bilayer graphene displays a set of interaction induced phases, such as superconductivity,[76,77] but also a quantized Hall effect in the absence of magnetic fields[79] known as the quantum anomalous Hall effect.

The above serves to emphasize how, without the constraints of translational invariance, topological phases can display a richer set of phenomena. Given the above, it is only natural to expect that amorphous solids can also display topological phases (see Figure 3(e)) with the additional benefit of being ubiquitous.

3.2. *Modeling amorphous topological matter*

Amorphous topological states were proposed independently by two groups at the dawn of 2017.[80,81] As a proof of principle, Agarwala and Shenoy[81] modeled an amorphous solid with a tight-binding hopping model using an uncorrelated uniform distribution of sites. They chose the Hamiltonian in the form

$$H = \sum_{i\alpha} \sum_{j\beta} (t(r) T_{\alpha\beta}(\hat{\mathbf{r}}) + \varepsilon_{\alpha\beta}\delta_{ij}) c^{\dagger}_{i\alpha} c_{j\beta}, \tag{5}$$

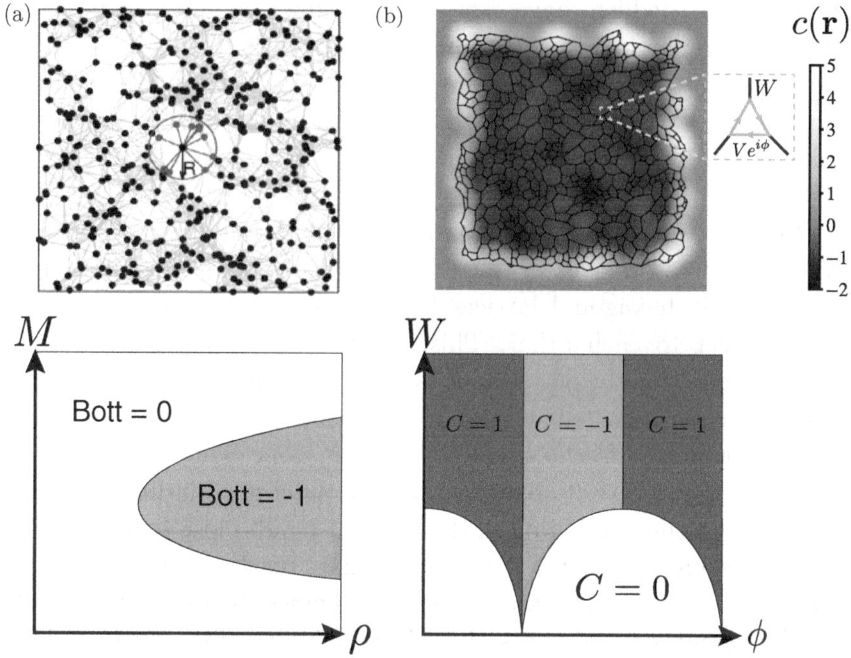

Fig. 4. **Examples of amorphous models of topological matter** (a) Random point set connected by hopping up to a radius R. The Chern insulator phase is diagnosed by localized edge states and the Bott index, Eq. (16), shown in the lower panel.[81] (b) Topological Weaire–Thorpe model with threefold coordination. The topological phase is diagnosed by a quantized local Chern marker, Eq. (15), in the bulk of the sample. The edge states contribute with a large and positive Chern marker such that the average over the sample is zero. The topological phase diagram (bottom) can be inferred analytically using amorphous symmetry indicators.[60]

that specifies the hopping $t(r)T_{\alpha\beta}(\hat{\mathbf{r}})$ between ij sites and $\alpha\beta$ orbitals with an on-site potential term $\varepsilon_{\alpha\beta}$. They imposed that the hopping amplitude is finite within a certain characteristic radius by choosing $t(r) = C\Theta(R-r)e^{-r/a}$ (see Figure 4(a)). The topological phase diagram is dictated by the angular part of the hopping, $T_{\alpha\beta}$. To define it, they were mainly inspired by two types of topological insulator models defined for crystal lattices: the three-dimensional $\mathbb{Z}_2$ time-reversal invariant topological insulator described earlier and the Chern insulator, a two-dimensional system with a quantized Hall conductivity. As a pedagogical exercise, we now describe the latter example in some detail.

The term Chern insulator refers to any two-dimensional insulator that has a finite Hall conductivity. This last property necessarily means that time-reversal symmetry is broken, either by an external magnetic field or intrinsic magnetic moments. The Chern insulator is a special kind of topological insulating state because although it cannot be deformed to an atomic insulator without closing the gap (as our definition of topological state requires), there is no symmetry that protects it. Semiclassically, the absence of time-reversal symmetry allows cyclotron orbits to form as in the Hall effect. Although seemingly localized in the bulk, at the edges, they result in skipping orbits that cannot back scatter. Quantum mechanically, this translates into an obstruction to find a (Wannierized) basis of localized orbitals if the Hall conductivity is finite as discussed in Section 2.2.

Since the Chern insulator does not require any symmetry, it is a natural starting point to construct a topological insulator. But how do we build such a model? On a periodic lattice, the Hall conductivity of a Chern insulator is given by the integral of the Berry curvature, Eq. (4), over the Brillouin zone, summed over all filled bands[82]

$$\sigma_{xy} = \frac{e^2}{h} \sum_{n \in \text{filled}} \int \frac{d^2k}{(2\pi)^2} \Omega_z^n(\mathbf{k}) = Ce^2/h. \tag{6}$$

Note that since in two dimensions, we only have two momenta, k_x and k_y, the Berry curvature only has one component, along z. The integer C is the Chern number, already encountered earlier. For the purposes of proving that it is indeed an integer, we shall restrict to two-band models, yet the statement is independent of the number of bands. Any two-band Hamiltonian defined in a periodic lattice has the general momentum space form $H(\mathbf{k}) = \mathbf{d}_\mathbf{k} \cdot \boldsymbol{\sigma} + \varepsilon_\mathbf{k}\sigma_0$ because the three Pauli matrices and the identity form a complete basis of 2×2 hermitian matrices. This implies that $\mathbf{d}_\mathbf{k}$ is a real vector function of momentum. $\varepsilon_\mathbf{k}$ is a different function of momentum, and they are all periodic. Given this form, it is well understood how to choose $\mathbf{d}_\mathbf{k}$ and $\varepsilon_\mathbf{k}$ to build a topological insulator with finite Hall conductivity.

The intuition is as follows. For such a two-band model, Eq. (6) can be rewritten as

$$\sigma_{xy} = \frac{e^2}{h}\int \frac{d^2k}{(2\pi)^2}\hat{d}_{\mathbf{k}} \cdot (\partial_{k_x}\hat{d}_{\mathbf{k}} \times \partial_{k_y}\hat{d}_{\mathbf{k}}), \tag{7}$$

where we defined the unit vector $\hat{d}_{\mathbf{k}} = \mathbf{d}_{\mathbf{k}}/|\mathbf{d}_{\mathbf{k}}|$. The mathematically inclined reader or those familiar with magnetic textures can recall that the integral is quantized to integers over a closed manifold, in this case, the Brillouin zone. This number, sometimes referred to as the skyrmion number, counts how many times the unit vector $\hat{d}_{\mathbf{k}}$ covers the surface of the sphere that it defines as we vary $\mathbf{k}$. Thus, it suffices to see how many times the vector $\hat{d}_{\mathbf{k}}$ reaches the north and south poles ($\hat{d}_{\mathbf{k}} = (0, 0, \pm 1)$).

We can now use the above arguments (or direct computation) to show that a simple model that interpolates between a trivial insulator and one with a finite Hall conductivity is

$$H = t_1 \sin(k_x)\sigma_x + t_1 \sin(k_y)\sigma_y + (M + 2t_1 - t_1 \cos(k_x) - t\cos(k_y))\sigma_z. \tag{8}$$

If M is large and positive, $\hat{d}_{\mathbf{k}}$ remains close to $(0, 0, 1)$ for all $\mathbf{k}$. Since $\hat{d}_{\mathbf{k}}$ does not cover the sphere, it defines a trivial insulator with $\sigma_{xy} = 0$, and thus, $C = 0$. If $0 < (M+2t) < 2$, however, we notice that $\hat{d}_{\mathbf{k}=(0,0)} = (0, 0, -1)$. As we move the momentum, the orientation of $\hat{d}_{\mathbf{k}}$ changes until it reaches $\hat{d}_{\mathbf{k}=(0,\pi)} = \hat{d}_{\mathbf{k}=(\pi,0)} = \hat{d}_{\mathbf{k}=(\pi,\pi)} = (0, 0, 1)$. Therefore, $\hat{d}_{\mathbf{k}}$ covers the sphere exactly once as we move $\mathbf{k}$ in the Brillouin zone, defining a Chern insulator with $\sigma_{xy} = e^2/h$, and thus, $C = 1$.

In real space, Eq. (8) can be defined on a square lattice with two orbitals per site

$$H = \frac{t_1}{2}\sum_i \left[c_i^\dagger \left(i\sigma_x - \sigma_z\right) c_{i+\hat{x}} + c_i^\dagger \left(i\sigma_y - \sigma_z\right) c_{i+\hat{y}} + \text{h.c.}\right] + (2t_1 + M)c_i^\dagger \sigma_z c_i, \tag{9}$$

where the creation and annihilation operators are two component vectors in orbital space (e.g., $c_i^\dagger = (c_{i,\alpha}^\dagger, c_{i,\beta}^\dagger)$) and $\hat{x}, \hat{y}$ are unit vectors.

Agarwala and Shenoy generalized these hoppings from a square lattice to a random point set. To do so, it is illustrative to write the on-site matrix $\varepsilon_{\alpha\beta}$ and hopping matrix in the x direction $T^{(x)}_{\alpha\beta}(\theta)$ connecting two sites i and j

$$\varepsilon_{\alpha\beta} = \begin{pmatrix} 2t_1 + M & 0 \\ 0 & -(2t_1 + M) \end{pmatrix},$$
$$T^{(x)}_{\alpha\beta} = \frac{t_1}{2}\begin{pmatrix} -1 & -i \\ -i & 1 \end{pmatrix}$$
$$\equiv T_{\alpha\beta}(\theta = 0). \tag{10}$$

The latter definition, where we introduced the polar angle θ, suggests how to generalize this hamiltonian to arbitrary θ, provided we identify the hoppings in x direction with $\theta = 0, \pi$ and the hoppings in y direction with $\theta = \pm\pi/2$. Hermiticity of the Hamiltonian demands that $T_{\alpha\beta}(\theta) = T^{*}_{\beta\alpha}(\theta + \pi)$. This results in

$$T_{\alpha\beta}(\hat{\boldsymbol{r}}) = \frac{1}{2}\begin{pmatrix} -t_1 + t_2 & -it_1 e^{-i\theta} + f(\theta) \\ -it_1 e^{i\theta} + f^{*}(\theta) & t_1 + t_2 \end{pmatrix}. \tag{11}$$

The phase factor $e^{-i\theta_{ij}} = (x_{ij} + iy_{ij})/|\mathbf{r}_{ij}|$, with $x_{ij} = x_i - x_j$ (respectively for y and $\mathbf{r}$), captures the angle dependence of the vector connecting two sites ij. The Hamiltonian in Eq. (9) is recovered when we exclude $f(\theta)$ and $t_2 = 0$, not present in Eq. (9), and only allow for vertical and horizontal hoppings. To break the particle–hole symmetry, Argawala and Shenoy[81] allowed the constant t_2 to be finite, and to take into account inter-orbital mixing, they incorporated an even function of the angle $f(\theta) = \lambda\left[\sin^2\theta(1+i) - 1\right]$. The latter effect also modifies the on-site energy as

$$\boldsymbol{\epsilon}_{\alpha\beta} = \begin{pmatrix} 2t_1 + M & (1-i)\lambda \\ (1+i)\lambda & -(2t_1 + M) \end{pmatrix}, \tag{12}$$

which completes the definition of the Agarwala and Shenoy model.[81]

With periodic boundary conditions, this model shows gaps as a function of M, as expected from our crystalline discussion. For open boundary conditions and intermediate values of M, the gaps become filled with states that are localized at the one-dimensional boundaries of the system, suggesting their topological origin. By computing topological markers, which are observables that signal topological phases for disordered systems described in the next section, they showed that these models could indeed host topological phases for high enough atomic densities, defined as $\rho = N_{\text{sites}}/V$, where N_{sites} is the number of sites and V is the volume (see Figure 4(a)).

The above recipe, is generalizable to representative models for all of the 10 strong topological symmetry classes.[81] Indeed, bulk gaps exist where metallic surface states of putative topological origin survive. We use putative because, as we will mention later when we discuss how to detect these phases, it is not known how topological markers generalize to all symmetries.

Concurrently, with the work by Argawala and Shenoy, Mitchell *et al.*[80] theoretically proposed and experimentally demonstrated the possibility of topological insulating phases in two dimensions using coupled gyroscopes. They demonstrated experimentally that several types of amorphous networks, built out of different point sets, including hyperuniform, jammed, quasi-crystalline, and uniformly distributed, can support edge states that propagate chirally along the edges of the sample.

Beyond the proof of principle, and remarkable experimental realization,[c] their findings showed that many, but not all networks, hosted topological phases. To define amorphous lattices, Mitchell *et al.* defined first a point set. The point sets were of four types: randomly distributed sites (as in the Argawala and Shenoy model), hyperuniform, jammed, and quasi-crystalline. Hyperuniform point[83] sets are defined as those sets where the number of points in a given observation window decays faster than the window's area. This is quantified by the structure factor $S(\mathbf{k}) = \frac{1}{N}\sum_i^N |\exp(i\mathbf{r}_i \cdot \mathbf{k})|^2$ for $k \neq 0$ demanding that it grows as $S(\mathbf{k}) \propto k^\alpha$ as $k \to 0$ with $\alpha > 0$. Jammed point sets are defined by minimizing a potential interaction (in this case a harmonic potential) between non-uniformly distributed (polydisperse) discs. For a given point

[c]We encourage the reader to view the supplementary videos in Ref. [80] for a fascinating visualization of the edge state dynamics.

set, Ref. [80] applied different connectivity rules to define different amorphous lattices. By connecting the nearest-neighbours of a given point, they arrived to an amorphous lattice formed by triangles of connected sites known as Delaunay triangulation. Connecting the centers of the Delaunay triangles, they defined the Voronized, threefold-coordinated lattice. Decorating each site in a Voronized lattice with a triangle, we arrive at a Kagome decoration of the Voronized lattice.

All but Delaunay triangulations showed gapped spectra and topological phases.[80] This observation was interpreted as the amorphous counterpart to the fact that gapped, topological phases need more than two bands to be realized. The Delaunay triangulation locally resembles a triangular lattice which, all sites being equal, is gapless. In contrast, the Voronoi tessellation locally resembles the honeycomb lattice which can host topological phases. As noted earlier on, in the amorphous community,[84] the local geometry of the lattice plays a crucial role in determining band gaps and consequently realizing topological phases. We will return to this point shortly.

The works by Agarwala and Shenoy,[81] and Mitchell *et al.*[80] opened up the door to topological phases in amorphous lattices in various physical systems. Soon after, amorphous photonic lattices with variable packing fractions formed by jammed point sets[85] and uniform point sets[86] were proposed to host chiral edge states. Amorphous two-dimensional topological superconductivity was proposed to emerge in a Shiba glass, arising from hybridizing states formed around magnetic moment on a superconducting surface with strong Rashba spin-orbit coupling.[87] Using tight-binding molecular dynamics, it was also proposed that liquid states can inherit topological properties upon melting a regular lattice.[88] Finally, starting from model tight-binding Hamiltonians, the approach of Agarwala and Shenoy described above serves to generalize amorphous topological phases to amorphous topological metals[89] and higher-order topological phases.[90]

Despite the above breakthroughs, many of the tight-binding models based on random point sets or exponentially decaying hopping matrices ignore that amorphous solids often do have a local environment very similar to that of a crystal with a fixed coordination number. In other words, amorphous matter is not entirely random. In covalently bonded amorphous solids, the local correlation is similar to that of the crystal. The coordination number remains fixed, nearest-neighbour distances are strongly peaked around the crystal bond lengths, and there are no dangling bonds.[8] Amorphous solids differ from the crystal in the absence of long-range correlation and the significant spread in bond angles, but the local environment remains fairly similar to the crystal.

The coordination number, in particular, conveys a significant amount of chemical and topological information, determining the local environment of a site. In the 1970s, this observation motivated Weaire and Thorpe to construct a model with fixed tetrahedral coordination to describe amorphous silicon.[84] For generic coordination z, it reads

$$H = \sum_{i,j\neq j'}^{z} V_{jj'}^{(i)} |i,j\rangle\langle i,j'| + \sum_{i\neq i',j}^{z} W_{ii'}^{(j)} |i,j\rangle\langle i',j|. \tag{13}$$

The index i labels the sites within a z coordinated lattice, and the j index labels the z orbitals within a site ($j = 1, 2, \ldots, z$); see Figure 4(b). Weaire and Thorpe chose the matrices $V_{jj'}^{(i)} = V$ and $W_{ii'}^{(j)} = W$ to be real and independent of the site i and orbital j, respectively. Using the resolvent method,[91] it is possible to show that for any generic V, W, and z, this model has band gaps where energy states are forbidden to appear. If the filling is such that it falls in one of these band gaps, the model describes an insulator. Although these models do not capture the fine spectral details of the density of states, they capture the general feature that covalently bonded amorphous are generically gapped.

Due to their gapped structure, it is then natural to consider the Weaire–Thorpe Hamiltonians as descriptions of topological insulating phases in amorphous solids. This route was explored recently[60] by allowing the intra-site coupling V to be complex by replacing it with $Ve^{i\phi}$ (see the inset in Figure 4(b)). Using the resolvent method, it is possible to prove that the model remains gapped. For a two-dimensional version without time-reversal symmetry, the model hosts different Hall insulating phases as a function of the magnitude and phase of the intra-site hopping (V and ϕ) and the inter-site hopping (W), proving that this model can indeed be topological for commensurate fillings. The authors noticed that the equivalence between orbitals and the fixed coordination of this model can be exploited further to reproduce the phase diagram without the need to calculate a topological invariant. We will return to this aspect in the next section when we discuss how to detect amorphous topological phases.

In addition to tight-binding models, more sophisticated numerical approaches are powerful tools to generate amorphous topological lattices. Density functional theory predicts[92] that the $\mathbb{Z}_2$ topological insulator character of crystal bismuthene,[93] a monolayer of bismuth recently synthesized, retains its topological character in its amorphous form. Molecular dynamics simulations have also been employed to model amorphous topological insulators. In its crystalline form, Sb_2Te_3 is a topological insulator,[94–96] and using molecular dynamics, its amorphous counterpart is predicted to be gapped.[97] However, the survival of topological properties was not discussed in this work. Supported by the experiments described in the following, molecular dynamics simulations indicate that a gap of 293 meV exists in amorphous Bi_2Se_3, similar in magnitude to that of the crystal ($\sim$300 meV) and consistent with the thermally activated resistivity curves.[98]

Finally, we note that there are other possible realistic scenarios to be explored deviating from continuous random network models. One intriguing direction is the fate of topological states under the occurrence of crystalline regions. For example, amorphous Si can have a substantial fraction of crystal structure.[99,100] Amorphous monolayer carbon has been observed[101] to have intercalated regions with perfect hexagonal order and regions described by a continuous random network.[102] These mixed crystal-amorphous order is captured under the crystallite model[103] which has not been explored in the context of topological matter. The effect of other possible characteristics of amorphous lattices, such as hyperuniformity,[83] also remains to be fully addressed in the context of topological phases despite encouraging first progress.[80]

3.3. *Physical properties and theoretical characterization of amorphous topological matter*

One of the difficulties that underlies amorphous topological matter is to signal its existence. Theoretically, diagnosing topological phases in the absence of crystalline symmetry is well developed due to the previously existing work on the survival of topology in the presence of disorder. Many of the tools developed in such context can be translated to amorphous matter, but, as we shall describe shortly, they also have their limitations.

3.3.1. *Local markers*

To address the robustness of topological phases to disorder some of the topological invariants that characterize them have been expressed in real space. These topological markers have made their way into aperiodic system and have been proven useful to characterize amorphous matter as well.[80,81,104]

Chern insulators are the simplest to diagnose. As we discussed earlier, Eq. (7) determines the Hall conductivity through an integral of the Berry curvature. Experimentally, a quantized Hall conductivity[82] or a quantized circular dichroism[105] will signal this type of topological insulator and can be measured independent of periodicity. Theoretically, in an amorphous structure, we cannot rely on the momentum space formulation of Eq. (6), and several alternatives have been developed.[104,106–108] Most of them can be obtained by Fourier transforming Eq. (6) into real space, leading to an expression in terms of the projector onto the occupied ($\hat{P}$) and unoccupied ($\hat{Q} = 1 - \hat{P}$) bands such that

$$C = -2\pi i \sum_{\text{all } \mathbf{r}_i} \langle \mathbf{r}_i | \left[\hat{Q}\hat{x}, \hat{P}\hat{y}\right] | \mathbf{r}_i \rangle \equiv \sum_{\text{all } \mathbf{r}_i} C_{xy}(\mathbf{r}_i), \tag{14}$$

where $C_{xy}(\mathbf{r}_i)$ is a local quantity known as the local Chern marker.[108] Interestingly, the local Chern marker can be written in terms of the single-particle density matrix $\rho_n(\mathbf{r}_i, \mathbf{r}_j) = \langle \mathbf{r}_i | n \rangle \langle n | \mathbf{r}_j \rangle$ as[106, 109]

$$C_{xy}(\mathbf{r}_i) = -4\pi \sum_{\mathbf{r}_j,\mathbf{r}_k} \sum_{\substack{E_l<0\\ E_m>0\\ E_n<0}} \mathrm{Im}\left[\rho_l(\mathbf{r}_i,\mathbf{r}_j)\rho_m(\mathbf{r}_j,\mathbf{r}_k)\rho_n(\mathbf{r}_k,\mathbf{r}_i)\right] x_j y_k. \tag{15}$$

If any pair of $\mathbf{r}_i$, $\mathbf{r}_j$, $\mathbf{r}_k$ is equal, the quantity inside the square brackets is purely real and the local Chern marker vanishes. The Chern number is obtained by summing over all possible triangles of sites, and a direct computation shows that triangles formed by nearest-neighboring and next-nearest-neighboring sites dominate the sum.[109] In this sense, even though the topological information is non-local, the local Chern marker is a quasi-local indicator of topology because it is to leading order given by a sum of local triangles.[80, 109]

The trace of the local Chern marker results in the Chern number upon summing over bulk sites, but it always vanishes exactly for finite systems. Notice that the local Chern marker is a trace over a commutator defined in a finite Hilbert space, which necessarily vanishes. Mathematically, this is reasonable, since topologically, the manifold of a system with open boundary conditions is flat and trivial, unlike a periodic system, which has the topology of a torus. In Figure 4(b), we show the local Chern marker for the topological Weaire–Thorpe Chern insulator model with parameters within the topological phase. The bulk is quantized to the Chern number of the periodic system ($C = -1$), while the boundary contributes with a large contribution of opposite sign so that the average over the whole system remains zero. In a trivial phase, we would have observed a vanishing Chern number both in the bulk and at the edge.

Interestingly, there is no clear relation between the original local Chern marker formula used by Kitaev,[106] essentially equivalent to Eq. (14), and the Kubo formula written in real space that determines the Hall conductivity.[104] They both seem to capture well the presence of a Hall insulator, and the survival of quantization in amorphous lattices can be proven rigorously, including spectral and mobility gaps.[104]

Beyond these mathematical subtleties, physically, the trace over the local Chern marker can be directly related to the frequency integrated circular dichroism, the total differential absorption between left and right circularly polarized light. This connection can be traced to Bennett and Stern[110] who observed that circular dichroism is determined by the Hall conductivity, later written as a trace over the local Chern marker.[105,111] Therefore, a finite Chern number implies a quantized circular dichroism, an observation that extends to amorphous systems.[60]

The Bott index[112] is an alternative topological marker to the local Chern marker. A finite Bott index signals the absence of a localized basis to represent the state. It avoids the subtleties of defining the position operators $\hat{x}, \hat{y}$ in periodic systems since it is defined through exponentials of position operators. Explicitly,

$$B = \frac{1}{2\pi}\mathrm{Im}\left\{\mathrm{Tr}[\log(WUW^{\dagger}U^{\dagger})]\right\}, \tag{16}$$

where $U = \hat{P}\exp(i\Theta)\hat{P}$, $W = \hat{P}\exp(i\Phi)\hat{P}$, and Θ and Φ are diagonal matrices that encode the position operators $\hat{x}$ and $\hat{y}$ rescaled to the interval $[0, 2\pi)$. The Bott index captures amorphous Chern insulator phases,[81] yet it is not straightforwardly connected to the local Chern marker, since no local version of the Bott index exists.

Topological markers are well developed for Chern insulators, but their generalization to other topological phases is challenging. As we reviewed in Section 2.2, there are many classes of topological insulators depending on the symmetries that protect them. For $\mathbb{Z}_2$ topological insulators in two-dimensions with conserved spin projection in the s_z direction, a spin-resolved version of the local Chern marker can be constructed.[113] In more general situations, spin-conservation is broken, but the $\mathbb{Z}_2$ topological insulator remains, and in this case, no local invariant exists. For three-dimensional $\mathbb{Z}_2$ topological insulators, it is possible to define a three-dimensional topological polarization θ that, as was mentioned in Section 2.2, should be quantized to $\theta = 0, \pi$. This quantity shows similar features to the one-dimensional polarization. It is only defined modulo 2π which causes troubles when formulating a closed real space expression.[114–117] A formula close to Kitaev's original local Chern marker proposal was developed recently.[118]

An interesting case where invariants can be efficiently computed is the case of mirror-topological insulators.[119] The eigenvalues can be labelled by their mirror eigenvalue allowing for their local Chern marker to be computed. These markers are promising options yet are still to be applied to amorphous systems. Additionally, the extension of local Chern markers to other topological phases, like superconductors, is yet to be developed.

3.3.2. *Responses to external fields*

As we introduced in Sections 2.1 and 2.2, topological phases are defined by quantized observables. By calculating or measuring these, it is possible to signal topology. A particularly well-known example in two dimensions is that of the Hall conductivity which is quantized to integers in units of fundamental constants. In three dimensions, a notable example occurs in $\mathbb{Z}_2$ topological insulators which host a topological magnetoelectric effect.[17,18]

The topological magnetoelectric effect is a rather unique response of three-dimensional $\mathbb{Z}_2$ topological insulators to external electromagnetic fields. We are accustomed to describing a medium by its dielectric permeability and magnetic permittivity. These enter the definitions of the displacement field $D_i = \varepsilon_{ij}E_j$ and the induction $H_i = \mu_{ij}^{-1}B_j$ in terms of the electric E_j and magnetic B_j fields, respectively. However, in magneto-electric materials, D_i and H_i are allowed to depend on B_i and E_i, respectively. To the lowest order in the fields, magnetoelectricity can be captured by a term $\alpha_{ij}E_iB_j$ in the effective free energy of the material. The tensor α_{ij} is the magnetoelectric tensor

$$\alpha_{ij} = \left.\frac{\partial P_j}{\partial B_i}\right|_{\mathbf{E}=0} = \left.\frac{\partial M_j}{\partial E_i}\right|_{\mathbf{B}=0}. \tag{17}$$

It accounts for the change in polarization P_j or magnetization M_j caused by a magnetic B_i or an electric field E_i, respectively. Soon after three-dimensional $\mathbb{Z}_2$ topological insulators were predicted, it was realized that there is an isotropic contribution to α_{ij} determined by the topological invariant introduced in Section 2.2 for this chapter,[17,18] $\theta = 0, \pi$, mod 2π. Specifically, the isotropic part of the magnetoelectric tensor is topological and defined by $\alpha_{ij}^{\text{iso}} = \theta(e^2/2\pi h)\delta_{ij}$.[114–117]

There are several physical consequences that can be traced back to α_{ij}^{iso}. They include a surface Hall effect[18,117] but also more exotic responses such as a quantized Kerr and Faraday rotation,[55] a repulsive Casimir effect,[120] and the Witten effect.[121] While all of them can signal how this topological state emerges even in the absence of translational invariance, only the Witten effect has been used so far as a theoretical diagnostic of putative amorphous $\mathbb{Z}_2$ topological insulators.[122] The Witten effect predicts that if a magnetic monopole is to be placed inside this type of topological insulator, it will bind a half-integer electric charge forming a composite particle known as a dyon, enabled by the magnetoelectric coupling α_{ij}^{iso}. By monitoring this charge as a function of model parameters, it is possible to map a topological phase diagram.[122]

Since the notion of topological marker has not been developed for all symmetry classes, the explicit calculation of topological responses can serve as a theoretical diagnostic of topological states. They have the benefit to generate specific physical predictions that can be used to discover topological phases, as we will discuss in Section 3.4.

3.3.3. *Symmetry indicators and effective Hamiltonians*

Although calculating the local Chern markers is possible, and calculating physical responses is necessary, these are not always computationally efficient to diagnose and classify topology. Developing a classification that can allow a systematic prediction of new materials requires a simple diagnosis of topological phases, which is currently lacking for amorphous systems. Two recent promising directions have emerged: the use of emergent symmetries and the definition of effective Hamiltonians.

Amorphous matter does not enjoy exact symmetries, in general, with the exception of time-reversal in non-magnetic systems and particle–hole symmetry in superconductors. However, it may enjoy average symmetries that can possibly be exploited. Additionally, even if the whole spectrum is not fully symmetric, special states can be exact eigenstates of a given symmetry. Is it possible to follow these to track how topology changes with different parameters? The answer is yes, at least for the Weaire–Thorpe models introduced in Eq. (13). By construction, the sites i are equivalent: upon permutation of orbitals j, the Hamiltonian remains invariant. This endows them with a mathematical structure that can be used to predict the evolution of spectral gaps as the parameters are varied.[91] The band edges remain exact eigenvalues of permutation of orbitals, and they can also be assigned a bonding or anti-bonding character. For a coordination z lattice, there are $2z$ band edges, which can be tracked as a function of parameters by solving exact inequalities,[60] monitoring when gap closings occur.

For the specific case of a threefold coordinated ($z = 3$) Weaire–Thorpe–Chern insulator model, the phase diagram can be reproduced by constructing a symmetry indicator that tracks the symmetry character of the band edges. Although there is for the moment no rigorous derivation for this approach, it can be justified by a parallelism to the theory of symmetry indicators in crystals.[1–3, 23–25] These are based on an idea that has been central to our discussion: in order to classify topological crystals, it is paramount to define what trivial means. In Section 2.1, we defined the atomic insulator for which the Wannier states are localized at atomic sites and which is for our purposes trivial. By defining the trivial limit, we defined a reference by which insulators can be compared, diagnosing topology through gap closings.

In the Weaire–Thorpe–Chern insulator, there are two trivial limits, $W = 0$ and $V = 0$, of decoupled sites and decoupled dimers, respectively. Taking the $W = 0$ as the reference trivial insulator, we can construct the following indicator:

$$C = \sum_{n \in \text{filled}} m_n - \sum_{n \in \text{filled}} m_n^{W=0} \mod 3, \tag{18}$$

where $m = 0, \pm 1$ label the band edges with eigenstates of the permutation symmetry with eigenvalue $\xi_m = e^{i2\pi m/z}$. The first term tracks band edge inversions, or in other words, when the gaps close. The second term acts as a reference for the trivial state. This formula exactly captures the phase diagram of the threefold coordinated ($z = 3$) Weaire–Thorpe–Chern insulator model as shown in Figure 4(b).

Another promising route to signal topology in amorphous systems is the use of an effective Hamiltonian.[59] The central idea is to project the real space Hamiltonian into a basis of plane waves with well-defined momentum $\mathbf{k}$, with the expectation that the topological information survives along the way. The effective Hamiltonian is defined as $H_{\text{eff}}(\mathbf{k}) = G_{\text{eff}}(\mathbf{k})^{-1} + E_F$ by projecting the single-particle Green's function onto plane wave states

$$G_{\text{eff}}(\mathbf{k}) = \langle \mathbf{k}|G|\mathbf{k}\rangle, \tag{19}$$

using Green's function $G = \lim_{\eta\to 0}(H - E_F + i\eta)^{-1}$ of the full Hamiltonian with E_F chosen to be in a gap. The key property of $H_{\text{eff}}(\mathbf{k})$ is that its gap closes when the gap of the full Hamiltonian closes,[59] thus retaining information about the topological state it describes.

The effective Hamiltonian approach has been successful to track gap closings in topological phases in quasi-crystals[59] and applied to the topological Weaire–Thorpe–Chern model.[60] However, in the latter case, $H_{\text{eff}}(\mathbf{k})$ does not recover the full phase diagram due to numerical instabilities, more pronounced in regions with a small gap. Should these problems be resolved, $H_{\text{eff}}(\mathbf{k})$ can constitute a promising tool to generalize the symmetry classifications of topological insulators to amorphous systems.

3.3.4. *Other probes*

We conclude this section by mentioning a few additional properties that remain to be explored in the context of topological amorphous matter. Recently, it was suggested that amorphous topological lattices may display distinct phase transitions compared to the crystalline counterparts. Strikingly, the critical exponent that determines how the correlation length ξ diverges as an amorphous Hall insulators consisting of growing domains reaches the percolation transition was reported to be different than previously found for two-dimensional electron gases in a magnetic field and Chern insulators.[123, 124] Further numerical studies will be helpful to establish solid evidence of this difference which could be, in principle, addressed in experiments.

Lastly, other approaches developed for disordered systems could be also important in the development and signaling of amorphous phases. The level statistics of the entanglement spectrum[125] or scattering matrix approaches[126] serve as alternatives to local markers numerically signal topological phases.

3.4. *Experimental realizations and promising platforms*

Translational invariance is heavily used to interpret experiments and identify topological states. A prime example is angle-resolved photoemission spectroscopy (ARPES), which generated the most explicit images of the energy–momentum conical dispersion of the surface Dirac cones of three-dimensional topological insulators, convincing the community that these topological insulators existed.[19, 20] For amorphous systems, the absence of momentum as a good quantum number suggests that typical tools sensitive to topological surface states, including ARPES, magnetic quantum oscillations, quasi-particle interference, scanning tunneling microscopy, transport phenomena, or optical responses should be revisited to scrutinize the fingerprints of amorphous topological states. Some encouraging progress suggests that this endeavour is not hopeless.

3.4.1. *Condensed matter systems*

Crystalline Bi_2Se_3 has been, since very early on, the paradigmatic example of a topological insulator.[94] It has a bulk gap of around 0.3eV, with a single two-dimensional Dirac surface state. The chemical potential of most samples lies above the gap turning this material into a metal. This fact is not a drawback for ARPES, which has been used to image the surface Dirac cone perfectly, but it is a drawback for applications which aim to exploit the robustness of its surface Dirac cone.

That topological properties can survive in non-crystalline solids was hinted by early experiments. For example, sputtered Bi_xSe_{1-x}[127] have been shown to present weak anti-localization effects, discussed in the following, associated with spin-momentum locking of a putative Dirac surface state.[128] Additionally, granular Bi_2Se_3 was shown to host topological surface states similar as those associated with the bulk crystal.[129]

In late 2019, amorphous Bi_2Se_3 was studied using transport, ARPES, and spin-resolved ARPES, showing signatures consistent with an amorphous topological insulator phase in the solid state.[98] High resolution transmission electron microscopy showed a diffraction pattern typical of amorphous matter, with well-defined coordination and inter-atomic distance, similar to the crystalline case, signaled by a diffuse but well-defined ring corresponding to the short-range ordering of nearest neighbors. The data showed no signs of nanocrystals or lattice fringes consistent with the amorphous nature of the samples. As anticipated in Section 3.2, these observations suggest that Weaire–Thorpe-type models, which fix coordination, are well suited to describe amorphous topological insulators.

As a function of temperature T, the $e^{-(T_0/T)^{1/4}}$ behavior of the conductivity observed in these samples is consistent with that expected from a variable range hopping model,[130] indicating that the bulk is largely Anderson localized. At low temperatures, when time-reversal symmetry is present, the spin-momentum locking of putative Dirac surface states suppresses back-scattering. The first quantum correction to the conductivity is positive and increases conductivity, a phenomenon that is known as weak anti-localization, which supports the robustness of the Dirac surface states (and other Dirac metals like graphene) to disorder.[131] When a magnetic field is applied, time-reversal symmetry is broken and back-scattering increases leading to a positive magnetoresistance. This phenomena has been observed in crystal topological insulators and taken as strong indication of the presence of surface states.[19,132] Upon applying a magnetic field to amorphous Bi_2Se_3, the sample showed weak anti-localization at small magnetic fields. The Hikami–Larkin–Nagaoka formula for weak anti-localization[133] indicated that the magnetoconductance can be well explained by two conducting channels at low temperatures. This is consistent with two surface conducting states and an Anderson localized bulk.[98]

Perhaps the most striking observation in this sample was that of a surface Dirac cone in ARPES and spin-resolved ARPES.[98] Although momentum is no longer a good quantum number, it is possible to represent the energy dispersion as a function of emission angle ϕ (see Figure 5(b)). The spectrum shows two nearly vertical features, consistent with a putative surface state, that merge with a high intensity region, consistent with a bulk valence band. By angle integrating the ARPES signal, the band gap was estimated to be close to 0.3 eV. The spin-resolved ARPES supports the topological state interpretation. Resembling a striking similarity with crystal Bi_2Se_3,[35] the experiment shows a clear spin asymmetry of the bulk and surface states. The putative surface band spin asymmetry is consistent with spin-momentum locking expected from Eq. (2).

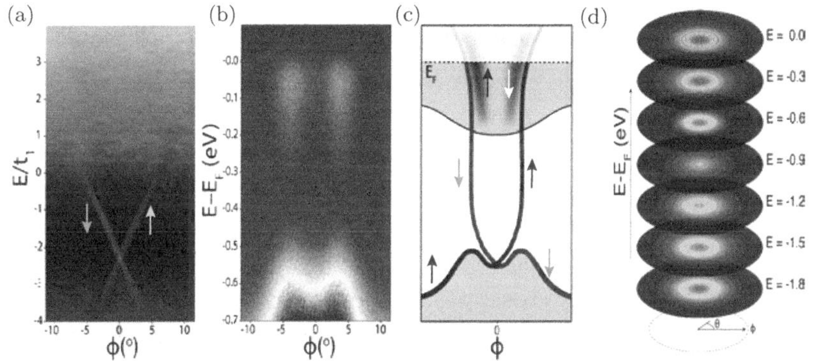

Fig. 5. **ARPES on Amorphous Bi_2Se_3** (a) An amorphous tight-binding model constructed with hoppings based on the model that describes crystalline Bi_2Se_3 shows a Dirac surface state as a function of detector angle ϕ. The arrows indicate the opposite spin projections of the two surface state branches. (b) and (d) show ARPES data at different photon energies as a function of the detection angles ϕ and θ. They evidence the presence of a spectral gap and crossing point, consistent with a surface Dirac cone. (c) The spin-ARPES results (not shown) support a schematic dispersion relation where spin projections reverse in the bulk bands compared to the in-gap surface state, resembling that observed in crystalline Bi_2Se_3.[98]

Although it might be striking that a spin polarized cone could be seen in photoemission, theoretical models support this picture. By calculating the spectral function of a model describing hoppings that lead a $\mathbb{Z}_2$ topological insulator on an amorphous lattice,[81] with fixed coordination,[98] a Dirac cone emerges with well-defined spin structure (see Figure 5(a)). It arises on top of blurry bulk states, defying the intuition that amorphous materials should only show only a noisy ARPES spectrum.

It is likely that other amorphous condensed matter systems can join the list of physical systems that present topological properties. Recent experiments observed that charge to spin conversion of sputtered amorphous Bi_xSe_{1-x} is consistent with the survival of topological properties.[134,135] Additionally, recent theoretical predictions offer alternatives to realize topological states in Shiba glasses emergent from randomly distributed magnetic impurities on the surface of a Rashba superconductor, which could realize amorphous topological superconductivity.[87] However, the difficulty in establishing the requirements to discover amorphous topological solids still precludes the task to predict and identify them in experiment. This impediment is not shared by synthetic systems, which we discuss to conclude.

3.4.2. *Synthetic systems*

The first and currently the only experimental demonstration of topological amorphous systems in synthetic matter was realized in arrays of coupled gyroscopes,[80] soon after the original proposal that amorphous states could exist.[81] Pairs of mechanical objects, in particular coupled gyroscopes, can be engineered to have non-reciprocal couplings that effectively break time-reversal symmetry and allow for a chiral propagating mode at their boundary. This was achieved in human size experiments at ETH Zurich where coupled pendula were designed to host a topological mode that propagated chiraly at the boundary, in analogy to a Chern insulator. Using a similar idea, Mitchell *et al.*[80] carried out the same principle on a significant number of amorphous lattices (see Section 3.2 for their description) of coupled gyroscopes. Their findings demonstrated experimentally that amorphous topological states can become an experimental reality for the first time.

Photonic lattices are another promising system to realize topological amorphous states.[136] Beyond the amorphous photonic Chern insulator proposal discussed above,[85] it is encouraging that the Weaire–Thorpe model has been recently realized in photonic topological lattices, achieving significant photonic band gaps.[83] Combining this realization with a periodically modulated waveguide in a third perpendicular direction effectively breaks time-reversal symmetry and acts like a magnetic field, experimentally realized in Ref. [137]. As for quasi-crystals,[66] this is a very promising route to realize the topological Weaire–Thorpe model proposed in Ref. [60].

It is reasonable to expect that the same synthetic platforms that have been successful in implementing topological band structures can serve to realize analogues of solid-state topological amorphous states.[104] Specifically, in addition to those mentioned earlier, acoustic systems, topo-electrical circuits, ultra-cold atoms trapped in optical lattices, and mechanical systems are all promising platforms to experimentally explore the properties of amorphous topological states.

4. Conclusions and Outlook

The potential of amorphous topological matter seems extraordinary. The lack of translational invariance is motivating to push the limits of our descriptions of topological matter and how to identify it in experiment. The ubiquity of amorphous solids in technology adds even more motivation to propose new device principles that combine the robustness of topology and the versatility of amorphousness.

The quest for a complete classification of topological amorphous solids that can aid discovery is in its infancy. Achieving this milestone can aid to discover topological insulators beyond the only known amorphous topological insulator in the solid state, amorphous Bi_2Se_3.[98] It can also aid to find amorphous topological matter that surpasses their crystal counterparts. For example, bismuth is superconducting at 6K in its amorphous form, a 10^4 increase from its crystalline counterpart. Extending this enhancement to amorphous superconducting materials with strong spin-orbit coupling can be the door to robust topological superconductivity.

Even more boldly, it is appealing to search for new topological phases that can only be realized in amorphous lattices. We have witnessed similar developments in quasi-crystals,[59] where an eight-fold rotational symmetry, absent in crystals, protects a higher order topological phase.

Many other questions remain to be addressed, such as the effect of interactions, the interplay between topological (or lattice) disorder and site disorder, or the nature of different topological phase transitions.[123] Together with the pressing need to discover new physical systems in both condensed and synthetic matter, it is hardly doubtful that this new field will thrive further, combining the efforts of two disconnected communities, those studying topological and amorphous states of matter.

Acknowledgments

The author is indebted to A. Cano, P. Corbae, S. Ciocys, E. Dresselhaus, Q. Marsal, O. Pozo, R. Queiroz, C. Repellin, B. Sbierski, and D. Varjas for stimulating discussions and related collaborations. The author especially thanks J. H. Bardarson and M. A. Ramos for the critical reading of this chapter and A. Cano for discussions on its structure. The author acknowledges financial support by the ANR under the grant ANR-18-CE30-0001-01 (TOPODRIVE) and the European Union Horizon 2020 research and innovation programme under grant agreement No 829044 (SCHINES).

References

1. F. Tang, H. C. Po, A. Vishwanath, and X. Wan. (2019). Comprehensive search for topological materials using symmetry indicators. *Nature*. **566**(7745), 486–489.
2. T. Zhang, Y. Jiang, Z. Song, H. Huang, Y. He, Z. Fang, H. Weng, and C. Fang. (2019). Catalogue of topological electronic materials. *Nature*. **566**(7745), 475–479.

3. M. G. Vergniory, L. Elcoro, C. Felser, N. Regnault, B. A. Bernevig, and Z. Wang. (2019). A complete catalogue of high-quality topological materials. *Nature.* **566**(7745), 480–485.
4. H. Watanabe, H. C. Po, and A. Vishwanath. (2018). Structure and topology of band structures in the 1651 magnetic space groups. *Sci. Adv.* **4**(8), eaat8685.
5. Y. Xu, L. Elcoro, Z. Song, B. J. Wieder, M. G. Vergniory, N. Regnault, Y. Chen, C. Felser, and B. A. Bernevig. (2020). High-throughput calculations of antiferromagnetic topological materials from magnetic topological quantum chemistry. *arXiv:2003.00012.*
6. L. Elcoro, B. J. Wieder, Z. Song, Y. Xu, B. Bradlyn, and B. A. Bernevig. (2020). Magnetic topological quantum chemistry. *arXiv e-prints.* art. arXiv:2010.00598.
7. D. Turnbull. (1969). Under what conditions can a glass be formed? *Contemp. Phys.* **10**(5), 473–488.
8. R. Zallen. (1998). *The Physics of Amorphous Solids.* Weinheim: Wiley.
9. Topology on top. (2016). *Nature Phys.* **12**(7), 615–615.
10. B. J. Wieder, B. Bradlyn, J. Cano, Z. Wang, M. G. Vergniory, L. Elcoro, A. A. Soluyanov, C. Fesler, T. Neupert, N. Regnault, and B. A. Bernevig. (2022). Topological materials discovery from crystal symmetry. *Nature Reviews Materials* **7**, 196.
11. R. Resta. (1994). Macroscopic polarization in crystalline dielectrics: The geometric phase approach. *Rev. Mod. Phys.* **66**, 899–915.
12. W. P. Su, J. R. Schrieffer, and A. J. Heeger. (1979). Solitons in polyacetylene. *Phys. Rev. Lett.* **42**, 1698–1701.
13. F. Schindler, B. Bradlyn, M. H. Fischer, and T. Neupert. (2020). Pairing obstructions in topological superconductors. *Phys. Rev. Lett.* **124**, 247001.
14. D. J. Thouless. (1984). Wannier functions for magnetic sub-bands. *J. Phys. C.* **17**(12), L325–L327.
15. A. A. Soluyanov and D. Vanderbilt. (2011). Wannier representation of z2 topological insulators. *Phys. Rev. B.* **83**, 035108.
16. C.-K. Chiu, J. C. Y. Teo, A. P. Schnyder, and S. Ryu. (2016). Classification of topological quantum matter with symmetries. *Rev. Mod. Phys.* **88**(3), 035005.
17. X. L. Qi, T. L. Hughes, and S.-C. Zhang. (2008). Topological field theory of time-reversal invariant insulators. *Phys. Rev. B.* **78**(19), 195424.
18. A. M. Essin, J. E. Moore, and D. Vanderbilt. (2009). Magnetoelectric polarizability and axion electrodynamics in crystalline insulators. *Phys. Rev. Lett.* **102**(14), 146805.

19. M. Z. Hasan and C. L. Kane. (2010). Colloquium: Topological insulators. *Rev. Mod. Phys.* **82**(4), 3045–3067.
20. X.-L. Qi and S.-C. Zhang. (2011). Topological insulators and superconductors. *Rev. Mod. Phys.* **83**, 1057–1110.
21. L. Fu and C. L. Kane. (2007). Topological insulators with inversion symmetry. *Phys. Rev. B.* **76**(4), 45302.
22. L. Fu. (2011). Topological crystalline insulators. *Phys. Rev. Lett.* **106**(10), 106802.
23. J. Kruthoff, J. de Boer, J. van Wezel, C. L. Kane, and R.-J. Slager. (2017). Topological classification of crystalline insulators through band structure combinatorics. *Phys. Rev. X.* **7**, 041069.
24. H. C. Po, A. Vishwanath, and H. Watanabe. (2017). Symmetry-based indicators of band topology in the 230 space groups. *Nature Commun.* **8**(1), 1–9.
25. H. C. Po. (2020). Symmetry indicators of band topology. *J. Phys. Condens. Matter.* **32**(26), 263001.
26. T. H. Hsieh, H. Lin, J. Liu, W. Duan, A. Bansil, and L. Fu. (2012). Topological crystalline insulators in the SnTe material class. *Nature Commun.* **3**(1), 982.
27. F. Schindler, Z. Wang, M. G. Vergniory, A. M. Cook, A. Murani, S. Sengupta, A. Y. Kasumov, R. Deblock, S. Jeon, I. Drozdov, H. Bouchiat, S. Guéron, A. Yazdani, B. A. Bernevig, and T. Neupert. (2018). Higher-order topology in bismuth. *Nature Phys.* **14**(9), 918–924.
28. R. M. Lutchyn, E. P. A. M. Bakkers, L. P. Kouwenhoven, P. Krogstrup, C. M. Marcus, and Y. Oreg. (2018). Majorana zero modes in superconductor–semiconductor heterostructures. *Nature Rev. Mater.* **3**(5), 52–68.
29. H. C. Po, H. Watanabe, and A. Vishwanath. (2018). Fragile topology and wannier obstructions. *Phys. Rev. Lett.* **121**, 126402.
30. Z. Song, L. Elcoro, N. Regnault, and B. A. Bernevig. (2019). Fragile phases as affine monoids: Full classification and material examples. *arXiv:1905.03262.* art. arXiv:1905.03262.
31. Z.-D. Song, L. Elcoro, and B. A. Bernevig. (2020). Twisted bulk-boundary correspondence of fragile topology. *Science.* **367**(6479), 794–797. ISSN 0036-8075.
32. A. Bouhon, A. M. Black-Schaffer, and R.-J. Slager. (2019). Wilson loop approach to fragile topology of split elementary band representations and topological crystalline insulators with time-reversal symmetry. *Phys. Rev. B.* **100**, 195135.
33. V. Peri, Z.-D. Song, M. Serra-Garcia, P. Engeler, R. Queiroz, X. Huang, W. Deng, Z. Liu, B. A. Bernevig, and S. D. Huber. (2020).

Experimental characterization of fragile topology in an acoustic metamaterial. *Science*. **367**(6479), 797–800.
34. K. C. Nowack, E. M. Spanton, M. Baenninger, M. König, J. R. Kirtley, B. Kalisky, C. Ames, P. Leubner, C. Brüne, H. Buhmann, L. W Molenkamp, D. Goldhaber-Gordon, and K. A. Moler. (2013). Imaging currents in HgTe quantum wells in the quantum spin Hall regime. *Nature Mater.* **12**(9), 787–791.
35. C. Jozwiak, J. A. Sobota, K. Gotlieb, A. F. Kemper, C. R. Rotundu, R. J. Birgeneau, Z. Hussain, D.-H. Lee, Z.-X. Shen, and A. Lanzara. (2016). Spin-polarized surface resonances accompanying topological surface state formation. *Nature Commun.* **7**, 13143.
36. N. P. Armitage, E. J. Mele, and A. Vishwanath. (2018). Weyl and Dirac semimetals in three-dimensional solids. *Rev. Mod. Phys.* **90**(1), 1443.
37. S. Mittal, V. V. Orre, G. Zhu, M. A. Gorlach, A. Poddubny, and M. Hafezi. (2019). Photonic quadrupole topological phases. *Nature Photonics.* **13**(10), 692–696.
38. G. E. Volovik. (2003). *The Universe in a Helium Droplet.* Oxford: Clarendon Press.
39. T. D. C. Bevan, A. J. Manninen, J. B. Cook, J. R. Hook, H. E. Hall, T. Vachaspati, and G. E. Volovik. (1997). Momentum creation by vortices in superfluid 3He as a model of primordial baryogenesis. *Nature.* **386**(6626), 689–692.
40. H. B. Nielsen and M. Ninomiya. (1981). Absence of neutrinos on a lattice: (I). Proof by homotopy theory. *Nucl. Phys. B.* **185**(1), 20–40.
41. F. de Juan, A. G. Grushin, T. Morimoto, and J. E. Moore. (2017). Quantized circular photogalvanic effect in Weyl semimetals. *Nature Commun.* **8**, 15995.
42. J. L. Mañes. (2012). Existence of bulk chiral fermions and crystal symmetry. *Phys. Rev. B.* **85**(15), 155118.
43. G. Chang, B. J. Wieder, F. Schindler, D. S. Sanchez, I. Belopolski, S.-M. Huang, B. Singh, D. Wu, T.-R. Chang, T. Neupert, S.-Y. Xu, H. Lin, and M. Z. Hasan. (2018). Topological quantum properties of chiral crystals. *Nature Mater.* **17**(11), 978–985.
44. B. Bradlyn, J. Cano, Z. Wang, M. G. Vergniory, C. Felser, R. J. Cava, and B. A. Bernevig. (2016). Beyond Dirac and Weyl fermions: Unconventional quasiparticles in conventional crystals. *Science.* **353**(6299), aaf5037.
45. G. Chang, S.-Y. Xu, B. J. Wieder, D. S. Sanchez, S.-M. Huang, I. Belopolski, T.-R. Chang, S. Zhang, A. Bansil, H. Lin, and M. Z. Hasan. (2017). Unconventional chiral fermions and large topological fermi arcs in RHSI. *Phys. Rev. Lett.* **119**, 206401.

46. P. Tang, Q. Zhou, and S.-C. Zhang. (2017). Multiple types of topological fermions in transition metal silicides. *Phys. Rev. Lett.* **119**, 206402.
47. N. B. M. Schröter, D. Pei, M. G. Vergniory, Y. Sun, K. Manna, F. de Juan, J. A. Krieger, V. Süß, M. Schmidt, P. Dudin, B. Bradlyn, T. K. Kim, T. Schmitt, C. Cacho, C. Felser, V. N. Strocov, and Y. Chen. (2019). Chiral topological semimetal with multifold band crossings and long Fermi arcs. *Nature Phys.* **15**(8), 759–765.
48. F. Flicker, F. de Juan, B. Bradlyn, T. Morimoto, M. G. Vergniory, and A. G. Grushin. (2018). Chiral optical response of multifold fermions. *Phys. Rev. B.* **98**(15), 155145. ISSN 2469-9950.
49. V. Peri, M. Serra-Garcia, R. Ilan, and S. D. Huber. (2019). Axial-field-induced chiral channels in an acoustic Weyl system. *Nature Phys.* **15**(4), 357–361.
50. T. Ozawa, H. M. Price, A. Amo, N. Goldman, M. Hafezi, L. Lu, M. C. Rechtsman, D. Schuster, J. Simon, O. Zilberberg, and I. Carusotto. (2019). Topological photonics. *Rev. Mod. Phys.* **91**(1), 634.
51. N. R. Cooper, J. Dalibard, and I. B. Spielman. (2019). Topological bands for ultracold atoms. *Rev. Mod. Phys.* **91**(1), 015005.
52. R.-P. Riwar, M. Houzet, J. S. Meyer, and Y. V. Nazarov. (2016). Multi-terminal Josephson junctions as topological matter. *Nature Commun.* **7**(1), 1–5.
53. C. L. Kane and T. C. Lubensky. (2014). Topological boundary modes in isostatic lattices. *Nature Phys.* **10**(1), 39–45
54. P. Delplace, J. B. Marston, and A. Venaille. (2017). Topological origin of equatorial waves. *Science.* **358**(6366), 1075–1077.
55. L. Wu, M. Salehi, N. Koirala, J. Moon, S. Oh, and N. P. Armitage. (2016). Quantized Faraday and Kerr rotation and axion electrodynamics of a 3D topological insulator. *Science.* **354**(6316), 1124–1127.
56. N. W. Ashcroft and N. D. Mermin. (1976). *Solid State Physics.* New York : Holt, Rinehart and Winston.
57. C. W. Groth, M. Wimmer, A. R. Akhmerov, J. Tworzydło, and C. W. J. Beenakker. (2009). Theory of the topological anderson insulator. *Phys. Rev. Lett.* **103**, 196805.
58. M. Brzezińska, A. M. Cook, and T. Neupert. (2018). Topology in the sierpiński-hofstadter problem. *Phys. Rev. B.* **98**, 205116.
59. D. Varjas, A. Lau, K. Pöyhönen, A. R. Akhmerov, D. I. Pikulin, and I. C. Fulga. (2019). Topological phases without crystalline counterparts. *Phys. Rev. Lett.* **123**, 196401.
60. Q. Marsal, D. Varjas, and A. G. Grushin. (2020). Topological Weaire-Thorpe models of amorphous matter. *Proceedings of the National Academy of Sciences* **117**(48), 30260–30265.

61. H. Jiang, L. Wang, Q.-f. Sun, and X. C. Xie. (2009). Numerical study of the topological anderson insulator in HgTe/CdTe quantum wells. *Phys. Rev. B.* **80**, 165316.
62. J. Li, R.-L. Chu, J. K. Jain, and S.-Q. Shen. (2009). Topological anderson insulator. *Phys. Rev. Lett.* **102**, 136806.
63. E. J. Meier, F. A. An, A. Dauphin, M. Maffei, P. Massignan, T. L. Hughes, and B. Gadway. (2018). Observation of the topological anderson insulator in disordered atomic wires. *Science.* **362**(6417), 929–933 ISSN 0036-8075.
64. Y. E. Kraus, Y. Lahini, Z. Ringel, M. Verbin, and O. Zilberberg. (2012). Topological states and adiabatic pumping in quasicrystals. *Phys. Rev. Lett.* **109**(10), 106402.
65. D.-T. Tran, A. Dauphin, N. Goldman, and P. Gaspard. (2015). Topological hofstadter insulators in a two-dimensional quasicrystal. *Phys. Rev. B.* **91**, 085125.
66. M. A. Bandres, M. C. Rechtsman, and M. Segev. (2016). Topological photonic quasicrystals: Fractal topological spectrum and protected transport. *Phys. Rev. X.* **6**(1), 011016.
67. I. C. Fulga, D. I. Pikulin, and T. A. Loring. (2016). Aperiodic weak topological superconductors. *Phys. Rev. Lett.* **116**(25), 257002.
68. H. Huang and F. Liu. (2018). Quantum spin Hall effect and spin Bott index in a quasicrystal lattice. *Phys. Rev. Lett.* **121**(12), 126401.
69. H. Huang and F. Liu. (2018). Theory of spin Bott index for quantum spin Hall states in nonperiodic systems. *Phys. Rev. B.* **98**(12), 125130.
70. R. Chen, C.-Z. Chen, J.-H. Gao, B. Zhou, and D.-H. Xu. (2020). Higher-order topological insulators in quasicrystals. *Phys. Rev. Lett.* **124**, 036803.
71. R. Chen, D.-H. Xu, and B. Zhou. (2019). Topological anderson insulator phase in a quasicrystal lattice. *Phys. Rev. B.* **100**, 115311.
72. D. Shechtman, I. Blech, D. Gratias, and J. W. Cahn. (1984). Metallic phase with long-range orientational order and no translational symmetry. *Phys. Rev. Lett.* **53**(20), 1951–1953.
73. A. Agarwala, S. Pai, and V. B. Shenoy. (2018). Fractalized metals. *arXiv:1803.01404.*
74. S. Pai and A. Prem. Topological states on fractal lattices. (2019). *Phys. Rev. B.* **100**, 155135.
75. M. Fremling, M. van Hooft, C. M. Smith, and L. Fritz. (2020). Existence of robust edge currents in sierpiński fractals. *Phys. Rev. Res.* **2**, 013044.
76. Y. Cao, V. Fatemi, A. Demir, S. Fang, S. L. Tomarken, J. Y. Luo, J. D. Sanchez-Yamagishi, K. Watanabe, T. Taniguchi, E. Kaxiras,

R. C. Ashoori, and P. Jarillo-Herrero. (2018). Correlated insulator behaviour at half-filling in magic-angle graphene superlattices. *Nature.* **556**(7699), 80–84.

77. Y. Cao, V. Fatemi, S. Fang, K. Watanabe, T. Taniguchi, E. Kaxiras, and P. Jarillo-Herrero. (2018). Unconventional superconductivity in magic-angle graphene superlattices. *Nature.* **556**(7699), 43–50.
78. R. Bistritzer and A. H. MacDonald. (2011). Moiré bands in twisted double-layer graphene. *Proc. Nat. Acad. Sci.* **108**(30), 12233–12237. ISSN 0027-8424.
79. M. Serlin, C. L. Tschirhart, H. Polshyn, Y. Zhang, J. Zhu, K. Watanabe, T. Taniguchi, L. Balents, and A. F. Young. (2020). Intrinsic quantized anomalous hall effect in a moiré heterostructure. *Science.* **367**(6480), 900–903. ISSN 0036-8075.
80. N. P. Mitchell, L. M. Nash, D. Hexner, A. M. Turner, and W. T. M. Irvine. (2018). Amorphous topological insulators constructed from random point sets. *Nature Phys.* **14**, 380–385.
81. A. Agarwala and V. B. Shenoy. (2017). Topological insulators in amorphous systems. *Phys. Rev. Lett.* **118**(23), 236402.
82. F. D. M. Haldane. (2004). Berry curvature on the fermi surface: Anomalous hall effect as a topological fermi-liquid property. *Phys. Rev. Lett.* **93**, 206602.
83. M. Florescu, S. Torquato, and P. J. Steinhardt. (2009). Designer disordered materials with large, complete photonic band gaps. *Proc. Nat. Acad. Sci.* **106**(49), 20658–20663.
84. D. Weaire and M. F. Thorpe. (1971). Electronic properties of an amorphous solid. i. a simple tight-binding theory. *Phys. Rev. B.* **4**, 2508–2520.
85. S. Mansha and Y. D. Chong. (2017). Robust edge states in amorphous gyromagnetic photonic lattices. *Phys. Rev. B.* **96**(12), 121405.
86. M. Xiao and S. Fan. (2017). Photonic Chern insulator through homogenization of an array of particles. *Phys. Rev. B.* **96**(10), 100202.
87. K. Pöyhönen, I. Sahlberg, A. Westström, and T. Ojanen. (2018). Amorphous topological superconductivity in a Shiba glass. *Nature Commun.* **9**(1), 2103.
88. G.-W. Chern. (2019). Topological insulator in an atomic liquid. *EPL (Europhysics Letters).* **126**(3), 37002.
89. Y.-B. Yang, T. Qin, D.-L. Deng, L.-M. Duan, and Y. Xu. (2019). Topological amorphous metals. *Phys. Rev. Lett.* **123**, 076401.
90. A. Agarwala, V. Juričić, and B. Roy. (2020). Higher-order topological insulators in amorphous solids. *Phys. Rev. Res.* **2**, 012067.

91. L. Schwartz and H. Ehrenreich. (1972). Comment on the tight-binding model for amorphous semiconductors. *Phys. Rev. B.* **6**, 4088–4090.
92. M. Costa, G. R. Schleder, M. B. Nardelli, C. Lewenkopf, and A. Fazzio. (2019). Toward realistic amorphous topological insulators. *Nano Lett.* **19**, 8941–8946.
93. F. Reis, G. Li, L. Dudy, M. Bauernfeind, S. Glass, W. Hanke, R. Thomale, J. Schäfer, and R. Claessen. (2017). Bismuthene on a SiC substrate: A candidate for a high-temperature quantum spin hall material. *Science.* **357**(6348), 287–290.
94. H. Zhang, C.-X. Liu, X. L. Qi, X. Dai, Z. Fang, and S.-C. Zhang. (2009). Topological insulators in Bi_2Se_3, Bi_2Te_3 and Sb_2Te_3 with a single Dirac cone on the surface. *Nat. Phys.* **5**(6), 438–442.
95. D. Hsieh, Y. Xia, D. Qian, L. Wray, F. Meier, J. H. Dil, J. Osterwalder, L. Patthey, A. V. Fedorov, H. Lin, A. Bansil, D. Grauer, Y. S. Hor, R. J. Cava, and M. Z. Hasan. (2009). Observation of time-reversal-protected single-dirac-cone topological-insulator states in bi_2te_3 and sb_2te_3. *Phys. Rev. Lett.* **103**, 146401.
96. C. Pauly, G. Bihlmayer, M. Liebmann, M. Grob, A. Georgi, D. Subramaniam, M. R. Scholz, J. Sánchez-Barriga, A. Varykhalov, S. Blügel, O. Rader, and M. Morgenstern. (2012). Probing two topological surface bands of Sb_2Te_3 by spin-polarized photoemission spectroscopy. *Phys. Rev. B.* **86**, 235106.
97. S. Caravati, M. Bernasconi, and M. Parrinello. (2010). First-principles study of liquid and amorphous Sb_2Te_3. *Phys. Rev. B.* **81**(1), 014201.
98. P. Corbae, S. Ciocys, D. Varjas, S. Zeltmann, C. H. Stansbury, M. Molina-Ruiz, Z. Chen, L.-W. Wang, A. M. Minor, A. G. Grushin, A. Lanzara, and F. Hellman. (2019). Evidence for topological surface states in amorphous Bi_2Se_3. *arXiv.org.*
99. J. M. Gibson, M. M. J. Treacy, T. Sun, and N. J. Zaluzec. (2010). Substantial crystalline topology in amorphous silicon. *Phys. Rev. Lett.* **105**(12), 119.
100. M. M. J. Treacy and K. B. Borisenko. (2012). The local structure of amorphous silicon. *Science.* **335**(6071), 950–953.
101. C.-T. Toh, H. Zhang, J. Lin, A. S. Mayorov, Y.-P. Wang, C. M. Orofeo, D. B. Ferry, H. Andersen, N. Kakenov, Z. Guo, I. H. Abidi, H. Sims, K. Suenaga, S. T. Pantelides, and B. Özyilmaz. (2020). Synthesis and properties of free-standing monolayer amorphous carbon. *Nature.* **577**(7789), 199–203.
102. V. Kapko, D. A. Drabold, and M. F. Thorpe. (2010). Electronic structure of a realistic model of amorphous graphene. *Phys. Status Solidi (b).* **247**(5), 1197–1200.

103. A. C. Wright. (2013). The great crystallite versus random network controversy: A personal perspective. *Int. J. Appl. Glass Sci.* **5**(1), 31–56.
104. C. Bourne and E. Prodan. Non-commutative Chern numbers for generic aperiodic discrete systems. *J. Phys. A.* **51**(23), 235202 (2018).
105. D. T. Tran, A. Dauphin, A. G. Grushin, P. Zoller, and N. Goldman. (2017). Probing topology by "heating": Quantized circular dichroism in ultracold atoms. *Sci. Adv.* **3**(8), e1701207.
106. A. Kitaev. (2005). Anyons in an exactly solved model and beyond. *Ann. Phys.* **321**(1), 2–111.
107. E. Prodan. (2010). Non-commutative tools for topological insulators. *New J. Phys.* **12**(6), 065003.
108. R. Bianco and R. Resta. (2011). Mapping topological order in coordinate space. *Phys. Rev. B.* **84**(24), 241106.
109. B. Irsigler, J.-H. Zheng, and W. Hofstetter. (2019). Microscopic characteristics and tomography scheme of the local chern marker. *Phys. Rev. A.* **100**, 023610.
110. H. S. Bennett and E. A. Stern. (1965). Faraday effect in solids. *Phys. Rev.* **137**(2A), A448–A461.
111. I. Souza and D. Vanderbilt. (2008). Dichroic f -sum rule and the orbital magnetization of crystals. *Phys. Rev. B.* **77**(5), 1–13.
112. T. A. Loring and M. B. Hastings. (2010). Disordered topological insulators via C*-algebras. *EPL (Europhysics Letters).* **92**(6), 67004.
113. A. Amaricci, L. Privitera, F. Petocchi, M. Capone, G. Sangiovanni, and B. Trauzettel. (2017). Edge state reconstruction from strong correlations in quantum spin hall insulators. *Phys. Rev. B.* **95**, 205120.
114. A. Malashevich, I. Souza, S. Coh, and D. Vanderbilt. (2010). Theory of orbital magnetoelectric response. *New J. Phys.* **12**(5), 053032.
115. S. Coh, D. Vanderbilt, A. Malashevich, and I. Souza. (2011). Chern-Simons orbital magnetoelectric coupling in generic insulators. *Phys. Rev. B.* **83**(8), 085108.
116. T. Olsen, M. Taherinejad, D. Vanderbilt, and I. Souza. (2017). Surface theorem for the Chern-Simons axion coupling. *Phys. Rev. B.* **95**(7), 075137.
117. T. Rauch, T. Olsen, D. Vanderbilt, and I. Souza. (2018). Geometric and nongeometric contributions to the surface anomalous Hall conductivity. *Phys. Rev. B.* **98**(11), 115108.
118. Z. Li and R. S. K. Mong. (2019). Local formula for the $\mathcal{F}_2$ invariant of topological insulators. *Phys. Rev. B.* **100**, 205101.
119. D. Varjas, M. Fruchart, A. R. Akhmerov, and P. M. Perez-Piskunow. (2020). Computation of topological phase diagram of disordered

$Pb_{1-x}Sn_xTe$ using the kernel polynomial method. *Phys. Rev. Res.* **2**, 013229.
120. A. G. Grushin and A. Cortijo. (2011). Tunable casimir repulsion with three-dimensional topological insulators. *Phys. Rev. Lett.* **106**(2), 793.
121. G. Rosenberg and M. Franz. (2010). Witten effect in a crystalline topological insulator. *Phys. Rev. B.* **82**(3), 035105.
122. P. Mukati, A. Agarwala, and S. Bhattacharjee. (2020) Topological and conventional phases of a three-dimensional electronic glass. *Phys. Rev. B.* **101**, 035142.
123. I. Sahlberg, A. Westström, K. Pöyhönen, and T. Ojanen. (2020) Topological phase transitions in glassy quantum matter. *Phys. Rev. Res.* **2**, 013053.
124. M. N. Ivaki, I. Sahlberg, and T. Ojanen. (2020) Criticality in amorphous topological matter — beyond the universal scaling paradigm. *arXiv:2006.05886*. art. arXiv:2006.05886.
125. E. Prodan, T. L. Hughes, and B. A. Bernevig. (2010). Entanglement spectrum of a disordered topological Chern insulator. *Phys. Rev. Lett.* **105**(11), 115501.
126. I. C. Fulga, F. Hassler, and A. R. Akhmerov. (2012). Scattering theory of topological insulators and superconductors. *Phys. Rev. B.* **85**, 165409.
127. M. DC, R. Grassi, J.-Y. Chen, M. Jamali, D. Reifsnyder Hickey, D. Zhang, Z. Zhao, H. Li, P. Quarterman, Y. Lv, M. Li, A. Manchon, K. A. Mkhoyan, T. Low, and J.-P. Wang. (2018). Room-temperature high spin-orbit torque due to quantum confinement in sputtered Bi_xSe_{1-x} films. *Nature Mater.* **17**(9), 800–807.
128. P. Sahu, J.-Y. Chen, J. C. Myers, and J.-P. Wang. (2018). Weak antilocalization and low-temperature characterization of sputtered polycrystalline bismuth selenide. *Appl. Phys. Lett.* **112**(12), 122402–6.
129. A. Banerjee, O. Deb, K. Majhi, R. Ganesan, D. Sen, and P. S. A. Kumar. (2017). Granular topological insulators. *Nanoscale.* **9**(20), 6755–6764.
130. R. M. Hill. (1976). Variable-range hopping. *Phys. Status Solidi (a).* **34**(2), 601–613.
131. J. H. Bardarson and J. E. Moore. (2013). Quantum interference and Aharonov–Bohm oscillations in topological insulators. *Rep. Progr. Phys.* **76**(5), 056501.
132. M. Brahlek, N. Koirala, N. Bansal, and S. Oh. (2015). Transport properties of topological insulators: Band bending, bulk metal-to-insulator transition, and weak anti-localization. *Solid State Commun.* **215–216**, 54–62. ISSN 0038-1098.

133. S. Hikami, A. I. Larkin, and Y. Nagaoka. (1980). Spin-orbit interaction and magnetoresistance in the two dimensional random system. *Progr. Theor. Phys.* **63**(2), 707–710.
134. M. DC, J.-Y. Chen, T. Peterson, P. Sahu, B. Ma, N. Mousavi, R. Harjani, and J.-P. Wang. (2019). Observation of high spin-to-charge conversion by sputtered bismuth selenide thin films at room temperature. *Nano Lett.* **19**(8), 4836–4844.
135. P. Sahu, J.-Y. Chen, X. Devaux, H. Jaffres, S. Migot, H. Dang, J.-M. George, Y. Lu, and J.-P. Wang. (2019). Room temperature high charge to spin conversion in amorphous topological insulator. *arXiv:1911.03323.* art. arXiv:1911.03323.
136. M. Rechtsman, A. Szameit, F. Dreisow, M. Heinrich, R. Keil, S. Nolte, and M. Segev. (2011). Amorphous photonic lattices: Band gaps, effective mass, and suppressed transport. *Phys. Rev. Lett.* **106**(19), 193904.
137. M. C. Rechtsman, J. M. Zeuner, Y. Plotnik, Y. Lumer, D. Podolsky, F. Dreisow, S. Nolte, M. Segev, and A. Szameit. (2013). Photonic Floquet topological insulators. *Nature.* **496**(7444), 196–200.
138. Jean-Noël Fuchs, Rémy Mosseri, and Julien Vidal. (2018). *Phys. Rev. B* **98**, 165427.
139. Jean-Noël Fuchs and Julien Vidal. (2016). *Phys. Rev. B* **94**, 205437.
140. X.-G. Wen. (2017). Colloquium: Zoo of quantum-topological phases of matter. *Rev. Mod. Phys.* **89**, 041004.

https://doi.org/10.1142/9781800612587_bmatter

Index